T0202802

Lecture Notes in Computer Science

Lecture Notes in Artificial Intelligence   **14195**

Founding Editor

Jörg Siekmann

Series Editors

Randy Goebel, *University of Alberta, Edmonton, Canada*
Wolfgang Wahlster, *DFKI, Berlin, Germany*
Zhi-Hua Zhou, *Nanjing University, Nanjing, China*

The series Lecture Notes in Artificial Intelligence (LNAI) was established in 1988 as a topical subseries of LNCS devoted to artificial intelligence.

The series publishes state-of-the-art research results at a high level. As with the LNCS mother series, the mission of the series is to serve the international R & D community by providing an invaluable service, mainly focused on the publication of conference and workshop proceedings and postproceedings.

Murilo C. Naldi · Reinaldo A. C. Bianchi
Editors

Intelligent Systems

12th Brazilian Conference, BRACIS 2023
Belo Horizonte, Brazil, September 25–29, 2023
Proceedings, Part I

 Springer

Editors
Murilo C. Naldi (ID)
Federal University of São Carlos
São Carlos, Brazil

Reinaldo A. C. Bianchi (ID)
Centro Universitario da FEI
São Bernardo do Campo, Brazil

ISSN 0302-9743   ISSN 1611-3349 (electronic)
Lecture Notes in Artificial Intelligence
ISBN 978-3-031-45367-0   ISBN 978-3-031-45368-7 (eBook)
https://doi.org/10.1007/978-3-031-45368-7

LNCS Sublibrary: SL7 – Artificial Intelligence

This Springer imprint is published by the registered company Springer Nature Switzerland AG
The registered company address is: Gewerbestrasse 11, 6330 Cham, Switzerland

Paper in this product is recyclable.

Preface

The 12th Brazilian Conference on Intelligent Systems (BRACIS 2023) was one of the most important events held in Brazil in 2023 for researchers interested in publishing significant and novel results related to Artificial and Computational Intelligence. The Brazilian Conference on Intelligent Systems (BRACIS) originated from the combination of the two most important scientific events in Brazil in Artificial Intelligence (AI) and Computational Intelligence (CI): the Brazilian Symposium on Artificial Intelligence (SBIA, 21 editions) and the Brazilian Symposium on Neural Networks (SBRN, 12 editions). The Brazilian Computer Society (SBC) supports the event, the Special Committee of Artificial Intelligence (CEIA), and the Special Committee of Computational Intelligence (CEIC). The conference aims to promote theoretical aspects and applications of Artificial and Computational Intelligence and exchange scientific ideas among researchers, practitioners, scientists, engineers, and industry.

In 2023, BRACIS took place in Belo Horizonte, Brazil, from September 25th to 29th, 2023, in the Campus of the Universidade Federal de Minas Gerais. The event was held in conjunction with two other events: the National Meeting on Artificial and Computational Intelligence (ENIAC) and the Symposium on Information and Human Language Technology (STIL).

BRACIS 2023 received 242 submissions. All papers were rigorously reviewed by an international Program Committee (with a minimum of three double-blind peer reviews per submission), followed by a discussion phase for conflicting reports. After the review process, 89 papers were selected for publication in three volumes of the Lecture Notes in Artificial Intelligence series (an acceptance rate of 37%).

The topics of interest included, but were not limited to, the following:

- Agent-based and Multi-Agent Systems
- Bioinformatics and Biomedical Engineering
- Cognitive Modeling and Human Interaction
- Combinatorial and Numerical Optimization
- Computer Vision
- Constraints and Search
- Deep Learning
- Distributed AI
- Education
- Ethics
- Evolutionary Computation and Metaheuristics
- Forecasting
- Foundation Models
- Foundations of AI
- Fuzzy Systems
- Game Playing and Intelligent Interactive Entertainment
- Human-centric AI

- Hybrid Systems
- Information Retrieval, Integration, and Extraction
- Intelligent Robotics
- Knowledge Representation and Reasoning
- Knowledge Representation and Reasoning in Ontologies and the Semantic Web
- Logic-based Knowledge Representation and Reasoning
- Machine Learning and Data Mining
- Meta-learning
- Molecular and Quantum Computing
- Multidisciplinary AI and CI
- Natural Language Processing
- Neural Networks
- Pattern Recognition and Cluster Analysis
- Planning and Scheduling
- Reinforcement Learning

We want to thank everyone involved in BRACIS 2023 for helping to make it a success: we are very grateful to the Program Committee members and reviewers for their volunteered contribution to the reviewing process; we would also like to thank all the authors who submitted their papers and laboriously worked to have the best final version possible; the General Chairs and the Local Organization Committee for supporting the conference; the Brazilian Computing Society (SBC); and all the conference's sponsors and supporters. We are confident that these proceedings reflect the excellent work in the artificial and computation intelligence communities.

September 2023   Murilo C. Naldi
                                                        Reinaldo A. C. Bianchi

Organization

General Chairs

Gisele Lobo Pappa	Universidade Federal de Minas Gerais, Brazil
Wagner Meira Jr.	Universidade Federal de Minas Gerais, Brazil

Program Committee Chairs

Murilo Coelho Naldi	Universidade Federal de São Carlos, Brazil
Reinaldo A. C. Bianchi	Centro Universitário FEI, Brazil

Steering Committee

Aline Paes	Universidade Federal Fluminense, Brazil
André Britto	Universidade Federal do Sergipe, Brazil
Anna H. R. Costa	Universidade de São Paulo, Brazil
Anne Canuto	Universidade Federal do Rio Grande do Norte, Brazil
Arnaldo Cândido Jr.	Universidade Estadual Paulista, Brazil
Felipe Meneguzzi	University of Aberdeen, UK
Filipe Saraiva	Universidade Federal do Pará, Brazil
Gina M. B. Oliveira	Universidade Federal Uberlandia, Brazil
Helida Santos	Universidade Federal do Rio Grande, Brazil
Leliane N. de Barros	Universidade de São Paulo, Brazil
Livy Real	B2W Digital, Brazil
Maria V. de Menezes	Universidade Federal do Ceará, Brazil
Marlo Souza	Universidade Federal da Bahia, Brazil
Renato Tinos	Universidade de São Paulo, Brazil
Ricardo Marcacini	Universidade de São Paulo, Brazil
Tatiane Nogueira	Universidade Federal da Bahia, Brazil
Thiago Pardo	Universidade de São Paulo - São Carlos, Brazil

Program Committee

Adenilton da Silva	Universidade Federal de Pernambuco, Brazil
Adriane Serapião	Universidade Estadual Paulista, Brazil
Adrião Duarte D. Neto	Universidade Federal do Rio Grande do Norte, Brazil
Alexandre Salle	Universidade Federal do Rio Grande do Sul, Brazil
Aline Neves	Universidade Federal do ABC, Brazil
Aline Paes	Universidade Federal Fluminense, Brazil
Alneu Lopes	Universidade de São Paulo - São Carlos, Brazil
Aluizio Araújo	Universidade Federal de Pernambuco, Brazil
Alvaro Moreira	Universidade Federal do Rio Grande do Sul, Brazil
Amedeo Napoli	LORIA, France
Ana Bazzan	Universidade Federal do Rio Grande do Sul, Brazil
Ana Carolina Lorena	Instituto Tecnológico de Aeronáutica, Brazil
Ana C. B. K. Vendramin	Universidade Tecnológica Federal do Paraná, Brazil
Anderson Soares	Universidade Federal de Goiás, Brazil
André Britto	Universidade Federal do Sergipe, Brazil
André P. L. F. Carvalho	Universidade de São Paulo - São Carlos, Brazil
André Rossi	Universidade Estadual Paulista, Brazil
André Ruela	Marinha do Brasil, Brazil
André Takahata	Universidade Federal do ABC, Brazil
Andrés E. C. Salazar	Universidade Tecnológica Federal do Paraná, Brazil
Anna H. R. Costa	Universidade de São Paulo, Brazil
Anne Canuto	Universidade Federal do Rio Grande do Norte, Brazil
Araken Santos	Universidade Federal Rural do Semi-árido, Brazil
Artur Jordão	Universidade de São Paulo, Brazil
Aurora Pozo	Universidade Federal do Paraná, Brazil
Bernardo Gonçalves	Universidade de São Paulo, Brazil
Bruno Masiero	Universidade Estadual de Campinas, Brazil
Bruno Nogueira	Universidade Federal de Mato Grosso do Sul, Brazil
Bruno Souza	Universidade Federal do Maranhão, Brazil
Bruno Veloso	Universidade Portucalense, Portugal
Carlos Ribeiro	Instituto Tecnológico de Aeronáutica, Brazil
Carlos Silla	Pontifícia Universidade Católica do Paraná, Brazil

Carlos Thomaz	Centro Universitário FEI, Brazil
Carlos A. E. Montesco	Universidade Federal de Sergipe, Brazil
Carlos E. Pantoja	Centro Federal de Educação Tecnológica - RJ, Brazil
Carolina P. de Almeida	Universidade E. do Centro-Oeste do Paraná, Brazil
Celia Ralha	Universidade de Brasilia, Brazil
Claudio Bordin Jr.	Universidade Federal do ABC, Brazil
Claudio Toledo	Universidade de São Paulo, Brazil
Cleber Zanchettin	Universidade Federal de Pernambuco, Brazil
Cristiano Torezzan	Universidade Estadual de Campinas, Brazil
Daniel Araújo	Universidade Federal do Rio Grande do Norte, Brazil
Daniel Dantas	Universidade Federal de Sergipe, Brazil
Danilo Perico	Centro Universitário FEI, Brazil
Danilo Sanches	Universidade Tecnológica Federal do Paraná, Brazil
Debora Medeiros	Universidade Federal do ABC, Brazil
Denis Mauá	Universidade de São Paulo, Brazil
Dennis B. Aranibar	Universidad Católica San Pablo, Peru
Diana Adamatti	Universidade Federal do Rio Grande, Brazil
Diego Furtado Silva	Universidade de São Paulo, Brazil
Donghong Ji	Wuhan University, China
Eder M. Gonçalves	Universidade Federal do Rio Grande, Brazil
Edson Gomi	Universidade de São Paulo, Brazil
Edson Matsubara	Universidade Federal de Mato Grosso do Sul, Brazil
Eduardo Costa	Corteva Agriscience, Brazil
Eduardo Goncalves	Escola Nacional de Ciências Estatísticas, Brazil
Eduardo Palmeira	Universidade Estadual de Santa Cruz, Brazil
Eduardo Spinosa	Universidade Federal do Paraná, Brazil
Edward H. Haeusler	Pontifícia Universidade Católica do R. de J., Brazil
Elaine Faria	Universidade Federal Uberlandia, Brazil
Elizabeth Goldbarg	Universidade Federal do Rio Grande do Norte, Brazil
Emerson Paraiso	Pontificia Universidade Catolica do Paraná, Brazil
Eric Araújo	Universidade Federal de Lavras, Brazil
Evandro Costa	Universidade Federal de Alagoas, Brazil
Fabiano Silva	Universidade Federal do Paraná, Brazil
Fábio Cozman	Universidade de São Paulo, Brazil
Felipe Leno da Silva	Lawrence Livermore National Lab., USA
Felipe Meneguzzi	University of Aberdeen, UK

Felix Antreich	Instituto Tecnológico de Aeronáutica, Brazil
Fernando Osório	Universidade de São Paulo, São Carlos, Brazil
Flavia Bernardini	Universidade Federal Fluminense, Brazil
Flavio Tonidandel	Centro Universitário FEI, Brazil
Flávio S. C. da Silva	Universidade de São Paulo, Brazil
Francisco Chicano	University of Málaga, Spain
Francisco De Carvalho	Universidade Federal de Pernambuco, Brazil
Gabriel Ramos	Universidade do Vale do Rio dos Sinos, Brazil
George Cavalcanti	Universidade Federal de Pernambuco, Brazil
Gerson Zaverucha	Universidade Federal do Rio de Janeiro, Brazil
Giancarlo Lucca	Universidade Federal do Rio Grande, Brazil
Gisele Pappa	Universidade Federal de Minas Gerais, Brazil
Gracaliz Dimuro	Universidade Federal do Rio Grande, Brazil
Guilherme Barreto	Universidade Federal do Ceará, Brazil
Guilherme Coelho	Universidade Estadual de Campinas, Brazil
Guilherme Derenievicz	Universidade Federal do Paraná, Brazil
Guilherme D. Pelegrina	Universidade Estadual de Campinas, Brazil
Guillermo Simari	Universidad Nacional del Sur in B. B., Argentina
Gustavo Giménez-Lugo	Universidade Tecnológica Federal do Paraná, Brazil
Heitor Gomes	Victoria University of Wellington, New Zealand
Helena Caseli	Universidade Federal de São Carlos, Brazil
Helida Santos	Universidade Federal do Rio Grande, Brazil
Heloisa Camargo	Universidade Federal de São Carlos, Brazil
Huei Lee	Universidade Estadual do Oeste do Paraná, Brazil
Isaac da Silva	Centro Universitário FEI, Brazil
Ivandré Paraboni	Universidade de São Paulo, Brazil
Ivette Luna	Universidade Estadual de Campinas, Brazil
Jaime S. Sichman	Universidade de São Paulo, Brazil
Jean Paul Barddal	Pontifícia Universidade Católica do Paraná, Brazil
João Papa	Universidade Estadual Paulista, Brazil
João C. Xavier-Júnior	Universidade Federal do RN (UFRN), Brazil
João Paulo Canário	Stone Co., Brazil
Jomi Hübner	Universidade Federal de Santa Catarina, Brazil
Jonathan Andrade Silva	Universidade Federal de Mato Grosso do Sul, Brazil
José Antonio Sanz	Universidad Publica de Navarra, Spain
José A. Baranauskas	Universidade de São Paulo, Brazil
Jose E. O. Luna	Universidad Católica San Pablo, Peru
Julio Nievola	Pontifícia Universidade Católica do Paraná, Brazil
Karla Roberta Lima	Universidade de São Paulo, Brazil

Karliane Vale	Universidade Federal do Rio Grande do Norte, Brazil
Kate Revoredo	Humboldt-Universität zu Berlin, Germany
Krysia Broda	Imperial College, UK
Laura De Miguel	Universidad Pública de Navarra, Spain
Leila Bergamasco	Centro Universitário FEI, Brazil
Leliane Nunes de Barros	Universidade de São Paulo, Brazil
Leonardo Emmendorfer	Universidade Federal do Rio Grande, Brazil
Leonardo Matos	Universidade Federal de Sergipe, Brazil
Leonardo T. Duarte	Universidade Estadual de Campinas, Brazil
Leonardo F. R. Ribeiro	Amazon, USA
Levy Boccato	Universidade Estadual de Campinas, Brazil
Li Weigang	Universidade de Brasilia, Brazil
Livy Real	B2W Digital, Brazil
Lucelene Lopes	Universidade de São Paulo - São Carlos, Brazil
Luciano Digiampietri	Universidade de São Paulo, Brazil
Luis Garcia	Universidade de Brasília, Brazil
Luiz H. Merschmann	Universidade Federal de Lavras, Brazil
Marcela Ribeiro	Universidade Federal de São Carlos, Brazil
Marcelo Finger	Universidade de São Paulo, Brazil
Marcilio de Souto	Université d'Orléans, France
Marcos Domingues	Universidade Estadual de Maringá, Brazil
Marcos Quiles	Universidade Federal de São Paulo, Brazil
Maria Claudia Castro	Centro Universitario FEI, Brazil
Maria do C. Nicoletti	Universidade Federal de São Carlos, Brazil
Marilton Aguiar	Universidade Federal de Pelotas, Brazil
Marley M. B. R. Vellasco	Pontifícia Universidade Católica do R. de J., Brazil
Marlo Souza	Universidade Federal da Bahia, Brazil
Marlon Mathias	Universidade de São Paulo, Brazil
Mauri Ferrandin	Universidade Federal de Santa Catarina, Brazil
Márcio Basgalupp	Universidade Federal de São Paulo, Brazil
Mário Benevides	Universidade Federal Fluminense, Brazil
Moacir Ponti	Universidade de São Paulo, Brazil
Murillo Carneiro	Universidade Federal de Uberlândia, Brazil
Murilo Loiola	Universidade Federal do ABC, Brazil
Murilo Naldi	Universidade Federal de São Carlos, Brazil
Myriam Delgado	Universidade Tecnológica Federal do Paraná, Brazil
Nuno David	Instituto Universitário de Lisboa, Portugal
Patrícia Tedesco	Universidade Federal de Pernambuco, Brazil
Paula Paro Costa	Universidade Estadual de Campinas, Brazil

Paulo Cavalin	IBM Research, Brazil
Paulo Pirozelli	Universidade de São Paulo, Brazil
Paulo Quaresma	Universidade de Évora, Portugal
Paulo Santos	Flinders University, Australia
Paulo Henrique Pisani	Universidade Federal do ABC, Brazil
Paulo T. Guerra	Universidade Federal do Ceará, Brazil
Petrucio Viana	Universidade Federal Fluminense, Brazil
Priscila Lima	Universidade Federal do Rio de Janeiro, Brazil
Priscila B. Rampazzo	Universidade Estadual de Campinas, Brazil
Rafael Giusti	Universidade Federal do Amazonas, Brazil
Rafael G. Mantovani	Universidade Tecnológica Federal do Paraná, Brazil
Rafael Parpinelli	Universidade do Estado de Santa Catarina, Brazil
Reinaldo A. C. Bianchi	Centro Universitario FEI, Brazil
Renato Krohling	UFES - Universidade Federal do Espírito Santo, Brazil
Renato Tinos	Universidade de São Paulo, Brazil
Ricardo Cerri	Universidade Federal de São Carlos, Brazil
Ricardo Marcacini	Universidade de São Paulo - São Carlos, Brazil
Ricardo Prudêncio	Universidade Federal de Pernambuco, Brazil
Ricardo Rios	Universidade Federal da Bahia, Brazil
Ricardo Suyama	Universidade Federal do ABC, Brazil
Ricardo A. S. Fernandes	Universidade Federal de São Carlos, Brazil
Roberta Sinoara	Instituto Federal de C., E. e T. de São Paulo, Brazil
Roberto Santana	University of the Basque Country, Spain
Rodrigo Wilkens	Université Catholique de Louvain, Belgium
Romis Attux	Universidade Estadual de Campinas, Brazil
Ronaldo Prati	Universidade Federal do ABC, Brazil
Rosangela Ballini	Universidade Estadual de Campinas, Brazil
Roseli A. F. Romero	Universidade de São Paulo - São Carlos, Brazil
Rosiane de Freitas R.	Universidade Federal do Amazonas, Brazil
Sandra Sandri	Instituto Nacional de Pesquisas Espaciais, Brazil
Sandro Rigo	Universidade do Vale do Rio dos Sinos, Brazil
Sílvia Maia	Universidade Federal do Rio Grande do Norte, Brazil
Sílvio Cazella	Universidade Federal de Ciências da S. de P. A., Brazil
Silvia Botelho	Universidade Federal do Rio Grande, Brazil
Solange Rezende	Universidade de São Paulo - São Carlos, Brazil
Tatiane Nogueira	Universidade Federal da Bahia, Brazil
Thiago Covoes	Wildlife Studios, Brazil
Thiago Homem	Instituto Federal de E., C. e T. de São Paulo, Brazil

Thiago Pardo	Universidade de São Paulo - São Carlos, Brazil
Tiago Almeida	Universidade Federal de São Carlos, Brazil
Tiago Tavares	Insper, Brazil
Valdinei Freire	Universidade de São Paulo, Brazil
Valerie Camps	Paul Sabatier University, France
Valmir Macario	Universidade Federal Rural de Pernambuco, Brazil
Vasco Furtado	Universidade de Fortaleza, Brazil
Vinicius Souza	Pontificia Universidade Católica do Paraná, Brazil
Viviane Torres da Silva	IBM Research, Brazil
Vladimir Rocha	Universidade Federal do ABC, Brazil
Washington Oliveira	Universidade Estadual de Campinas, Brazil
Yván Túpac	Universidad Católica San Pablo, Peru
Zhao Liang	Universidade de São Paulo, Brazil

Additional Reviewers

Alexandre Alcoforado
Alexandre Lucena
Aline Del Valle
Aline Ioste
Allan Santos
Ana Ligia Scott
Anderson Moraes
Antonio Dourado
Antonio Leme
Arthur dos Santos
Brenno Alencar
Bruna Zamith Santos
Bruno Labres
Carlos Caetano
Carlos Forster
Carlos José Andrioli
Caroline Pires Alavez Moraes
Cedric Marco-Detchart
Cinara Ghedini
Daiane Cardoso
Daniel da Silva Junior
Daniel Guerreiro e Silva
Daniela Vianna
Diego Cavalca
Douglas Meneghetti
Edilene Campos

Edson Borin
Eduardo Costa Lopes
Eduardo Max
Elias Silva
Eliton Perin
Emely Silva
Estela Ribeiro
Eulanda Santos
Fabian Cardoso
Fabio Lima
Fagner Cunha
Felipe Serras
Felipe Zeiser
Fernando Pujaico Rivera
Guilherme Mello
Guilherme Santos
Israel Fama
Javier Fumanal
Jefferson Oliva
João Fabro
João Lucas Luz Lima Sarcinelli
Joelson Sartori
Jorge Luís Amaral
José Angelo Gurzoni Jr.
José Gilberto Medeiros Junior
Juan Colonna

Leandro Lima
Leandro Miranda
Leandro Stival
Leonardo da Silva Costa
Lucas Alegre
Lucas Buzuti
Lucas Carlini
Lucas da Silva
Lucas Pavelski
Lucas Queiroz
Lucas Rodrigues
Lucas Francisco Pellicer
Luciano Cabral
Luiz Celso Gomes Jr.
Maëlic Neau
Maiko Lie
Marcelino Abel
Marcella Martins
Marcelo Polido
Marcos José
Marcos Vinícius dos Santos Ferreira
Marisol Gomez
Matheus Rocha
Mária Minárová
Miguel de Mello Carpi
Mikel Sesma
Murillo Bouzon
Murilo Falleiros Lemos Schmitt

Newton Spolaôr
Odelmo Nascimento
Pablo Silva
Paulo Rodrigues
Pedro Da Cruz
Rafael Berri
Rafael Gomes Mantovani
Rafael Krummenauer
Rafael Orsi
Ramon Abílio
Richard Gonçalves
Rubens Chaves
Sarah Negreiros de Carvalho Leite
Silvio Romero de Araújo Júnior
Tatiany Heiderich
Thiago Bulhões da Silva Costa
Thiago Carvalho
Thiago Dal Pont
Thiago Miranda
Thomas Palmeira Ferraz
Tiago Asmus
Tito Spadini
Victor Varela
Vitor Machado
Weber Takaki
Weverson Pereira
Xabier Gonzalez

Contents – Part I

Resource Allocation and Planning

Rules and Feature Extraction

IA and Education

Agent Systems

Explainability

IA Models

Contents – Part II

Deep Learning Applications

Reinforcement Learning and GAN

Machine Learning Analysis

Contents – Part III

Language and Models

Graph Neural Networks

Pattern Recognition

AI Applications

Best Paper

Embracing Data Irregularities in Multivariate Time Series with Recurrent and Graph Neural Networks

Marcel Rodrigues de Barros[1]([✉]), Thiago Lizier Rissi[1],
Eduardo Faria Cabrera[1], Eduardo Aoun Tannuri[1], Edson Satoshi Gomi[1],
Rodrigo Augusto Barreira[2], and Anna Helena Reali Costa[1]

[1] Universidade de São Paulo, São Paulo, SP, Brazil
{marcel.barros,thiago.rissi,eduardofcabrera,eduat,gomi,anna.reali}@usp.br
[2] Petróleo Brasileiro S.A., Rio de Janeiro, Brazil
barreira@petrobras.com.br

Abstract. Data collection in many engineering fields involves multivariate time series gathered from a sensor network. These sensors often display differing sampling rates, missing data, and various irregularities. To manage these issues, complex preprocessing mechanisms are required, which become coupled with any statistical model trained with the transformed data. Modeling the motion of seabed-anchored floating platforms from measurements is a typical example for that. We propose and analyze a model that uses both recurrent and graph neural networks to handle irregularly sampled multivariate time series, while maintaining low computational cost. In this model, each time series is represented as a node in a heterogeneous graph, where edges depict the relationships between each measured variable. The time series are encoded using independent recurrent neural networks. A graph neural network then propagates information across the time series using attention layers. The outcome is a set of updated hidden representations used by the recurrent neural networks to create forecasts in an autoregressive manner. This model can generate forecasts for all input time series simultaneously while remaining lightweight. We argue that this architecture opens up new possibilities as the model can be integrated into low-capacity systems without needing expensive GPU clusters for inference.

Keywords: Multivariate Time Series · Machine Learning · Artificial Intelligence · Graph Neural Networks · Recurrent Neural Networks · Data Imputation

We gratefully acknowledge the support from ANP/PETROBRAS, Brazil (project N. 21721-6), CNPq (grants 310085/2020-9 and 310127/2020-3), CAPES (finance code 001) and C4AI-USP (FAPESP grant 2019/07665-4 and IBM Corporation).

M. C. Naldi and R. A. C. Bianchi (Eds.): BRACIS 2023, LNAI 14195, pp. 3–17, 2023.
https://doi.org/10.1007/978-3-031-45368-7_1

1   Introduction

Although the study of Multivariate Time Series (MTS) spans over 40 years, this area remains relatively unexplored due, in part, to its distinct nature compared to conventional statistical theory. Classical methods primarily rely on random samples and independence assumptions, whereas MTS are characteristically highly correlated [19].

Despite this, the field has recently garnered attention from Machine Learning (ML) researchers, motivated by the recent achievements in token sequence modeling, particularly in tasks related to Natural Language Processing (NLP). Models like the Transformer [17] have generated groundbreaking results in tasks such as question answering and text classification, thereby prompting the question: Can such successes be replicated in the context of time series?

For the majority of applications, ML algorithms currently underperform when compared to traditional statistical methods for univariate time series [9]. One of the main hurdles is that deep learning models, largely responsible for the ML revolution in NLP, are particularly susceptible to overfitting in univariate scenarios. However, multivariate problems exhibiting complex interactions are not easily modeled by classical approaches. This makes it an appealing area for ML research exploration.

This work introduces a cost-effective encoder-decoder architecture founded on Recurrent and Graph Neural Networks (RNNs and GNNs), coupled with vector time representation capable of forecasting MTS. Moreover, this proposed model can process data from sensors with differing sampling rates and missing data profiles without resorting to data imputation techniques. Our dataset consists of motion data from Floating Production, Storage, and Offloading units (FPSOs)—floating sea platforms used for oil extraction. Moored to the seafloor, FPSOs display intricate oscillatory behavior influenced by environmental factors such as wind, currents, and waves.

The key contributions of this work are fourfold:

- We demonstrate that time encoding enhances forecasting results for complex oscillatory dynamics, especially when large missing data windows are present.
- We provide evidence that heterogeneous graph neural networks can effectively propagate information between abstract representations of sensor data.
- We define a comprehensive end-to-end system capable of modeling FPSO hydrodynamic behavior at low computational costs and without the need for elaborate data preprocessing routines.
- Additionally, we argue that the proposed architecture can be successful in a wide range of applications.

In light of the above, this article is organized as follows: Section 2 presents the foundational concepts underpinning this work and provides a synopsis of how data irregularities can affect common ML architectures. Section 3 outlines the proposed architecture, while Sect. 4 offers a more detailed description of the dataset and experimental setups. Sections 5 and 6, respectively, present the experimental results and offer some conclusions, as well as directions for future work.

2   Background

This section introduces grounding concepts for this work, namely: MTS definitions and terminology, RNNs and GNNs main characteristics as means to restrict the search space for the parameter optimization process, and how FPSO motion forecast context is centered in dealing with data irregularities.

2.1   Multivariate Time Series

Formally, we can define a multivariate time series context as a (finite) set of time series $\mathcal{S} = \{\mathbf{Z}_\mathbf{t}^\mathbf{i}\}_{i=1}^S$. Each element of this set is a sequence of events of length T_z, $\mathbf{Z}_\mathbf{t} = [Z_1, Z_2, \cdots, Z_{T_z}]$, such that $Z_t \in \mathbb{R}^{K_z}$. Note that we choose to represent a time series as a boldface matrix $\mathbf{Z}_\mathbf{t}$ with T_z observations of K_z features. This work treats each feature vector Z_t as an event associated with a scalar timestamp representing t.

The set $\mathcal{S}$ is defined over χ which is the input space of the model. Let's suppose that, in the simplest case, one wishes to forecast the features associated with a single future event Z_{T_z+d} from a time series $\mathbf{Z}_\mathbf{t}^\mathbf{i}$, but making use of all information contained in $\mathcal{S}$.

In other words, in this scenario we wish to approximate an unknown function $f : \chi \to \mathbb{R}^{K_z}$ such that $f(\mathcal{S}) = Z_{T_z+d}$ for all $\mathcal{S}$. We call the approximated function $\tilde{f} \in \mathcal{F}$.

An important concept in this work is the choice of function class $\mathcal{F}$ that contains the function $\tilde{f}$. The typical deep learning approach is to use a parametrized function class $\mathcal{F} = \{f_\theta \in \Theta\}$. An adequate parametrization guarantees the existence of a function $\tilde{f}$ that is arbitrarily close to the true function f [6]. This is called the interpolating regime i.e. $\tilde{f}(\mathcal{S}) = f(\mathcal{S})$ for all $\mathcal{S}$. Usually, the stochastic gradient descent algorithm is used to optimize the parameters by minimizing a loss function $L(\tilde{f}(\mathcal{S}), f(\mathcal{S}))$.

However, each data point in a time series context typically has little to no information since the process is mainly driven by strong correlations and white noise processes. In this scenario, simple deep neural networks such as Multilayer Perceptrons (MLPs) rapidly pick-up spurious patterns and are unable to generalize to unseen data. Thus, to learn these complex interactions from the data available it is necessary to reduce the size of $\mathcal{F}$ by restricting it.

2.2   RNNs and GNNs as Means to Restrict $\mathcal{F}$

Many successful deep learning models can be described as restricted optimization problems in function classes that represent some kind of symmetry. For example, Convolutional Neural Networks, which is the main architecture for image processing tasks, implement translational symmetry through the convolution operation on grids. Other analogous interpretations can be derived for gated RNNs, Transformers, and others [3].

In this work, we apply gated RNNs, which exhibit *time warping* symmetry invariance [16]. This symmetry is defined as any monotonically increasing differentiable mapping $\tau : \mathbb{R}^+ \to \mathbb{R}^+$ between times. Being invariant to this symmetry means that gated RNNs can associate patterns in time series data even if they are distorted in time as long as order is preserved.

We also apply GNNs to propagate information between different $\mathbf{Z}_t^i$. The GNN architecture [14] can be described by the message-passing mechanism:

$$\mathbf{x}_i^{(k)} = \gamma^{(k)} \left(\mathbf{x}_i^{(k-1)}, \square_{j \in \mathcal{N}(i)} \, \phi^{(k)} \left(\mathbf{x}_i^{(k-1)}, \mathbf{x}_j^{(k-1)} \right) \right), \tag{1}$$

where $\mathbf{x}_i^{(k)} \in \mathbb{R}^D$ is the new representation of the features of node i at layer k, $\phi^{(k)}$ is a differentiable *message function* that produces a hidden representation that is sent from node j to node i. $\square$ denotes a differentiable, permutation invariant *aggregation function* that aggregates all messages from neighbour nodes. $\gamma^{(k)}$ is a differentiable function that updates the node's hidden representation. Once again the crucial fact is the permutation invariance that restricts the function class $\mathcal{F}$ with the fact that isomorphic graphs have the same representation.

There are several variants of GNNs and this work does not intend to detail their differences. A comprehensive survey can be found at [21]. In this work we use the Heterogeneous Graph Attention Network architecture. That means we consider the graph that models the MTS as being heterogeneous, i.e., as having different types of nodes and edges. This may become clearer in Sect. 3 that details how these RNNs and a GNN are combined in an end-to-end system.

2.3   FPSO Dynamics and Irregularities in MTS

While the contributions of this work apply to a general MTS context, it was developed to address the challenge of maintaining stable positions of floating platforms in deep water. FPSO units have been employed for years to safely extract offshore oil reserves. These floating platforms enable the extraction and storage of crude oil while being held in place using mooring lines anchored to the ocean floor. Mooring systems play a critical role in ensuring the positional stability, personnel safety, and smooth functioning of various operations on a platform, such as extraction, production, and oil offloading.

The constant exposure of floating structures to offshore environmental conditions like waves, currents, and wind leads to continuous stresses and strains on the platforms. Consequently, the structural integrity and mooring lines of these platforms degrade over time. Understanding and modeling the complex dynamics of the floating platform allows for real-time detection of unexpected oscillatory patterns, which could assist in assessing the mooring lines' integrity. In this work, we analyze four variables: UTMN and UTME, which are north and east coordinates from a GPS, heading, which measures the rotation movement around the vertical axis, and roll, which measures the rotation along the longitudinal axis of the platform. These four variables are the most significant to characterize the platform movement.

The ever-increasing amount of data generated by sensors can be used to train ML algorithms for this task, but the same environment characteristics that deteriorate the mooring system also affect the sensors which are reflected in the measurements. Typical sensor data from FPSOs suffers from long-missing data windows, different polling rates between sensors, and other deformities.

Several data imputation approaches to achieve grid-like data structure have been proposed [5]. A regular grid structure allows for the usage of simple RNNs or even MLPs [13], but also causes the model behavior to be highly dependent on the data imputation technique used. Furthermore, achieving grid-like structure in complex sensor networks becomes rapidly unfeasible due to missing data points and shifted time series as shown in Fig. 1. Our proposal aims to offer a simple and robust solution that does not rely in any imputation process.

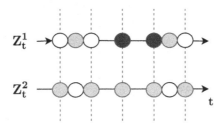

Fig. 1. White balls denote observations from sensors. $\mathbf{Z}_t^1$ is represented by a single feature and has two missing data points, denoted in dark gray, while $\mathbf{Z}_t^2$ also has one feature, but a different frequency, and is shifted. Red balls denote data that would have to be imputed to achieve a grid-like structure. (Color figure online)

3   Architecture

The proposed architecture consists of a set of single-layered RNNs, a single layer of a Heterogeneous Graph Attention Network [2,18], a time encoding module, and a linear projection layer.

Each input time series is processed by a different RNN instance into a fixed-size representation as shown in Fig. 2. In this work, we use Gated Recurrent Units (GRUs) [4] which is a proven gated RNN architecture. This is a design choice that allows both T_z and K_z to vary between time series since RNNs can process arbitrary sequence lengths and each RNN instance has its width predefined based on the number of features of its input time series.

A central aspect in this work is the encoding of timestamps. Instead of feeding RNNs with only feature values, we also encode the time scalar for each observation using the positional encoding mechanism from [17]. The process can be visualized in Fig. 3.

More formally, the time encoder can be defined as a function $T : \mathbb{R}^{T_z} \rightarrow \mathbb{R}^{T_z \times \mathcal{T}}$ that encodes each scalar $t - t_0$, where t_0 represents the instant of inference, into a representation of size $\mathcal{T}$. Each RNN encoder can be defined as $E^i : \mathbb{R}^{T_z \times (K_z + \mathcal{T})} \rightarrow \mathbb{R}^{\mathcal{H}}$ that encodes a sequence of T_z vectors with K_z features enriched by $\mathcal{T}$ time features into a single hidden representation of size $\mathcal{H}$.

Once each input time series has now a fixed-size representation, the information is shared between nodes using the message-passing mechanism defined in

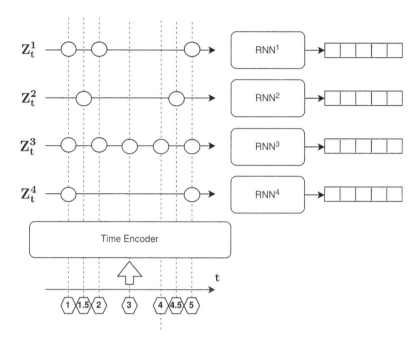

Fig. 2. Encoding each time series with a different RNN instance provides flexibility to handle irregularities without the need for data imputation. Each time series is encoded into a fixed-size representation.

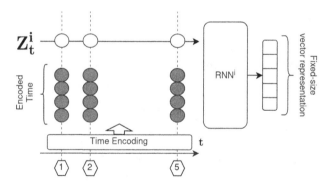

Fig. 3. Detailed view of encoding of a single time series. An RNN can ingest a sequence of arbitrary length T_z. Instead of just feeding it with K_z features (1 in this example), the model adds the encoded timestamps as additional information. RNNs generate an output for each element in the input sequence, but our architecture only makes use of the last as a fixed-size representation of the whole input sequence.

Eq. 1. Note that this expression denotes a homogeneous message passing. In a pure heterogeneous case, a different pair of functions ϕ and γ would be trained for each edge type. That means for a graph with N nodes there would be N^2

pairs of learnable parametrized message-passing functions. To prevent overfitting due to this quadratic increase of parametrized functions the regularization technique proposed in [15] is applied. In it instead of training a new pair of functions for each new edge type a fixed set of function pairs is trained and each edge type applies a different learnable linear combination of this set of *bases*.

The structures of ϕ and γ are determined by the choice of GNN architecture, this work applies the Graph Attention Network v2 [2] that calculates attention masks for the incoming messages of each node. This allows nodes to incorporate the most relevant neighborhood information dynamically.

The GNN part of the encoder can be defined as $\mathcal{G} : (\mathbb{R}^{N \times \mathcal{H}}, \mathcal{E}) \to \mathbb{R}^{N \times \mathcal{H}}$ that updates the hidden representation of each node $i \in N$ based on the set of edges $\mathcal{E}$. A fully-connected graph representation for our problem can be visualized in Fig. 4.

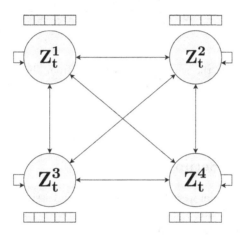

Fig. 4. Heterogeneous GNN takes as input the fixed-size representation of each time series and propagates the information through the graph using the message-passing mechanism. Outputs are the updated vector representations for each node.

Finally, the decoding process uses the same RNNs used for encoding, but now in an autoregressive configuration coupled with a linear layer. The initial hidden state $h_{t_0}^i$ is given by the updated hidden representation obtained from $\mathcal{G}$. Each forecasted data point y_{t+1}^i is obtained by projecting the hidden representation that is iteratively updated. During this process, we also enrich the input features for the RNN with the encoded representations of the distance to the target timestamp $(t + 1) - t_0$. The whole decoding process can be visualized in Fig. 5. Formally, each decoding step $\mathcal{D}^i : (\mathbb{R}^{\mathcal{H}}, \mathbb{R}^{K_z}, \mathbb{R}) \to (\mathbb{R}^{\mathcal{H}}, \mathbb{R}^{K_z})$ receives a hidden state h_t^i, the last forecasted label y_t^i and a target time $t+1$. It outputs an updated state h_{t+1}^i and its related output forecast y_{t+1}^i.

It is important to note that each module-specific implementation could be replaced. Adopting alternatives for time encoding such as Time2Vec [8], different RNN architectures, such as LSTMs [7], or different GNN architectures would not change the grounding principles behind this work but could bring further performance gains.

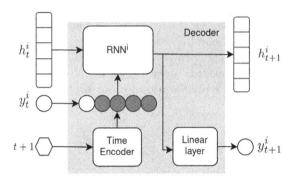

Fig. 5. The decoder uses the same RNN instances from the encoder. It takes the updated representations as its initial state and uses a linear projection to produce y_{t+1}^i in an autoregressive manner.

4 Experimental Setup

Three variants of the model were trained: a complete version including the timestamp encoding method and a fully connected graph (named full model); a second version without the timestamp encoding strategy (named no time enconding model); and a third version with a totally disconnected graph (named disconnected graph model). With these comparisons, we intend to evaluate the effectiveness of GNN in propagating information between abstract representations of sensor data and the improvement in regression performance provided by timestamp encoding.

For each model, 3 sets of tests were evaluated, varying the proportion of missing data points: 40%, 20% and 0%, in order to verify the model's robustness in addressing the irregularities that usually occur in real sensor data. Long windows of 5% missing data were randomly inserted into the test dataset before inference. Often, missing data in sensor networks appear in the form of long missing data windows instead of isolated missing data points [1].

The dataset is obtained from numerical simulations that applies partial differential equations to model environmental conditions acting onto a 3D model of an existing FPSO. The simulator generates three-hour MTS with 1 s frequency for 7003 different environmental conditions, of which we separated 50% for training, 20% for validation and 30% to test. This represents 14,006 h of simulated data.

In this work validation data is only used for training convergence plots, i.e., no hyperparameter optimization was performed. All experiments used the following settings: $T = 50, H = 200$. For the Heterogeneous GNN, the number of bases for regularization is 6 and the number of attention heads is 16. A single 10% dropout layer is applied to GNN outputs during training. Optimizing these hyperparameters may be beneficial, but it is not in the scope of this work.

To prepare the data, we applied a differentiation in each time series such that $ds_t = s_t - s_{t-1}$. This is a common technique in autoregressive models and is especially relevant for GPS data, since FPSO initial offset can vary widely with environmental conditions. In addition, we applied a Z-Score normalization to the differential time series.

The instant of inference is denoted t_0. The model builds the input MTS S by filtering measurements within a certain context window c in seconds so that $t_0 - c \leq t < t_0$. This architecture allows for each time series to adopt a different context window length c based on its characteristics. Longer context windows of $1800\,s$ are chosen for UTMN and UTME since their signals present lower frequencies, while shorter $600\,s$ windows are enough to characterize the spectrum of heading and roll motions, which mostly consist of high frequencies. Such as the input, the model builds the target MTS S_o by filtering the measurements within a certain forecast window d in seconds so that $t_0 \leq t < t_0 + d$. For all the four variables we adopted forecast windows of $d = 100\,s$.

The loss function $L(S_o, \hat{S}_o)$ used for training was based on the Index of Agreement [20] that provides a fit measurement between two time series that falls within the unit interval. This metric had been used successfully in the ML for MTS context by [10] since it provides a common range for all time series losses. For a MTS context it can be precisely defined as:

$$L(S_o, \hat{S}_o) = 1 - IoA = 1 - \frac{1}{S}\sum_{i=1}^{S}\left(\frac{\sum_t (\mathbf{Z^i} - \hat{\mathbf{Z}}^i)^2}{\sum_t (|\hat{\mathbf{Z}}^i - \bar{\mathbf{Z}}^i| + |\mathbf{Z^i} - \bar{\mathbf{Z}}^i|)^2}\right) \qquad (2)$$

In each training epoch, the model chooses randomly a t_0 from the three-hour MTS for doing the forecasting. With this, the model avoids becoming specialized in forecasting at some specifics inference times. For testing, we determined a set of inference times beforehand so each model can be consistently compared.

All models were trained for 2000 epochs. For the full model that corresponds to 22 h of training in a single NVIDIA RTX 3060Ti GPU. This inexpensive training is possible since the model is only four layers deep and contains 8.35M parameters taking up only 33.4 MB of disk space. The inference can be run in a consumer-level CPU in around 0.2 s while consuming 100 MB of memory.

5   Results and Analysis

Figure 6 displays each model's results for the time series in a randomly selected environmental condition and t_0, for different levels of missing data. The full model proved to be resistant to the increase of missing data points, and managed to maintain a consistent performance as the prediction horizon increased.

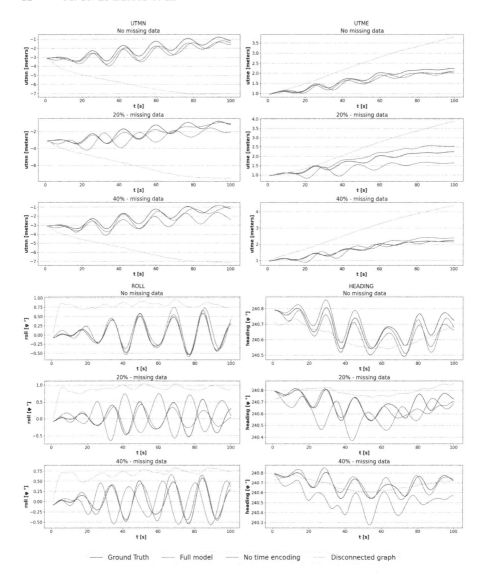

Fig. 6. Illustration of the three models predictions (full model, no time encoding model, and disconnected model) with distinct missing data proportions for the four time series (UTMN, UTME, Roll, and Heading) in the same random environmental condition, considering a forecast window of 100 s.

Additionally, the model using disconnected graphs performed worst than both the full model and no-time encoding model, since in several cases the model was not able to correctly learn the movement tendency, and produced forecasts with significant offsets. In other cases, it failed to represent high frequency components of the MTS.

From this qualitative view it is noticeable that the full model also outperforms the no-time encoding model, especially as missing data windows increase. This is the hypothesized behavior since the time encoding provides a way for the RNN to recognize a time gap and adjust for it. Thus, the model was able to better extrapolate temporal patterns of motion.

Box plots in Fig. 7 present the evolution of the regression's mean absolute error in the forecast window using the full model. For better visualization we omit the outliers, limiting the whiskers to comprehend the range of 5% to 95% percentiles. It is observed that as the temporal distance increases the error increases, as expected.

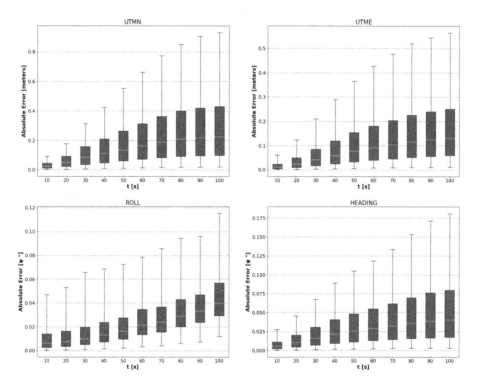

Fig. 7. Boxplots of the absolute error versus the time distance to target in the forecast window for all the test environmental conditions. The whiskers are limited to comprehend the range of 5% to 95% percentiles.

Numerical results are presented in Table 1, displaying absolute errors' statistics calculated upon the last predicted point in the inference horizon of 100 s. The results confirm the analysis for Fig. 6, showing that the full model outperformed the other two models in the vast majority of the cases, except for the 25% and 50% percentiles for the *Roll* series without missing data, that were better predicted by the no-time encoding model.

Table 1. Table containing all numeric results calculated for the last predicted point in the inference horizon for all different forecast windows and environments conditions. The 25, 50 and 75 columns refer to the absolute error that represents the respective percentiles, while the max column refers to the maximum error obtained in the forecast of any environmental condition and t_0.

	UTMN				UTME			
	25	50	75	max	25	50	75	max
No missing data								
No time encoding	0.13	0.30	0.56	6.28	0.07	0.18	0.37	7.39
Disconnected model	3.57	4.75	6.25	25.7	1.71	2.30	3.09	15.1
Full model	**0.10**	**0.22**	**0.43**	**4.06**	**0.06**	**0.13**	**0.25**	**4.18**
20% missing data								
No time encoding	0.40	0.96	2.07	14.37	0.29	0.89	2.50	20.2
Disconnected model	3.24	4.67	6.28	33.2	1.86	2.50	3.40	17.5
Full model	**0.21**	**0.47**	**0.91**	**12.6**	**0.13**	**0.30**	**0.63**	**14.4**
40% missing data								
No time encoding	0.52	1.23	2.54	15.8	0.40	1.14	2.73	23.0
Disconnected model	3.02	4.58	6.30	38.0	1.95	2.63	3.57	17.2
Full model	**0.27**	**0.62**	**1.20**	**15.6**	**0.18**	**0.43**	**0.90**	**12.2**
	Roll				Heading			
	25	50	75	max	25	50	75	max
No missing data								
No time encoding	**0.02**	**0.03**	0.08	3.34	0.03	0.07	0.13	1.70
No graph connectivity	0.36	0.64	0.99	5.75	0.13	0.25	0.42	4.92
Full model	0.03	0.04	**0.06**	**2.03**	**0.02**	**0.04**	**0.08**	**0.97**
20% missing data								
No time encoding	**0.03**	0.07	0.20	4.98	0.07	0.16	0.30	3.11
No graph connectivity	0.36	0.66	1.03	6.76	0.11	0.22	0.39	5.55
Full model	**0.03**	**0.05**	**0.09**	**3.02**	**0.03**	**0.07**	**0.14**	**2.34**
40% missing data								
No time encoding	0.04	0.09	0.27	4.59	0.09	0.19	0.35	3.41
No graph connectivity	0.35	0.66	1.04	7.23	0.10	0.21	0.38	4.96
Full model	**0.03**	**0.05**	**0.10**	**2.44**	**0.03**	**0.09**	**0.18**	**2.14**

Here it is emphasized the resilience of the full model to the increase in missing data. This occurs because missing windows are recognized by the time encoding mechanism and missing information can be filled in by neighbouring nodes in a latent space, allowing for an "intrinsic data imputation" to occur. This effect of feature completion in GNNs was previously described by [12].

All three models displayed forecast outliers where errors were several times larger than the 95% percentile. Preventing or recognizing these outliers is an important task to enable these type of models for operating in mission critical tasks. Nevertheless, the full model also exhibited fewer and less extreme outliers.

To conclude, the evolution of the IoA during training calculated over the validation dataset is provided in Fig. 8. The results are evidence that both graph connectivity and time encoding act together to improve convergence speed and stability during training.

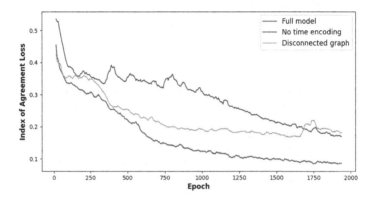

Fig. 8. Evolution of Index of Agreement loss during training. This plot denotes IoA loss running mean for all time series for the last 30 epochs calculated over the validation dataset.

6   Conclusion

Results provide evidence that heterogeneous GNNs can effectively propagate information between abstract representations of sensor data, exploiting the relationship between them. This work also demonstrates that GNNs can be associated with RNNs and time encoders to tackle missing data windows in MTS.

This general architecture is a simple, cost-effective, yet robust solution for restricting the function class $\mathcal{F}$ and allowing the stochastic gradient optimization to converge faster and more accurately in complex multivariate time series contexts. We argue, based on the results and conveniences of this model, that this architecture could be successful in a multitude of applications making it a strong one-fits-all model in the MTS context.

As future research directions, we note that other works have been successful on modeling MTS as abstract networks. We believe that approaches treat each single measurement as a different node such as [22] can be combined with latent graph inference techniques, allowing for dynamic inference of an underlying optimized graph topology.

Creating specialized functions that are able to store relevant information from hidden states into persistent vector representations for nodes between different inference times such as proposed by [11] is also an interesting direction for long-term predictions.

Finally, we believe a more comprehensive study on the choice of loss function and how known MTS losses affect this architecture could bring other insights on its capabilities and weaknesses.

This work was financed in part by ANP/PETROBRAS, Brazil (project N. 21721-6). We also gratefully acknowledge partial support from CNPq (grants 310085/2020-9 and 310127/2020-3).

References

1. Adhikari, D., Jiang, W., Zhan, J.: Imputation using information fusion technique for sensor generated incomplete data with high missing gap. Microprocess. Microsyst., 103636 (2021). https://doi.org/10.1016/j.micpro.2020.103636
2. Brody, S., Alon, U., Yahav, E.: How attentive are graph attention networks? CoRR abs/2105.14491 (2021). https://arxiv.org/abs/2105.14491
3. Bronstein, M.M., Bruna, J., Cohen, T., Velickovic, .P.: Geometric Deep Learning Grids, Groups, Graphs, Geodesics, and Gauges. CoRR abs/2104.13478, 160 (2021)
4. Cho, K., et al.: Learning phrase representations using RNN encoder-decoder for statistical machine translation, September 2014. https://doi.org/10.48550/arXiv.1406.1078. arXiv:1406.1078 [cs, stat]
5. Emmanuel, T., Maupong, T., Mpoeleng, D., Semong, T., Mphago, B., Tabona, O.: A survey on missing data in machine learning. J. Big Data 8(1), 140 (2021). https://doi.org/10.1186/s40537-021-00516-9
6. Hartman, E.J., Keeler, J.D., Kowalski, J.M.: Layered neural networks with Gaussian hidden units as universal approximations. Neural Comput. 2(2), 210–215 (1990). https://doi.org/10.1162/neco.1990.2.2.210
7. Hochreiter, S., Schmidhuber, J.: Long short-term memory. Neural Comput. 9, 1735–80 (1997). https://doi.org/10.1162/neco.1997.9.8.1735
8. Kazemi, S.M., et al.: Time2Vec: learning a vector representation of time (2019). https://doi.org/10.48550/arXiv.1907.05321. arXiv:1907.05321 [cs]
9. Makridakis, S., Spiliotis, E., Assimakopoulos, V.: Statistical and machine learning forecasting methods: concerns and ways forward. PLoS ONE 13(3), e0194889 (2018). https://doi.org/10.1371/journal.pone.0194889. https://journals.plos.org/plosone/article?id=10.1371/journal.pone.0194889. Publisher: Public Library of Science
10. Netto, C.F.D., et al.: Modeling oceanic variables with dynamic graph neural networks, June 2022. https://doi.org/10.48550/arXiv.2206.12746. arXiv:2206.12746 [cs]
11. Rossi, E., Chamberlain, B., Frasca, F., Eynard, D., Monti, F., Bronstein, M.: Temporal graph networks for deep learning on dynamic graphs. arXiv:2006.10637 [cs, stat], October 2020. arXiv: 2006.10637
12. Rossi, E., Kenlay, H., Gorinova, M.I., Chamberlain, B.P., Dong, X., Bronstein, M.: On the unreasonable effectiveness of feature propagation in learning on graphs with missing node features, May 2022. https://doi.org/10.48550/arXiv.2111.12128. arXiv:2111.12128 [cs]

13. Saad, A.M., et al.: FPSO mooring line failure detection based on predicted motion. In: Proceedings of the ASME 2021 40th International Conference on Ocean, Offshore and Arctic Engineering. American Society of Mechanical Engineers Digital Collection, October 2021. https://doi.org/10.1115/OMAE2021-62413

14. Scarselli, F., Gori, M., Tsoi, A.C., Hagenbuchner, M., Monfardini, G.: The graph neural network model. IEEE Trans. Neural Netw. **20**(1), 61–80 (2009). https://doi.org/10.1109/TNN.2008.2005605. http://ieeexplore.ieee.org/document/4700287/

15. Schlichtkrull, M., Kipf, T.N., Bloem, P., van den Berg, R., Titov, I., Welling, M.: Modeling relational data with graph convolutional networks. In: Gangemi, A., et al. (eds.) ESWC 2018. LNCS, vol. 10843, pp. 593–607. Springer, Cham (2018). https://doi.org/10.1007/978-3-319-93417-4_38

16. Tallec, C., Ollivier, Y.: Can recurrent neural networks warp time? In: International Conference on Learning Representation 2018, February 2018. https://openreview.net/forum?id=SJcKhk-Ab

17. Vaswani, A., et al.: Attention is all you need. In: Guyon, I., et al. (eds.) Advances in Neural Information Processing Systems, vol. 30. Curran Associates, Inc. (2017)

18. Veličković, P., Cucurull, G., Casanova, A., Romero, A., Lió, P., Bengio, Y.: Graph attention networks. arXiv:1710.10903 [cs, stat], February 2018. http://arxiv.org/abs/1710.10903

19. Wei, W.W.S.: Multivariate Time Series Analysis and Applications. Wiley, p. 528 (2019)

20. Willmott, C.J.: On the validation of models. Phys. Geogr. **2**(2), 184–194 (1981). https://doi.org/10.1080/02723646.1981.10642213. Publisher: Taylor & Francis _eprint

21. Wu, Z., Pan, S., Chen, F., Long, G., Zhang, C., Yu, P.S.: A comprehensive survey on graph neural networks. IEEE Trans. Neural Netw. Learn. Syst. **32**(1), 4–24 (2021). https://doi.org/10.1109/TNNLS.2020.2978386. https://ieeexplore.ieee.org/document/9046288/

22. Zhang, X., Zeman, M., Tsiligkaridis, T., Zitnik, M.: Graph-guided network for irregularly sampled multivariate time series, March 2022. https://doi.org/10.48550/arXiv.2110.05357. http://arxiv.org/abs/2110.05357 [cs]

Regulation and Ethics of Facial Recognition Systems: An Analysis of Cases in the Court of Appeal in the State of São Paulo

Cristina Godoy B. de Oliveira[1][(✉)], Otávio de Paula Albuquerque[2],
Emily Liene Belotti[1], Isabella Ferreira Lopes[3], Rodrigo Brandão de A. Silva[4],
and Glauco Arbix[4]

[1] Faculty of Law of Ribeirão Preto, University of São Paulo (USP), São Paulo, Brazil
{cristinagodoy,emilybelotti}@usp.br
[2] School of Arts, Sciences and Humanities, USP, São Paulo, Brazil
otavioalbuquerque@usp.br
[3] Institute of Mathematical and Computer Sciences of São Carlos, USP, São Paulo, Brazil
isabella.lopes6@usp.br
[4] Department of Sociology, USP, São Paulo, Brazil
{brandao-cs,garbix}@usp.br

Abstract. Context: The use of Artificial Intelligence (AI) in various sectors of the economy is already a reality in Brazil. Consequently, since 2019, the number of cases in the Judiciary involving AI has increased. Cases involving facial recognition systems (FRS) for contracting bank credit are increasing annually, so it is necessary to analyze how the Judiciary handles the issues. Problem: Why is the São Paulo Court of Appeal ruling in favor of banks in all cases involving taking out credit through facial recognition technology? Methodology and Methods: Data were collected and processed using automated computer programs. The qualitative analysis used the analytical, comparative and monographic methods. Results: The Court of Appeal of São Paulo considers it difficult to deceive an AI system, therefore, the burden of proof is on the author, even if there is a consumer relationship. That is, the decisions are contrary to the general rule of the Code of Consumer Protection in Brazil, which consists of reversing the burden of proof in consumer relations when one of the parties is underprivileged. Contributions and Solutions: The research points to the path of jurisprudence in cases involving the contracting of credit through FRS, and the Judiciary is deciding against the bank's customers, dispensing with the production of evidence by the banking sector. Therefore, it is necessary to alert the National Council of Justice and the Central Bank regarding this situation so that it is disciplined adequately since the FRS is fallible and does not guarantee the absence of fraud.

This work was carried out at the Center for Artificial Intelligence (C4AI-USP), with support by the São Paulo Research Foundation (FAPESP grant #2019/07665-4) and by the IBM Corporation. This study was financed in part by the Coordenação de Aperfeiçoamento de Pessoal de Nível Superior - Brasil (CAPES) - Finance Code 001.

M. C. Naldi and R. A. C. Bianchi (Eds.): BRACIS 2023, LNAI 14195, pp. 18–32, 2023.
https://doi.org/10.1007/978-3-031-45368-7_2

Keywords: Facial Recognition Systems and Legal Cases · Artificial Intelligence and Law · AI and Jurisprudence

1   Introduction

The regulation of Artificial Intelligence (AI) has garnered the attention of various stakeholders worldwide. In Brazil, the subject gained visibility between 2019 and 2021 when three bills concerning the regulation of AI began to be discussed in the National Congress. In 2022, the discussion on the regulation of AI in the country became even more intense due to the creation, by the Federal Senate, of a commission of jurists responsible for gathering information to create a regulatory framework for AI. Amidst these topics, many researchers have overlooked the fact that the legal regulation of AI depends, among other elements, on decisions by the Judiciary regarding disputes involving the different applications of the technology in question, rather than solely relying on general laws that directly deal with regulation.

In addition to this gap in the discussions on the legal regulation of AI, there is another gap in the discussions on the use of AI by the Brazilian Public Power: it is still not completely clear to researchers how AI applications have been used by national public offices. In the case of the Brazilian Judiciary, Salomão *et al.* [22] attempted to map AI use cases. Investigations such as this are valuable but do not provide us with a clear understanding of how this Power actually deals with AI in concrete situations where the technology in question may have caused objective harm to specific individuals.

To partially address these issues, our research question is: When analyzing cases involving facial recognition systems, does the Judiciary of the State of São Paulo evaluate the transparency of these systems and their cybersecurity level against fraud? Guided by this question, we applied web scraping techniques to collect data from the Court of Justice web portal and analyzed the arguments presented by the Judiciary Power to justify its decisions involving the use of facial recognition (FR) systems for credit loan contracts. We also assessed how clearly the use of such technologies is understood by both the Judiciary Power and the contracting party of the credit.

We chose to study facial recognition systems because, as Daly [6] points out, they are an example of an AI application that tangibly affects individuals' lives, allowing them to conceive the technology in a more concrete manner, even though it may sometimes seem abstract. The choice of the State of São Paulo is justified by its status as the most prosperous state in the country, potentially making it the state with the most frequent use of AI applications. Finally, we discuss the challenges of transparency in AI, as we believe that ethical AI is impossible without it.

In addition to this introductory section, the article comprises three additional sections. Section 2 presents the methodology used to scrape data on the decisions of the Judiciary in the State of São Paulo and includes the methodology employed to analyze the collected data. Section 3 presents and discusses the obtained results, highlighting the dependence of facial recognition systems on sensitive personal data and noting that the Judiciary of the State of São Paulo has given little importance to the requirement of free and informed consent for the treatment of such data. Section 4 concludes the article.

2   Materials and Methods

An exploratory research was conducted to investigate the factors that influence the decision-making of the State of São Paulo's Court of Appeal regarding subjects related to Artificial Intelligence [10]. The study was based on data collected from the decisions stored in the portal of services (E-SAJ) of the Court of Appeal portal[1] using a Web Scraping technique, which consists of extracting online data to obtain content of interest in a systematic and structured manner [11].

In this research, most of the data collection and treatment procedures were automated using computer programs. The technologies used can be defined into two categories: (1) Web scraping tools and (2) data treatment coding.

2.1   Web Scraping Tools and Frameworks

To scrape the data from the Court of Appeal processes, an Application Programming Interface (API) named TJSP was used. It is a community open-source application published on GitHub [8]. The main goals of the TJSP tool are to collect, organize, and structure data from process judgments of the first and second instances of São Paulo's Court of Appeal, using different methods for various data extraction proposals. This tool was developed using the programming language R, and the integrated development environment (IDE) called RStudio was used to manipulate the code and instantiate the API for our research, defining the terms and authenticity properties.

The API was used to extract data from the Court of Appeal web portal proceedings by searching for all decision documents containing the term "Artificial Intelligence" and was improved to deliver the data in a structured and organized file. After defining the search terms, the following steps were taken: (1) Download the corresponding HTML pages with all the processes from the São Paulo's State Court of Appeal web portal; (2) Based on the obtained HTML files, identify the information about the processes, returning information such as class, subject, rapporteur, court district, judging body, date of judgment, date of publication, case, summary, and code of the decision; (3) Store the retrieved information in a data frame, where each column represents specific information and each row represents a process; (4) Convert the data frame into a spreadsheet for qualitative and quantitative analysis.

The collected decisions in this document refer to 190 (one hundred ninety) matching results, comprising data from 2012 to 2023, related to the most general terms in the AI context: "inteligência artificial" and "artificial intelligence"; "aprendizado de máquina" and "machine learning"; "aprendizado profundo" and

---

[1] São Paulo's State Court of Justice web portal - Jurisprudence consultations - Full consultation: https://esaj.tjsp.jus.br/cjsg/consultaCompleta.do?f=1.

"deep learning". The search included the terms searched in Portuguese and English, respectively, and in singular and plural. These results correspond to the first decision document registered on the portal on june 24th, 2010, to the latest scraping round performed on April 20th, 2023.

2.2   Data Treatment

Between the information retrieved from the decisions, there are different types of data, including integral, date-time, and character strings. Given the goal of our study and to help facilitate the qualitative analysis, we applied different data treatment techniques[2], divided into pre- and post-processing approaches. These approaches were applied to the "Jurisprudence" column, which contains the decision text. Cleaning techniques were developed using the Python programming language and deployed in the Google Colab IDE to take advantage of the performance benefits offered by the platform and make them available to the research team.

Data Pre-processing. The pre-processing and cleaning of data are the most critical and time-consuming parts of a web scraping project. To obtain a good visualization of the data collection, it usually requires multiple iterations of cleaning, transforming, and visualizing the data before it is in a suitable form [27]. There are different techniques, ranging from the simplest to the most advanced, to preprocess a data collection, i.e., prepare, clean, and transform data from an initial form to a structured and context-specific form. In our research, we applied data mining techniques to clean and search for repetitive or redundant parts in the "Jurisprudence" column content to improve the qualitative analysis.

The decision content follows a standard structure containing repetitive or redundant sections. These sections include: (1) titles, headers, references, and terms that are identical for all documents, such as "Judiciary Power," "Court of Appeal of the State of São Paulo," "Electronic Signature," "Decision," among others; (2) appeal and vote numbers, which differ for each decision but will not be used in the analysis. These numbers are identified by being preceded by terms such as "appeal n," "civil appeal n," "appeal.," "record:," "vote n," and others; and (3) document page identifiers, such as current page numbers, referenced section page numbers, or total page numbers of the decisions. These identifiers are preceded by sections such as "(fls.," "(fl." Using the previously mentioned cleaning techniques, these expressions were removed from the results spreadsheets.

Data Post-processing. Once the data cleaning stage is complete, we move on to post-processing the information. In order to derive insights from the data, it is imperative that the data is presented in a format that is readily analyzable. This requires the data to be not only clean and consistent but also well-structured, aiding in addressing inconsistencies in scraped data [14]. This stage involves structuring the collected data according to the research context, or in other

---

[2] Project coding and data treatment methods applied to this work: https://github.com/isabellalopes-usp/tjsp_data_treatment.

words, highlighting and grouping the most important parts of each decision to streamline and aid in the accuracy of manual work done by analysts.

In the context of data post-processing, automated procedures were developed to improve the qualitative analysis by identifying relevant terms/contents that could make it easier to discover the topic addressed in each process and determine whether or not it should be part of the ongoing research analysis scope. (1) From each decision, all paragraphs in which the searched term appears were selected and added in a new column; (2) It was identified that some decisions refer to others using the number of the respective process. Thus, for each decision, references to its process number in other documents were searched, and this information was also stored in a new column of the results spreadsheet. Another procedure developed was the grouping of decisions based on rapporteurs, also using Python modules for data manipulation, this improve made it more practical and efficient to perform a qualitative analysis of the results in a critical and embracing way.

Also, aiming to continuing collecting data, to improve the useful information, and to make the research dynamic by keeping the results up to date, a periodicity of repetition of the scraping procedures was established. For this, other methods were developed in Python programs. They are: (1) A results counting mechanism, which compares the number of terms with corresponding matches and the number of matches for each term, comparing these numbers with the respective ones from the last scraping performed. These data are inserted into a new spreadsheet and updated every month; (2) In each results spreadsheet, the new scraped processes, compared to the ones from previous scraping, are marked with the character '*', so that analysts can easily identify new results, avoiding redundancies.

2.3   Qualitative Analysis Method

The main procedural methods adopted were analytical, comparative, and, especially, monographic. This was revealed through a longitudinal study of the specificities of each case, investigating the factors that influenced decision-making in each unique event, based on sectoral analysis and making generalizations about similar events. This article examined all decisions of the State of São Paulo Court of Appeal that contain the following six terms: "inteligência artificial" and "artificial intelligence," "aprendizado de máquina" and "machine learning," and "aprendizado profundo" and "deep learning." During the search, 190 items were retrieved, out of which 41 were deeply analyzed for the purpose of this ongoing research, totaling 21.58% of the results. These analyses focused specifically on banking matters, with particular attention to disputes involving the validity of digital payroll loan contracts carried out by facial biometrics. The first decision analyzed was issued and published on August 30, 2021, and the last decision (according to data scraping conducted in April 2023) was on April 20, 2023.

Using the monographic method, this research will present the standard arguments issued by each of the rapporteurs in the decisions referred to, justifying the validity of the loan contracts. Firstly, the existence of a legal relationship between the bank and the consumer will be analyzed, as well as the alleged lack of voluntary consent. This will provide a generalized framework of the views adopted by judges regarding facial recognition technology (FRT) and its use in digital banking contracts in Brazil.

Finally, the analysis of the Appeal Court's decisions was divided into 13 subcategories: (i) case number; (ii) class (type of appeal analyzed in the decision); (iii) rapporteur; (iv) judicial district; (v) judging body; (vi) subject; (vii) trial date; (viii) publication date; (ix) disputing parties; (x) reasoning; (xi) final decision; (xii) context of the use of the term "Inteligência Artificial" (A.I.); (xiii) jurisprudence.

3 Results and Discussion

From the web portal of the State of São Paulo's Court of Justice, 190 decisions covering a span of 12 years (from 2012 to 2023) were analyzed. The results showed that 177 decisions involved the term "Inteligência Artificial," 4 decisions referred to the term "Artificial Intelligence" (from October 2022 to April 2023), 3 decisions related to "Aprendizado de Máquina" (from August 2022 to February 2023), and 6 decisions mentioned "Machine Learning" (from September 2018 to November 2022). However, no decisions were found for the terms "Deep Learning" or "Aprendizado Profundo" through web scraping on the Appellate Court's website.

Figure 1 illustrates the exponential growth trend in the number of decisions containing these six terms. The graph covers the period from 2012 to 2023, showing relative stability until 2019 and a significant increase from 2020 onwards. Several factors contribute to this trend, including: (1) The enactment of the "Lei Geral de Proteção de Dados Pessoais" (LGPD), Law No. 13,709, on August 14, 2018, which came into force in Brazil in September 2020. This has led to an increase in disputes related to the implementation of AI systems in various sectors of society, such as banking, service provision through applications, and compliance with the new standards. (2) As mentioned in the "qualitative analysis method" section, 37.3% of the "Inteligência Artificial" results are related to banking matters, including disputes involving credit card operations, ATM fraud, nighttime bank robberies with threats for consecutive withdrawals, phone or SMS scams where scammers pretend to be AI virtual assistants to obtain passwords/personal data, and payroll loan cases involving facial recognition (the focus of this article). The occurrence of facial recognition-related cases is also expected to grow rapidly, as depicted in Fig. 1.

The apparent drop in 2023 does not indicate an actual decrease in the number of decisions. It is important to note that this research only compiled data up until April of that year. The growth trend remains high, as the number of decisions analyzed up until April 2023 alone has already more than doubled the entire year of 2019.

Our results will focus solely on decisions related to the term "Artificial Intelligence," totaling 41 in number. The other terms do not involve disputes regarding the validity of digital payroll loan contracts carried out through facial biometrics. In fact, they do not pertain to banking matters at all. These decisions cover topics such as "health plans," "higher education," "contractual readjustment," "business management," "indemnity for moral damages," and "provision

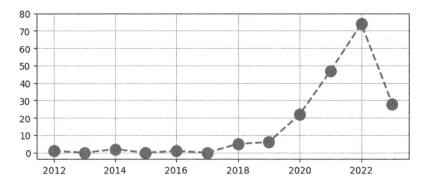

Fig. 1. The Number of Occurrences Related to 6 Term searched "Inteligência Artificial","Artificial Intelligence", "Aprendizado de Máquina", "Machine learning", "Aprendizado Profundo" and "Deep learning", between 2012 and 2023.

of transport application services," among others. None of the appellants in these cases are banking parties.

Out of the 41 decisions analyzed, in 40 of them, it was determined that a legal relationship existed between the contracting bank and the customer. Therefore, the validity of the electronic payroll loan agreement was confirmed by the judges who recognized the legitimacy of the debt and the legality of the bank. Furthermore, the Court of Appeal acknowledged that there is no unenforceability of the loan refinancing agreement. In cases where the bank files an appeal, it is granted in terms of the contract's validity and the absence of unenforceability for material and moral damages (when claimed by the plaintiffs). When the judgment is unfavorable to the plaintiff, their appeal is dismissed. The existence of the loan and the allegation of bad faith litigation (when claimed) are recognized or partially dismissed (removing the bad faith litigation while maintaining the validity of the loan).

Figure 2, it is possible to point out that the exponential growth trend in the occurrence of payroll loan cases involving facial biometrics over time. The graph covers the period from 2020 to 2023. However, the registered cases in the facial biometrics contexts emerge from 2021. Based on these data, it can be inferred that such an event occurs in this way, since the regulation of facial recognition technology (FRT) is still quite incipient in the country, both in terms of the legislative framework and its implementation. In the LGPD, for example, there is not even a mention of the term "facial recognition" or "facial biometrics", restricting itself only to classifying in its art. 5, item II, as "sensitive" biometric data (relating to health, sex life, genetic data, religious/political/philosophical group, etc.). In Brazil, there are only three bills dealing with FRT: PL 2392/22, which prohibits the use of FRT for identification purposes in the public and private sectors without a prior report on the impact on people's privacy; PL 3069/22, which regulates the use of automated facial recognition by public security forces in criminal investigations or administrative procedures; and PL 2338/23, which regulates artificial intelligence and allows the use of facial recognition systems for public security purposes in public spaces only with legal and judicial authorizations. Therefore, deprived of a normative framework that minimally regulates not only the FRT implementation procedures in the country but also the rights and duties of consumers and banks regarding such technology, judges end up issuing judicial decisions that could be more grounded, complete, and divergent.

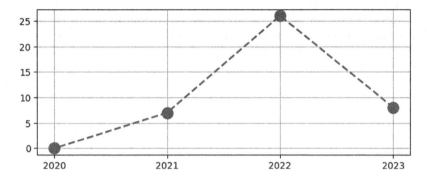

Fig. 2. The Number of Occurrences Related to Facial Recognition Between 2020 and 2023.

Meanwhile, the four main arguments presented by judges to justify the existence of the digital payroll loan agreement involving facial recognition technology (FRT) are listed below:

(i) Non-challenge of the documents presented by the bank. The plaintiff did not challenge his photo displayed by the bank (defendant) nor the receipts for withdrawing the money used by the plaintiff, nor did he require the production of an expert report at the appropriate procedural time to prove possible tampering with the evidence presented by the opposing party.

(ii) Article 3 of Normative Instruction No. 28/2008 of the INSS/PRES allows the contracting of a loan to take place electronically. Therefore, the signature is not a requirement for the existence or even the validity of the contract, as it would only serve to confirm the expression of will. This is indeed essential and indispensable and was granted by the author when he agreed to collect his facial biometrics for the specific purpose of formalizing the contract. Therefore, a contract carried out electronically dispenses with the formal rigors of a standard contract, precisely to the extent that the electronic form can be carried out anywhere, as long as the consumer consents.

(iii) In consumer relations, according to the Brazilian consumer protection code (CDC), there is a reversal of the burden of proof: presentation, by the bank, of a bank credit note (signed by the plaintiff through the capture of facial biometrics through 'selfie', similar to its ID photo); proof of transfer of the contracted amount made to the applicant's account, confirmation of address at the same domicile declared by the plaintiff, digital signature, presentation of original documents, geolocation, digital contract indicating that it was signed by mobile application, telephone model cell phone used for hiring, in addition to time and IP address.

(iv) All decisions are based on a judgment that considered: "As for electronic hiring and facial biometrics, I understand that such a form of hiring is perfectly appropriate, given the technological resources available and widely used [...] "The electronic medium is valid. Falsifying a signature is something quite common, unfortunately, but falsifying a person's face to "fool" the artificial intelligence, even if it is possible, requires more sophisticated fraud, and the burden of proving the occurrence of this fraud was on the author, even if it were it is a consumer relationship, since it is a constitutive fact of the right alleged by itself."

3.1   Consent

Payroll loan contracts have a history of problems in Brazil. In 2020, complaints about such credit doubled [3], totaling 20,564 records, half of the total in 2019 (39,012 records), according to records available in the government database. Likewise, the Consumer Protection and Defense Foundation (PROCON) in São Paulo recorded a 50% increase in complaints against banking institutions [21]. Idec - Brazilian Institute of Consumer Protection - evaluated more than 300 reports of complaints sent on the consumer portal, which show, in general, the easy access of banks to confidential bank data of consumers with the INSS (National Institute of Social Security), such as confirmation of bank operations without the direct involvement of the person responsible for the account or the release of the payroll loan without the consumer's consent [4].

Such data can be explained by a conjuncture of phenomena, among them, the "esteira invertida", in which fraudsters—usually bank correspondents[3] or financial market operators—deposit an uncontracted loan in the account of the retiree or pensioner, without their authorization, to receive up to 6% of the transaction amount as commission. The fraudster voluntarily chooses victims who have previously carried out payroll loans, reusing the personal data of these consumers. Thus, in a situation of vulnerability, such retirees often end up using the money without knowing its origin. Therefore, the argument used by the reporting judges that the plaintiff did not challenge the withdrawal receipts presented by the bank does not hold up, as it is entirely plausible that the retiree, a hypo sufficient consumer in relation to the financial institution, really believed that the money in his account was his, not knowing that he had been the victim of fraudulent activity.

The Payroll Law determines that it can only be deducted from the benefit "when expressly authorized by the beneficiary" (Law no. 8213, art. 5, inc.V). If there is no express authorization, the discount will be undue. Likewise, article 39, item III of the Consumer Defense Code expressly prohibits the supplier of

---

[3] Individuals who acquire certification to work in the sale of banking services, with emphasis on credit operations and credit card granting.

products or services from sending to the consumer, without prior request, any product or taking advantage of the condition of ignorance, social situation, age or health to sell its products or services (article 39, item IV). Such devices refer to the "consent" provided for in the LGPD (Brazilian General Law for the Protection of Personal Data, Law No. 13.709/2018), which is inspired by the General Data Protection Regulation of the European Union (GDPR): natural person holder of the data, or by their legal guardian, and must be expressed clearly and unequivocally, in writing or not".

Consent must be informed, that is, before obtaining it, it is essential to provide information to data subjects so that they can understand what they agree with, for example: the identity of the person responsible for processing the data, the purpose of each of the processing operations for which consent is sought, what types of data will be analyzed, etc. [26]. Article 4, n⁰ 11 of the GDPR clearly states that consent requires an "unequivocal positive statement or act" from the data subject, which means that it has to be given using positive action or explicit statement, mainly in situations where there is a severe risk to data protection and, therefore, it is appropriate to have a high level of security, such as biometric data and/or financial information - both used in payroll loan contracts. Therefore, such an explicit manifestation may be through contracts or electronic forms, but it is necessary to stick to adhesive contracts with pre-signed adhesion conditions, in which the data subject's intervention is necessary to prevent acceptance.

Thus, the argument of the reporting judge that "contracts executed in electronic form dispense with the formal rigors of a standard contract, precisely to the extent that the electronic form can be carried out anywhere" does not hold, since the generalized acceptance of general conditions does not match the aforementioned definition of unequivocal, explicit and informed consent. Likewise, even though Article 107 of the Brazilian Civil Code (CCB) provides for freedom of form for contracting, using the electronic contract which dispenses with a traditional signature, any new technology used must prove the explicit demonstration of consent, guaranteeing the reliability of the protection of consumer data, ensuring unequivocal fraud prevention mechanisms—deficient characteristics in the use of facial biometrics, presented below.

3.2   Facial Biometrics

The General Data Protection Regulation (GDPR) clarifies what biometric data is in its Art. 4th [7]:

> Article 4 - Definitions For the purposes of this Regulation, it is understood by: [...] "Biometric data", personal data resulting from a specific technical treatment relating to the physical, physiological or behavioral characteristics of a natural person that allows or confirm the unique identification of that natural person, namely facial images or dactyloscopic data;

In the Brazilian legal framework, Decree No. 10,046/2019, which "provides for governance in data sharing within the scope of the federal public administration and establishes the Base Citizen Register and the Central Data Governance Committee" defines biometric attributes [1] as:

Art. 2nd For the purposes of this Decree, it is considered: [...] II - biometric attributes - measurable biological and behavioral characteristics of the natural person that can be collected for automated recognition, such as the palm of the hand, fingerprints, retina or iris of the eyes, face shape, voice and way of walking.

Biometric processing corresponds to the processing of biometric data. It can "be referred to interchangeably as recognition, identification, authentication, detection or other related terms, as well as (often opaque) ways of collecting and storing biometric data, even if the data is not processed immediately" [19]: Generally speaking, a facial recognition system operates by using biometrics to capture a person's facial features, comparing the information to a database. They usually work through similar steps: a image of the person's face is captured from a photo or video; then, the facial recognition software using a machine learning algorithm, verifies a series of facial characteristics such as the distance between the eyes, the curvature of the mouth, etc., to extract "facial signatures" based on the identified facial characteristics patterns. This signature would then be compared to a database of pre-registered faces. Finally, there is the determination, in which the verification of the analyzed face can occur [16].

From a technical point of view, facial recognition technology (FRT) is a subcategory within Artificial Intelligence (AI). FRT is less accurate than, for example, fingerprinting, mainly when used in real-time or in large databases [9]. Several factors influence the probability/accuracy of a match, such as: the quality of image, represented by the physical characteristics, external accessories usage, equipment and format usage, and environment conditions; quality of the dataset, size and proportion of the data; and the quality of the model by the choose of threshold, parameters, and fine-tuning [20, 23].

Determining the accuracy level of facial recognition can be challenging and result in both false positives (a face is mistaken for another in a database where it shouldn't be) and false negatives (a face doesn't match in a database in which it should be registered). Based on machine learning nature, the algorithms will never bring a definitive result, only probabilities. This means that identical twins can be misidentified. In addition, cutouts of ethnic, gender or age groups can also generate false results.

Debora Lupton [15], based on the studies of Donna Haraway [13] and Annemarie Mol [18], explains that digital data is never a "tabula rasa" (superficial), but must be understood and experienced as something generated by structures, through processes of different characteristics, but above all, perhaps, social and cultural. Artificial Intelligence technology is not created"by itself", nor is it devoid of the structural constraints that shape it. It was found that, on average, only 10 to 20% of the group responsible for developing artificial intelligence technologies is made up of women [25]—the data refer to the largest technology companies in the USA, a global hub in the development of this field. Thus, darker people, ethnic groups, women, transgender people and people with physical disabilities are more likely to incur false negatives or positives due to less inaccuracy than other technologies (fingerprinting) and the low diversification of facial databases.

Furthermore, the quality of facial recognition (FR) models can be measured by their error rate, that is, the number of times such models fail to compare images of person's face. Known technically as False Non-Match Rate, the error rate varies according to the types of images presented to FR algorithms. It tends to be smaller when the comparison is based on images taken in controlled environments, where variables such as the position of the face and the incidence of light can be manipulated. International visa photos are an example of this type of image. Compared to images obtained in real situations, the error rate of FR systems tends to be higher. In this context, it is worth mentioning that, "As of 2022, the top-performing models on all of the [. . .] datasets [that integrate the National Institute of Standards and Technology (NIST)'s Face Recognition Vendor Test], with the exception of WILD Photos, each posted an error rate below 1%, and as low as a 0.06% error rate on the VISA Photos dataset" [17].

In this way, the justification given by the judge rapporteur for recognizing the validity of the FRT is highly generic and devoid of foundation. Faking the FRT does not necessarily require more sophisticated fraud, and automatic erroneous matching may occur depending on the accuracy level of a specific recognition system. Not limited to occasional errors, like any other computational technologies, biometric data is also at risk of violation and misuse, either by outsiders (hackers) or insiders, employees who may use the pre-collected data for their own use [5]. An example of insiders is the banking correspondents of payroll loans, which can improperly renew the loans or activate them without the consumer's authorization, or even the bank attendants themselves who use the wide range of personal data of consumers stored by the institution to defraud loans.

In Brazil, studies such as the one by NIST are lacking. Therefore, we need to find out what are, in general, the error rates of the systems used in the country. For this reason, Brazilian banks must be able not only to inform the public about the error rates of the systems they develop and use, but also whether they are as low as those identified by the US agency. Brazilian banks must also be able to inform if the error rates of their systems are equally distributed among different social groups or if they are concentrated in specific ones. This requirement is necessary because recent studies have shown that different FR systems make more errors when confronted with the faces of non-white people [2,12,24]. Finally, and most importantly, any institution that decides to use FR systems must have clear protocols for using the outputs generated by the technology. In the case of taking out credit, for example, banks should investigate with users whether they are aware that automated facial recognition is a crucial step in the process.

4   Conclusions and Future Work

In the context of payroll loans, it is clear that the motivation and grounds for decisions are very generic, incomplete or even erroneous. The fact that the judge uses facial recognition as an infallible means of proof of hiring is wrong, since in addition to several factors influencing its accuracy, the algorithms do not bring a definitive result, based on alternative solutions for situations that do not appear or are not frequent in their databases. Likewise, presenting the digital contract as evidence indicating that it was signed by mobile application, cell phone model used for contracting, in addition to time and IP address, is not sufficient to demonstrate the existence of free consent. The geolocation mechanism, for example, is susceptible to fraudulent mechanisms both for the one based on IP (VPNs, proxies, tor, tunneling) and for the one based on GPS (through fake GPS applications). As for the electronic signature, despite art. 107 of the Brazilian Civil Code provides for free forms of contract, an electronic contract presupposes the same free, explicit and informed consent provided for in the LGPD and the GDPR.

Therefore, when analyzing disputes involving hiring through facial recognition, it is recommended that the judge try to base his decisions on the greatest possible number of consolidated references, both in the area of science and technology, as well as in the area of sociology and public policies, given that this is a field still undergoing implementation and adaptation in the most diverse instances of society.

From this, it is noted that it is still necessary to improve the standards that encompass facial recognition technology. Therefore, within the scope of financial institutions, it is recommended that the Brazilian Central Bank regulate this matter and define the measures that other banks must implement to guarantee the reliability, security, transparency, responsibility, diversity and effectiveness of the FRT systems, in addition to defining essential minimum conditions for the use of such technology in its service contracts, guaranteeing unequivocal, explicit and informed consent to its customers/consumers regarding the new facial biometrics technologies.

Finally, the CNJ (National Court of Justice), according to Article 102 of Resolution 67 of 2009, may edit normative acts, resolutions, ordinances, instructions, administrative statements and recommendations. Additionally, resolutions and statements will have binding force on the Judiciary Power once approved by the majority of the CNJ Plenary. Thus, it is recommended that the CNJ pass a resolution that governs the criteria to be met by magistrates for the production of evidence in cases involving AI systems. Transparency must be guaranteed in all cases, and it is necessary not to assume that the AI is infallible. In any case related to AI, the absence of production of evidence concerning to the existence of fraud should be accepted and the review of the decisions made by the AI should be enforced. If this situation in the Judiciary continues, we will watch the weakening of due process of law and human dignity.

This research carried out the web scrap data using umbrella terms from the field of Artificial Intelligence both in English and Portuguese such as "artificial intelligence", "inteligência artificial", "machine learning", "aprendizado de máquina", "deep learning" and "aprendizado profundo", which may have limited the amount of returned decisions. As future work, we will analyze a list of terms related to applied artificial intelligence and its subareas, and their applications.

References

1. Brasil: Decreto n° 10.046, de 9 de outubro de 2019: Dispõe sobre a governança no compartilhamento de dados no âmbito da administração pública federal e institui o cadastro base do cidadão e o comitê central de governança de dados. Diário Oficial da República Federativa do Brasil (2019). https://www.planalto.gov.br/ccivil_03/_ato2019-2022/2019/decreto/D10046.htm
2. Buolamwini, J., Gebru, T.: Gender shades: intersectional accuracy disparities in commercial gender classification. In: Conference on Fairness, Accountability and Transparency, pp. 77–91. PMLR (2018)
3. Camargo, C.H., et al.: Um ano de pandemia da covid-19: diversidade genética do sars-cov-2 no brasil. BEPA. Boletim Epidemiológico Paulista **18**(207), 12–33 (2021)
4. Camarotto, M.: Reclamações explodem e governo decide olhar de perto o crédito consignado (2020). https://valorinveste.globo.com/produtos/credito/noticia/2020/10/16/reclamacoes-explodem-e-governo-decide-olhar-de-perto-o-credito-consignado.ghtml. Accessed 1 May 2023
5. Cheng, L., Liu, F., Yao, D.: Enterprise data breach: causes, challenges, prevention, and future directions. Wiley Interdisc. Rev. Data Mining Knowl. Disc. **7**(5), e1211 (2017). https://doi.org/10.1002/widm.1211
6. Daly, A.: Everyday AI ethics: from the global to local through facial recognition. OSF Preprints (2022). https://doi.org/10.31219/osf.io/bhm3w
7. European Union: EUR-LLex: access to European union law. micro-, small- and medium-sized enterprises: definition and scope (2016). https://eur-lex.europa.eu/legal-content/PT/TXT/?uri=celex:32016R0679. Accessed 1 May 2023
8. Filho, J.J.: Pacote TJSP: Coleta e organização de dados do tribunal de justiça de são paulo (2020). https://github.com/jjesusfilho/tjsp. Accessed 20 Nov 2022
9. Garvie, C.: The perpetual line-up: Unregulated police face recognition in America. Center on Privacy & Technology, Georgetown Law (2016)
10. Gil, A.C.: Como elaborar projetos de pesquisa, 4th edn. Editora Atlas, São Paulo (2002)
11. Glez-Peña, D., Lourenço, A., López-Fernández, H., Reboiro-Jato, M., Fdez-Riverola, F.: Web scraping technologies in an API world. Brief. Bioinform. **15**(5), 788–797 (2014). https://doi.org/10.1093/bib/bbt026
12. Hao, K.: A us government study confirms most face recognition systems are racist (2019). https://www.technologyreview.com/2019/12/20/79/ai-face-recognition-racist-us-government-nist-study/. Accessed 10 May 2023
13. Haraway, D.J.: The companion species manifesto: dogs, people, and significant otherness. Prickly Paradigm Press, Chicago (2003)
14. Lan, H.: COVID-scraper: an open-source toolset for automatically scraping and processing global multi-scale spatiotemporal COVID-19 records. IEEE Access **9**, 84783–84798 (2021). https://doi.org/10.1109/ACCESS.2021.3085682
15. Lupton, D.: Digital companion species and eating data: implications for theorising digital data-human assemblages. Big Data Soc. **3**(1), 1–5 (2016). https://doi.org/10.1177/2053951715619947

16. Lynch, J.: Face off: law enforcement use of face recognition technology. Elect. Front. Found. (EFF) (2020). https://doi.org/10.2139/ssrn.3909038
17. Maslej, N., et al.: Artificial intelligence index report 2023. Technical report, Institute for Human-Centered AI - Stanford University, Stanford (2023). https://aiindex.stanford.edu/wp-content/uploads/2023/04/HAI_AI-Index-Report_2023.pdf
18. Mol, A.: I eat an apple on theorizing subjectivities. Subjectivity **22**, 28–37 (2008). https://doi.org/10.1057/sub.2008.2
19. Montag, L., Mcleod, R., De Mets, L., Gauld, M., Rodger, F., Pełka, M.: The rise and rise of biometric mass surveillance in the EU: a legal analysis of biometric mass surveillance practices in Germany, The Netherlands, and Poland. EDRi (European Digital Rights) and EIJI (Edinburgh International Justice Initiative) (2021)
20. Phankokkruad, M., Jaturawat, P.: Influence of facial expression and viewpoint variations on face recognition accuracy by different face recognition algorithms. In: 2017 18th IEEE/ACIS International Conference on Software Engineering, Artificial Intelligence, Networking and Parallel/Distributed Computing (SNPD), pp. 231–237 (2017). https://doi.org/10.1109/SNPD.2017.8022727
21. Procon: Procon tem reclamações de aposentados que não pediram empréstimos consignados (2020). https://globoplay.globo.com/v/8970817/. Accessed 1 May 2023
22. Salomão, L.F., outros: Inteligëncia artificial: Tecnologia aplicada á gestão dos conflitos no âmbito do poder judiciário brasileiro (2020). https://ciapj.fgv.br/sites/ciapj.fgv.br/files/estudos_e_pesquisas_ia_1afase.pdf. Accessed 17 April 2023
23. Sharif, M., Naz, F., Yasmin, M., Shahid, M.A., Rehman, A.: Face recognition: a survey. J. Eng. Sci. Technol. Rev. **10**(2), 1–12 (2017)
24. da Silva, T.: Visão computacional e racismo algorítmico: branquitude e opacidade no aprendizado de máquina. Revista da Associação Brasileira de Pesquisadores/as Negros/as (ABPN) **12**(31), 428–448 (2020)
25. Simonite, T.: AI is the future-but where are the women (2018). https://www.wired.com/story/artificial-intelligence-researchers-gender-imbalance. Accessed 30 April 2023
26. União Europeia: Grupo de trabalho para o artigo 29° para a proteção de dados (2017), https://edpb.europa.eu/about-edpb/more-about-edpb/article-29-working-party_pt. Accessed 1 May 2023
27. VanderPlas, J.: Python for Data Science Handbook. O'Reilly Media, Sebastopol (2016)

A Combinatorial Optimization Model and Polynomial Time Heuristic for a Problem of Finding Specific Structural Patterns in Networks

Igor M. Sampaio$^{(\boxtimes)}$ 🄳 and Karla Roberta P. S. Lima 🄳

University of Sao Paulo - USP, Sao Paulo, Brazil
{igor.ms,ksampaiolima}@usp.br

Abstract. The development of tools based on robust mathematical models to deal with real-world, computationally intractable problems has increasingly aligned with combinatorial optimization and integer linear programming techniques due to enormous technological advances and the real need to deal with large volumes of data. In this paper, the Maximum Tropical Path Problem on graphs (MTPP), which is known to be NP-hard, was investigated. It is a problem of searching for specific structural patterns in networks that can represent various applications, among them biological interactions, such as metabolic, neurological or protein interaction networks. The main result of this work consists of a polynomial-time heuristic algorithm that, according to experimental results, finds good solutions in practice. Furthermore, an integer linear programming model was developed and implemented so that it could be compared with the heuristic presented. To conclude, an empirical analysis was performed through computational experiments on both random and real-word instances to evaluate the results presented in this paper.

Keywords: Optimization Combinatorial · Integer linear programming · Graphs

1 Introduction

The tropical path search problem in graphs represents a search problem for specific structural patterns in networks. Different types of pattern search have been presented in the literature, taking into account the peculiarities of each application; the one used in this work originates from the problem of occurrence of the *motifs* in graphs and was proposed by Lacroix et al. (2006) [10]. The initial problem was investigated in the context of metabolic network analysis [5,9], in which the set of reactions involved in the synthesis and degradation of certain molecules is represented through a network and the *motifs* correspond to the "modules" into which this network can be decomposed in order to facilitate the interpretation of the functions that each of these modules plays in cellular metabolism.

This work was partially supported by grant 2021/07080-6, São Paulo Research Foundation (FAPESP) and by the Graduate Support Program (Proap)-CAPES.

M. C. Naldi and R. A. C. Bianchi (Eds.): BRACIS 2023, LNAI 14195, pp. 33–47, 2023.
https://doi.org/10.1007/978-3-031-45368-7_3

In graph theory terms, the problem of searching for colored motifs in graphs consists of finding a subgraph whose set of colors matches the multicolored set given as input.

Contributions. This work aims to present results in the line of combinatorial optimization for a problem with applications in bioinformatics that represents the problem of searching for structural patterns in graphs. Specifically, the main contributions of this work are as follows:

– polynomial time heuristic algorithm for the MTPP that transforms a simple graph instance into a cactus graph instance and subsequently runs an algorithm that finds the path of interest.
– formulation of an integer linear programming (ILP) model for, given a graph colored at the vertices, finding a path that maximizes the number of colors used.
– tests and analysis of the quality and performance of the proposed algorithm based on the ILP model through large instances, both real-world and random.

This paper is organized as follows: Sect. 2 presents some initial concepts and a brief review of the literature on the tropical path problem in graphs; Sect. 3 presents the ILP model formulation; Sect. 4 presents the proposed heuristic algorithm; Sect. 5 presents the experimentation and results; and Sect. 6 presents the concluding remarks.

2   Concepts and Related Work

A *simple graph* G is an ordered pair (V, E), where V is a finite set of elements called *vertices* and E is a set of elements called *edges*, where each edge is an unordered pair of distinct vertices (u, v). If $u, v \in V$ is an edge, then u and v are *adjacent*. A *path* in a graph G is a sequence of distinct vertices $P = (v_1, v_2, \ldots, v_k)$, such that $v_i v_{i+1} \in E$ for $i = 1, \ldots, k-1$. A graph G is *connected*, if for any pair of distinct vertices u and v, there is a path from u to v in G. A *coloring* of a graph $G = (V, E)$ is a function $C : V \rightarrow \mathcal{C}$, where $\mathcal{C}$ is a set of colors. The coloring defined here corresponds to a simple assigning of colors to the vertices of the graph, without any restrictions. A *color graph* G^c consists of a graph G and a coloring C of the vertices of G. Given a graph G colored at the vertices, a path P in G is said to be *tropical* if each color, assigned to some vertex in the input graph, appears at least once in the P. A *cactus graph* is a connected graph in which any two simple cycles have at most one vertex in common; equivalently, each edge belongs to at most one simple cycle. A cactus G in which a vertex is defined as a root is called a *rooted cactus*. The root of a cactus G is denoted by $root(G)$.

The problem of interest consists of, given a colored graph of vertices, finding a path whose vertices use the greatest number of distinct colors. This problem will be referred to as *maximum tropical path problem (MTPP)*. Note that any solution to the *MTPP* is optimal if the path found is tropical and not every optimal solution is a tropical path (Fig. 1). Furthermore, if each vertex in the input graph has a color distinct from the others, MTPP reduces to the classical Hamiltonian path problem. Cohen et al. (2017) [4] proved that MTPP is NP-hard even on DAGs (acyclic directed graph), cactus graphs, and interval graphs, and it can be solved in polynomial time for some specific classes, such as trees, block graphs, proper interval graphs, bipartite chain graphs and threshold graphs. Chapelle et al. (2017) [2] proposed exact algorithms for the problem of searching for a tropical subset of minimum size. Anglès et al. (2018) [1] studied tropical subsets in graphs from a structural and algorithmic point of view and obtained results on the line of inapproximability. In general the problem of searching for tropical subgraphs in vertex colored graphs has been explored extensively for some classes of graphs; for more details see [1,3,6].

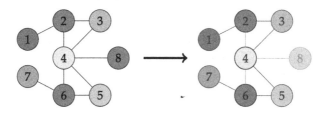

Fig. 1. An optimal solution with 7 vertices and 6 colors. The instance has 7 colors. See that in the solution it is allowed to repeat colors.

3   An IP Model for MTPP

In this section, the integer linear programming model developed for the MTPP problem is presented. The initial purpose of developing an integer linear programming model for the MTPP was to explore the combinatorial structures of the problem and obtain algorithms involving branch-and-bound and branch-and-cut techniques, for example. However, the model was satisfactorily used to directly obtain the optimal solution for large instances to be compared to the heuristics proposed in this paper.

To formulate the MTPP as an integer linear program, an instance of this problem is considered to consist of a graph colored at the vertices $G^C = (G, C)$. In the presented formulation, each edge $(u, v) \in E(G)$ is replaced by two directed arcs uv and vu. Variables X_{uc}, Y_{uv}, and Z_c, $\forall u \in V, \forall uv \in A$, and $\forall c \in C$ are used and are associated with vertices, arcs, and colors, respectively. The variable $X_{u,c} = 1$ indicates that the vertex u of color c is in the solution, the variable $Y_{uv} = 1$ indicates that the arc uv is in the solution, and the variable $Z_c = 1$ imposes that the color c is in the solution. Additionally, the variable $l_v \geq 0 \ \forall v \in V$ is created, which allows us to number (label) the vertices along the path in increasing order. This variable is used in the constraint proposed by Miller, Tucker, and Zemlin (1960) [12] to eliminate circuits in the solution.

3.1   ILP for General Graphs

In summary, the objective function maximizes the number of colors that should compose the solution. The constraint (1) prevents the variable Z_c from assuming value 1 without at least one vertex of color c being in the solution. Given a vertex v, the constraints (2) and (3) guarantee that if a vertex v is in the solution, the number of incoming and outgoing arcs of v will be at most 1, respectively; The constraint (4) makes use of the function l_v that allows to number the vertices along the path in increasing order. This inequality is essential to avoid cycles in the solution. The equality (5) guarantees that the difference between the number of vertices and the number of edges that make up the solution is exactly 1. The constraints (6) and (7) guarantee the integrality of variables.

$$\max \sum_{c \in C} Z_c$$

s. t.

$$Z_c \leq \sum_{u \in V} X_{uc}, \qquad\qquad \forall c \in C \qquad\qquad (1)$$

$$\sum_{c \in C} X_{vc} - \sum_{u \in \delta^-(v)} Y_{uv} \geq 0, \qquad\qquad \forall v \in V \qquad\qquad (2)$$

$$\sum_{c \in C} X_{vc} - \sum_{w \in \delta^+(v)} Y_{vw} \geq 0, \qquad\qquad \forall v \in V \qquad\qquad (3)$$

$$(l_v - l_u) \geq Y_{uv} - (1 - Y_{uv})N, \qquad\qquad \forall uv \in A \qquad\qquad (4)$$

$$\sum_{u \in V} \sum_{c \in C} X_{uc} - \sum_{u \in V} \sum_{v \in V} Y_{uv} = 1, \qquad\qquad (5)$$

$$X_{uc}, Y_{uv} \in \{0, 1\}, \qquad\qquad \forall u \in V, \forall uv \in A, \forall c \in C \qquad (6)$$

$$Z_c \in \{0, 1\}, l_v \geq 0; \qquad\qquad \forall v \in V, \forall c \in C \qquad\qquad (7)$$

Note that the proposed formulation has a linear amount of inequalities added to the model as the (1) constraints add $|\mathcal{C}|$ inequalities to the model, while the (2) and (3) constraints each add $|V|$ inequalities, and the (4) constraints add $2|A|$ inequalities. Last, the (5) adds only one inequality to the model, totaling a set of $|\mathcal{C}| + 2|V| + 2|A| + 1$ linear inequalities.

Theorem 1. *The restricted solution x of the integer solutions (x, y, z) for the formulation from (1) to (7) defines a path on the graph (G, C) that uses $|C'|$ colors, where $C' \subseteq C$.*

Proof. For any feasible integer solution (x, y, z) of the formulation, let $V_P = \{u : X_{uc} = 1\}$, $A_P = \{uv : Y_{uv} = 1\}$, and $C_P = \{c : Z_c = 1\}$. Note that if a vertex $u \in V_P$, then $c \in C_P$. Let C' be the set formed by the colors of vertices where $X_{uc} = 1$. If an arc $uv \in A_P$, by the constraint (2), $\sum_{c \in C} X_{vc} = 1$, and by the constraint (3), $\sum_{c \in C} X_{uc} = 1$, so $u, v \in V_P$. By the same conditions imposed in constraints (2) and (3), for each $u \in V_P$, there is at most one outgoing arc in u and at most one incoming arc in u. Note that the subgraph $G = (V_P, A_P, C_P)$ is well defined but not necessarily connected. If an arc $uv \in A_P$, by the constraint (4) $Y_{uv} = 1$ and the difference between the labels of vertices u and v has to be at least one, which prohibits circuits, finally, the connectivity of G is established by adding the constraint (5) and thus x indicates a path that uses C' colors, which completes the proof.

Although not done in this paper, one way to evaluate the strength of the ILP model would be to compare the value of the integer optimal solution to the value of the optimal solution of the model without the integrality conditions. This is the integrality gap.

4   Polynomial-Time Heuristic for MTPP

It is known that MTPP is NP-hard for simple graphs in general, especially for DAGs, cactus graphs, and interval graphs. The inherent intractability of MTPP encourages the development of heuristics that are designed to find bounds, in particular lower bounds, for the problem.

In this section, it is presented a polynomial-time heuristic called the *Tropical-Cactus* algorithm for finding maximum tropical paths in vertex-colored graphs. For this the algorithm makes use of the algorithm presented in Sect. 4.2. Existing literature does not provide any exact, approximation, or heuristic algorithms for the MTPP. Tropical-Cactus algorithm is inspired by a linear-time algorithm proposed by Markov et al. (2012) [11] for calculating the longest paths in cactus graphs. Cleverly, the algorithm finds a path whose number of colors empirically comes very close to an optimal solution. Additionally, an auxiliary algorithm, presented in Sect. 4.1 called the *Cactus-Graph* algorithm was developed to construct a cactus graph from a simple graph.

4.1   Cactus-Graph Algorithm

Given a simple graph G, the Cactus-Graph algorithm extracts from G a cactus graph G' that will be the input of the Tropical-Cactus algorithm. To perform this process, it is necessary to find *long cycles* in the graph. Given an integer $k \geq 3$, a cycle in an undirected graph is called *long* if it has at least k edges. The Cactus-Graph algorithm uses as a subroutine the algorithm presented in [7] that finds a path of length $exp(\Omega(\sqrt{log\ L}))$, where L is the length of the longest path (the details of this algorithm will be omitted here). This subroutine will be called in this paper $FindCycle(G, v)$ and receives as input a cactus graph G and a vertex v.

Extraction Step of a Cactus Graph from a Graph G: Given a simple graph G, any vertex $v \in G$, and a positive integer $k \geq 3$, the constructed graph is defined by A_k. Two types of components form A_k. A component in A_k is either a *singular component* or a *circuit component*. A *singular component* corresponds

---

Algorithm 1: Cactus-Graph

Input: G is a vertex-colored graph and v is a random vertex in the graph.
Output: A_k.

1 $C = FindCycle(G, v)$;
2 **if** $C \neq \emptyset$ **then**
3 $\quad$ $A_k \leftarrow C - v$;
4 $\quad$ **forall** *component* $H \in G - C$ **do**
5 $\quad\quad$ Choose a vertex $u \in H$ neighbor of $w \in C$;
6 $\quad\quad$ $V(A_k) \leftarrow u$;
7 $\quad\quad$ $E(A_k) \leftarrow (u, w)$;
8 $\quad\quad$ **if** $V(H) \backslash u \neq \emptyset$ **then**
9 $\quad\quad\quad$ Cactus-Graph(H, u);

10 **else**
11 $\quad$ Choose a vertex $u \in G$ neighbor of v;
12 $\quad$ $V(A_k) \leftarrow u$;
13 $\quad$ $E(A_k) \leftarrow (v, u)$;
14 $\quad$ **if** $G - v \backslash u \neq \emptyset$ **then**
15 $\quad\quad$ Cactus-Graph($G - v, u$);

---

to a single vertex $u \in G$ and defines a vertex in A_k, while a *circuit component* corresponds to a circuit $C \in G$ and defines a circuit in A_k.

The Fig. 2 demonstrates an example of transforming a simple graph G into a cactus graph A_k, where k is the minimum circuit size. Note that the set of vertices circled by the red line represents a *circuit component* in the graph A_k and has a length of 9. The vertices circled by the blue line represent *singular components* in the graph A_k. Note that vertex 7 could be a neighbor of $2, 6$, or 11 in the graph A_k, and in a second execution of the algorithm, a circuit of different size could be found.

The first vertex to become part of A_k is the given vertex v, and after the execution of the algorithm, v may be a singular component or become part of a circuit-component in A_k. It is also generally defined in this work that the minimum cycle size k is equal to 3, as such variation did not present a significant change in the final results of the paths found.

Note that the Cactus-Graph algorithm is linear, and at each step of the algorithm either a cycle or a vertex is extracted from G which in turn is connected to another cycle or another vertex, maintaining the properties of a cactus graph.

4.2 Labelling Algorithm

Definition 1. *Given a vertex u in a cactus graph G that is root, the children of u are all vertices adjacent to u or that are in a cycle that contains u.*

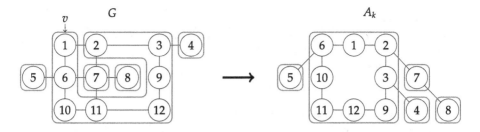

Fig. 2. Example of transforming G into A_k for $k \geq 3$

Definition 2. *Given a cactus graph G and a vertex $u \in V(G)$, F^u denotes the set of all vertices that are children of u, and $H^u \subseteq F^u$ denotes the subset of F^u consisting only of the children of u that belong to some cycle containing u.*

Definition 3. *A vertex u is a bridge vertex if u has degree two, belongs to a cycle, and is the child of some other visited vertex.*

Definition 4. *A vertex u', child of a root u, has no children if u' is a bridge vertex or a leaf vertex.*

The Fig. 3 demonstrates a cactus graph rooted in u, and presents an example of the formation of the sets F^u and H^u. In the figure, F^u is highlighted by the red line and, in this example, is formed by the vertices $F^u = v_1, v_2, v_3, v_4, v_5, v_6, v_7$, while H^u is highlighted by the blue line and, in this example, is formed by the vertices $H^u = v_3, v_4, v_5, v_6, v_7$.

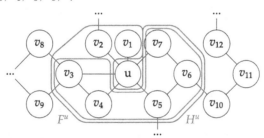

Fig. 3. A rooted cactus graph at u.

Step of Calculating the Vertex Labels of a Cactus Graph: By associating an integer value to each vertex of a cactus graph, one obtains a *labeled cactus graph*. In this algorithm, two labels are associated with each vertex $u \in V$: the partial label and the total label, both defined below.

Definition 5. *The value of the **partial** label of u ($r_p(u)$) represents the largest number of colors on a path in a cactus subgraph rooted at u.*

Definition 6. *The value of the **total** label of u ($r_t(u)$) represents the largest number of colors on a path that contains u and is contained in cactus subgraphs rooted at u.*

Algorithm 2: Label-Calculation

Input: (G, C, u) vertex-colored cactus graph and u a root vertex of the graph.
Output: a labeled cactus graph G^r.

1 **if** u *it is a leaf or a bridge* **then**
2     $r_p(u) \leftarrow 1$; $r_t(u) \leftarrow 1$; $L(u) \leftarrow c(u)$
3 **else**
     /* $F^u = v_1, v_2, \ldots, v_d$ children of u */
4   **forall** $v_i \in F^u$ **do**
5   Label-Calculation(G, C, v_i);
6   **forall** $v_i \in F^u$ **do**
7   **if** $v_i \in H^u$ **then**
          /* Let $\{w_1, w_2, \ldots, v_i, \ldots, w_t\} \subseteq H^u$, $P_1 = \{w_1, \ldots, v_i\}$ e
              $P_2 = \{v_i \ldots w_t\}$ */
8           $C(P_1) \leftarrow \{L(v_i) \dot{\cup} c(u) \dot{\cup} \{c(w_j) \mid j : 1 \ldots v_i\}\}$;
9           $C(P_2) \leftarrow \{L(v_i) \dot{\cup} c(u) \dot{\cup} \{c(w_j) \mid j : v_i \ldots t\}\}$;
10   **if** $|C(P_1)| \geq |C(P_2)|$ **then**
11             $CF_{v_i} \leftarrow C(P_1)$;
12   **else**
13             $CF_{v_i} \leftarrow C(P_2)$;
14   **else**
15          $CF_{v_i} \leftarrow L(v_i) \dot{\cup} c(u)$;
16     $L(u) \leftarrow CF_{v_i}$, where $|CF_{v_i}| \geq |CF_{v_j}| \; \forall v_i \neq v_j \; e \; v_i, v_j \in F^u$;
17     $r_p(u) \leftarrow |L(u)|$;
18     $r_t(u) \leftarrow \max(|CF_{v_i} \dot{\cup} CF_{v_j}|) \mid \forall v_i \neq v_j \; e \; v_i, v_j \in F^u$;
19 **return** G^r;

To compute the label of the vertices of the cactus graph, each vertex $u \in V$ receives a color list $L(u)$ that will initially contain the color of u . This list L, after execution of the algorithm, will contain the set of colors of the vertices of a path of a subgraph with a root at u that contains the greatest number of colors. The label of each vertex, except the leaf and bridge vertices which have labels = 1, is calculated based on the color lists of its children.

Note that the Label-Calculation algorithm labels either leaf-type or bridge-type vertices first, and so on until it reaches the root. The algorithm Label-Calculation is executed for all vertices of the set F^u, children of the root u. The vertex label u is calculated only after all vertices in F^u already have a label. It is easy to see that the algorithm can be adapted to return the vertices of the best path found in linear time.

4.3 Tropical-Cactus Algorithm

Given any vertex u defined as a root of G, the main Tropical-Cactus algorithm uses the auxiliary Label-Calculation algorithm to label all vertices of the cactus graph G. For each root $u' \neq u$, the Tropical-Cactus algorithm is run, and the vertices of G are given a label.

In short, the Tropical-Cactus algorithm returns the largest found value of the labels $r_t(u)$ that is calculated for each $u' \in V$ possible root of G. By the definition of $r_t(u)$, this is equivalent to the largest number of colors of a found path that passes through u.

Algorithm 3: Tropical-Cactus

Input: A pair (G, C), where G is a cactus graph found by the *Cactus-Graph algorithm* and C is the coloring of G.
Output: $|C^*|$, the largest number of colors found on a path in the input graph.
1 **forall** *root* $\in V$ **do**
2    $\lfloor$ Label-Calculation$(G, C, root)$;
3 **return** $\max(r_t(root))$

The following result shows the complexity of the Tropical-Cactus algorithm. The idea in this context is to show that the execution of the algorithm takes polynomial time.

Lemma 1. *The Label-Calculation algorithm runs in linear time* $\Theta\left(|V\left(G\right)|\right)$.

Proof. The algorithm is initially run for the root vertex u. In line 4, the algorithm is called recursively for each child u' of u, which will either be a bridge or a leaf, or there will be a cactus subgraph rooted in u', of which u is no longer part, thus the total number of recursive calls of the algorithm is the number of vertices in the graph. Additionally, note that the operations in lines 1 and 2 and lines 6–18 take $O(1)$ time, which shows that the algorithm has linear complexity in input size.

Proposition 1. *The Tropical-Cactus algorithm runs in polynomial time* $\Theta\left(|V\left(G\right)|^2\right)$

Proof. Note that the Tropical-Cactus algorithm makes $|V\left(G\right)|$ calls to the Label-Calculation algorithm, one call for each vertex that will be taken as the root of the graph. By the Lemma 1, the *Label-Calculation* algorithm runs on $\Theta\left(|V\left(G\right)|\right)$ and therefore the result follows.

5   Experimental Results

This section presents an analysis of the experimental results performed in this work. In general terms, the performance of the heuristic algorithm was evaluated in relation to the integer linear programming model for MTPP in general graphs, and in cactus graphs. Although the problem is NP-hard in all these classes, the idea was to obtain the optimal solution for instances of various dimensions, in terms of number of vertices, edges, and colors, and compare it with the result of the heuristic algorithm for this same set of instances. For this purpose, instances were generated randomly and also obtained from real-world data sets.

Implementation Details. The computational experiments were carried out on a PC with the following specifications: Intel(R) Core (TM) i7-8550U CPU 1.99 GHz processor, 8 GB of RAM, and Windows 10 Home operating system version 20H2. For the implementation of the LP models, the commercial optimization software Gurobi Optimizer version 9.5.0 was used. The Gurobi parameters were configured with their respective default values. To consolidate the results, all instances were run 10 times, and the final results of each instance were defined by the average of the valid results.

Benchmark Instances. Real-world instances were obtained from a dataset repository [8]. In some instances, adaptations were necessary due to the different characteristics of the networks in relation to the problem studied. The adaptations included removing isolated vertices, parallel edges, and loops.

To study the impact of varying the number of colors in an instance, three random colorings were set for all networks that did not have a previous coloring. To explore the behavior of both the heuristic algorithm and the ILP model in terms of the number of vertices, edges, and colors, random graphs were generated. For each n vertices, instances were generated with the following variations in the number of edges: $n + (n/10)$, $2n$, $(n * (n-1))/4$, and $(n * (n-1))/3$. Additionally, for each number of vertices, three variations in the number of colors were generated, representing 10%, 50%, and 90% of the total number of vertices. It was noted that the ILP model exhibited performance limitations in instances with around 800 vertices, although, an optimal integer solution was obtained for isolated instances with more than 1200 vertices. Therefore, for comparative purposes, instances with 200, 400, 600, and 800 vertices were considered.

Forty-eight random instances will be presented in this analysis, which are identified in the Tables 2 and 3. The tables use an index for each variation in the number of vertices, colors, and edges. The index follows the format A_{ikj}, where i represents the variation in the number of vertices, k represents the variation in the number of colors, and j represents the variation in the number of edges. The instance identification table is grouped by the number of vertices $|V|$, and each set contains the columns: Id, number of colors $|\mathcal{C}|$, and number of edges $|E|$. Since the cactus graph has a limited number of edges compared to a traditional graph, two variations on the number of edges were defined for each cactus graph of n vertices and k colors (Table 1).

5.1   Computational Results

The results of the computational experiments for the ILP model and the heuristic algorithm are presented in the following tables: Table 4, Table 5 and Table 6 and

Table 1. Identification of real-world instances for graphs in general

Id	Nome	$|V|$	$|E|$	$|\mathcal{C}|$	Id	Nome	$|V|$	$|E|$	$|\mathcal{C}|$
R_1	Les Miserables$_1$	77	254	8	R_{12}	CPAN Authors$_4$	839	2112	310
R_2	Les Miserables$_2$	77	254	39	R_{13}	EuroSIS$_1$	1285	6462	13
R_3	Les Miserables$_3$	77	254	70	R_{14}	EuroSIS$_2$	1285	6462	22
R_4	Primary$_1$	236	5899	11	R_{15}	Diseasome	1419	2738	25
R_5	Primary$_2$	238	5539	11	R_{16}	Yeast$_1$	2284	6646	228
R_6	C.Eleg$_1$	297	2148	30	R_{17}	Yeast$_2$	2284	6646	1142
R_7	C.Eleg$_2$	297	2148	149	R_{18}	Yeast$_3$	2284	6646	2056
R_8	C.Eleg$_3$	297	2148	268	R_{19}	CPAN Distributions$_1$	2719	5016	272
R_9	CPAN Authors$_1$	839	2112	32	R_{20}	CPAN Distributions$_2$	2719	5016	1360
R_{10}	CPAN Authors$_2$	839	2112	60	R_{21}	CPAN Distributions$_3$	2719	5016	2448
R_{11}	CPAN Authors$_3$	839	2112	68	–	–	–	–	–

Table 2. Identification of the random instances for graphs in general

200			400			600			800		
Id	$|\mathcal{C}|$	$|E|$	Id	$|\mathcal{C}|$	$|E|$	Id	$|\mathcal{C}|$	$|E|$	Id	$|\mathcal{C}|$	$|E|$
A_{111}	20	220	A_{211}	40	440	A_{311}	60	660	A_{411}	80	880
A_{112}	20	400	A_{212}	40	800	A_{312}	60	1200	A_{412}	80	1600
A_{113}	20	9950	A_{213}	40	39900	A_{313}	60	89850	A_{413}	80	159800
A_{114}	20	13266	A_{214}	40	53200	A_{314}	60	119800	A_{414}	80	213066
A_{121}	100	220	A_{221}	200	440	A_{321}	300	660	A_{421}	400	880
A_{122}	100	400	A_{222}	200	800	A_{322}	300	1200	A_{422}	400	1600
A_{123}	100	9950	A_{223}	200	39900	A_{323}	300	89850	A_{423}	400	159800
A_{124}	100	13266	A_{224}	200	53200	A_{324}	300	119800	A_{424}	400	213066
A_{131}	180	220	A_{231}	360	440	A_{331}	540	660	A_{431}	720	880
A_{132}	180	400	A_{232}	360	800	A_{332}	540	1200	A_{432}	720	1600
A_{133}	180	9950	A_{233}	360	39900	A_{333}	540	89850	A_{433}	720	159800
A_{134}	180	13266	A_{234}	360	53200	A_{334}	540	119800	A_{434}	720	213066

Table 3. Identifying random instances of cactus graphs

Id	$|V|$	$|\mathcal{C}|$	$|E|$	$|Cycles|$	Id	$|V|$	$|\mathcal{C}|$	$|E|$	$|Cycles|$
A^c111	200	20	209	10	A^c311	600	60	629	30
A^c112	200	20	219	20	A^c312	600	60	659	60
A^c121	200	100	209	10	A^c321	600	300	629	30
A^c122	200	100	219	20	A^c322	600	300	659	60
A^c131	200	180	209	10	A^c331	600	540	629	30
A^c132	200	180	219	20	A^c332	600	540	659	60
A^c211	400	40	419	20	A^c411	800	80	839	40
A^c212	400	40	439	40	A^c412	800	80	879	80
A^c221	400	200	419	20	A^c421	800	400	839	40
A^c222	400	200	439	40	A^c422	800	400	879	80
A^c231	400	270	419	20	A^c431	800	720	839	40
A^c232	400	270	439	40	A^c432	800	720	879	80

present the following nomenclatures, from left to right: the instance (Id); the CPU time spent, in seconds, to obtain the integer optimal solution ($T_{ILP}(s)$); the value of the integer optimal solution (Sol_{ILP}); the CPU time spent, in seconds, of the heuristic algorithm ($T_{Alg}(s)$); the value of the solution returned by the heuristic algorithm (Sol_{Alg}); and finally the difference in percentage between the integer optimal solution and the approximate solution of the heuristic algorithm (GAP_{Sol}).

Table 4. Results of the computational experiments on random instances for graphs in general

200						400					
Id	$T_{ILP}(s)$	Sol_{ILP}	$T_{Alg}(s)$	Sol_{Alg}	GAP_{Sol}	Id	$T_{ILP}(s)$	Sol_{ILP}	$T_{Alg}(s)$	Sol_{Alg}	GAP_{Sol}
A_{111}	1.32	20	0.06	20	0.00%	A_{211}	19.35	40	1.07	39	2.50%
A_{112}	1.72	20	0.02	20	0.00%	A_{212}	10.24	40	0.12	40	0.00%
A_{113}	1.71	20	0.02	20	0.00%	A_{213}	12.56	40	0.14	40	0.00%
A_{114}	2.63	20	0.02	20	0.00%	A_{214}	14.67	40	0.37	40	0.00%
A_{121}	6.11	61	0.20	48	21.31%	A_{221}	37.65	111	1.32	89	19.82%
A_{122}	1.45	100	0.37	83	17.00%	A_{222}	48.19	198	6.89	166	16.16%
A_{123}	3.25	100	0.29	95	5.00%	A_{223}	201.56	200	1.86	184	8.00%
A_{124}	9.05	100	0.46	97	3.00%	A_{224}	47.13	200	2.50	198	1.00%
A_{131}	3.99	67	0.16	49	26.87%	A_{231}	6.06	140	1.13	104	25.71%
A_{132}	2.25	171	0.57	123	28.07%	A_{232}	329.34	343	5.87	238	30.61%
A_{133}	4.03	180	0.24	164	8.89%	A_{233}	511.23	360	2.17	317	11.94%
A_{134}	6.25	180	0.33	172	4.44%	A_{234}	319.54	360	2.81	335	6.94%
600						800					
Id	$T_{ILP}(s)$	Sol_{ILP}	$T_{Alg}(s)$	Sol_{Alg}	GAP_{Sol}	Id	$T_{ILP}(s)$	Sol_{ILP}	$T_{Alg}(s)$	Sol_{Alg}	GAP_{Sol}
A_{311}	41.19	60	3.43	58	3.33%	A_{411}	395.11	80	6.95	74	7.50%
A_{312}	43.88	60	0.29	60	0.00%	A_{412}	46.12	80	1.78	80	0.00%
A_{313}	39.08	60	0.36	60	0.00%	A_{413}	110.94	80	0.77	80	0.00%
A_{314}	41.91	60	0.41	60	0.00%	A_{414}	144.46	80	0.98	80	0.00%
A_{321}	44.42	168	4.97	135	19.64%	A_{421}	372.76	234	6.45	151	35.47%
A_{322}	134.80	292	31.09	239	18.15%	A_{422}	760.51	393	26.05	299	23.92%
A_{323}	361.62	300	7.74	292	2.67%	A_{423}	2273.05	400	24.10	393	1.75%
A_{324}	582.18	300	10.43	296	1.33%	A_{431}	66.43	269	5.51	166	38.29%
A_{331}	42.78	209	2.80	128	38.76%	A_{432}	2504.11	672	28.37	417	37.95%
A_{332}	1827.28	501	21.66	348	30.54%	A_{433}	1914.03	720	26.53	638	11.39%
A_{333}	2955.88	540	9.07	490	9.26%	A_{434}	2168.50	720	39.87	679	5.69%

From an individual analysis of the ILP model, it was noted that even though some instances exceeded the proposed execution time limit of 60 min, over 77% of the executed instances obtained optimal integer solutions in less than seven minutes. Some isolated tests obtained an optimal integer solution for instances of up to 1250 vertices, 1125 colors, and 1375 edges.

From the perspective of random graph instances in general, the heuristic algorithm found the integer optimal solution for 28.26% of the instances; 86.96%

Table 5. Results of the computational experiments on real-world instances for graphs in general

Id	$T_{ILP}(s)$	Sol_{ILP}	$T_{Alg}(s)$	Sol_{Alg}	GAP_{Sol}	Id	$T_{ILP}(s)$	Sol_{ILP}	$T_{Alg}(s)$	Sol_{Alg}	GAP_{Sol}
R_1	0.72	8	0.00	8	0.00%	R_9	5.33	31	4.43	22	29.03%
R_2	2071.25	32	0.02	20	37.50%	R_{10}	16.65	56	4.41	36	35.71%
R_3	143.73	50	0.02	26	48.00%	R_{11}	16.83	67	4.35	45	32.84%
R_4	1.23	11	0.02	11	0.00%	R_{12}	31.33	143	4.56	62	56.64%
R_5	1.10	11	0.02	11	0.00%	R_{13}	22.69	13	5.28	13	0.00%
R_6	4.49	30	0.17	30	0.00%	R_{14}	70.93	22	5.37	22	0.00%
R_8	1183.24	254	1.77	197	22.44%	R_{16}	1291.21	228	254.77	218	4.39%

of the instances presented a gap, that is, a difference, in percent, between the integer optimal solution presented by the ILP model and the solution of the heuristic algorithm, of less than 30%; and 60.87% of the instances presented a gap of less than 10%. The worst case had a gap of 38.16%. It was also noted that, on average, the solution of the heuristic algorithm diverges from the optimal solution presented by the ILP model by 11.37%.

Table 6. Results of the computational experiments for cactus graphs

Id	$T_{ILP}(s)$	Sol_{ILP}	$T_{Alg}(s)$	Sol_{Alg}	GAP_{Sol}
A^c111	29.52	19	0.154	19	0.00%
A^c112	6.30	20	0.005	20	0.00%
A^c121	8.23	33	0.090	33	0.00%
A^c122	23.42	38	0.180	37	2.63%
A^c131	11.67	35	0.122	35	0.00%
A^c132	18.37	35	0.121	35	0.00%
A^c221	123.91	41	0.360	41	0.00%
A^c222	666.00	48	0.374	48	0.00%
A^c231	19.60	48	0.383	48	0.00%
A^c232	32.26	51	0.378	46	9.80%
A^c321	1791.34	56	0.848	56	0.00%
A^c331	1519.61	49	0.824	49	0.00%

For the real instances, the heuristic algorithm found the integer optimal solution in 42.86% of the instances; 64.29% of the instances presented a gap less than 30%; and 50% of the instances presented a gap smaller than 10%. The worst case presented a solution gap of 56.64%. For this data set, the solution of the heuristic algorithm diverges, on average, by 19.04% from the optimal solution presented by the ILP model.

For the random cactus graph instances, the heuristic algorithm found the optimal solution for 83.33% of the instances; 100% of the instances presented a gap less than 10%. The worst-case instance had a gap of 9.80%, and on average, this gap was 1.04%.

We noticed an improved behavior of the heuristic algorithm on graphs where the number of possible circuits is limited; moreover, in real-world instances, the heuristic algorithm performed better than in random instances, which concludes that in general, the algorithm is a good tool to obtain lower bounds for the MTPP.

6   Concluding Remarks

In this article, a contribution to the state of the art of the problem of tropical paths in graphs is proposed from the perspective of combinatory optimization. Specifically, an integer linear optimization model for MTPP has been developed. This model contains, in order, $O(nk)$ variables, being n the number of vertices, and k the number of colors, and $O(n)$ linear inequalities. Due to being an NP-hard problem, its execution is impractical for large instances. Additionally, a polynomial-time heuristic algorithm was also presented, which, as far as it is known, is the first result of this line of research. The differential of the optimization model was the association of two important inequalities that together guaranteed both the connectivity and the absence of circuits in the solution, unlike the commonly used models in which an exponential number of subsets of graphs are exploited for this purpose. The proposed heuristic algorithm is a recursive algorithm that, for a given random root, finds the path with the largest number of colors containing that root. All vertices are visited as roots, and the best path in these terms is returned. The proposed heuristics were tested on a benchmark consisting of 93 instances, among them random and real-world, of various dimensions. The results were compared with the respective entire optimal solutions of the model, and the experimental results pointed out that the average percentage of the difference between the optimal solution and the heuristic algorithm solution is less than 11%, which proves experimentally the strength of the proposed heuristic algorithm. In practice, when faced with computationally intractable problems, it is natural to sacrifice optimality in favor of computational time to obtain a competitive solution. In this regard, the recursive algorithm proved to be an advantageous tool for obtaining good solutions for the MTPP.

References

1. Anglès d'Auriac, J.A., et al.: Tropical dominating sets in vertex-coloured graphs. J. Disc. Algor. **48**, 27–41 (2018). https://doi.org/10.1016/j.jda.2018.03.001. https://www.sciencedirect.com/science/article/pii/S1570866718300595

2. Chapelle, M., Cochefert, M., Kratsch, D., Letourneur, R., Liedloff, M.: Exact exponential algorithms to find tropical connected sets of minimum size. Theor. Comput. Sci. **676**, 33–41 (2017). https://doi.org/10.1016/j.tcs.2017.03.003. https://www.sciencedirect.com/science/article/pii/S0304397517301883

3. Cohen, J., Manoussakis, Y., Phong, H., Tuza, Z.: Tropical matchings in vertex-colored graphs. Electron. Notes Disc. Math. **62**, 219–224 (2017). https://doi.org/10.1016/j.endm.2017.10.038, https://www.sciencedirect.com/science/article/pii/S1571065317302779

4. Cohen, J., Italiano, G.F., Manoussakis, Y., Nguyen, K.T., Pham, H.P.: Tropical paths in vertex-colored graphs. In: Gao, X., Du, H., Han, M. (eds.) COCOA 2017. LNCS, vol. 10628, pp. 291–305. Springer, Cham (2017). https://doi.org/10.1007/978-3-319-71147-8_20

5. Deville, Y., Gilbert, D.R., van Helden, J., Wodak, S.J.: An overview of data models for the analysis of biochemical pathways. Brief. Bioinf. **4**(3), 246–259 (2003). https://doi.org/10.1093/bib/4.3.246

6. Foucaud, F., Harutyunyan, A., Hell, P., Legay, S., Manoussakis, Y., Naserasr, R.: The complexity of tropical graph homomorphisms. Disc. Appl. Math. **229**(C), 64–81 (2017). https://doi.org/10.1016/j.dam.2017.04.027

7. Gabow, H.N., Nie, S.: Finding a long directed cycle. ACM Trans. Algor. (TALG) **4**(1), 1–21 (2008)

8. Gephi: Conjuntos de dados gephi (2009). https://github.com/gephi/gephi/wiki/Datasets. Accessed 27 Sept 2021

9. Kelley, B.P., et al.: Conserved pathways within bacteria and yeast as revealed by global protein network alignment. Proc. Natl. Acad. Sci. **100**(20), 11394–11399 (2003). https://doi.org/10.1073/pnas.1534710100. https://www.pnas.org/content/100/20/11394

10. Lacroix, V., Fernandes, C.G., Sagot, M.F.: Motif search in graphs: application to metabolic networks. IEEE/ACM Trans. Comput. Biol. Bioinf. **3**(4), 360–368 (2006). http://doi.ieeecomputersociety.org/10.1109/TCBB.2006.55

11. Markov, M., Ionut Andreica, M., Manev, K., Tapus, N.: A linear time algorithm for computing longest paths in cactus graphs. Serdica J. Comput. **6**(3), 287p–298p (2012)

12. Miller, C.E., Tucker, A.W., Zemlin, R.A.: Integer programming formulation of traveling salesman problems. J. ACM (JACM) **7**(4), 326–329 (1960)

Efficient Density-Based Models for Multiple Machine Learning Solutions over Large Datasets

Natanael F. Dacioli Batista$^{(\boxtimes)}$ ⓘ, Bruno Leonel Nunes ⓘ, and Murilo Coelho Naldi ⓘ

Department of Computing Science, Federal University of São Carlos (UFSCar) Rod, Washington Luís, km 235 - PBOX 676, São Carlos, SP, Brazil
{natanael.batista,brunoleonel}@estudante.ufscar.br, naldi@ufscar.br

Abstract. Unsupervised and semi-supervised machine learning is very advantageous in data-intensive applications. Density-based hierarchical clustering obtains a detailed description of the structures of clusters and outliers in a dataset through density functions. The resulting hierarchy of these algorithms can be derived from a minimal spanning tree whose edges quantify the maximum density required for the connected data to characterize clusters, given a minimum number of objects, $MinPts$, in a given neighborhood. $CORE\text{-}SG$ is a powerful spanning graph capable of deriving multiple hierarchical solutions with different densities with computational performance far superior to its predecessors. However, density-based algorithms use pairwise similarity calculations, which leads such algorithms to an asymptotic complexity of $O(n^2)$ for n objects in the dataset, impractical in scenarios with large amounts of data. This article enables hierarchical machine learning models based on density by reducing the computational cost with the help of Data Bubbles, focusing on clustering and outlier detection. It presents a study of the impact of data summarization on the quality of unsupervised models with multiple densities and the gain in computational performance. We provide scalability for several machine learning methods based on these models to handle large volumes of data without a significant loss in the resulting quality, enabling potential new applications like density-based data stream clustering.

Keywords: Density-based Data Models · Unsupervised Learning · Clustering · Data Sumarization · Big Data

1 Introduction

Massive amounts of data are generated daily [3,23]. Unsupervised machine learning is advantageous in such scenarios as it finds significant relationships between the data and does not require labels, as the last can be rare or expensive in big data scenarios [2]. Among the various unsupervised techniques, data clustering algorithms are worth mentioning, especially those based on density, as

M. C. Naldi and R. A. C. Bianchi (Eds.): BRACIS 2023, LNAI 14195, pp. 48–62, 2023.
https://doi.org/10.1007/978-3-031-45368-7_4

they can find clusters and outliers without any information external to the data. Furthermore, these algorithms are not bound by format constraints for clusters and do not require a prior definition of several clusters like parametric clustering algorithms. Its applications are present in high-impact journals across various research areas, including human behavior [9], chemistry [14], genetics [19], robotics [21], and others.

Density-based algorithms estimate that objects of a dataset are chainlike connected if there is a minimum number of objects in their neighborhood (known as the smoothing parameter $MinPts$) to characterize a region dense enough to form a cluster. Otherwise, objects are considered outliers. Given a value of $MinPts$, it is possible to establish the minimum density and similarity necessary to connect any pair of objects in the data set, forming a complete graph G in which each node represents an object and each edge a measure of the density of mutual reach that connects them. For example, the HDBSCAN [6] algorithm, state-of-the-art in clustering by density, derives a hierarchy between objects and their clusters from the Minimum Spanning Tree (MST) of the complete graph G. This enables transmission of clusters by density, detection, and evaluation of outliers, and other helpful information in various areas such as passing bridge approval [8], cancer research [15], shipping systems [16], among others. Later, Gertrudes $et\ al.$ [10] demonstrated that several other unsupervised and semi-supervised algorithms could also be explained from a perspective based on instances or variations of utilizing this MST over G, considering the resulting minimum density of $MinPts$. However, you may need to find the most appropriate value for $MinPts$, which can be done by analyzing multiple $MSTs$ over G, with each MST requiring an asymptotic computational complexity of $O(n^2. \log n)$ for a dataset containing n objects.

In some cases, it is necessary to analyze different minimum densities, which can be computationally prohibitive given the re-execution of the MST extraction algorithm of the complete graph G. Recently, Neto $et\ al.$ [18] proposed the replacement of the graph G by a much smaller graph called $CORE\text{-}SG$ which has as its main characteristic to contain all the $MSTs$ of G for any value of $MinPts$ less than a given maximum value $MinPts_{max}$. An extraction of MST from $CORE\text{-}SG$ has an asymptotic complexity of $O(n.\log n)$, with $n >> MinPts_{max}$, which is a very significant reduction compared to the extraction over G. $CORE\text{-}SG$ is very advantageous, as it allows the analysis of results for multiple values of $MinPts$ at the computational cost of obtaining a single result since the construction of the $CORE\text{-}SG$ requires the MST over G with density estimated as $MinPts_{max}$.

In scenarios where the amount of data generated is massive, using $CORE\text{-}SG$ to extract results with different densities is even more computationally advantageous. However, in these scenarios, its construction is prohibitive due to the intrinsic complexity of density estimation methods using pairwise calculations, which remains $O(n^2)$. Because of their complexity, density-based methods might be limited to smaller datasets. Breunig $et\ al.$ [4] proposed the use of Data Bubbles, a data summarization method data for hierarchical clustering by density, which makes it possible to preserve the quality of the result obtained and, at the same time, considerably increase the computational performance of the algorithm. Furthermore, Data Bubbles are advantageous when compared to other feature vectors (Clustering Feature - CF) as they were explicitly designed by some of the authors of HDBSCAN* [6] to be used accurately in the calculation of reachability distances based on density.

Contributions. This work enables the construction of the $CORE\text{-}SG$ on large datasets, which allows for the direct application of a variety of unsupervised and semi-supervised algorithms in tasks like clustering, outlier detection, classification, and others [10]. Density-based algorithms are known for the quality of their results, and their computational cost is traditionally limited to $O(n^2)$. This work proposes summarization of the data in a significantly smaller amount of Data Bubbles and, subsequently, the construction of the $CORE\text{-}SG$ on the Data Bubbles. Once built, $CORE\text{-}SG$ can be used to obtain hierarchical results extracted with different values for its density parameter with computational performance $O(n)$, which enables its application in large volumes of data. Far as the authors are aware, this is the only work in the literature that allows the construction of multiple hierarchical models with different density estimates from a single model of summarized data.

The present work analyzes the impact of summarization on the quality and performance of hierarchical models based on densities to measure to what extent it can obtain significant results without degenerating these models. The presented results, in addition to benefiting a variety of existing algorithms in the literature and organized in [10], also enables the potential for new density-based applications that include scenarios with continuous data streams.

2   Related Works

Data summarization has been successfully used to enable machine learning tasks on large data sets or data streams, which are potentially infinite. For hierarchical clustering, one of the best-known algorithms is BIRCH [24], which builds and maintains a hierarchy of structures called *Clustering Features* (*CF*). Each *CF* stores the number of summarized data, the linear and square sum of the data. The BIRCH algorithm structure of *CFs* forms a tree of subclusters and is built incrementally, as objects are inserted sequentially in the *CF* that best represents it. Parameters define when new *CFs* are added to the tree to increase the hierarchy. Using *CFs* statistics improves the algorithm's performance without significantly impacting the quality of the result.

Traditional *CF* summarize data in such a way as to lose information about its original placement and its nearest neighbors, creating a distortion when represented only by its central objects. This loss of information impacts the quality of density-based algorithms. To solve this problem, Data Bubbles (*DBs*) [4] were proposed to add information to *CF* necessary for the calculation of density estimates in the region covered by *DB*. This information is a "representative" vector to indicate the placement of *DB*; its radius and estimation of average distances between objects within the *DB*, essential for density calculations. *DBs* have been successfully used in a variety of jobs [13,20,26] related to new summarization approaches and density calculations. Additionally, *DBs* were proposed for application in hierarchical algorithms, as was done with the OPTICS [1] in [5] where the use of *DBs* obtained a great increase in computational performance, maintaining the quality of the final result. The potential of applying *DBs* for analysis and visualization of large databases is also discussed in [5].

A characteristic of *DBs* is that they are incremental, which means they can be updated with new data. This characteristic is explored in [17], where *DBs* are used to provide incrementality for the method that the authors call Incremental and Effective Data Summarization (*IEDS*), designed for large volumes of data dynamic, where new data is continually add in a hierarchical clustering tree. The IEDS method speeds up the incremental construction of summarized data using triangular inequalities for inserting and deleting *DBs*. Furthermore, the algorithm analyzes the *DBs* to keep those with high-quality compressed data.

However, when initializing the summarization, defining which data will be summarized in each *DB* is necessary. This choice can generate different and unsatisfactory results, mainly if data separated by low-density regions are summarized in the same *DB*. To avoid the performance of the algorithm not so sensitive to the compression rate and the location of the seeds that initialize the *DBs*, a method of data summarization and initialization of *DBs* is proposed in [25]. In it, a sampling based on density for selecting a subset of data to be used in the construction of *DBs* is used, thus avoiding the choice of data separated by regions of low density for the same *DB*.

Recently, *DBs* have been used as summary frameworks along with *framework* MapReduce for data summarization and [20] scalability. They were applied to enable the scalable version of HDBSCAN* [6] to summarize data (and even other *DBs*) into a hierarchy of clusters by density. This approach is similar to the one proposed in this article, as it enables the application of a hierarchical algorithm based on density over large amounts of data. However, unlike the approach used in [6], our work considers the summarization of a structure that results in multiple hierarchies with different densities and not just a single one, which is quite advantageous when one does not have prior knowledge of the fittest values of the density parameters.

3   Proposed Method

Our proposal is divided into two steps: summarizing the data and obtaining hierarchies with multiple densities. The original *DB* [4] uses the parameter ϵ, which is incompatible with HDBSCAN* and *CORE-SG*. Therefore, we propose an adaptation and new definitions to enable the joint application of *DB* and *CORE-SG*.

3.1   Density-Based Summarization

Given a dataset X with n objects of d dimensions, an integer value for the density parameter $MinPts$ that defines the number of neighbors needed for an object to be considered core, our method starts by applying a $O(n)$ computational complexity method to divide X into m subsets, where $m << n$. Clustering into m in subsets can be adopted in this step or a guided initialization [25].

Definition 1 - *Data Bubbles: The data subset $X_i \subseteq X$ summarized in a DB which is defined as a tuple $B_i = (rep_i, n_i, extent_i, nnDist_i)$, where rep_i is the position of the representative of X_i, n_i is the $|X_i|$ cardinality, $extent_i$ is a natural number that defines an extended radius around the rep_i that contains most objects of X_i, and $nnDist_i$ is a function that given an integer value $k \leq MinPts < n_i$ that denotes the estimated average distance of the k-nearest neighbor within the set of objects X_i. In a Euclidean space, the statistics linear and quadratic sums of the objects in X_i, LS_i, and SS_i, respectively, can be used to calculate the characteristics of B_i. With these statistics, $rep_i = \frac{LS_i}{n_i}$, $extent_i = \sqrt{\frac{2 \cdot n_i \cdot SS_i - 2 \cdot LS_i^2}{n_i \cdot (n_i - 1)}}$ is calculated and the expected distance from the k-nearest neighbor is $nnDist_i(k) = (\frac{k}{n_i})^{\frac{1}{d}} \cdot extension_i$.*

Definition 2 - *Distance between DBs: If B_i and B_j are two DBs, the direct distance between B_i and B_j is defined as:*

$$dist(B_i, B_j) = \begin{cases} 0, \ if \ B_i = B_j, \\ \\ \begin{aligned} &dist(rep_i, rep_j) - (extend_i + extend_j) \\ &\quad + nnDist_i(1) + nnDist_j(1), \\ &if \ dist(rep_i, rep_j) - (extend_i + extend_j) \geq 0, \end{aligned} \\ \\ max(nnDist_i(1), nnDist_j(1)), \ otherwise. \end{cases} \tag{1}$$

In other words, if $B_i = B_j$, their distance will be zero. If they do not overlap, their distance is given by the distance between their representatives minus their radii plus their expected distances to the nearest neighbor. Finally, if the *DBs* overlap, the distance is the maximum between their expected nearest neighbor distances. The described cases are illustrated in Fig. 1.

Definition 3 - *Core distance from a DB: Given a value of $MinPts < n$, the function $NN(B_i, k)$ that returns the k-th closest DB to B_i or B_i if $k = 0$, a function that returns the minimum number of neighbors of B_i needed to summarize MinPts objects, i.e., $SNN(B_i, MinPts) = \arg\min_k \sum_{j=0}^{k} n_j \geq MinPts \mid B_j = NN(B_i, j)$; the core distance of B_i is:*

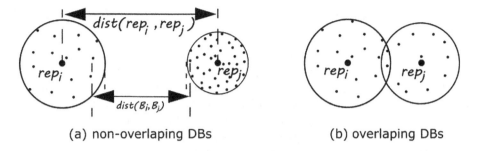

(a) non-overlaping DBs   (b) overlaping DBs

Fig. 1. Distance between Data Bubbles [4].

$$core_{MinPts}(B_i) = \begin{cases} nnDist_i(MinPts), \ if \ SNN(B_i, MinPts) = 0 \\ \\ dist(rep_i, rep_k) - extend_k + \\ nnDist_k(MinPts - \sum_{l=0}^{k-1} n_l | B_l = NN(B_i, l)), otherwise \end{cases} \tag{2}$$

where $B_k = NN(B_i, SNN(B_i, MinPts))$ is the kth nearest neighbor of B_i. The core distance is the minimum distance from a DB that summarizes at least $MinPts$ objects. It is important to point out that $core_a(B_i) \geq core_b(B_i) \ \forall \ a \geq b$, i.e., increasing the value of $MinPts$ cannot reduce the core distance of a DB. We can calculate the reachability distances used in density estimates using the distance between two DBs and their core distances.

Definition 4 - *Mutual Reachability Distance (MRD) between DBs: Given the DBs B_i and B_j, the minimum number of objects $MinPts$, the mutual reachability distance between the bubbles B_i and B_j, i.e., the distance that make them density reachable within each other is:*

$$mrd_{MinPts}(B_i, B_j) = max(core_{MinPts}(B_i), core_{MinPts}(B_j), dist(B_i, B_j)) \tag{3}$$

3.2   *CORE-SSG*: CORE Summarized Spanning Graph

The *MRD* reflects the minimum density needed to connect two *DBs*. With it, we can estimate the distance that connects all pairs of *DBs* in the data set by density, forming a virtual graph G that represents all the density relationships between the data, given a minimum value $MinPts$ of points needed to obtain the desired density.

Definition 5 - *Mutual Reachability Graph G: Given a minimum density value $MinPts$, the $G_{MinPts} = (V, E)$ is a complete graph in which the set of vertices*

V *represent the DBs and the set of edges is defined as* $E = \{e(B_i, B_j) \mid B_i, B_j \in V$ *with weights* $w(e) = mrd_{MinPts}(B_i, B_j)\}$.

The main idea behind density-based hierarchical machine learning algorithms is to extract a minimal graph that connects all objects in the dataset by density [6,10]. That is, directly or indirectly, these algorithms obtain the MST_{MinPts} of the graph G_{MinPts} and, with it, build a hierarchy that allows obtaining clusters, identification of outliers and applications. However, different $MinPts$ values tend to derive different results, which would require multiple extractions of $MSTs$ at a high computational cost. To avoid this, given an upper bound for $MinPts$ defined here as $MinPts_{Max}$, we can replace the complete graph G with $O(n^2)$ edges by the graph $CORE\text{-}SG_{MinPts_{Max}}$ with $O(n)$ edges when applied to a dataset with n objects, if $MinPts \le MinPts_{Max} << n$ [18], which is a natural assumption.

However, obtaining the $CORE\text{-}SG$ of G over n objects using a traditional MST extraction algorithm is $O(n^2 \cdot log\, n)$, prohibitive for large values of n. In the present work, we summarize the data into DBs and consider the complete graph G over m DBs, which makes the cost to build the $CORE\text{-}SG_{MinPts_{Max}}$ $O(m^2 \cdot log\, m)$ and the cost of each extraction of MST_{MinPts} $O(m \cdot log\, m)$. In a scenario with a large volume of data, it is natural to consider that $MinPts_{Max} << m << n$ and, therefore, to limit the necessary computational cost for our proposal to $O(n)$. In scenarios where this assumption is invalid, the data can be used normally, and summarization is unnecessary.

As it is applied over DBs, the summarized $CORE\text{-}SG$ will be referred to here as CORE Summarized Spanning Graph ($CORE\text{-}SSG$) and has definitions and properties different from its original version [18]. Next, we will present its new concepts and proprieties.

Definition 6 - *Minimum Nearest Neighbor Graph* $MNNG_{MinPts}$: *Given a minimum density value* $MinPts$, *the graph* $MNNG_{MinPts} = (V, E)$ *has the generated DBs as a set of vertices* V *over the data and the set of edges* $E = \{e(B_i, B_j) \mid B_i, B_j \in V \wedge B_j = NN(B_i, k) \; \forall k = [0, SNN(B_i, MinPts)]\}$. *This graph connects the DBs to guarantee a minimum number of MinPts connected summarized objects. We see that* $MNNG_{MinPts}$ *has two important properties, as shown in Lemmas 1 and 2.*

Lemma 1: $\forall\, a \le b,\; MNNG_a \subseteq MNNG_b$.

Proof: Since $MNNG_a = (V_a, E_a)$, $MNNG_b = (V_b, E_b)$ and $V_a = V_b = V$, it is necessary to prove that $E_a \subseteq E_b$ so that Lemma 1 be true. Assuming that $E_a \nsubseteq E_b$, there must be an edge in E_a that is not in E_b. By definition, $SNN(B_i, a) \le SNN(B_i, b) \; \forall\, a \le b, B_i \in V$. Therefore, the set of integer values of the interval $[0, SNN(B_i, a)]$ is contained in $[0, SNN(B_i, b)]$, which makes false the statement that there is an edge E_a that is not in E_b.

Lemma 2: $\forall\, a \le b\; :\; mrd_a(B_i, B_j) = max(core_a(B_i), core_a(B_j)) \Rightarrow e(B_i, B_j) \in MNNG_b$.

Proof: Assuming that $mrd_a(B_i, B_j) = \max(core_a(B_i), core_a(B_j))$, given the definition of the MRD in Definition 4, it follows that $core_a(B_i) \geq dist(B_i, B_j)$ or $core_a(B_j) \geq dist(B_i, B_j)$, or both. Hence, at least one of these DBs must be in the neighborhood of the other, i.e., $B_j = NN(B_i, k), k = [0, SNN(B_i, a)]$ or/and $B_i = NN(B_j, k), k = [0, SNN(B_j, a)]$ must hold and implies that $e(B_i, B_j) \in MNNG_a$. Since $a \leq b$, it follows by Lemma 1 that $MNNG_a \subseteq MNNG_b$ and, accordingly, $e(B_i, B_j) \in MNNG_b$.

Definition 7 - *Summarized CORE-SG (CORE-SSG): Given a maximum value for the minimum density parameter $MinPts_{Max}$, the $CORE\text{-}SSG_{MinPts_{Max}}$ $= MNNG_{MinPts_{Max}} \cup MST_{MinPts_{Max}}$ is a non-directional graph that contains the edges of all MST_{MinPts} of G_{MinPts} for each value of $MinPts <$ $MinPts_{Max}$. This feature allows replacing G by $CORE\text{-}SSG_{MinPts_{Max}}$ in the process of extracting a MST_{MinPts}, simply replacing the weight of the edges $e(B_i, B_j)$ in $CORE\text{-}SSG_{MinPts_{Max}}$ by $w(e) = mrd_{MinPts}(B_i, B_j)$ to $MinPts <$ $MinPts_{Max}$.*

Theorem 1: *For two values p and q for the density parameter $MinPts$, let M_p be the set of all possible $MSTs$ of G_p, and let MST_q be any MST of G_q, so $\forall\, p < q: \exists MST_p \in M_p: MST_p \subseteq MST_q \cup MNNG_q$.*

Proof: Let's consider an edge $e(B_i, B_j)$ in a $MST_p^* \in M_p$ that connects two subgraphs I and J. The edge weight $e(B_i, B_j)$ is defined by $mrd_p(B_i, B_j)$, which is the maximum between $dist(B_i, B_j)$ and their core distances $core_p(B_i)$ and $core_p(B_j)$, according to Definition 4. Due to the cutting property of MSTs, $e(B_i, B_j)$ must have the lowest weight among all the edges that connect the sets I and J in G_p. There are two scenarios:

(1) $e(B_i, B_j) \in MST_q$
(2) $e(B_i, B_j) \notin MST_q$

In scenario (1), since $e(B_i, B_j) \in MST_q$, it trivially follows that $e(B_i, B_j) \in MST_q \cup MNNG_q$. In scenario (2), MST_q must have a different edge $e(C_i, C_j)$ that connects the DBs C_i and C_j that are contained in the graphs I and J, respectively, having the lowest weight among the edges that connect these graphs ($MSTs$ are connected graphs). Thus, the following two statements must be true (i) $mrd_p(B_i, B_j) \leq mrd_p(C_i, C_j)$ (otherwise $e(B_i, B_j)$ could not be the edge with the smallest weight connecting I and J in MST_p^*) and (ii) $mrd_q(B_i, B_j) \geq mrd_q(C_i, C_j)$ (otherwise $e(C_i, C_j)$ could not be the least weighted edge connecting I and J in MST_q). We saw in Definition 3 that by increasing the value of $MinPts$, the core distances $core_{MinPts}(\cdot)$ can only grow and the distance $dist(\cdot, \cdot)$ between two DBs is constant. Therefore, the weight of an edge, given by $mrd_{MinPts}(\cdot, \cdot)$, cannot decrease when the value of $MinPts$ increases. Consequently, the mutual reachability distances (edge weights), defined in Eq. 3, can only remain the same or increase. Therefore, when $MinPts$ increases from p to q ($p < q$ by assumption in the theorem), one of the following must be true:

(a) $mrd_p(B_i, B_j) < mrd_q(B_i, B_j)$
(b) $mrd_p(B_i, B_j) = mrd_q(B_i, B_j)$

If (a) is true, as the distance $dist(\cdot, \cdot)$ between two DBs does not depend on $MinPts$, the mutual reachability distance $mrd_q(B_i, B_j)$ must be determined by the core distance of B_i or B_j, i.e., $mrd_q(B_i, B_j) = max(core_q(B_i), core_q(B_j))$. From Lemma 2 with $MinPts = q$ it then follows that $e(B_i, B_j) \in MNNG_q$ and therefore $e(B_i, B_j) \in MST_q \cup MNNG_q$. If (b) is true, then from statements (i) and (ii) above, we have $mrd_q(C_i, C_j) \le mrd_q(B_i, B_j) = mrd_p(B_i, B_j) \le mrd_p(C_i, C_j)$. However, since $p < q$ and mrd_{MinPts} cannot decrease when $MinPts$ increases, it follows that $mrd_q(C_i, C_j) = mrd_p(B_i, B_j) = mrd_p(C_i, C_j)$. In this case, replacing $e(B_i, B_j)$ in MST_p^* with $e(C_i, C_j)$ will result in another $MST_p' \in M_p$, as the total weight of MST_p^* and MST_p' are the same. MST_p^* includes the edge $e(C_i, C_j)$ and by assumption $e(C_i, C_j) \in MST_q$, and therefore $e(C_i, C_j) \in MST_q \cup MNNG_q$.

It follows that, given a MST of G_q, MST_q, and a MST of G_p, $MST_p^* \in M_p$, with $q > p$, we can construct a weighted graph MST_p' as follows: (1) include all edges and edge weights $e(B_i, B_j) \in MST_p$ that belong to $MST_q \cup MNNG_q$ (according to scenarios (1) or (2)-a); (2) replace all edges $e(B_i, B_j) \in MST_i$ that do not belong in $MST_q \cup MNNG_q$ with edges $e(C_i, C_j)$ that must exist in $MST_q \cup MNNG_q$(as scenario (2)-b), connecting the same two subsets of points as $e(B_i, B_j)$ and having the same edge weight as $e(B_i, B_j)$. This graph MST_p' is a MST of G_p, that is, $MST_p' \in M_p$ and $MST_p' \subseteq MST_q \cup MNNG_q$. Thus there exists a $MST_p \in M_p$ such that $MST_p \subseteq MST_q \cup MNNG_q$.

Theorem 1 supports that the complete graph $G_{MinPts_{max}}$ can be replaced by $MST_{MinPts_{max}} \cup MNNG_{MinPts}$ when calculating additional results of $MSTs$ with $MinPts < MinPts_{max}$, with potential for greater execution time and memory savings.

4   Experiments

In this section, we evaluate the use of $CORE\text{-}SSG$ for the clustering task to illustrate its utility, although it can be used in various applications [10]. The $CORE\text{-}SSG$[1] was implemented in Python/Cython, and the experiments were performed on computer with 16GB of RAM and 8 processing cores. The experimental evaluations are divided into three parts: assessment of the quality of the hierarchies and extracted partitions from $CORE\text{-}SSG$; the clustering solutions from $CORE\text{-}SG$ and $CORE\text{-}SSG$ are compared; an analysis of the runtime and speed-up of the compared algorithms is made.

4.1   Evaluation of the Quality Loss Resulted from $CORE\text{-}SSG$ Summarization

Three bi-dimensional datasets with 38k, 43k, and 49k objects were synthesized for the quality evaluation experiments and divided into 13, 10, and 12 clusters

---
[1] Implementation in: https://github.com/natanaelbatista99/CORE-SSG.

with noise. After that, each dataset was divided into subsets of the same volume so that the number of subsets totalized $n/10$, i.e., 10% of the datasets' sizes. Then, the *DBs* were created over these subsets, and a *CORE-SSG* with $MinPts = 200$ was built. From the $CORE\text{-}SSG_{200}$, we extracted density-based *MSTs* for *MinPts* ranging from 200 to 2 in decreasing steps of 2. Using the extracted *MSTs*, we built one hundred HDBSCAN* clustering solutions and calculated their pairwise similarity using the HAI index [12]. Based on their similarities, we have chosen four of the most representative clustering solutions, i.e., we selected four solutions that are most similar to the others and different within themselves, according to the HAI index. These four hierarchical clustering results are presented in the clustering trees of Fig. 2, where the most prominent clusters, according to the FOSC framework [7], are circled in red. The FOSC framework uses the concept of excess of mass to extract the most prominent partition from a hierarchical clustering solution. As done in the HDBSCAN* article, we assumed the minimum cluster size value equals *MinPts*.

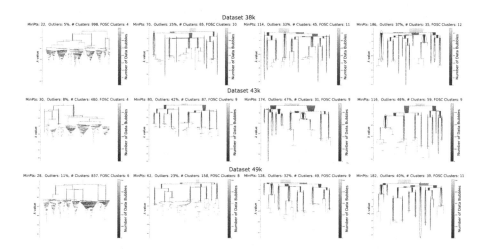

Fig. 2. Cluster hierarchies of *CORE-SSG*'s most representative solutions.

The clustering hierarchies presented in Fig. 2 accurately reflect the behavior of HDBSCAN* clustering for a range of *MinPts* values. For small *MinPts* values, the hierarchy is full of small, highly detailed clusters that are hard to visualize but can be analyzed using cluster validation indexes [22]. In contrast, as *MinPts* values increase, the hierarchies comprise a smaller number of more robust clusters that reflect denser structures. The results perfectly mimic what is expected from density-based clustering algorithms. The difference resides in the fact that *CORE-SSG* does not have the hierarchical structure of the summarized data, i.e., within the *DBs*. In most cases, the summarized data has a level of detail of minor importance for large volumes of data. However, if the user requires the structure within the *DBs*, the *CORE-SG* can be applied over the summarized data of each *DBs* independently, and the results (which are graphs) combined. If the summarized data is still extensive, the *CORE-SSG* can be applied recursively.

Although a hierarchy of clusters is a more complete and richer way of representing the data structure, they can be too complex and extensive for very large datasets. In such scenarios, the automatic extraction of a partition of clusters may be preferred. FOSC is an excellent [7] choice for this job. In Fig. 3, partitions selected (circled) by FOSC in Fig. 2 are presented. It is possible to note that FOSC successfully extracted prominent clusters in the datasets and correctly separated noise. Additionally, different *MinPts* values resulted in different cluster selections, where small values favor the selection of broader clusters that resides in the higher levels of the hierarchy, as they are more present in the hierarchy than their several smaller nested child clusters. In contrast, a higher *MinPts* value prevents smaller cluster structures at the bottom of the hierarchy, favoring the selection of denser clusters nested into the broader ones. The results explicit the importance of the analysis of multiple results with different *MinPts* values, as all of them have the potential to be considered "the best" and validate the main objective of the present work. Furthermore, summarizing the data did not impact the choice of clusters made by FOSC, as the framework has detected the most prominent ones seamlessly for different *MinPts* values.

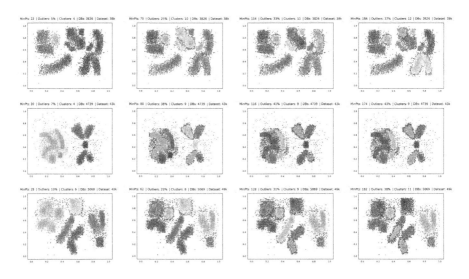

Fig. 3. Cluster chart by representative *MinPts* in Datasets.

4.2 Comparison Between *CORE-SG* and *CORE-SSG* Partitions

As the hierarchies of *CORE-SG* and *CORE-SSG* are not comparable due to summarization, we compared the FOSC extracted partitions between *CORE-SG* and

CORE-SSG to check if the prominent clusters are similar. We used the Adjusted Rand Index (ARI) [11] to obtain the similarities between the extracted partitions where one indicates that the partitions are identical and zero is adjusted to reflect randomness. Figure 4 presents the ARI heatmap resulting from the pairwise comparison among partitions from *CORE-SG* and *CORE-SSG* with different $MinPts$ values. *CORE-SG* partitions are presented in the horizontal axis, and *CORE-SSG* partitions are in the vertical axis. The heatmaps show that, for small $MinPts$ values, the *CORE-SG* has a much more detailed partition, and *CORE-SSG* tends to unify small clusters, which is expected because the information within the *DBs* is summarized and cannot be split. That is reflected in the darker area over small $MinPts$ values. For large datasets, small values of $MinPts$ tend to over-fragment the clustering structure into a hard-to-interpret solution. According to the ARI, for medium and high values of $MinPts$, most heatmaps present bright areas that reflect a similarity over 0.85. This result shows that the main cluster structure of the data is present in the results of both graphs, sometimes for distinct values of $MinPts$, which also support the importance of a multi-density analysis of the dataset, the objective of this paper.

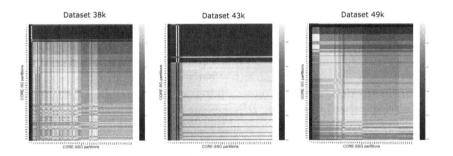

Fig. 4. Comparing partitions with and without *DBs* in Datasets.

4.3 Runtime Analysis

To assess the computational performance of *CORE-SSG* in relation to *CORE-SG*, we created datasets with a variety of sizes, dimensions, and different levels of summarization, as described in Table 1. To evaluate each characteristic separately, default values were fixed and presented in bold in Table 1. Additionally, we also considered a variety of $MinPts_{max}$ values used to build the graphs. The runtime necessary to obtain the graphs, *CORE-SG* and *CORE-SSG*, added to the time needed for the hierarchical density solutions extraction, are presented in Fig. 5.

Table 1. Dataset characteristics and experimental setup.

Variables	Values
$MinPts_{max}$	100, 150, **200**, 250, 300
#objects	100k, 250k, **500k**, 750k, 1M
#dimension	2, **4**, 6, 8
#summariation	5%, **10%**, 15%

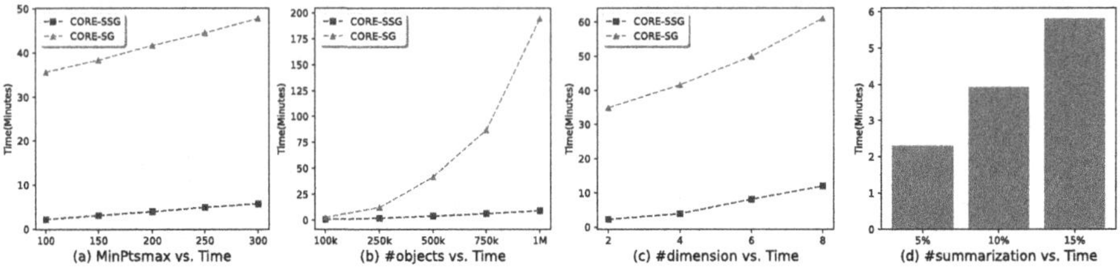

Fig. 5. Runtime to build the graphs and extract the hierarchical density solutions for *CORE-SG* and *CORE-SSG*.

Figure 5 (a) shows that the parameter *MinPts* has a linear correlation with the processing time needed to run the experiment for both graphs, but *CORE-SSG* presented a speedup close to 20 times better than *CORE-SG*. Considering the dataset size n and $m << n$, Fig. 5 (b) clearly shows the quadratic behavior of *CORE-SG* as the dataset size increases up to one million and how the computational cost is amortized when using *DBs*, i.e., for *CORE-SSG*. The present results support the main research hypothesis of this work: data summarization enables scalability for density-based machine learning solutions, clustering algorithms in particular. Summarization also had an impact on the algorithms' total runtime when the data's dimensionality increases, as shown in Fig. 5 (c), with speedups up to 15 times. Last but not least, the total time needed for the data summarization, presented in Fig. 5 (d), reflects the linear increase of the computational time related to the proportion of *DBs* used, which reflects the expected behavior for our proposal and shows that summarization requires a small fraction of the total runtime.

5   Conclusion and Future Work

The main goal of the present research, enabling the use of *CORE-SG* over larger datasets, was achieved as we showed that a single *CORE-SSG* permits the accurate extraction of multiple density-based hierarchical solutions for different densities parameters at a subquadratic computational cost. That was shown theoretically, with the proof that *CORE-SSG* holds the main characteristics of *CORE-SG*, and empirically through experiments. Moreover, the experiments also supported the importance of the analysis of multiple density-based results, as datasets can have nested density structures with different characteristics, each structure with its relative importance, which justifies the application of *CORE-SG* in a pleura of machine learning tasks [10], including clustering.

The most prominent clusters (according to the FOSC framework) present in the original version of *CORE-SG* are also present in our *CORE-SSG* solution, obtaining a significant computational cost reduction. The downside of our approach is the loss of small density-based structures summarized inside the *DBs*, which were rarely considered prominent by the FOSC framework as they connect a very small fraction of the dataset. Thus, our solution is indicated for large volumes of data. Even so, if these small structures are indeed important for the application, we recommend the use of *CORE-SG* (or *CORE-SSG*) over the data within each *DBs*, followed by the combination of the resulted graphs.

In addition to the direct benefits of our findings for the machine learning applications previously referred in [10], we believe that *CORE-SSG* has much potential to innovate in scenarios with continuous data streams. That is because parameters of density-based applications for data streams usually must be set (fixed) as an algorithm input, which may be hard to define and require changes during the evolution of the data (concept drift). *CORE-SSG* enables the extraction of different density-based parameters, even multiple results, from a single induced model, which can be analyzed online to adapt the model for different density levels and concept drift. Indeed, this is a future work worth exploring.

Acknowledgement. The authors would like to thank CNPq MAI/DAI program, CAPES, and FAPESP for their financial support. We also thank Prof. Jörg Sander for his insight and support.

References

1. Ankerst, M., Breunig, M.M., Kriegel, H.P., Sander, J.: Optics: ordering points to identify the clustering structure. SIGMOD Rec. **28**(2), 49–60 (1999)
2. Barlow, H.: Unsupervised Learning. Neural Comput. **1**(3), 295–311 (1989)
3. Blazquez, D., Domenech, J.: Big data sources and methods for social and economic analyses. Technol. Forecast. Soc. Chang. **130**, 99–113 (2018)
4. Breunig, M.M., Kriegel, H.P., Kröger, P., Sander, J.: Data bubbles: quality preserving performance boosting for hierarchical clustering. In: Proceedings of the 2001 ACM SIGMOD International Conference on Management of Data, pp. 79–90 (2001)
5. Breunig, M.M., Kriegel, H.-P., Sander, J.: Fast hierarchical clustering based on compressed data and OPTICS. In: Zighed, D.A., Komorowski, J., Żytkow, J. (eds.) PKDD 2000. LNCS (LNAI), vol. 1910, pp. 232–242. Springer, Heidelberg (2000). https://doi.org/10.1007/3-540-45372-5_23
6. Campello, R.J.G.B., Moulavi, D., Zimek, A., Sander, J.: Hierarchical density estimates for data clustering, visualization, and outlier detection. ACM Trans. Knowl. Discov. Data **10**(1), 1–51 (2015)
7. Campello, R.J., Moulavi, D., Zimek, A., Sander, J.: A framework for semi-supervised and unsupervised optimal extraction of clusters from hierarchies. Data Min. Knowl. Disc. **27**, 344–371 (2013)

8. Cheema, P., Alamdari, M.M., Chang, K., Kim, C., Sugiyama, M.: A drive-by bridge inspection framework using non-parametric clusters over projected data manifolds. Mech. Syst. Signal Process. **180**, 109401 (2022)

9. Djonlagic, I., et al.: Macro and micro sleep architecture and cognitive performance in older adults. Nat. Hum. Behav. **5**(1), 123–145 (2021)

10. Gertrudes, J.C., Zimek, A., Sander, J., Campello, R.J.G.B.: A unified view of density-based methods for semi-supervised clustering and classification. Data Min. Knowl. Discov. **33**(6), 1894–1952 (2019)

11. Hubert, L., Arabie, P.: Comparing partitions. J. Classif. **2**, 193–218 (1985)

12. Johnson, D., Xiong, C., Gao, J., Corso, J.: Comprehensive cross-hierarchy cluster agreement evaluation. ACM TKDD. **10**, 1–51 (2013)

13. Liu, B., Shi, Y., Wang, Z., Wang, W., Shi, B.: Dynamic incremental data summarization for hierarchical clustering. In: Yu, J.X., Kitsuregawa, M., Leong, H.V. (eds.) WAIM 2006. LNCS, vol. 4016, pp. 410–421. Springer, Heidelberg (2006). https://doi.org/10.1007/11775300_35

14. Miccio, L.A., Schwartz, G.A.: Mapping chemical structure-glass transition temperature relationship through artificial intelligence. Macromolecules **54**(4), 1811–1817 (2021)

15. Minussi, D.C., et al.: Breast tumours maintain a reservoir of subclonal diversity during expansion. Nature **592**(7853), 302–308 (2021)

16. Murray, B., Perera, L.P.: An AIS-based deep learning framework for regional ship behavior prediction. Reliab. Eng. Syst. Saf. **215**, 107819 (2021)

17. Nassar, S., Sander, J., Cheng, C.: Incremental and effective data summarization for dynamic hierarchical clustering. In: Proceedings of the 2004 ACM SIGMOD International Conference on Management of Data, pp. 467–478. SIGMOD 2004, Association for Computing Machinery, New York, NY, USA (2004)

18. Neto, A.C.A., Naldi, M.C., Campello, R.J.G.B., Sander, J.: Core-SG: efficient computation of multiple MSTS for density-based methods. In: 2022 IEEE 38th International Conference on Data Engineering (ICDE), pp. 951–964 (2022)

19. Norman, T.M., et al.: Exploring genetic interaction manifolds constructed from rich single-cell phenotypes. Science **365**(6455), 786–793 (2019)

20. dos Santos, J.A., Syed, T.I., Naldi, M.C., Campello, R.J., Sander, J.: Hierarchical density-based clustering using MapReduce. IEEE Trans. Big Data **7**(1), 102–114 (2019)

21. Savoie, W., et al.: A robot made of robots: emergent transport and control of a smarticle ensemble. Sci. Robot. **4**(34), eaax4316 (2019)

22. Vendramin, L., Campello, R.J., Hruschka, E.R.: Relative clustering validity criteria: a comparative overview. Statist. Anal. Data Mining ASA Data Sci. J. **3**(4), 209–235 (2010)

23. Zerhari, B., Lahcen, A.A., Mouline, S.: Big data clustering: Algorithms and challenges. In: Proceedings of International Conference on Big Data, Cloud and Applications (BDCA-5) (2015)

24. Zhang, T., Ramakrishnan, R., Livny, M.: Birch: an efficient data clustering method for very large databases. SIGMOD Rec. **25**(2), 103–114 (1996)

25. Zhang, Y., Cheung, Y., Liu, Y.: Quality preserved data summarization for fast hierarchical clustering. In: 2016 International Joint Conference on Neural Networks (IJCNN), pp. 4139–4146 (2016)

26. Zhou, J., Sander, J.: Data bubbles for non-vector data: Speeding-up hierarchical clustering in arbitrary metric spaces. In: Freytag, J.C., Lockemann, P., Abiteboul, S., Carey, M., Selinger, P., Heuer, A. (eds.) Proc. 2003 VLDB Conf., pp. 452–463. Morgan Kaufmann, San Francisco (2003)

Exploring Text Decoding Methods
for Portuguese Legal Text Generation

Kenzo Sakiyama[1]([✉])(iD), Raphael Montanari[1](iD), Roseval Malaquias Junior[1](iD),
Rodrigo Nogueira[2](iD), and Roseli A. F. Romero[1](iD)

[1] Universidade de São Paulo - ICMC, São Paulo, Brazil
{kenzosakiyama,rmontanari,roseval}@usp.br, rafrance@icmc.usp.br
[2] Universidade de Campinas - Neuralmind, Campinas, Brazil
rfn@unicamp.br

Abstract. In recent years, there has been considerable growth in the volume of legal proceedings in Brazil. In this context, there is a lot of potential in using recent advances in Natural Language Processing to automate tasks and analysis in the legal domain. In this article, we investigate text decoding methods for automating the writing of keyphrases, a sequence of key terms present in documents used in courts throughout Brazil. For this purpose, a text-to-text framework based on generative Transformers is used to generate keyphrases and evaluate three decoding techniques: greedy, top-K, and top-p. Since the keyphrases are designed to improve retrieval tasks, we evaluated keyphrases generated by the decoding methods in legal document retrieval. Traditional retrieval methods (TF-IDF and BM25) were used to evaluate the quality of the generated keyphrases. The results obtained (in terms of IR metrics) were statistically significant, and they indicate that greedy decoding generates high-quality keyphrases for the dockets used in this work, providing keyphrases close to the ones generated by human specialists.

Keywords: Text Generation · Information Retrieval · Legal Texts

1 Introduction

Based on data from the National Council of Justice - *Conselho Nacional de Justiça* (CNJ) [11], there were 77.3 million cases in transit in the Brazilian judiciary at the end of 2021, indicating a 10.4% increase from the previous year. The analysis of such cases contributes to the slowness of the Brazilian legal system due to the human effort required to both write and analyze the legal cases. In this context, dockets consist of a special type of document, that aims to provide a summary of a legal case. They are used in courts all around Brazil and are designed to provide a summarised representation of judicial decisions. Figure 1 presents an example of a docket.

The dockets usually follow a pre-defined structure composed of two components: keyphrases and enumerated paragraphs. The keyphrases consist of a

M. C. Naldi and R. A. C. Bianchi (Eds.): BRACIS 2023, LNAI 14195, pp. 63–77, 2023.
https://doi.org/10.1007/978-3-031-45368-7_5

PROCESSUAL CIVIL. AGRAVO EM RECURSO ESPECIAL. ACÓRDÃO RECORRIDO. FUNDA-
MENTAÇÃO CONSTITUCIONAL. REVISÃO. IMPOSSIBILIDADE.
1. À luz do art. 105, III, da Constituição Federal, o recurso especial não serve à revisão da fundamentação
constitucional.
2. Na hipótese dos autos, o acórdão a quo foi proferido com fundamento constitucional, uma vez que aponta
na direção da violação pela lei estadual da proibição do efeito confiscatório da multa tributária e da propor-
cionalidade da penalidade (art. 150, IV, da CF).
3. Agravo interno desprovido.

Fig. 1. Example of a docket. Keyphrases are highlighted in bold text.

header present at the beginning of the docket and are composed of sequences
of capitalised key terms that highlight the key subjects present in the document. This header is created to improve the search and retrieval of jurispru-
dences (precedents) [7]. The enumerated paragraphs discuss the themes (or topics) present in the document.

By analysing the form and linguistic style present in the keyphrases, it's
possible to note similarities between the writing of keyphrases and two Natural
Language Processing (NLP) tasks: summarization and key terms extraction.
However, keyphrases are not written in a fluid and natural manner such as
summaries. In addition, most of the terms present in their text are not present
in the remainder of the docket which originated the keyphrase, which makes it
difficult to treat its writing as an extractive task.

Given the predictable structure and availability of dockets, it would be possible to prepare input-output pairs in order to generate keyphrases using the
enumerated paragraphs as inputs, by employing a supervised approach. Transformers, such as GPT [19], were already proven effective in various text-to-text
generative Natural Language Processing (NLP) tasks [6] (such as translation,
question answering and summarization). Also, the availability of pre-trained language models [3,25,33] presents a lot of opportunities to automate NLP tasks.

Thus, in this work, we aim to investigate the usage of state-of-the-art generative Transformers to automate the writing of keyphrases. Specifically, we seek
to investigate text decoding methods in order to generate keyphrases that aid
retrieval in the legal domain. This study is unprecedented in Brazil and can be
used to automate keyphrase generation in courts around the country. At last,
the main contributions of this work are:

1. Investigation of a novel approach to generate keyphrases from Brazilian dockets, using a sequence-to-sequence Transformer;
2. Comparison of three different text decoding methods for the proposed task
   (greedy and sampling methods);
3. Quantitative and qualitative analysis of the generated keyphrases.

This paper is organised as follows. Section 2 presents related works. Section 3
discussed the methodology applied for the keyphrase generation. Next, Sect. 4
presents and discusses the obtained results. At last, Sect. 5, presents conclusions
and future works.

2   Related Works

In this Section, we will present studies related to the main objectives of this proposal. The Transformer [29] consists of a deep neural network architecture that achieved the state of the art in several NLP tasks. It consists of an encoder-decoder architecture used originally for translation. However, the context-aware representations generated by the model can be used for a large variety of tasks.

Following the success of the Generative Pretrained Transformers (GPT) models [19, 25, 33], there is a predominance of decoder-only models in NLP tasks that can be approached as text generation (such as question answering and summarization) [6]. In addition, recent studies showed the great potential in using such models in zero and few-shot scenarios [2]. Other studies [20, 31] investigate text generation using the full Transformer architecture (encoder-decoder) for some NLP tasks. The T5 [20] Transformer proposes the unification of a series of NLP tasks in a single text-to-text framework and Xue *et al.* [31] expanded the original work to add multilingual support.

Although the presented Transformer approaches for text generation are different in terms of architecture and scale (number of parameters), they all deal with common issues concerning the quality of the artificially generated text. Generated texts are often simplistic, inconsistent, or end up being repetitive [8]. There is also the possibility of hallucination, generating contradictory texts, meaningless and without foundation or evidence [10].

In order to mitigate the challenges (repetitive and predictable texts), there were initiatives aimed at making text generation non-deterministic [4,8]. Such proposals arise as an alternative to simpler text generation methods (also called greedy decoding), arguing that choosing most probable words (or tokens) is one of the main causes of repetitive texts.

Another example of a study aimed at mitigating repetitive texts is the work from Su *et al.* [28]. Proposed in 2022, the contrastive-search consists of a modification in the choosing words (or tokens) predicted by a textual generator, which aims to increase the variability of the text while maintaining its coherence. For this purpose, the authors suggest penalising, during decoding or unsupervised training of the language model, the softmax scores of the most likely tokens by their similarity to other tokens within the context. The importance given to the similarity is controlled by a parameter alpha.

At last, we will present examples of studies employing Transformers to generate text in the legal domain. Keyphrases such as the ones used in this work are exclusive to Brazil and, for the best of our knowledge, this is the first in depth study of decoding methods for brazillian keyphrase generation. Feijo and Moreira [5] and Yoon *et al.* [32] applied Transformer models to summarise rulings from the Brazilian Supreme Court and Korean legal cases, respectively. Peric *et al.* [16] proposed the use of Transformers to generate opinions about legal cases originating in the *U.S Circuit Court*, by employing an encoder-decoder architecture.

Huang *et al.* [9] proposed a solution to automate the Legal Judgment Prediction (PJL) subtasks using the T5 text-to-text framework. At last, Althammer *et al.* [1] investigated the use of summaries (generated by Transformer) as part of an information retrieval pipeline for the legal domain as part of the 2021 Competition on Legal Information Extraction/Entailment (COLIEE).

3   Methodology

The methodology used in this work is composed of: I) Data Collection and Preprocessing, II) Keyphrase Generator Training, III) Decoding Methods Evaluation and IV) Qualitative Analysis. These components will be discussed below.

3.1   Data Collection and Preprocessing

In 2022, the Brazilian *Supremo Tribunal de Justiça* (STJ) - Supreme Court of Justice made available the *Dados Abertos*[1] platform. The platform consists of a public website, sharing legal decisions from various courts in Brazil. The published documents comprise a large variety of topics in Brazil's legal domain, such as crimes in general, commerce, taxes, etc. We collected a total of 712,161 documents from the platform in August 2022.

After the data collection, we extracted the dockets from the documents metadata and preprocessed the text of the decisions. We removed duplicated examples and removed URLs from the text. 111,964 dockets remained after the preprocessing described. With the remaining examples, we extracted the keyphrases and enumerated paragraphs from the dockets, identifying and extracting capitalised sentences present in the header of the collected decisions. By extracting the inputs (enumerated paragraphs) and expected outputs (keyphrases), the original keyphrases (written by specialists) compose the reference set used for supervised training and evaluation.

As a final preprocessing step, we divided the corpus (111,964 examples) into training (70%), validation (10%), and test (20%) splits. From the examples of the training set, we observed that enumerated paragraphs and keyphrases have a mean of 203.26 and 55.84 space-separated tokens, respectively. We used the splits to train and evaluate a supervised deep learning text generator.

3.2   Keyphrase Generation

In this section is described the methodology employed for keyphrase training and generation.

Transformers for Text Generation. Based on the dockets collected, we noted that most of the terms in the keyphrases are not directly present in the dockets. By further analysing examples from the validation set, we noted that only ∼10% of the terms present in the keyphrases are in fact present in the input text. Thus, we decided to approach writing keyphrases as generation rather than extraction

---

[1] https://dadosabertos.web.stj.jus.br/.

of text. For this purpose, a sequence-to-sequence (or text-to-text) Transformer model was chosen.

We choose PTT5 [3] as our keyphrase generator. PTT5 was pretrained in a large Brazilian Portuguese corpus (*brWaC* [30]) with 2.7 billion tokens and the *base* version of the model (220M parameters) was used in our experiments. We experimented with other state-of-the-art multilingual generative Transformer models (such as mT5 [31], BLOOM [25] and OPT [33]), but the Portuguese model (PTT5) performed better. Previous works [3,22,27] observed that models pretrained for the task language tend to outperform multilingual models on the same tasks, and the same trend was observed in our experiments.

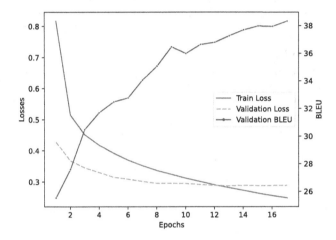

Fig. 2. Train and validation losses obtained for the PTT5 model are shown in the left y-axis. The plot also shows the validation BLEU scores in the right y-axis.

Training Details. We fixed the input (enumerated paragraphs) and output (keyphrases) sizes to 512 and 256 sentence-piece tokens, respectively. We padded shorter sequences of tokens and truncated longer sequences (to the maximum length). We fine-tuned the method PTT5 using a fixed learning rate of 1×10^{-3}, batch size equal to 256 and 20 maximum training epochs.

For the sequence-to-sequence training, the cross-entropy loss function was adopted. The BLEU score [17] metric was considered to evaluate the text generation quality. We used early-stopping during training, monitoring the BLEU metric in the validation set. The training process is stopped after two epochs without improving the BLEU score. For evaluation, we repeated the training process with five different seeds (1000, 2000, 3000, 4000 and 5000) and obtained a 37.254±0.783 BLEU score. The best performing model achieved 38.607 BLEU on the test set. The fine-tuning was done using the *HuggingFace*[2] library, and a Tesla P100 GPU with 16 GB VRAM.

---

[2] huggingface.co/.

Figure 2 shows the train and validation losses for the best execution of the PTT5 model in addition to the validation BLEU scores. The model was trained for 17 epochs, totaling 5219 iterations.

3.3 Decoding Methods Evaluation

For the evaluation of the generated text and to compare the decoding methods, we have concatenated the generated keyphrases to their original document and used a real use case of retrieval to extract IR metrics. We opted to use an IR task to evaluate the generated keyphrases (created using different decoding methods) to evaluate them in their intended use: improving retrieval tasks. The details of the evaluations will be presented to follow.

Decoding Methods. Decoding techniques are used to guide neural text generation, in order to generate meaningful and coherent text. The methods are used to generate human-readable text from the internal representations of language models. In this work, we evaluate three decoding methods: greedy, top-k [4] and top-p [8]. Top-k and top-p are sampling decoding methods, that, during text generation, sample tokens from finite sets. A brief description of the methods will be presented next.

- **Greedy:** greedy decoding always selects the most probable token (highest softmax score) during generation.
- **Top-K:** consists of filtering the most probable K tokens at a given instant, and redistributing their probabilities among them before sampling.
- **Top-p:** limits the set of selectable tokens to a set of more probable tokens whose summed probabilities are lower than the established threshold p. Note that the number of tokens that can be chosen is dynamic, since the probability distributions vary at each instant.

Task Formulation. We have used the themes (categorical information), present in the dockets' metadata, to simulate a retrieval task in which a specialist seeks to retrieve documents similar to a query document using a search engine. The themes are unique identifiers, that are mapped to common questions of law. Thus, by using the binary relevance definition: given a query document Q, the relevant documents R to Q must have the same theme as Q. Note that, in the real scenario, the documents consists in dockets containing both keyphrases and enumerated paragraphs.

From the collected decisions, only 801 have themes. These documents were removed from the training set and used to prepare query - relevant document pairs for IR evaluation. The query set consists of dockets whose themes occur at least twice. From those, we prepared 482 query - relevant document pairs (pairs of same theme documents).

To prepare the final retrieval corpus, we combined the test set presented in Sect. 3.1 with the dockets with themes and obtained a total of 23,194 documents.

We increased the retrieval corpus to make the retrieval task more challenging. In the worst case, the documents without a theme may introduce false negatives (documents with the same theme of the query, but considered non-relevant), hindering the IR metrics.

Experimental Setup. Two different experiments were performed during the IR evaluation and they are described in the following.

1. *Studying Sampling-based Decoding Methods*: this experiment aims to investigate the generation of multiple keyphrases from a single docket using sampling decoding. By concatenating multiple keyphrases to a single docket, we expect to see improvements in the IR metrics since we are adding more text variations.

   We generated up to 10 keyphrases for each example in the search corpus, using top-K and top-p sampling, and concatenated them to the original input (enumerated paragraphs) to generate the IR metrics. We repeated the text generation five times with seeds of different random numbers (1,000, 2,000, 3,000, 4,000, and 5,000) and aggregated the results for comparisons. The effects of the K of the top-K, and the p of the top-p sampling methods were also evaluated in this experiment, varying the values of both K and p.

   We choose K and p from the following sets of values: $K \in \{15, 50, 100\}$ and $p \in \{0.1, 0.5, 0.9\}$. Note that for this experiment, we are not interested in determining the best number of repetitions, nor the best value for K or p. The goal is to investigate the effect of the parameters on the proposed IR task, but the results may indicate the best parameter ranges.

2. *Decoding Methods Comparison*: in order to compare the decoding methods, we extracted IR metrics for dockets using keyphrases generated using top-K and top-p sampling. We used the generated ones in place of the originals in this experiment. For reference, we also evaluated IR metrics considering documents with and without the original keyphrases (for both query and corpus documents) and using simple greedy decoding.

   We choose to use only one keyphrase generated by each method based on the results of the previous experiment and to evaluate the decoding methods in similar scenarios. For this experiment, we used the following parameters for the sampling-based decoding methods: $K = 15$ and $p = 0.9$ (based on the performances obtained in the previous experiment).

The experiments with sampling decoding methods were inspired by the work *doc2query* [14]. In this work, for each example in a corpus, the authors generated several queries related to the example's content using a sequence-to-sequence Transformer model. The authors used top-K sampling in order to generate several queries per example. Then the queries were concatenated to the input documents in order to improve IR metrics. Considering both experiments with sampling methods, we used contrastive-search with $alpha = 0.6$, based on the original paper [28]. We choose the K and p values based on previous works with top-K and top-p sampling [8,14].

Information Retrieval Methods and Metrics. To evaluate the proposed IR task, we choose two traditional methods: TF-IDF and BM25 [21]. The methods were chosen due to their popularity in search engines (such as Lucene[3]) and competitive performance [18,23]. Previous works [12,13,15] also discussed that sparse representation methods (such as the chosen ones) tend to perform better in similar tasks in the legal domain.

As an additional preprocessing for the IR methods, the documents were tokenized and Portuguese stop-words and punctuation were removed. For TF-IDF, we utilized a vocabulary size of 10,000 tokens (that appeared at least three times), and n-grams from 1 to 3. To sort documents during retrieval using TF-IDF, we used the cosine similarity between queries and documents. Considering BM25, the documents were sorted by the probability ranking principle, estimating the relevance of a document to the presented query. The additional preprocessing was done using spaCy[4] and sklearn[5]. For BM25, we used the implementation and default parameters from *rank-bm25*[6].

At last, we evaluated the performance in the proposed IR task using two traditional IR metrics: Mean Reciprocal Rank (MRR) and Recall. The metrics were chosen by their use in similar works in the legal domain [24,26]. We used a threshold of 10 documents (top-10 ranked documents) to compute the metrics. According to Russel *et al.* [24], law professionals tend to analyze, for the most part, up to 50 documents in their searches. Therefore, we are evaluating an even more challenging scenario than the described by the authors.

3.4   Qualitative Analysis

As a final analysis, for all decoding methods evaluated (greedy, top-K and top-p), we sampled examples generated using all methods and performed a qualitative analysis on them. For this analysis, we compared the generated keyphrases to the references and discussed the similarities between them and the effect of the sampling methods.

4   Results and Discussions

These Sections discuss the results obtained for each experiment described in Sect. 3.3. In all experiments, we aim not to compare the retrieval methods (TF-IDF and BM25), but to use them to evaluate the quality of the generated keyphrases using different decoding methods.

4.1   Studying Sampling-Based Decoding Methods

Tables 1 and 2 present the IR metrics obtained, varying both the number of repetitions and K e p parameters. The metrics consist in the mean of five different executions (using five different seeds).

---

[3] https://lucene.apache.org/.
[4] https://spacy.io/.
[5] https://scikit-learn.org.
[6] https://pypi.org/project/rank-bm25/.

Table 1. Top-k experiments evaluation metrics. N represents the number of samples included at the beginning of each docket.

| | (a) TF-IDF experiments. | | | | | | | (b) BM25 experiments. | | | | | |
| | k=15 | | k=50 | | k=100 | | | k=15 | | k=50 | | k=100 | |
N	MRR@10	R@10	MRR@10	R@10	MRR@10	R@10	N	MRR@10	R@10	MRR@10	R@10	MRR@10	R@10
1	0.826	0.804	0.824	0.811	0.824	0.809	1	0.861	0.882	0.857	0.880	0.856	0.881
2	0.820	0.796	0.818	0.798	0.817	0.799	2	0.857	0.879	0.857	0.876	0.856	0.875
3	0.815	0.793	0.812	0.789	0.812	0.790	3	0.858	0.875	0.856	0.874	0.856	0.873
4	0.811	0.784	0.812	0.785	0.812	0.783	4	0.859	0.878	0.856	0.875	0.856	0.874
5	0.809	0.779	0.810	0.780	0.810	0.780	5	0.858	0.876	0.857	0.872	0.854	0.872
6	0.808	0.776	0.810	0.782	0.810	0.783	6	0.858	0.875	0.854	0.873	0.852	0.873
7	0.810	0.781	0.806	0.783	0.805	0.782	7	0.859	0.874	0.855	0.872	0.855	0.872
8	0.807	0.781	0.809	0.784	0.809	0.784	8	0.860	0.874	0.855	0.870	0.853	0.870
9	0.806	0.777	0.810	0.784	0.808	0.783	9	0.860	0.873	0.855	0.871	0.854	0.869
10	0.804	0.778	0.810	0.784	0.809	0.787	10	0.861	0.871	0.857	0.871	0.856	0.871

Table 2. Top-p experiments evaluation metrics. N represents the number of samples included at the beginning of each docket.

| | (a) TF-IDF experiments. | | | | | | | (b) BM25 experiments. | | | | | |
| | p=0.1 | | p=0.5 | | p=0.9 | | | p=0.1 | | p=0.5 | | p=0.9 | |
N	MRR@10	R@10	MRR@10	R@10	MRR@10	R@10	N	MRR@10	R@10	MRR@10	R@10	MRR@10	R@10
1	0.821	0.815	0.823	0.813	0.826	0.807	1	0.854	0.877	0.855	0.880	0.859	0.883
2	0.787	0.768	0.804	0.790	0.820	0.797	2	0.842	0.866	0.852	0.880	0.858	0.880
3	0.761	0.724	0.788	0.774	0.810	0.790	3	0.842	0.862	0.853	0.875	0.860	0.881
4	0.743	0.703	0.781	0.763	0.809	0.785	4	0.832	0.853	0.852	0.872	0.858	0.878
5	0.727	0.684	0.776	0.748	0.805	0.779	5	0.828	0.845	0.852	0.866	0.855	0.876
6	0.717	0.672	0.772	0.740	0.806	0.781	6	0.827	0.836	0.851	0.866	0.857	0.876
7	0.713	0.661	0.768	0.734	0.803	0.780	7	0.823	0.826	0.851	0.865	0.858	0.874
8	0.707	0.649	0.764	0.733	0.804	0.777	8	0.821	0.823	0.853	0.864	0.859	0.876
9	0.701	0.640	0.760	0.724	0.805	0.777	9	0.819	0.822	0.853	0.862	0.859	0.876
10	0.697	0.637	0.758	0.719	0.801	0.774	10	0.816	0.818	0.853	0.861	0.861	0.873

When carrying out this experiment, the expectation was to observe a logarithmic growth as more different keyphrases were concatenated to the dockets (similar to *doc2query* [14]), since we are using more variations of keyphrases. However, this result was not observed in any of the evaluated metrics. Contrary to expectations, in the worst cases, there was a decay in the metrics as new variations were added to the input texts for both top-K and top-p decoding methods. The decay is more noticeable for TF-IDF method, with reductions between 2% (top-K) and 12% (top-p) in all observed metrics. The mentioned behaviors were observed for all evaluated K and p values.

The worst performances were observed in increasing repetitions for $p = 0.1$ (top-p experiments). The most probable explanation is that the low p value is too restrictive, reducing the set of selectable tokens. Hence, the top-p tends to generate similar keyphrases with low text variation (more similar or equal keyphrases). This way, the repetitive text hindered the performance of both IR methods evaluated.

By increasing the K and p values, we increase the variability of the generated text, since the tokens to be predicted are chosen from a larger set. A positive effect on the metrics was also expected due to the possibility of adding more discriminative terms in the generated keyphrases, which is beneficial for the evaluated sparse methods. However, we observed deterioration of the performance and, at the best case, similar metrics by varying the K and p values. We suspect that even with the increase in variability, the generated keyphrases remained similar to each other, resulting in addition of repetitive texts to the dockets.

The conclusion from these results is that there is no evidence that using more keyphrases (by using sampling decoding) is beneficial to the evaluated task. Also, there is no benefit in using K values above 15, and p values lower than 0.9. We will discuss the results of the sampling methods further in Sect. 4.3.

4.2 Decoding Methods Comparison

In Table 3 is presented the results comparing decoding methods. For both TFIDF and BM25, a single keyphrase using greedy was generated for both top-K and top-p decoding. We adopted $K = 15$ and $p = 0.9$ for the top-K and top-p decoding, respectively, based on the results from the previous experiment. Table 3 also presents the results obtained performing the proposed retrieval task with and without the original (reference) keyphrases for comparisons.

Table 3. IR metrics obtained for each decoding method evaluated. Superscript characters denote a pairwise statistically significant difference, according to a paired T-test (p-value < 0.05).

(a) TF-IDF experiments.	MRR@10	R@10
a) Without	0.806	0.790
b) Generated greedy	0.822^a	0.815^a
c) Generated top-K	0.828^a	0.810^a
d) Generated top-p	0.825^a	0.811^a
Original	0.825^a	0.832^a

(b) BM25 experiments.	MRR@10	R@10
a) Without	0.819	0.878
b) Generated greedy	0.854^a	0.877
c) Generated top-K	0.863^a	0.890
d) Generated top-p	0.859^a	0.880
Original	0.879^a	0.916^a

We observe that the keyphrases are, indeed, beneficial to retrieval tasks by comparing the metrics obtained by using documents with and without the keyphrases. For both metrics, we observed statistically significant differences. Since both sparse methods (TF-IDF and BM25) benefit from the existence of discriminative terms in the documents, these results were already expected. Note that the metrics obtained using the original keyphrases act as an upper bound to our experiments.

Considering the TF-IDF retrieval, we observed an increment in all metrics by using the generated texts (compared to not using any). The differences in all metrics are statistically significant (see Table 3a). For the BM25 method, we observe similar results (see Table 3b). However, no significant differences were observed when considering the R@10 metric. Note that by using generated keyphrases, there is the possibility of introducing false positives (false similar) and false negatives (false non-similar) in the search corpus, originated by noisy keyphrases.

The IR metrics obtained by the BM25 method suggest that the method was sensitive to these noisy examples.

By comparing the decoding methods, we note small increments for the sampling methods in relation to greedy decoding. However, considering a paired T-Test using a threshold of 5%, there is no significant difference between the decoding methods. Thus, there is not enough evidence to reject the null hypothesis (metrics have the same mean) by observing the comparisons between the metrics of all three decoding approaches. Therefore, there is no evidence that justifies choosing to sample decoding methods over a simpler greedy decoding approach, considering the proposed task.

4.3   Qualitative Analysis

Reference: AGRAVO REGIMENTAL NO RECURSO ORDINÁRIO EM HABEAS CORPUS. IMPETRAÇÃO SUBSTITUTIVA DE REVISÃO CRIMINAL. NULIDADE. OITIVA DE TESTEMUNHAS. AUSÊNCIA DE INTIMAÇÃO PELO JUÍZO DEPRECADO. DESNECESSIDADE. PREJUÍZO NÃO COMPROVADO. NO MAIS, NÃO ENFRENTAMENTO DOS FUNDAMENTOS DA DECISÃO AGRAVADA. SÚMULA 182/STJ. AGRAVO DESPROVIDO.
Generated: AGRAVO REGIMENTAL NO RECURSO ORDINÁRIO EM HABEAS CORPUS. CRIME DE ESTUPRO DE VULNERÁVEL. NULIDADE. INTIMAÇÃO DA EXPEDIÇÃO DA CARTA PRECATÓRIA. DESNECESSIDADE. NO MAIS, NÃO ENFRENTAMENTO DOS FUNDAMENTOS DA DECISÃO AGRAVADA. SÚMULA 182/STJ. AGRAVO DESPROVIDO.

Reference: DIREITO TRIBUTÁRIO. EMBARGOS DE DECLARAÇÃO NO AGRAVO INTERNO NO AGRAVO EM RECURSO ESPECIAL. APURAÇÃO DO VALOR VENAL DE IMÓVEL NÃO PREVISTO ORIGINALMENTE NA PLANTA GENÉRICA DE VALORES. REPERCUSSÃO GERAL RECONHECIDA. DEVOLUÇÃO À ORIGEM PELO TEMA 1.084. PRECEDENTES DO SUPREMO TRIBUNAL FEDERAL.
Generated: PROCESSUAL CIVIL. EMBARGOS DE DECLARAÇÃO NO AGRAVO INTERNO NO AGRAVO EM RECURSO ESPECIAL. AÇÃO DE COBRANÇA. IPTU. IMÓVEL NÃO PREVISTO NA PLANTA VALOR VENAL. REPERCUSSÃO GERAL RECONHECIDA. TEMA 1.084 COM REPERCUSSÃO GERAL RECONHECIDA. DEVOLUÇÃO À ORIGEM.

Reference: AGRAVO REGIMENTAL NO HABEAS CORPUS. TRÁFICO DE DROGAS. PRISÃO PREVENTIVA. PERICULUM LIBERTATIS. CRISE MUNDIAL DE COVID-19. NÃO DEMONSTRADA A EXISTÊNCIA DE RISCO À INTEGRIDADE FÍSICA DO ENCARCERADO. AGRAVO REGIMENTAL NÃO PROVIDO.
Generated: AGRAVO REGIMENTAL NO HABEAS CORPUS. TRÁFICO DE DROGAS. PRISÃO PREVENTIVA. PERICULUM LIBERTATIS. AGRAVO REGIMENTAL NÃO PROVIDO.

Fig. 3. Examples generated using greedy decoding.

Greedy Decoding. Examples of keyphrases generated by PTT5, using greedy decoding, are presented in Fig. 3. We can note that with BLEU scores close to 40%, although being generated by a model trained in a modest training set (less than 100K examples), the generated keyphrases do not present spelling and lexical errors. They captured the writing style of the reference keyphrases and are very similar to the keyphrases written by humans.

A comparison between the number of tokens of the original and the generated keyphrases using greedy decoding is shown in Fig. 4. It is possible to observe that, although the distributions presented by the two histograms are similar, the generated keyphrases have a higher concentration of examples below 60 tokens. The average of space-separated tokens of the generated keyphrases is lower than the average of the tokens presented by the references (42.34 compared to 48.28).

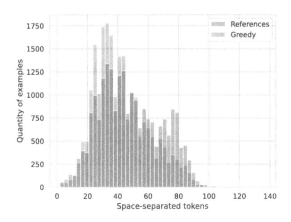

Fig. 4. Histogram comparing the number of space-separated tokens between the reference keyphrases and the ones generated using greedy decoding.

Therefore, we identified that the keyphrases generated with greedy decoding tend to have a smaller length (in tokens) than the originals. We also observed the same pattern for the keyphrases generated using sampling decoding.

Sampling Decoding. In Fig. 5 is shown keyphrase examples generated using top-K and top-p decoding. We used $K = 15$ and $p = 0.9$ based on the results from the previous analysis. From the examples, it is possible to note that the main effect of using sampling is the generation of paraphrases of the original keyphrase. We also observe examples of reordering of the phrases present in the keyphrases. Hence, the generated keyphrases tend to be similar to each other. The described behaviours are justified by the working of language models based on Transformers since, during text generation, they tend to generate tokens that appear in similar contexts.

By using sampling-based methods, we observed an increase in text variability. However, the possibility of the model generating text not related to the input also increases, which may have harmed the IR methods studied. In addition, when concatenating multiple variations of keyphrases similar to each other, we added many repeated terms to the documents, which may influence negatively the sparse IR methods evaluated.

In addition to the justifications presented, the amount of training data may also have affected the sampling methods. Although the results for greedy generation were better, the lack of variability in the training examples (due to their small size), may have harmed the decoding using top-K and top-p sampling.

Original: PROCESSUAL CIVIL E TRIBUTÁRIO. AGRAVO INTERNO NO RECURSO ESPECIAL. EXECUÇÃO FISCAL. NOMEAÇÃO DE BENS À PENHORA. RECUSA INJUSTIFICADA DA FAZENDA. REVISÃO DAS CONCLUSÕES ADOTADAS NA ORIGEM QUE DEMANDA REEXAME DE PROVA. SÚMULA 7/STJ. AGRAVO INTERNO DA FAZENDA DO ESTADO DE SÃO PAULO DESPROVIDO.

1) Top-K: PROCESSUAL CIVIL E TRIBUTÁRIO. AGRAVO INTERNO NO RECURSO ESPECIAL. EXECUÇÃO FISCAL. PENHORA. ORDEM LEGAL. NOMEAÇÃO. RECUSA JUSTIFICADA. REVISÃO. IMPOSSIBILIDADE. SÚMULA 7/STJ. AGRAVO INTERNO DA FAZENDA DO ESTADO DE SÃO PAULO A QUE SE NEGA PROVIMENTO.

2) Top-K: TRIBUTÁRIO. AGRAVO INTERNO NO RECURSO ESPECIAL. EXECUÇÃO FISCAL. PENHORA. NOMEAÇÃO. RECUSA JUSTIFICADA. POSSIBILIDADE. REVISÃO. INCIDÊNCIA DA SÚMULA 7/STJ. AGRAVO INTERNO DA FAZENDA DO ESTADO DE SÃO PAULO DESPROVIDO.

3) Top-p: TRIBUTÁRIO. AGRAVO INTERNO NO AGRAVO EM RECURSO ESPECIAL. EXECUÇÃO FISCAL. PENHORA DE BEM NOMEADO NA PRIMEIRA OPORTUNIDADE. RECUSA PELO FISCO EXEQUENTE NÃO JUSTIFICADA. POSSIBILIDADE. REVISÃO. SÚMULA 7/STJ. AGRAVO INTERNO DA FAZENDA DO ESTADO DE SÃO PAULO DESPROVIDO.

4) Top-p: TRIBUTÁRIO E PROCESSUAL CIVIL. AGRAVO INTERNO NO AGRAVO EM RECURSO ESPECIAL. EXECUÇÃO FISCAL. PENHORA. NOMEAÇÃO DE PRECATÓRIOS. ORDEM LEGAL. NEGATIVA DE PRESTAÇÃO JURISDICIONAL. NÃO OCORRÊNCIA. REEXAME DE FATOS E PROVAS. INVIABILIDADE. AGRAVO INTERNO DA FAZENDA DO ESTADO DE SÃO PAULO DESPROVIDO.

Fig. 5. Examples of keyphrases generated by top-K ($K = 15$) and top-p ($p = 0.9$) decoding.

5   Conclusion and Future Works

In this paper, we successfully trained a sequence-to-sequence Transformer to generate keyphrases and investigated three different text decoding methods. The results showed us that the keyphrases bring significant increments to IR metrics when used in combination with the dockets. This result was observed for all the keyphrases evaluated: the references and the generated ones (using greedy, top-K, and top-p decoding). Although we have evaluated different parameters and concatenated multiple variations of keyphrases generated using sampling decoding (top-K and top-p), the simpler greedy decoding performed similarly to these methods. We presented and discussed possible justifications for such behaviour, and the results suggest that greedy decoding is enough for keyphrase generation considering legal dockets.

As future works, we intend to experiment pre-training language models on legal documents in order to improve keyphrase generation. Furthermore, we aim to improve the quality of the training by collecting more dockets from more sources around Brazil. At last, this work can also be used to inspire other works aiming to automatise text writing in the legal domain.

Acknowledgement. This study was supported by the Coordenação de Aperfeiçoamento de Pessoal de Nível Superior - Brazil (CAPES) - Finance Code 001. We thank CEMEAI for granting access to the Euler cluster for the experiments. Also, this work is partially funded by Fundação de Amparo à Pesquisa do Estado de São Paulo (FAPESP), grant 2022/01640-2. We would like also to thank INCT (CAPES Concessão 88887.136349/2017-00, CNPQ 465755/2014-3 and FAPESP 2014/50851-0) for the support.

References

1. Althammer, S., Askari, A., Verberne, S., Hanbury, A.: Dossier@ coliee 2021: leveraging dense retrieval and summarization-based re-ranking for case law retrieval. arXiv preprint arXiv:2108.03937 (2021)
2. Brown, T., et al.: Language models are few-shot learners. Adv. Neural. Inf. Process. Syst. **33**, 1877–1901 (2020)
3. Carmo, D., Piau, M., Campiotti, I., Nogueira, R., Lotufo, R.: PTT5: pretraining and validating the T5 model on Brazilian Portuguese data. arXiv preprint arXiv:2008.09144 (2020)
4. Fan, A., Lewis, M., Dauphin, Y.N.: Hierarchical neural story generation. CoRR abs/1805.04833 (2018). http://arxiv.org/abs/1805.04833
5. Feijo, D., Moreira, V.: Summarizing legal rulings: comparative experiments. In: Proceedings of the International Conference on Recent Advances in Natural Language Processing (RANLP 2019), pp. 313–322 (2019)
6. Floridi, L., Chiriatti, M.: GPT-3: its nature, scope, limits, and consequences. Mind. Mach. **30**(4), 681–694 (2020)
7. Guimarães, J.A.C., Santos, J.C.G.: A ementa jurisprudencial como resumo informativo em um domínio especializado: aspectos estruturais. Braz. J. Inf. Sci. **10**(3), 32–43 (2016)
8. Holtzman, A., Buys, J., Forbes, M., Choi, Y.: The curious case of neural text degeneration. CoRR abs/1904.09751 (2019). http://arxiv.org/abs/1904.09751
9. Huang, Y., Shen, X., Li, C., Ge, J., Luo, B.: Dependency learning for legal judgment prediction with a unified text-to-text transformer. arXiv preprint arXiv:2112.06370 (2021)
10. Ji, Z., et al.: Survey of hallucination in natural language generation. ACM Comput. Surv. **55**(12), 1–38 (2022)
11. de Justiça CNJ, C.N.: Conselho nacional de justiça - justiça em números (2023). https://www.cnj.jus.br/pesquisas-judiciarias/justica-em-numeros/. Accessed 08 May 2023
12. Lima, J.P., Costa, J.A., Araújo, D.C.: Comparison of feature extraction methods for Brazilian legal documents clustering. In: 2021 IEEE Latin American Conference on Computational Intelligence (LA-CCI), pp. 1–5. IEEE (2021)
13. Mandal, A., Ghosh, K., Ghosh, S., Mandal, S.: Unsupervised approaches for measuring textual similarity between legal court case reports. Artif. Intell. Law **29**(3), 417–451 (2021). https://doi.org/10.1007/s10506-020-09280-2
14. Nogueira, R., Yang, W., Lin, J., Cho, K.: Document expansion by query prediction. arXiv preprint arXiv:1904.08375 (2019)
15. Pedroso, D.D.S.C., Ladeira, M., de Paulo Faleiros, T.: Does semantic search performs better than lexical search in the task of assisting legal opinion writing? In: 2019 18th IEEE International Conference on Machine Learning and Applications (ICMLA), pp. 680–685. IEEE (2019)
16. Peric, L., Mijic, S., Stammbach, D., Ash, E.: Legal language modeling with transformers. In: Proceedings of the Fourth Workshop on Automated Semantic Analysis of Information in Legal Text (ASAIL 2020) held online in conjunction with 33rd International Conference on Legal Knowledge and Information Systems (JURIX 2020) 9 December 2020, vol. 2764. CEUR-WS (2020)
17. Post, M.: A call for clarity in reporting bleu scores. arXiv preprint arXiv:1804.08771 (2018)

18. Pradeep, R., et al.: H2oloo at TREC 2020: when all you got is a hammer... deep learning, health misinformation, and precision medicine. Corpus **5**(d3), d2 (2020)
19. Radford, A., Wu, J., Child, R., Luan, D., Amodei, D., Sutskever, I., et al.: Language models are unsupervised multitask learners. OpenAI Blog **1**(8), 9 (2019)
20. Raffel, C., et al.: Exploring the limits of transfer learning with a unified text-to-text transformer. J. Mach. Learn. Res. **21**(140), 1–67 (2020)
21. Robertson, S.E., Walker, S.: Okapi/Keenbow at TREC-8. In: TREC, vol. 8, pp. 151–162. Citeseer (1999)
22. Rosa, G.M., Bonifacio, L.H., de Souza, L.R., Lotufo, R., Nogueira, R.: A cost-benefit analysis of cross-lingual transfer methods. arXiv preprint arXiv:2105.06813 (2021)
23. Rosa, G.M., Rodrigues, R.C., Lotufo, R., Nogueira, R.: Yes, BM25 is a strong baseline for legal case retrieval. arXiv preprint arXiv:2105.05686 (2021)
24. Russell-Rose, T., Chamberlain, J., Azzopardi, L.: Information retrieval in the workplace: a comparison of professional search practices. Inf. Process. Manag. **54**(6), 1042–1057 (2018)
25. Scao, T.L., et al.: BLOOM: a 176B-parameter open-access multilingual language model. arXiv preprint arXiv:2211.05100 (2022)
26. Souza, E., et al.: Assessing the impact of stemming algorithms applied to Brazilian legislative documents retrieval. In: Anais do XIII Simpósio Brasileiro de Tecnologia da Informação e da Linguagem Humana, pp. 227–236. SBC (2021)
27. Souza, F., Nogueira, R., Lotufo, R.: BERTimbau: pretrained BERT models for Brazilian Portuguese. In: Cerri, R., Prati, R.C. (eds.) BRACIS 2020. LNCS (LNAI), vol. 12319, pp. 403–417. Springer, Cham (2020). https://doi.org/10.1007/978-3-030-61377-8_28
28. Su, Y., Lan, T., Wang, Y., Yogatama, D., Kong, L., Collier, N.: A contrastive framework for neural text generation. arXiv preprint arXiv:2202.06417 (2022)
29. Vaswani, A., et al.: Attention is all you need. In: Advances in Neural Information Processing Systems, vol. 30 (2017)
30. Wagner Filho, J.A., Wilkens, R., Idiart, M., Villavicencio, A.: The brWaC corpus: a new open resource for Brazilian Portuguese. In: Proceedings of the Eleventh International Conference on Language Resources and Evaluation (LREC 2018) (2018)
31. Xue, L., et al.: MT5: a massively multilingual pre-trained text-to-text transformer. arXiv preprint arXiv:2010.11934 (2020)
32. Yoon, J., Junaid, M., Ali, S., Lee, J.: Abstractive summarization of Korean legal cases using pre-trained language models. In: 2022 16th International Conference on Ubiquitous Information Management and Communication (IMCOM), pp. 1–7. IEEE (2022)
33. Zhang, S., et al.: OPT: open pre-trained transformer language models. arXiv preprint arXiv:2205.01068 (2022)

Community Detection for Multi-label Classification

Elaine Cecília Gatto[1]([✉]) [ID], Alan Demétrius Baria Valejo[1] [ID],
Mauri Ferrandin[2] [ID], and Ricardo Cerri[1] [ID]

[1] Department of Computer Science, Federal University of São Carlos,
Rodovia Washington Luis, km 235, São Carlos, SP, Brazil
`elainegatto@estudante.ufscar.br`, {`alanvalejo,cerri`}`@ufscar.br`
[2] Federal University of Santa Catarina, Rua João Pessoa, Blumenau 2750,
Santa Catarina, Brazil
`mauri.ferrandin@ufsc.br`

Abstract. Exploring label correlations is one of the main challenges in multi-label classification. The literature shows that prediction performances can be improved when classifiers learn these correlations. On the other hand, some works also argue that the multi-label classification methods cannot explore label correlations. The traditional multi-label local approach uses only information from individual labels, which makes it impractical to find relationships between them. In contrast, the multi-label global approach uses information from all labels simultaneously and may miss more specific relationships that are relevant. To overcome these limitations and verify if improving the prediction performances of multi-label classifiers is possible, we propose using Community Detection Methods to model label correlations and partition the label space into partitions between the local and global ones. These partitions, here named hybrid partitions, are formed of disjoint clusters of correlated labels, which are then used to build multi-label datasets and train multi-label classifiers. Since our proposal can generate several hybrid partitions, we validate all of them and choose the one that is considered the best. We compared our hybrid partitions with the local and global approaches and an approach that generates random partitions. Although our proposal improved the predictive performance of the used classifier in some datasets compared with other partitions, it also showed that, in general, independent of the approach used, the classifier still has difficulties learning several labels and predicting them correctly.

Keywords: Multi-Label Partitions · Multi-Label Learning · Label Correlations · Label Communities · Label Problems

1 Introduction

Multi-label classification is the task of simultaneously assigning multiple labels to an instance. Thus, a classifier must be induced to predict a set of labels for new instances. As real-world applications, we can mention recognition of sensor activities [13], protein function prediction [39], and drug-target interactions [20].

A multi-label problem can be solved mainly through local and global approaches. The local approach divides the original multi-label dataset into many binary ones, so labels are learned individually. One classifier per label is trained, and the individual predictions are gathered for the final evaluation. Any classification algorithm can be used, and traditional methods are Binary Relevance [25,36] and Classifier Chains [23]. The global approach creates new methods or modifies existing ones to deal with all classes of the problem at the same time. Therefore, only one multi-label classifier is trained and is responsible for predicting all labels of an instance. Traditional methods in this approach include Back Propagation Multi-Label Learning (BPMLL) [37] and Multi-Label k-Nearest Neighbors (MLkNN) [38].

Among the challenges of multi-label classification, we can highlight class imbalance [33], high dimensionality [3], and label correlations [4]. Learning label correlations can help to predict labels that would probably not be predicted using traditional methods. Works in the literature have shown that modeling label correlations is an important aspect of multi-label classification and can help to build better classification models [12,19].

The literature presents different strategies to model label correlations, and community detection methods are one of them [7] [32]. The motivation to use them is that the network topology can encode interactions of the data systematically and find relationships between them [17,31]. Community detection methods are also graph partitioning algorithms: they divide vertices into clusters to minimize the number of edges between them. Roughly speaking, a community (cluster) is a set of vertices with many edges inside and a few edges outside the community, a desired characteristic for this study [31]. In multi-label problems, labels can be considered vertices and the correlations between them, the edges. Hence, we can build a graph of correlated labels, which can then be partitioned to find label communities.

In this paper, we verify if the performance of a classifier can be improved, and if it can better learn label correlations, using a different approach from the conventional global and local ones. We propose an approach that generates and validates several partitions from the label space - which explores the label correlations - and chooses the best one. Such partitions are composed of disjoint correlated label clusters and are called Hybrid Partitions.

Figure 1 presents our idea of multi-label partitions, where squares are the partitions, circles are label clusters, and diamonds are labels. Figure 1a shows a global partition (global approach) composed of only one cluster with all labels. We induce one global classifier for global partitions, and this classifier simultaneously learns and outputs predictions for all labels. Figure 1b shows a local partition (local approach) composed of one cluster for each label. One local classifier per cluster is induced in local partitions, also with train and test sets, and then we gather those predictions to make the final evaluation for the dataset. Finally, Fig. 1c shows an example of a hybrid partition, where correlated labels are clustered. We need to induce a global-based classifier for clusters with more than one label, in order to learn correlations between those labels. In contrast, we induce a local-based classifier for clusters with a single label. The final prediction is then obtained combining the predictions performed by all global and local-based classifiers.

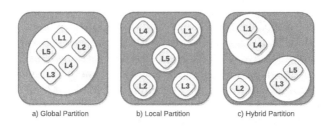

a) Global Partition   b) Local Partition   c) Hybrid Partition

Fig. 1. Illustration of the Global, Local and Hybrid Partitions.

To get consistent and coherent results, we must use the same classifier in all clusters; otherwise, we cannot observe how much the classifier learned. Therefore, we emphasize that using a multi-label classifier with local and global versions is necessary for our strategy since we want to compare the improvements from hybrid partitions in relation to local, global, and even random generated partitions. Therefore, the same classifier must be used in all partitions.

As an example, it does not make sense to use a global-based classifier such as ML-C4.5 [8] since it cannot handle clusters containing a single label. In our strategy, using Binary Relevance in clusters with more than one label is not possible, as we would be re-dividing the problem and learning separate labels when we want to learn the clustered labels. Accordingly, the classifier used in our strategy has to deal with all types of partitions.

For this reason, here we used the Clus Framework [34], which induces binary and multi-label decision trees based on Predictive Clustering Trees (PCT), and is considered to be one of the state-of-the-art methods in the literature. We train a binary PCT for each label and combine their outputs to form the final multi-label prediction for the local partition. For the global partition, only one multi-label PCT is necessary. To test our hybrid partitions, a set of multi-label PCTs, or a combination of binary and multi-label PCTs, can be applied, depending on how many labels are in the partition clusters. The individual outputs are then combined to form the final multi-label prediction.

As already pointed out by Basgalupp et al. [1], finding a suitable partition in a set with n labels is a hard task, whose difficulty increases with the number of labels. Considering a set with n labels, the number of possible partitions consisting of k separate and non-empty subsets is given by $\sum_{k=0}^{n} \left\{ {n \atop k} \right\}$ [16]. Thus, the number of possible partitions drastically increases with the number of labels. As an example, with $n = 12$ labels, 44.213.597 partitions can be generated.

Our experiments were conducted using 20 benchmark multi-label datasets from five application domains with different characteristics. First, we used Jaccard Index and Rogers-Tanimoto similarity measures to model the label correlations, which are used to build the label co-occurrence graphs. Then, we applied 7 community detection methods that partition the graphs and generate several hybrid partitions. Next, those hybrid partitions were validated, and the best one tested. Finally, the performance results were compared with the global, local, and random generated partitions using the same base classifier.

Our results showed that our approach could improve the classifier's prediction, be competitive with the local and random generated partitions, and outperform the global approach. However, the results also showed that independent of the partitioning used, the overall performance did not improve much, i.e., we noticed that the average performance remained competitive for most methods and datasets. Thus, the valuable insight we had with our results is that the multi-label classification methods need to improve because regardless of the partitioning used or if the correlations were (or not) explored, we cannot state with absolute certainty that they are correctly learning the labels.

The remainder of this paper is organized as follows: Sect. 2 presents some related studies; Sect. 3 presents our proposed approach; Sect. 4 explains how we conducted our experiments; Sect. 5 presents and discusses our results; and finally, Sect. 6 presents our conclusions and future works.

2   Related Work

A related and similar study was presented by Melo and Paulheim [15]. The authors used local and global feature selection approaches with binary relevance to compare flat and hierarchical problems. They showed that when using multi-label transformation methods, the overall predictive performance of the local feature selection approach was better and superior to the global. However, the authors also elucidated that, in some cases, the results of the best general performance between adaptation or transformation methods are inconclusive. Although the authors used the instance space and feature selection, the results presented go in the same direction as our work.

Rivolli et al. [25] presented a hypothesis that base algorithms can have a more substantial influence than the binary transformation strategies on the predictive performance of multi-label models. The authors demonstrated that despite the performance improvements, all investigated strategies and base algorithms could not predict some labels or mispredict labels for many datasets. Therefore, the conclusions are aligned with our study.

On the problem of learning label correlations, Luaces et al. [14] demonstrated that the level of dependency between labels in benchmark datasets is deficient, so the methods cannot explore the correlations. That was one of the reasons that led them to build a synthetic dataset generator, combining synthetic data with benchmark datasets to better analyze the behavior of multi-label classification methods. They pointed out the difficulty of extracting useful conclusions using only benchmark datasets.

In Gatto et al. [12], the authors modeled label correlations using the Jaccard index and partitioned the label space using Agglomerative Hierarchical Clustering, which was able to find hybrid partitions. They used average, complete, and single linkage as an agglomerative method to build dendrograms and cut them into levels. Different hybrid partitions were generated and validated using the Clus framework, which can deal with local and global partitions. The highest Macro-F1 performance was used to choose the more suitable hybrid partition, compared with local, global, and random generated partitions. The authors concluded that the results from hybrid partitions were competitive, or better in some cases, compared to the local and global approaches. Similar to our work, the authors, in the same direction as Melo and Paulheim [15], observed that the local approach was superior to the global one in most cases.

Also trying to find partitions in-between the global and local ones, the study presented by Basgalupp et al. [1] investigated whether it was possible to obtain better results by learning multi-target models on several partitions of the targets. They used a Genetic Algorithm (GA) and an exhaustive search to find alternate partitions between the global and local ones. They tested those partitions with random forests and decision trees in multi-target regression and multi-label problems and then compared them with the local and global ones. An oracle experiment was also conducted, proving that it is possible to find partitions beyond traditional ones and obtain superior performances. Although the global approach led to better results for the multi-label case with decision trees, the results showed no statistically significant differences among local, global, and alternate partitions, which matches our conclusions. In our work, we also found hybrid partitions that led to better results than the local and global ones for some datasets, and the conclusions from this study are similar to the ones from Melo and Paulheim [15].

From the revised works we can find many open research questions in Multi-Label Classification. Our objective here is not to answer or find solutions for all of them. Instead, we focused on finding partitions between global and local ones and verifying whether they could improve the classifier's predictions.

3   Proposed Approach

Our approach consists of the following steps: 1) data preprocessing, which divides the dataset into train, validation, and test sets; 2) modeling label correlations, where we build graphs computing Jaccard Index and Rogers-Tanimoto similarity measures on the label space of the train set; 3) applying community detection methods to find hybrid partitions; 4) building the hybrid partitions and the corresponding datasets for each partition; 5) validate the hybrid partitions and choose the best one using the silhouette coefficient; and 6) test the best hybrid partition with the test set. Figure 2 shows an overview of our approach. The complete process is presented in the following subsections.

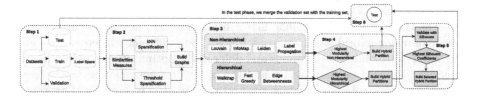

Fig. 2. Overview of the Proposed Approach.

3.1 Steps 1 and 2: Preprocess and Model Label Correlations

Step 1 divides the dataset into the train, validation, and test sets and also splits the label space from the train set, while step 2 models label correlations. According to Silva and Zhao [31], we need a similarity or dissimilarity matrix to build a network from vector-based data, which makes it possible to establish links between pairs of labels with weights according to that matrix. However, links with small weights can lead to poor results and be considered noises that could provide wrong information to the machine learning algorithm. To avoid that, sparsification must be applied, a preprocess that cuts those edges and can improve the learning stage.

One traditional sparsification is based on the concept of k nearest neighbors, where the cuts occur in the links that are not part of the neighborhood. Another type allows a threshold that cuts some percentage of the links [31]. We use both types in our study. We choose Jaccard Index (Eq. 1) and Rogers-Tanimoto (Eq. 2) as similarities measures to build label co-occurrence graphs, as they are well-known and popular measures [6,11,35].

In Eqs. 1 and 2, a stands for the number of times labels l_p and l_q occur together, while d stands for the number of times those two labels never occur together. Variables b and c indicate the number of times labels l_p and l_q occur alone. Each measure is applied in the label space of the train set, resulting in similarity matrices. Each matrix represents a complete graph modeled with the label correlations, where an edge (similarity value) connects every pair of distinct vertices (labels).

$$J = \frac{a}{a+b+c} \qquad (1) \qquad Ro = \frac{a+d}{a+2(b+c)+d} \qquad (2)$$

3.2 Step 3: Apply Community Detection Methods

In this step, we applied seven community detection methods to each graph. The methods are divided into two categories: a) *Hierarchical*: Waktrap [21], Fast Greedy [9] and Edge Betweenness [18]; and b) *Non-Hierarchical*: Louvain [2], InfoMap [27], SpinGlass [24] and Label Propagation [22]. Hierarchical methods provide dendrograms that can be used to construct several hybrid partitions. The procedure is similar to hierarchical clustering but uses different techniques to agglomerate vertices: *Walktrap*, which is based on random walks; *Fast Greedy* based on modularity gain, and *Edge Betweenness* based on the number of shortest paths through the edge. Unlike hierarchical methods, non-hierarchical ones provide only one hybrid partition, constructed using different concepts. *Louvain* and *Infomap* are multilevel methods and differ in the optimization function used: the former is based on FastGreedy and the latter on map equation. *SpinGlass* is based on Potts Spin Glass, which uses rotation models to perform clustering, while in *Label Propagation*, each vertex is assigned to the most frequent label among its neighbors.

3.3   Step 4: Build Hybrid Partitions

After applying community detection methods, only one is chosen to generate hybrid partitions. This choice is performed using a modularity criterion that measures the separation among vertices, quantifying the density of links within communities compared to links between communities [18,31]. Finally, the method with the highest modularity value is chosen. As an example, consider that between non-hierarchical methods, Louvain obtained the highest modularity value. Hence, each identified community can be considered a cluster of correlated labels. Then, we build a corresponding dataset for each cluster using the train instances assigned to the labels of the cluster, induce a classifier, and test it in the test set. The same occurs with hierarchical methods, but as they generate more than one hybrid partition, we must validate all generated partitions from the chosen community detection method and then choose only one hybrid partition to test.

3.4   Step 5: Validate Hybrid Partitions

This step is executed only in the partitions generated by hierarchical community detection methods since these generate several hybrid partitions. Each hybrid partition created using hierarchical methods in step 4 is now evaluated in a validation dataset. Here we used the silhouette coefficient [28,30] cluster validation measure. It measures the quality of partitions based on the proximity between, in this case, the labels of a particular cluster and the distance between them and the closest cluster. The partition with the highest silhouette coefficient is then evaluated in a test dataset.

3.5   Step 6: Testing Hybrid Partitions

Finally, the last step is to test the best-validated hybrid partition. As already explained, the Clus Framework was used in our experiments, configured with the hyper-parameters listed below. For a detailed explanation of Clus and its hyper-parameters, the reader can consult the Clus webpage[1].

---

[1] https://dtai.cs.kuleuven.be/clus/.

- General: Compatibility = MLJ08
- Attributes: ReduceMemoryNominalAttrs = yes and Weights = 1
- Tree: Heuristic = VarianceReduction and
  FTest = [0.001,0.005,0.01,0.05,0.1,0.125]
- Model: MinimalWeight = 5.0
- Output: WritePredictions = Test

4   Experimental Setup

We evaluated our approach using 20 multi-label datasets from five different application domains, which were chosen to diversify the domain and number of labels. Their characteristics are summarized in Table 1 with the name of each dataset, domain, the total number of instances, attributes, and labels. *Cardinality* is the average of labels per instance; *Density* is the average frequency of labels, and *TCS* (Theoretical Complexity Score) [5] is a measure that computes how difficult it is to learn a predictive model from the dataset.

Table 1. Characteristics of the Datasets.

Name	Domain	Instances	Attributes	Labels	Cardinality	Density	TCS
birds	Audio	645	279	19	1.01	0.05	13.40
cal500	Music	502	242	174	26.04	0.15	15.60
cellcycle	Biology	3757	255	178	2.19	0.01	16.79
derisi	Biology	3725	241	178	2.20	0.01	16.58
eisen	Biology	2424	244	165	2.34	0.01	16.40
emotions	Music	593	78	6	1.87	0.31	9.36
EukaPseAAC	Biology	7766	462	22	1.15	0.05	13.90
flags	Image	194	26	7	3.39	0.48	8.88
gasch1	Biology	3764	351	178	2.19	0.01	17.59
GnegativeGO	Biology	1392	1725	8	1.05	0.13	12.47
GpositiveGO	Biology	519	916	4	1.01	0.25	10.15
langlog	Text	1460	1079	75	1.18	0.02	16.95
medical	Text	978	1494	45	1.25	0.03	15.63
pheno	Biology	1591	234	165	2.24	0.01	16.00
PlantGO	Biology	978	3103	12	1.08	0.09	13.99
scene	Image	2407	300	6	1.07	0.18	10.18
seq	Biology	3919	656	178	2.14	0.01	18.62
VirusPseAAC	Biology	207	446	6	1.22	0.20	10.71
yeast	Biology	2417	117	14	4.24	0.30	12.56
Yelp	Text	10806	676	5	1.64	0.33	11.58

Our source code and all necessary materials to replicate the experiments are freely available[2]. We executed our experiments within the 10-fold cross-validation strategy, using an iterative stratification algorithm proposed for multi-label data [29]. Our proposal was compared with three approaches to generate partitions: local, global, and random. Global and local partitions were already explained in Sect. 1. Random partitions were generated similarly to hybrid partitions. The main difference is in Step-2: given an adjacency matrix[3] obtained from the label space, k-NN sparsification is executed with a random chosen k value.

To evaluate the results, we focused mainly on two multi-label measures: a) *Macro-F1* (MaF1), which is the harmonic mean of the Macro-Precision (MaP) and Macro-Recall (MaR), and b) *Missing Label Problem* (MLP). MaF1 is a good measure for multi-label problems since it considers the individual performances in each class. Furthermore, its analysis is complemented by MLP, which calculates the proportion of labels that are never predicted [26].

Analyzing the results with the MLP measure is important since traditional multi-label measures cannot correctly identify when some individual labels are not correctly predicted for any given instance. However, it is possible to identify the predictive performance of each label for each instance individually. Thus, MLP is an important measure to identify labels the classifiers could not learn and thus are never predicted. Equations 3, 4, 5, and 6 show the evaluation measures used in the experiments where: $\mathcal{L}$ is the set of all labels in the label space, and tp_j, fp_j and fn_j are, respectively, the number of true positives, false positives, and true negatives for class y_j.

$$MLP = \frac{1}{l} \sum_{j=1}^{|\mathcal{L}|} I\Big((tp_j + fp_j) == 0\Big) \quad (3)$$

$$MaP = \frac{1}{|\mathcal{L}|} \sum_{j=1}^{|\mathcal{L}|} \frac{tp_j}{tp_j + fp_j} \quad (4)$$

$$MaR = \frac{1}{|\mathcal{L}|} \sum_{j=1}^{|\mathcal{L}|} \frac{tp_j}{tp_j + fn_j} \quad (5)$$

$$MaF1 = \frac{2 \times MaP \times MaR}{MaP + MaR} \quad (6)$$

5   Results, Analysis and Discussion

We generated 20 different variations of our proposal (Table 2). Considering k-NN sparsification, we experimented three values for k ($k = 1$, $k = 2$ and $k = 3$). Considering threshold sparsification, we considered two alternatives: $T0$, which cuts only the self-loops from the graphs, and $T1$, which cuts 10% of edges from the graphs. Table 2 also shows the global and local methods and two variations of random methods, one using hierarchical community detection methods and another using non-hierarchical community detection methods.

5.1   Community Detection Methods Selected

For the Rogers-Tanimoto similarity measure, the most chosen hierarchical community detection methods were Edge Betweenness for threshold sparsification

---

[2] https://github.com/cissagatto/Bracis2023/.

[3] Computes the number of instances classified in all pairs of labels.

and WalkTrap for $k-$NN sparsification. In contrast, for Jaccard Index, Walktrap was the most chosen in both sparsification types. In the case of the non-hierarchical methods, InfoMap was the most chosen community detection method for both similarity measures and sparsification types. Finally, for Random Communities, the hierarchical method most chosen was WalkTrap, while InfoMap was the most chosen for non-hierarchical methods.

5.2 Hybrid Partitions Selected

For *birds, emotions, EukaryotePseAAC*, and *yeast* datasets, the best partitions chosen were close to the local partition. In contrast, *eisen, GnegativeGO, Gpos-*

Table 2. Characteristics of the Compared Methods and Variations.

N.	Method's Variation	Similarity Measure	Community Detection Method	Sparsification Method
1	Global (G)	none	none	none
2	Local (L)	none	none	none
3	H.Ra	Random	Hierarchical	k-NN (k is random)
4	NH.Ra	Random	Non Hierarchical	k-NN (k is random)
5	H.J.K1	Jaccard	Hierarchical	k-NN: k=1
6	H.Ro.K1	Rogers	Hierarchical	k-NN: k=1
7	H.J.K2	Jaccard	Hierarchical	k-NN: k=2
8	H.Ro.K2	Rogers	Hierarchical	k-NN: k=2
9	H.J.K3	Jaccard	Hierarchical	k-NN: k=3
10	H.Ro.K3	Rogers	Hierarchical	k-NN: k=3
11	H.J.T0	Jaccard	Hierarchical	Threshold: T0
12	H.Ro.T0	Rogers	Hierarchical	Threshold: T1
13	H.J.T1	Jaccard	Hierarchical	Threshold: T0
14	H.Ro.T1	Rogers	Hierarchical	Threshold: T1
15	NH.J.K1	Jaccard	Non Hierarchical	k-NN: k=1
16	NH.Ro.K1	Rogers	Non Hierarchical	k-NN: k=1
17	NH.J.K2	Jaccard	Non Hierarchical	k-NN: k=2
18	NH.Ro.K2	Rogers	Non Hierarchical	k-NN: k=2
19	NH.J.K3	Jaccard	Non Hierarchical	k-NN: k=3
20	NH.Ro.K3	Rogers	Non Hierarchical	k-NN: k=3
21	NH.J.T0	Jaccard	Non Hierarchical	Threshold: T0
22	NH.Ro.T0	Rogers	Non Hierarchical	Threshold: T1
23	NH.J.T1	Jaccard	Non Hierarchical	Threshold: T0
24	NH.Ro.T1	Rogers	Non Hierarchical	Threshold: T1

tiviGO, *scene*, and *Yelp* datasets had the best partitions chosen close to the global partition. Considering the *Langlog* dataset, the partition most frequently chosen was similar to the local. In contrast, in *pheno* dataset, there is a balance between partitions near the global or local ones. Finally, the best partitions for the other datasets vary, but a partition close to the global partition configuration was more frequently chosen. Our experiments observed that the best hybrid partitions generated by the chosen community detection methods are similar to the global partition. This can be one reason our performance results are competitive compared with other partitions, overcome the global, and are not superior to the local ones for some datasets.

5.3   Classifier's Performance

Figure 3 presents the number of wins, losses, and ties considering the 24 methods and 20 datasets. In the figure, blue indicates wins, purple indicates losses, and green indicates ties. Considering the performances over the 20 datasets in both MaF1 and MLP measures, the local approach tends to obtain the best results, followed by the results obtained with the hybrid partitions. On the other hand, the global approach obtained the worst results. Furthermore, the Theoretical Complexity Score (Table 1) presents high values for all datasets except emotions and flags. This indicates that it is hard to learn label correlations in most of the datasets.

The classifier's performance results[4], in several datasets, using random partitions outperformed its performance when using global partitions. When comparing *H.Ra* and *NH.Ra* with *Lo*, the random partitions led to competitive results. Regarding the sparsification methods, *k*-NN obtained slightly better performances when compared to threshold sparsification.

The clusters of the partitions found by the community detection methods for each dataset were similar and sometimes the same. By all indications, our method is not significantly affected by sparsification with *k*-NN. We also notice that the results from graphs built with the Jaccard index and Hierarchical methods outperformed the ones with the Rogers-Tanimoto similarity, while in Non-Hierarchical methods, Rogers-Tanimoto were generally better than the Jaccard index. Finally, in a method-by-method pairwise comparison, the local partitions performed better than all other partitions on an average of 16 datasets. Meanwhile, hybrid and random partitions achieved the best results on an average of 10 datasets.

For the MLP measure, the resulting values for many datasets are about 0.9. As the best value for this measure is 0.0, a value near 1.0 indicates that not all labels were correctly learned nor predicted by the classifier. The pheno dataset had the worst result when using the local partition, 0.977, which means that the classifier did not correctly predict 97% of all labels. In contrast, at least 40% of the labels were predicted by the other methods, indicating that the correlated labels were helpful in the learning process.

---

[4] For the sake of space and better presentation, the tables with all the performance values were not included in this paper but are available in the repository.

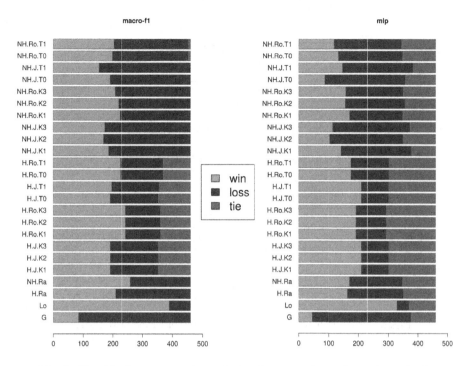

Fig. 3. Graphical representations of wins, losses, and ties for each method.

Considering the MaF1 average values over the 20 datasets, most of the methods obtained performance values between 0.3 and 0.2, which is considered low. The best results (close to 1.0) were obtained in the datasets GpositiveGO and GnegativeGO, which have few labels. Similar results can also be found in related studies, where the authors showed that multi-label classifiers could not correctly learn and predict the labels and correlations. Finally, our results show that the traditional local and global approaches cannot correctly learn label correlations since random generated partitions led to very similar results in many cases.

5.4 Statistical Tests

To verify the statistical significance of our results, we executed the Friedman test ($\alpha = 0.05$), followed by the Nemenyi post-hoc test. They are adequate when comparing many classifiers over multiple datasets [10]. The respective Friedman p-values were $5.75e$-09 and $9.27e$-14, and the critical distances are shown in Figs. 4 and 5. No statistically significant differences were detected between the results obtained with the local approach and those obtained with the random partitions generated with non-hierarchical community detection methods. There were also no significant differences when comparing the global partition and the random partitions generated with hierarchical community detection methods.

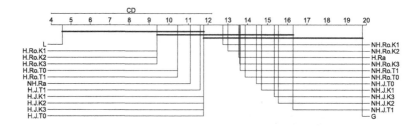

Fig. 4. Critical Diagram for the MaF1 Results.

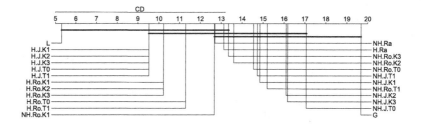

Fig. 5. Critical Diagram for the Missing Label Problem Results.

When comparing local and hybrid partitions from hierarchical community methods, there were no significant differences, meaning that hybrid partitions led to competitive results. The same occurs with global partitions and hybrid partitions from non-hierarchical community methods. Given that random generated partitions led to results equal or superior to the local and global approaches, we can conclude that local and global-based methods still need many improvements. Thus, our results show that the conventional and widely-used local and global approaches may not correctly learn label correlations.

6   Conclusions and Future Works

In this paper, we presented an approach that aims to improve multi-label classifiers' performances by partitioning the multi-label dataset into in-between local and global partitions. These so-called hybrid partitions are obtained with community detection methods by learning label correlations in the label space. First, we modeled label correlations with the Jaccard Index and the Rogers-Tanimoto similarity measures and used the resulting similarity matrices to model graphs. Next, we applied k-NN and *threshold* sparsification methods to prune edges from the graphs. We then applied seven community detection methods, three hierarchical and four non-hierarchical.

For the non-hierarchical methods, we generated hybrid partitions directly using a training set and tested them with a test set. For hierarchical methods, we obtained dendrograms that were cut into levels to generate several hybrid partitions, also using a training set. Then, we validated those partitions with a silhouette coefficient using a validation set. Finally, the hybrid partition with the highest coefficient value was tested.

Our results showed that our proposed approach could obtain better or more competitive results in several datasets when compared to conventional local and global approaches. We also compared the results obtained by the global and local approaches with the ones obtained using random generated partitions. We could verify if the partitioning improved the prediction power of the classifier. We concluded that there is no vast improvement besides our competitive results, and independently of the partitioning used. According to the results obtained when using the Missing Label Problem measure, this probably occurred because most labels were not learned by the classifier, even by traditional approaches. Therefore, we can conclude that the local and global approaches still need many improvements and may not correctly learn label correlations.

In future works, we intend to explore other multi-label evaluation measures, such as Wrong Label Problem and Constant Label Problem [26], and also use other base classifiers to test the generated partitions. With this, we will be able to compare the results from different classifiers, verify possible improvements, and recommend the best classifier and label partition for a given dataset.

Acknowledgments. This study was financed in part by the Coordenação de Aperfeiçoamento de Pessoal de Nível Superior - Brasil (CAPES) - Finance Code 001. The authors also thank the Brazilian research agencies FAPESP and CNPq for financial support.

References

1. Basgalupp, M., Cerri, R., Schietgat, L., Triguero, I., Vens, C.: Beyond global and local multi-target learning. Inf. Sci. **579**, 508–524 (2021)
2. Blondel, V.D., Guillaume, J.L., Lambiotte, R., Lefebvre, E.: Fast unfolding of communities in large networks. J. Stat. Mech. Theory Exp. **2008**, P10008 (2008)
3. Bogatinovski, J., Todorovski, L., Džeroski, S., Kocev, D.: Comprehensive comparative study of multi-label classification methods. Expert Syst. Appl. **203**, 117215 (2022)
4. Chang, W., Yu, H., Zhong, K., Yang, Y., Dhillon, I.S.: A modular deep learning approach for extreme multi-label text classification. CoRR abs/1905.02331 (2019)
5. Charte, F., Rivera, A., del Jesus, M.J., Herrera, F.: On the impact of dataset complexity and sampling strategy in multilabel classifiers performance. In: Hybrid Artificial Intelligent Systems, pp. 500–511. Springer International Publishing (2016)
6. Choi, S., Cha, S., Tappert, C.C.: A survey of binary similarity and distance measures. J. Systemics Cybern. Inform. **8**, 43–48 (2010)
7. Chu, Y., et al.: DTI-MLCD: predicting drug-target interactions using multi-label learning with community detection method. Briefings in Bioinform. **22**, bbaa205 (2020)
8. Clare, A., King, R.D.: Predicting gene function in saccharomyces cerevisiae. In: Proceedings of the European Conference on Computational Biology (ECCB 2003), September 27–30, 2003, Paris, France, pp. 42–49 (2003)

9. Clauset, A., Newman, M.E.J., Moore, C.: Finding community structure in very large networks. Phys. Rev. E **70**, 066111 (2004)

10. Demšar, J.: Statistical comparisons of classifiers over multiple data sets. J. Mach. Learn. Res. **7**, 1–30 (2006)

11. Garg, A., Enright, C.G., Madden, M.G.: On asymmetric similarity search. In: 2015 IEEE 14th International Conference on Machine Learning and Applications (ICMLA) (2015)

12. Gatto, E.C., Ferrandin, M., Cerri, R.: Exploring label correlations for partitioning the label space in multi-label classification. In: 2021 International Joint Conference on Neural Networks (IJCNN) (2021)

13. Lin, S.C., Chen, C.J., Lee, T.J.: A multi-label classification with hybrid label-based meta-learning method in internet of things. IEEE Access **8**, 42261–42269 (2020)

14. Luaces, O., Díez, J., Barranquero, et al., J.: Binary relevance efficacy for multilabel classification. Progress in Artificial Intelligence (2012)

15. Melo, A., Paulheim, H.: Local and global feature selection for multilabel classification with binary relevance an empirical comparison on flat and hierarchical problems (2017)

16. Mezo, I.: The r-bell numbers. J. Integer Sequences **14**, A11 (2011)

17. Mittal, R., Bhatia, M.P.S.: Classification and comparative evaluation of community detection algorithms. Archives of Computational Methods in Engineering (2020)

18. Newman, M.E.J., Girvan, M.: Finding and evaluating community structure in networks. Phys. Rev. E **69**, 026113 (2004)

19. Nguyen, T.T., Nguyen, T.T.T., Luong, A.V., Nguyen, Q.V.H., Liew, A.W.C., Stantic, B.: Multi-label classification via label correlation and first order feature dependance in a data stream. Pattern Recogn. **90**, 35–51 (2019)

20. Pliakos, K., Vens, C., Tsoumakas, G.: Predicting drug-target interactions with multi-label classification and label partitioning. IEEE/ACM Trans. Comput. Biol. Bioinform. **18** 1596–1607 (2021)

21. Pons, P., Latapy, M.: Computing communities in large networks using random walks (long version) (2005)

22. Raghavan, U.N., Albert, R., Kumara, S.: Near linear time algorithm to detect community structures in large-scale networks. Phys. Rev. E **76**, 036106 (2007)

23. Read, J., Pfahringer, B., Holmes, G., Frank, E.: Classifier chains: a review and perspectives. J. Artif. Intell. Res. **70**, 683–718 (2021)

24. Reichardt, J., Bornholdt, S.: Statistical mechanics of community detection. Phys. Rev. E **74**, 016110 (2006)

25. Rivolli, A., Read, J., Soares, C., Pfahringer, B., de Leon Ferreira de Carvalho, A.C.P.: An empirical analysis of binary transformation strategies and base algorithms for multi-label learning. Machine Learning (2020)

26. Rivolli, A., Soares, C., Carvalho, A.C.P.d.L.F.d.: Enhancing multilabel classification for food truck recommendation. Expert Systems (2018)

27. Rosvall, M., Bergstrom, C.T.: Maps of random walks on complex networks reveal community structure. Proceedings of the National Academy of Sciences 105 (2008)

28. Rousseeuw, P.: Silhouettes: a graphical aid to the interpretation and validation of cluster analysis. J. Comput. Appl. Math. **20**, 53–65 (1987)

29. Sechidis, K., Tsoumakas, G., Vlahavas, I.: On the stratification of multi-label data. In: Gunopulos, D., Hofmann, T., Malerba, D., Vazirgiannis, M. (eds.) ECML PKDD 2011. LNCS (LNAI), vol. 6913, pp. 145–158. Springer, Heidelberg (2011). https://doi.org/10.1007/978-3-642-23808-6_10

30. Shahapure, K.R., Nicholas, C.: Cluster quality analysis using silhouette score. In: 2020 IEEE 7th International Conference on Data Science and Advanced Analytics (DSAA) (2020)

31. Silva, T.C., Zhao, L.: Machine Learning in Complex Networks. Springer Publishing Company, Incorporated (2016)

32. Szymański, P., Kajdanowicz, T., Kersting, K.: How is a data-driven approach better than random choice in label space division for multi-label classification? Entropy 18 (2016)

33. Tahir, M.A.U.H., Asghar, S., Manzoor, A., Noor, M.A.: A classification model for class imbalance dataset using genetic programming. IEEE Access **7**, 71013–71037 (2019)

34. Vens, C., Struyf, J., Schietgat, L., Džeroski, S., Blockeel, H.: Decision trees for hierarchical multi-label classification. Mach. Learn. **73**, 185–214 (2008)

35. Warrens, M.J.: Similarity coefficients for binary data: Properties of coefficients, coefficient matrices, multi-way metrics and multivariate coefficients. Master's thesis, Leiden University (2008)

36. Zhang, M.L., Li, Y.K., Liu, X.Y., Geng, X.: Binary relevance for multi-label learning: an overview. Front. Comput. Sci. **12**, 191–202 (2018)

37. Zhang, M.L., Zhou, Z.H.: Multi-label neural networks with applications to functional genomics and text categorization. IEEE Trans. Knowl. Data Eng. **18**, 1338–1351 (2006)

38. Zhang, M.L., Zhou, Z.H.: Ml-knn: a lazy learning approach to multi-label learning. Pattern Recogn. **40**, 2038–2048 (2007)

39. Zhou, J.P., Chen, L., Guo, Z.H., Hancock, J.: Iatc-nrakel: an efficient multi-label classifier for recognizing anatomical therapeutic chemical classes of drugs. Bioinformatics **36**, 1391–1396 (2020)

Logic and Heuristics

A Monte Carlo Algorithm
for Time-Constrained General Game
Playing

Victor Scherer Putrich[1]([✉]), Anderson Rocha Tavares[2], and Felipe Meneguzzi[1,3]

[1] Pontifical Catholic University of Rio Grande do Sul, Porto Alegre, Brazil
`Victor.Putrich@edu.pucrs.br`
[2] Federal University of Rio Grande do Sul, Porto Alegre, Brazil
`artavares@inf.ufrgs.br`
[3] University of Aberdeen, Aberdeen, Scotland
`felipe.meneguzzi@abdn.ac.uk`

Abstract. General Game Playing (GGP) is a challenging domain for AI
agents, as it requires them to play diverse games without prior knowledge. In this paper, we develop a strategy to improve move suggestions
in time-constrained GGP settings. This strategy consists of a hybrid
version of UCT that combines Sequential Halving and $UCB_{\sqrt{}}$, favoring
information acquisition in the root node, rather than overspend time on
the most rewarding actions. Empirical evaluation using a GGP competition scheme from the Ludii framework shows that our strategy improves
the average payoff over the entire competition set of games. Moreover,
our agent makes better use of extended time budgets, when available.

Keywords: General Game Playing · Monte Carlo · Regret

1 Introduction

General Game Playing (GGP) is a research area focused on developing intelligent
agents capable of playing a wide variety of games without prior knowledge of
any specific game being played [5]. GGP agents receive the rules of potentially
unknown games and must play them effectively. This prevents the creation of
game-specific heuristics.

The Upper Confidence for Trees (UCT) [7] algorithm has been effectively
utilized in GGP environments. UCT is based on building a search tree using
Monte Carlo Tree Search (MCTS). MCTS employs Monte Carlo simulations to
iteratively build a game tree.

UCT guarantees asymptotic optimal convergence. However, it may spend too
much time on the most promising choice so far, instead of attempting other less
explored options to potentially find a better one. This results in UCT taking too
long to produce high-quality recommendations in certain scenarios. A significant
challenge in GGP is designing algorithms that can efficiently find solutions in a

© The Author(s), under exclusive license to Springer Nature Switzerland AG 2023
M. C. Naldi and R. A. C. Bianchi (Eds.): BRACIS 2023, LNAI 14195, pp. 97–111, 2023.
https://doi.org/10.1007/978-3-031-45368-7_7

timely manner, particularly in competitive contexts, where the time required to find a solution is critical to the agent's performance.

In this paper, we tackle the problem of GGP with scarce time resources. Specifically, we focus on the following question: Is it UCT the best option for GGP environments with strict time constraints?

In response, we develop $UCT_{\sqrt{SH}}$[1] (presented at Sect. 4). Our algorithm is based on a hybrid scheme for building game-trees though MCTS [15], which is presented at Sect. 3. $UCT_{\sqrt{SH}}$ aims to be more exploratory in the root node than UCT, avoiding repeatedly probing the currently most promising action, and using the budget to find potentially better alternatives.

We conduct two distinct experiments (Sect. 5). First we present the Prize Box Selection experiment, which is a simplified Multi Armed Bandit (MAB) problem to compare how selection polices allocate their resources under scenarios with high and low reward variance. The second experiment aims to measure the agents' performance relative to UCT under time constraints. For this purpose, we use the Ludii GGP environment [12]. Specifically, we use the Kilothon tournament, one of the tracks of Ludii's GGP competition[2]. Such international competitions have a crucial role in motivating GGP research [14].

The main contributions of this paper are as follows: (i) the $UCT_{\sqrt{SH}}$ algorithm, a new decision-making method that attempts to use less budget on the greedy choice to favour less-explored ones; (ii) the Clock Bonus Time (*cbt*) approach, which enhances the estimation of thinking time in a GGP environment; (iii) the Prize Box Selection experiment, which highlights the resilience of $UCT_{\sqrt{SH}}$ allocation criteria on decision-making problems with high and low variance, compared to UCB_1 and other selection policies (examined in our work at Sect. 2 and 3); and (iv) empirical evidence of the improved performance of $UCT_{\sqrt{SH}}$ over UCT on a GGP environment, suggesting its effectiveness as a selection policy.

2   Monte Carlo Methods

Monte Carlo techniques employ random sampling via simulations to acquire information to address problems that are otherwise intractable. In game-playing, Monte Carlo methods can be used to evaluate a game-tree node by computing the expected outcome of its actions by sampling a sufficient large number of random completions of a game, also called *playouts*. This section presents useful concepts regarding Monte Carlo methods, that are useful throughout the remainder of the paper.

2.1   Regret on Bandit Problem

Multi-Armed Bandit (MAB) problems [1] represent a category of decision-making scenarios in which the outcomes of chosen actions are unknown. Imagine

---

a casino slot machine (bandit) with k distinct arms, each with its own reward and probability of winning. The gambler's objective is to plan a strategy that maximizes their overall profit in a previously unknown bandit. The challenge lies in determining the number of times to pull each arm to maximize returns (exploitation) while learning rewards and probability distributions (exploration).

One way to measure performance in the Multi-Armed Bandit problem is via regret, defined as the difference in the reward obtained from the pulled versus the optimal arm. We use two important measures of regret adapted from the definitions of Pepels [10] and Bubeck [3]. Specifically, we use cumulative and simple regret from Definitions 1 and 2.

Definition 1. *Cumulative regret is the regret accumulated over a number n of arm pulls. Let $\mu^\star$ be the expected reward of the best arm, μ_i be the expected reward of the arm pulled in the i-th trial. Then, the cumulative regret R_n can be defined as:*

$$R_n = \sum_{i=1}^{n}(\mu^\star - \mu_i) \tag{1}$$

An alternative experimental setup involves just finding the optimal arm without the need to maximize the reward during this process. Then, the gambler makes a final pull in this arm. Simple regret then measures the sub-optimality of this final pull, as per Definition 2.

Definition 2. *Simple regret is the difference between the expected reward of the best arm $\mu^\star$ and the expected reward of the arm of the final pull μ_n:*

$$r_n = \mu^\star - \mu_n \tag{2}$$

Bubeck et al. [3] showed that there is a trade-off between minimizing cumulative and simple regret. Specifically, they found that a smaller upper bound on cumulative regret (R_n) leads to a larger lower bound on simple regret (r_n), meaning that when an algorithm performs well in terms of cumulative regret (worst-case scenario), it is likely to have a higher minimum simple regret (best-case scenario), and vice-versa.

2.2   Upper Confidence Bound

Upper Confidence Bound (UCB) [1] is a selection policy in MAB problems and MCTS. The policy optimizes cumulative regret over time at a logarithmic rate over the number of trials performed. A widely adopted variant, UCB$_1$, is favored for its simplicity and its ability to consistently deliver robust performance outcomes.

UCB$_1$ establishes a statistical confidence interval for the estimated mean action-value.

The UCB$_1$ equation, adapted from Auer et al. [1], is presented in Eq. 3:

$$\text{UCB}_1(s, a) = Q_{s,a} + \sqrt{\frac{2 \ln N_s}{n_{s,a}}} \ .$$

$$(3)$$

Here, $Q_{s,a}$ is the mean reward from action a, when selected from state s. N_s is the number of visits on state s, while $n_{s,a}$ is the number of times action a has been selected in state s. The square-root term quantifies the upper-confidence bound, i.e., the uncertainty in the estimate of taking action a on state s. It serves as a bonus to encourage exploring less visited actions.

UCB$_1$ offers a desirable property: the discovery process can be interrupted at any time, providing an estimate of each option's quality based on collected samples. This anytime property allows for more flexibility in managing computational resources.

2.3   Sequential Halving

Sequential Halving [6] is a flat, non-exploiting approach[3] for the MAB problem. The algorithm evenly distributes a pre-defined *budget* (representing the number of simulations to be conducted) among all actions and progressively eliminates the bottom half in terms of performance. While effective at reducing simple regret compared to UCB$_1$, it reduces cumulative regret at slower rates. Sequential Halving cannot be interrupted prematurely and offers less robust assurances regarding the quantity of suboptimal selections made [6].

Algorithm 1 The Sequential Halving algorithm (adapted from [6])

1: **function** SEQUENTIALHALVING(s, $\mathcal{B}$)
2:     **Input:** state s, budget $\mathcal{B}$
3:     **Output:** Recommended action
4:     $v_{root} \leftarrow \langle s, \text{ACTIONS}(s) \rangle$
5:     $k \leftarrow |\text{CHILDREN}(v_{root})|$
6:     **while** $k > 1$ **do**
7:         $b \leftarrow \lceil \frac{\mathcal{B}}{k \times \log_2 |\text{CHILDREN}(v_{root})|} \rceil$
8:         **for** $v' \in \text{HEAD}(\text{CHILDREN}(v_{root}), k))$ **do**
9:             $Q(v') \leftarrow Q(v') + \text{SIMULATE}(v', b)$
10:             $N(v') \leftarrow N(v') + b$
11:         SORT($\text{CHILDREN}(v_{root}), k$)
12:         $k = \lceil k/2 \rceil$
13:     **return** $\text{CHILDREN}(v_{root})[0]$

Algorithm 1 depicts Sequential Halving, employing a tree-like structure for compatibility with subsequent algorithms. Each node, denoted as v, retains the following data: a state s, total reward Q, number of visits N, and a list of child nodes originated from v (produced by the CHILDREN function if it's not

---

[3] In contrast with exploiting policies, that allocate most resources to the most promising choice, non-exploiting policies allocate resources uniformly among choices, iteratively discarding the poorly-performing ones.

a leaf node). We use k to restrict the quantity of possible actions, and the HEAD function gives the first k children of v_{root}. The Sequential Halving formula in Line 7 divides the budget $\mathcal{B}$ by the number of times v_{root}'s children can be halved, given by $\log_2$ CHILDRENv_{root}. To distribute the budget over the remaining children in the current iteration, $\mathcal{B}$ is also divided by k.

2.4   Monte Carlo Tree Search

Monte Carlo Tree Search (MCTS) [13] employs Monte Carlo simulations to iteratively build a game tree. MCTS is designed to progressively converge on the best action as it gathers more statistical information about the domain.

MCTS is based on two principles: (1) with sufficient time, the sampled average reward from random simulations converges to the true state value, and (2) previous samples can guide future searches.

Algorithm 2 Pseudocode for MCTS algorithm (adapted from [13])

1: **function** MCTS(s, $\mathcal{R}$)
2:    **Input:** State s, Resource $\mathcal{R}$
3:    **Output:** Recommended action
4:    $v_{root} \leftarrow \langle s, \text{ACTIONS}(s) \rangle$
5:    **while** $r \leq \mathcal{R}$ **do**
6:       $v_k \leftarrow$ SELECTION(v_{root}, π)
7:       $v_{k+1} \leftarrow$ EXPANSION(v_k)
8:       $rw \leftarrow$ SIMULATION(v_{k+1}, π_Δ)
9:       BACKPROPAGATION(v_{k+1}, rw)
10:      update r
11:   **return** RECOMMEND(v_{root})

Algorithm 2 outlines the MCTS process, starting instantiating a root node, denoted as v_{root}. A node v consists of a state s, the list of applicable actions in s, the parent node, a list of children, and the Q and N values for cumulative reward and visits count, respectively.

The search process involves the following four steps:

- *Selection*: beginning at tree root, the selection phase traverses the tree using a *tree policy* (π) that guides the search towards promising nodes, until finding a node with untried actions.
- *Expansion*: a node is expanded by applying a random untried action to its state, resulting in a new child node. This new node is initialized with the new state, a list of applicable actions, an empty child list, and its parent reference.
- *Simulation*: a playout evaluates the potential reward r of the new node. This is done by following a *default policy* (π_Δ), which usually applies random actions until reaching a terminal state.
- *Backpropagation*: each node from v_{k+1} up to the root are updated: their Q is updated by rw and N increases by 1.

The search process continues until the algorithm uses up a specified resource $\mathcal{R}$, which can be time or a number of iterations. RECOMMEND function selects a move according to one of three criteria: *Max Child*, with the highest Q value; *Robust Child*, with the highest N value; or *Max-Robust Child*, combining both Q and N values.

The most popular tree policy for the selection phase is the Upper Confidence Bound (UCB$_1$, Sect. 2.2), which considers each node as an individual MAB problem. When used in MCTS, the algorithm is called Upper Confidence Bound Applied to Trees (UCT). MCTS and UCT exhibit an anytime property, allowing them to recommend useful actions even if the search execution is interrupted.

3   Alternatives to UCT

Simple regret (SR) minimization is strictly related to choosing a child node from the root at the recommendation phase, and the cumulative regret (CR) is related to the searching process. UCB$_1$ has optimal bounds on cumulative regret recommendation, but it is penalized in terms of simple regret. At the root node, sampling in MCTS/UCT typically focuses on finding the best move with high confidence. Once UCB$_1$ identifies such a move, it continues to spend time on it, possibly resulting in low information gain [15].

In time-sensitive situations, not considering other options and continuing with the current best choice may be a potential flaw that could be improved. Bubeck et. al [3] shows that UCB$_1$ exhibits a slow decrease in terms of simple regret, with the best-case scenario being a polynomial rate decrease. This can be problematic when fast recommendations are required. Karning et. al [6] suggest that the more exploratory policies have better bounds on simple regret minimization, like the Sequential Halving algorithm (Sect. 2.3).

3.1   UCB$_{\sqrt{}}$ and SR+CR

Tolpin and Shimony [15] modify UCB$_1$'s policy into UCB$_{\sqrt{}}$. This policy adjusts the UCB$_1$ formula using a quicker-growing sublinear function, leading to a faster increase in the uncertainty bonus. The new policy changes the $\ln N_s$ term in UCB$_1$ to $\sqrt{N_s}$, aiming to narrowing the gap between the selections of non-greedy nodes.

Tolpin and Shimony also point out that nodes closer to the root and those deeper in the tree have different goals. The former is more crucial for move recommendations. As a result, the search strategy near the root should prioritize reducing simple regret more quickly, while nodes deeper in the tree should aim to match the value of taking the optimal path, aligning more with cumulative regret minimization.

The Simple Regret plus Cumulative Regret (SR+CR) scheme proposed by Tolpin and Shimony integrates two different policies to strike a balance between minimizing simple regret and cumulative regret. They introduced two specific algorithms, both of which combine the UCT policy with more exploratory strategies. The first one, UCB$_{\sqrt{}}$+UCT, operates by applying the UCB$_{\sqrt{}}$ policy at the root node and UCB$_1$ to all child nodes. Their second algorithm, $\frac{1}{2}$-greedy+UCT, operates by randomly selecting a node at the root with a 50% probability.

3.2 Hybrid Monte Carlo Tree Search

Hybrid Monte Carlo Tree Search (H-MCTS) [10] is a SR+CR algorithm that combines the Sequential Halving Applied on Trees (SHOT) algorithm [4] with UCT. SHOT is a recursive adaptation of Sequential Halving, used for constructing game-trees using a non-exploiting policy. In H-MCTS, SHOT is used not only at the root node but also deeper within the tree to address simple regret minimization.

The proposed method switches from UCT to SHOT when the computational budget spent in the node achieves a certain threshold, after considering changing the policy to be safe (when a subset of good moves are already identified and evaluated). The algorithm shifts its focus from minimizing cumulative regret to minimizing simple regret after beginning to expand SHOT within UCT regions that have been sufficiently visited. Since the computational budget per node is initially small, the simple regret tree remains shallow, as SHOT eliminate nodes from selection, the budget spent increases, causing the SHOT tree to grow deeper.

H-MCTS outperforms UCT for various exploration coefficients [10] and is highly effective in games with large branching factors, as it prunes low-promising nodes and directs the search towards the most promising areas.

4 Improving UCT in Time-Restricted Scenarios

In this Section, we present a new SR+CR algorithm, and a new method to calculate the amount of time the agent should spend in each move, assuming a time constraint for the entire game.

4.1 UCT$_{\sqrt{SH}}$

Although H-MCTS is promising at balancing simple and cumulative regret, it requires a predefined budget for the SHOT portion, which is not possible to estimate for previous unknown domains. Furthermore, by neglecting the exploitation of nodes, the agent becomes prone to excessive resource allocation in unpromising regions.

To enhance the performance of UCT under a GGP environment with rigid time constraints, we propose a different SR+CR method, using Sequential Halving and UCB$_{\sqrt{}}$, as shown in Algorithm 3.

Algorithm 3 Pseudocode of UCT$_{\sqrt{\text{SH}}}$ algorithm

1: **function** UCT$_{\sqrt{\text{SH}}}(s, \mathcal{R})$
2:    **Input:** State s, Resource $\mathcal{R}$
3:    **Output:** Recommended action
4:    start r
5:    $v_{root} \leftarrow \langle s, \text{ACTIONS}(s) \rangle$
6:    $n \leftarrow |\text{CHILDREN}(v_{root})|$
7:    $h \leftarrow 1; k \leftarrow n$
8:    **while** $r \leq \mathcal{R}$ **do**
9:        **if** $k > k_{min}$ and $r > (\mathcal{R}\frac{h}{\log_2 n})$ **then**
10:            SORT(v_{root}, k)
11:            $h \leftarrow h + 1; k \leftarrow \text{MAX}(k_{min}, k/2)$
12:        $ch \leftarrow \text{CHILDREN}(v_{root})$
13:        $v_s \leftarrow \underset{v \in \text{HEAD}(ch,k)}{\arg\max} \; \pi_{UCB_{\sqrt{}}}(v)$
14:        $v_k \leftarrow \text{SELECTION}(v_s, \pi_{UCB_1})$
15:        $v_{k+1} \leftarrow \text{EXPANSION}(v_k)$
16:        $rw \leftarrow \text{SIMULATION}(v_{k+1}, \pi_\Delta)$
17:        BACKPROPAGATION(v_{k+1}, rw)
18:        update r
19:    RECOMMEND$(\text{CHILDREN}(v_{root}))$

UCT$_{\sqrt{\text{SH}}}$ prioritize simple regret minimization at root node by combining UCB$_{\sqrt{}}$ with Sequential Halving eliminations, and the cumulative regret component uses UCT. In UCT$_{\sqrt{\text{SH}}}$, the aim of Sequential Halving is not to converge to the single best move, but rather to limit the number of children to search, which allows UCB$_{\sqrt{}}$ to explore the most promising areas.

We set a lower limit, k_{min}, on the number of child nodes necessary for the elimination process to occur. The halving operation is executed by dividing k by two. During the child selection process at the root, we use k to constrain the selection to the first k child nodes. Upon reaching a new elimination point, the algorithm arranges the root's child nodes in descending order according to their anticipated reward.

We employ an iterative methodology (presented at line 11) to ascertain when to halve the number of children. We compute a ratio representing the fraction of halving stages already completed. For that, we divide the halve counter h by the maximum number of halving operations $\log_2 n$. We compute the resource allocation required for the next halving operation multiplying this ratio with the total resource $\mathcal{R}$. After the used resource r surpasses this value, we increment h by one. This method ensures that the same portion of $\mathcal{R}$ is equally divided across all halving stages.

A key distinction from traditional MCTS lies in the separate treatment of root selection. The root selection, depicted at line 13, iterates over the first k-th children, by calling the HEADch, k function. The selected child is the one which maximizes $\pi_{UCB_{\sqrt{}}}$. Notice that rather than eliminating moves based on the number of visits a node has, we determine when to apply eliminations based on $\mathcal{R}$, which can be time or number of simulations.

4.2   Clock Bonus Time

GGP agents play games without prior knowledge, usually with a time limit for the entire game. Agents must allocate time for each move. A common strategy is to use a predefined fixed time budget, which can lead to inefficient time management, either by exhausting the time on long games, or leaving unused time in shorter games. We propose a method for estimating the time to spend on each move in a GGP environment, using a certain number of simulations during the search to gather information about the game itself. Our model employs a minimum thinking time, and for games where the agent can spend more time, it provides a thinking time bonus. The Clock Bonus Time (cbt) formula is as follows:

$$cbt = \max(\tau_{min}, \min(\tau_{max}, G/\overline{m})) - \tau_{min} \ . \tag{4}$$

In Eq. 4, G is the total time budget, τ_{min} and τ_{max} are the minimum and maximum allowed thinking times per move, respectively. The bonus is given by G divided by the estimated number of moves left to finish the game $\overline{m}$, which we compute using playouts. The max and min functions ensure that the agent performs at least the minimum thinking time and avoids overestimating the time it has. We then discount the new time by τ_{min} because it is a bonus increased to the minimum thinking time. One way to integrate cbt with MCTS consists of calling cbt after half of τ_{min} has passed, which is when $r \geq \mathcal{R}/2$.

5   Experiments

To evaluate $UCT_{\sqrt{SH}}$, we conduct two experiments. First, we examined the agent's decision-making in two different scenarios of reward variance. This design simulate game situations where making a suboptimal choice significantly affects the outcome (high variance), as well as those where suboptimal choices have a milder effect and are less harmful (low variance). However, in these latter scenarios, a series of misjudgments due to lack of confidence in the most rewarding decision can potentially lead to an overall loss.

The second test examine our agent's performance within a practical context. We utilized a game competition called Kilothon, hosted within the Ludii environment. This competition served as a benchmark on $UCT_{\sqrt{SH}}$ performance in a GGP environment.

5.1   Prize Box Selection Experiment

The Prize Box Selection Experiment is a simplified MAB where there are K boxes containing a deterministic amount of money. The money for each box is pre-selected from a Gaussian distribution $N(\mu, \sigma)$. We test different policies for a given number of trials and number of boxes, recording how often the policy selects each box during the experiment.

We compare UCB$_1$, Sequential Halving, UCB$_{\sqrt{}}$, and UCB$_{\sqrt{sh}}$ (root policy of UCT$_{\sqrt{SH}}$), at a scenario with high and low variance in reward's distributions. Both scenarios with 10000 trials for distribute across 30 boxes. In the low variance case, we set $\mu = 0.3$ and $\sigma = 0.05$, limited to $[-0.5, 0.5]$. For the high variance case, we set $\mu = 0.3$ and $\sigma = 0.5$, limited to $[-1,1]$.

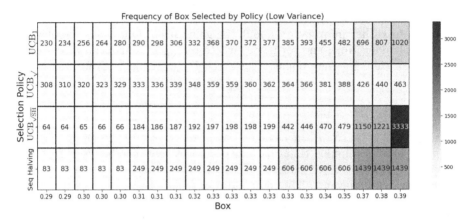

Fig. 1. Box selection frequency under low variance reward distribution.

Figure 1 depicts the low variance scenario, where the boxes are arranged in descending order, displaying only the 20 boxes with the highest rewards (i.e., boxes 1–10 have the lowest reward and are omitted).

In this scenario, UCB$_{\sqrt{}}$ presents the most dispersed trials among all boxes, with UCB$_1$ following a similar pattern. The use of eliminations in this specific scenario guides the policies that adopt them towards more focused selections, UCB$_{\sqrt{sh}}$ and Sequential Halving concentrated a higher quantity of resources on a smaller subset of boxes than UCB$_1$ and UCB$_{\sqrt{}}$.

In the high variance test at Fig. 2, both UCB$_1$ and UCB$_{\sqrt{sh}}$ exhibit a stronger preference for the highest rewarded box. While Sequential Halving does not change its selection distribution no matter the reward distribution is, due to its non-exploiting nature. UCB$_{\sqrt{}}$ and Sequential Halving share more similar frequencies of selection between them, indicating their preference for exploration over UCB$_1$. Using UCB$_{\sqrt{sh}}$ avoids overspending trials on the best box in high-variance scenarios, which can be a desirable characteristic from the perspective of simple regret minimization, although it has a clear preference for the best rewarding box.

Our analysis emphasizes that while exploiting (i.e. allocating budget to the most promising option) is valuable for reward maximization, exploration is crucial in game playing as it leads to the rapid discovery of beneficial moves. In this context, UCB$_{\sqrt{sh}}$ displays a particularly desirable quality of enhanced exploration in our experiments.

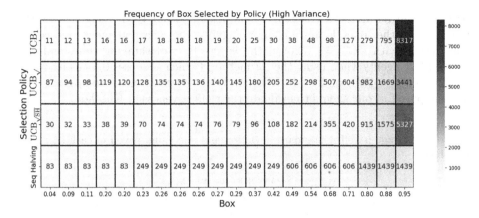

Fig. 2. Box selection frequency under high variance reward distribution.

Essentially, the primary objective is to identify and execute the optimal move in the game. The frequency of selecting the best move during the search is not of primary concern. Furthermore, $UCB_{\sqrt{sh}}$ appears to be less sensitive to varying rewards distributions. This resilience stems from its ability to not overlook exploitation in high variance scenarios, and to focus resources on low variance situations through the application of Sequential Halving eliminations.

5.2 GGP Competition Experiment

General Game Playing (GGP) is a research area focused on developing intelligent agents capable of playing a wide variety of games without prior knowledge of any specific game being played [5]. Ludii is a system for general game research, which has contributed significantly to the field. Games are implemented using Ludii's Game Description Language (GDL). Ludii's GDL is robust and straightforward, it allows researchers and game designers to create new games and even reproduce historical ones [12].

Ludii hosted a GGP competition, where games chosen for the competition were turn-based, adversarial, sequential, and fully observable, including deterministic and stochastic games. Kilothon was one of the competition tracks, where participants play 1094 games against an implementation of UCT algorithm, native from Ludii (which we will refer to as *Kilothon agent*). Each agent has a strict one-minute time limit to play each game in its entirety. When the one-minute time limit is reached, the agent must resort to random moves until the game concludes.

The Kilothon agent uses a fixed thinking time of 0.5 s per move, and incorporates two modifications to the pure UCT algorithm: Tree Reuse enables the agent to store the search tree from previous plays and reusing it in the future, and Open Loop [11] for dealing with stochastic games.

Experimentation. As a baseline, we implemented our UCT version, without tree reuse neither open loop, to compete against the Kilothon agent, and compare results with $UCT_{\sqrt{SH}}$. For both, we use 0.5 s of thinking time, and we differentiate agents when using the *cbt* method (which also uses 0.5 s as its τ_{min} and 2 s as τ_{max}). We conduct 10 Kilothon trials for each agent, computing the average payoff of our agents to evaluate their effectiveness in Kilothon.

Table 1. Average payoff ± standard deviation and maximum payoff of each agent in Kilothon.

AGENT	PAYOFF ± s.d	MAX
$UCT_{\sqrt{SH}}^{cbt}$	0.1512 ± 0.0176	0.1984
UCT^{cbt}	0.0813 ± 0.0334	0.1489
$UCT_{\sqrt{SH}}$	0.0672 ± 0.0196	0.1019
UCT	-0.0063 ± 0.0168	0.0157

Table 1 presents the average payoff ± standard deviation of our tested agents across all games, along with the maximum payoff achieved by each of them. Our results highlights the performance of $UCT_{\sqrt{SH}}$ method over UCT, which achieves better scores than UCT including after adding the *cbt* method. $UCT_{\sqrt{SH}}^{cbt}$ had the highest score, that could achieve second place in the official competition, where the first place achieved 0.231, and the second 0.031.

The *board* games category, contains the vast majority of games in Kilothon contest. Ludii board games are classified according the following classifications [2,8]: **hunt**, where a player controls more pieces and aims to immobilize the opponent; **race**, where the first to complete a course, with moves controlled by dice or other random elements, wins; **sow or mancala**, where players sow seeds to specific positions and capture opponent seeds; **space**, where players place and/or move pieces to achieve a specific pattern, with possibility of blocks and captures; and **war**, where the goal is to control territory, immobilize or capture all opponent's pieces.

Figure 3 showcases the winning rate of our agents under the five board game categories. The win ratio is computed as win/(win+loss), not including draws.

$UCT_{\sqrt{SH}}$ outperforms UCT in sow (+6%), space (+7%), and war (+4%) games. Against the Kilothon agent, $UCT_{\sqrt{SH}}$ secures the respective win ratios in hunt, race, sow, space and war, respectively: 56%, 53%, 57%, 55%, 51% . Sow and space games shows the highest variability among agents, with the highest scores achieved by $UCT_{\sqrt{SH}}^{cbt}$ of 69% and 63%. Both these games display significant performance boosts via the *cbt* method for UCT^{cbt} and $UCT_{\sqrt{SH}}^{cbt}$, both surpassing a 60% win rate. Our evaluations reveal that the $UCT_{\sqrt{SH}}$ strategy, especially with the *cbt* method, outperforms baseline UCT. The $UCT_{\sqrt{SH}}^{cbt}$ agent had the highest score, showcasing its improvement over the baseline.

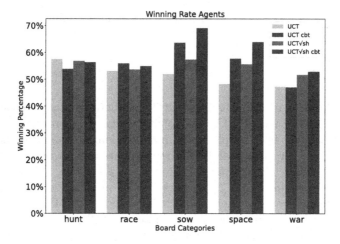

Fig. 3. Winning percentage in board categories for each agent, after running 10 Kilothon tournaments each.

GGP Subset: Five Board Games. While Kilothon competition encompassed an extensive variety of games, we examine a subset that fell within the board game category. To this end, we selected games that were also part of a study conducted by Pepels [9].

We compare $\mathrm{UCT}_{\sqrt{\mathrm{SH}}}$ vs UCT, where both agents had 0.5, 1, and 2 s of thinking time for each move. Each experiment running over 1000 matches. Table 2 showcases the results.

Table 2. Win rate of $\mathrm{UCT}_{\sqrt{\mathrm{SH}}}$ against UCT.

Game	0.5 s/move	1 s/move	2 s/move
Pentalath	$51.7\% \pm 3\%$	$64.9\% \pm 2\%$	$66.8\% \pm 2\%$
AtariGo	$57.0\% \pm 3\%$	$63.6\% \pm 2\%$	$71.9\% \pm 2\%$
NoGo	$61.0\% \pm 3\%$	$66.9\% \pm 2\%$	$79.6\% \pm 2\%$
Breakthrough	$48.8\% \pm 3\%$	$56.2\% \pm 3\%$	$66.9\% \pm 2\%$
Amazons	$46.9\% \pm 3\%$	$60.0\% \pm 3\%$	$70.1\% \pm 2\%$
Overall	$53.0\% \pm 3\%$	$62.3\% \pm 2\%$	$71.0\% \pm 2\%$

Our study reveals the advantages of $\mathrm{UCT}_{\sqrt{\mathrm{SH}}}$ over UCT across various game domains. $\mathrm{UCT}_{\sqrt{\mathrm{SH}}}$ makes a better use of the time budget, as its win rate over UCT increases significantly when the budget increases. In Pentalath, the win rate began at 51.7% and rose to 66.8% as the thinking time increased, while NoGo achieved the highest final win rate of 79.6%. Both Breakthrough and Amazons exhibited a significant increase in win rates, escalating from less than 50% to 66.9% and 70.1%, respectively. The overall result shows that $\mathrm{UCT}_{\sqrt{\mathrm{SH}}}$ improve its advantage over UCT after both had increased its thinking time, achieving win rates of 53% up to 71%.

6   Conclusion

This work addresses two drawbacks in UCT, the base method of most general game playing (GGP) agents: (i) UCT exploitation factor guarantees asymptotic optimality but suffers from simple regret minimization; (ii) the use of fixed time-budget for search per move may be an overestimate and an underestimate for long and short games, respectively.

For (i), we introduce $UCT_{\sqrt{SH}}$, a new MCTS method, which foregoes the asymptotic optimality in exchange for a timely response. $UCT_{\sqrt{SH}}$ uses the simulation budget more exploratively than traditional UCT, since during the search time the goal is to find the best possible move and return it to the game. For (ii), we present the Clock Bonus Time (*cbt*) strategy to better allocate the search time per move, given a fixed time budget to play the entire game.

We use two experiments to empirically evaluate $UCT_{\sqrt{SH}}$ against UCT. The Prize Box Experiment indicates that $UCT_{\sqrt{SH}}$ is less sensible to changes in the distribution of rewards as UCB_1 and $UCB_{\sqrt{}}$ do. These latter two have a more spread-out allocation when rewards have minimal variation. However, when there's a lot of variation in the rewards, $UCT_{\sqrt{SH}}$ tends to favor the best option, although not as consistently as UCB_1, which almost always selects the top choice.

Although it may appear that constantly selecting the known best option would maximize rewards, as proposed in Tolpin and Shimony's study [15], they argue differently. They suggest that policies that promote more exploration at the root level can actually lead to faster identification of better moves.

In the Kilothon GGP competition, our method exceeded the performance of the baseline UCT. The implementation of the *cbt* strategy more than doubled the scores for both agents. $UCT_{\sqrt{SH}}^{cbt}$ could achieve the second place in the official Kilothon competition, according to the final competition results. This achievement is remarkable, considering our agent relies solely on Monte Carlo simulations and does not utilize any other enhancements or parallelism.

Our study also showed that $UCT_{\sqrt{SH}}$ uses increased time budgets significantly better than UCT. Moreover, our findings suggest that using 0.5 s of thinking time, as the default UCT Kilothon agent does, the time constraint imposed might be too unrealistic for agents to play in many game domains.

References

1. Auer, P., Cesa-Bianchi, N., Fischer, P.: Finite-time analysis of the multiarmed bandit problem. Mach. Learn. **47**(2), 235–256 (2002)
2. Brice, W.C.: A history of board-games other than chess. by h. j. r. murray. oxford: Clarendon press, 1952. pp. viii 287, 86 text figs. 42s. J. Hellenic Stud. **74**, 219–219 (1954). https://doi.org/10.2307/627627

3. Bubeck, S., Munos, R., Stoltz, G.: Pure exploration in finitely-armed and continuous-armed bandits. Theoret. Comput. Sci. **412**(19), 1832–1852 (2011)
4. Cazenave, T.: Sequential halving applied to trees. IEEE Trans. Comput. Intell. AI Games **7**(1), 102–105 (2014)
5. Genesereth, M., Love, N., Pell, B.: General game playing: overview of the AAAI competition. AI Mag. **26**(2), 62–62 (2005)
6. Karnin, Z., Koren, T., Somekh, O.: Almost optimal exploration in multi-armed bandits. In: International Conference on Machine Learning, pp. 1238–1246. PMLR (2013)
7. Kocsis, L., Szepesvári, C., Willemson, J.: Improved monte-carlo search. Univ. Tartu, Estonia, Tech. Rep. **1**, 1–22 (2006)
8. Parlett, D.: The Oxford history of board games. Oxford University Press, Oxford (1999)
9. Pepels, T.: Novel selection methods for monte-carlo tree search. Master's thesis, Department of Knowledge Engineering, Maastricht University, Maastricht, The Netherlands (2014)
10. Pepels, T., Cazenave, T., Winands, M.H., Lanctot, M.: Minimizing simple and cumulative regret in monte-carlo tree search. In: Workshop on Computer Games, pp. 1–15. Springer (2014)
11. Perez Liebana, D., Dieskau, J., Hunermund, M., Mostaghim, S., Lucas, S.: Open loop search for general video game playing. In: Proceedings of the 2015 Annual Conference on Genetic and Evolutionary Computation, pp. 337–344 (2015)
12. Piette, É., Soemers, D.J.N.J., Stephenson, M., Sironi, C.F., Winands, M.H.M., Browne, C.: Ludii - the ludemic general game system. In: Giacomo, G.D., Catala, A., Dilkina, B., Milano, M., Barro, S., Bugarín, A., Lang, J. (eds.) Proceedings of the 24th European Conference on Artificial Intelligence (ECAI 2020). vol. 325, pp. 411–418. IOS Press (2020)
13. Świechowski, M., Godlewski, K., Sawicki, B., Mańdziuk, J.: Monte carlo tree search: A review of recent modifications and applications. Artificial Intelligence Review, pp. 1–66 (2022)
14. Świechowski, M., Park, H., Mańdziuk, J., Kim, K.J.: Recent advances in general game playing. Sci. World J. **2015** (2015)
15. Tolpin, D., Shimony, S.: Mcts based on simple regret. In: Proceedings of the AAAI Conference on Artificial Intelligence, vol. 26.1, pp. 570–576 (2012)

α-MCMP: Trade-Offs Between Probability and Cost in SSPs with the MCMP Criterion

Gabriel Nunes Crispino$^{(\boxtimes)}$, Valdinei Freire , and Karina Valdivia Delgado

Universidade de São Paulo, São Paulo, Brazil
gcrispino@alumni.usp.br, {valdinei.freire,kvd}@usp.br

Abstract. In Stochastic Shortest Path (SSP) problems, not always the requirement of having at least one policy with a probability of reaching goals (probability-to-goal) equal to 1 can be met. This is the case when dead ends, states from which the probability-to-goal is equal to 0, are unavoidable for any policy, which demands the definition of alternate methods to handle such cases. The α-strong probability-to-goal priority is a property that is maintained by a criterion if a necessary condition to optimality is that the ratio between the probability-to-goal values of the optimal policy and any other policy is bound by a value of $0 \leq \alpha \leq 1$. This definition is helpful when evaluating the preference of different criteria for SSPs with dead ends. The Min-Cost given Max-Prob (MCMP) criterion is a method that prefers policies that minimize a well-defined cost function in the presence of unavoidable dead ends given policies that maximize probability-to-goal. However, it only guarantees α-strong priority for $\alpha = 1$. In this paper, we define α-MCMP, a criterion based on MCMP with the addition of the guarantee of α-strong priority for any value $0 \leq \alpha \leq 1$. We also perform experiments comparing α-MCMP and GUBS, the only other criteria known to have α-strong priority for $0 \leq \alpha \leq 1$, to analyze the difference between the probability-to-goal of policies generated by each criterion.

Keywords: Decision making under uncertainty · Markov Decision Processes · Stochastic Shortest Path

1 Introduction

Markov Decision Processes (MDPs) [11] constitute the main theoretical framework for modeling probabilistic planning problems, in which an agent interacts with an environment by applying actions with stochastic outcomes, to optimize a given objective function. In this context, Stochastic Shortest Path (SSP) [1] problems model scenarios in which an agent needs to minimize the expected accumulated cost to goal. The solution to an SSP is a policy, a mapping between states to actions. Conventionally, SSPs assume that there is at least one policy that has a probability of 1 to reach goals when followed by the agent.

M. C. Naldi and R. A. C. Bianchi (Eds.): BRACIS 2023, LNAI 14195, pp. 112–127, 2023.
https://doi.org/10.1007/978-3-031-45368-7_8

Realistically this is a strong requirement. Environments commonly have dead ends, defined as states from which goals can not be reached. When these states exist and are unavoidable, the conventional requirement for solving SSPs is not well-defined. This requires the definition of alternate criteria to address this particular case.

Several criteria were proposed in the literature in this context. The MAX-PROB criterion [7] chooses policies that maximize the probability of reaching goals (probability-to-goal). The S^3P [14] and iSSPUDE [6] criteria also prefer policies that maximize probability-to-goal, then choose the ones that minimize cost measures only considering histories that reach the goal. The MCMP criterion [16], on the other hand, chooses policies that maximize probability-to-goal, but then it minimizes a different cost measure that does not ignore histories that have dead-end states. This can lead to more natural optimal policies when compared to S^3P and iSSPUDE since dead ends are taken into account. Also, since it can be formulated as a linear program, efficient state-of-the-art methods can be used to solve this criterion.

Other criteria make trade-offs between probability-to-goal and cost measures, such that they do not only prefer policies that maximize probability-to-goal. The fSSPUDE [6] criterion uses a finite penalty to give up, such that the agent can pay this value to exit the process at any step. A discount factor, commonly used in infinite horizon MDPs [11], can also be used in SSPs for making such trade-offs [15]. The GUBS criterion [4] was proposed as an alternative for criteria that make trade-offs between probability-to-goal and cost measures. Among other features, GUBS maintains good theoretical properties, like the α-strong probability-to-goal priority [2], which guarantees a lower bound on the ratio of probability-to-goal values considering pairs of policies, for a given value of $0 \leq \alpha \leq 1$.

Between the criteria that were mentioned, GUBS is the only one that guarantees α-strong probability-to-goal priority for $0 \leq \alpha \leq 1$. However, we can modify constraints in the MCMP linear programming formulation such that a new criterion, which also guarantees α-strong probability-to-goal priority for $0 \leq \alpha \leq 1$, is obtained.

In this paper, we thus propose α-MCMP, a criterion derived from modifying a constraint in MCMP's linear programming formulation such that it guarantees α-strong probability-to-goal priority for $0 \leq \alpha \leq 1$. We also perform experiments to evaluate the differences in the preference of policies of α-MCMP and GUBS.

The structure of this work is defined as follows: Sect. 2 contains the background for this work, outlining definitions and properties of SSPs and alternate criteria to solve them in the presence of unavoidable dead ends; Sect. 3 introduces α-MCMP and a proof that it maintains the α-strong probability-to-goal priority property, as well as some differences between this new criterion and GUBS; Sect. 4 describes the empirical evaluation that was performed; and Sect. 5 has the conclusion of the present work.

2   Background

The following subsections will cover definitions and properties that will be used throughout this paper.

2.1   Stochastic Shortest Path (SSP) Problems

A Stochastic Shortest Path (SSP) problem [1] is an indefinite-horizon stochastic process in which at time t the agent is at a state s_t and can choose an action a_t that will lead to s_{t+1} given a probability distribution on s_t and a by paying a cost of c_t.

Definition 1. *An SSP [1] is a tuple* $\mathcal{M} = \langle \mathcal{S}, s_0, \mathcal{A}, P, c, \mathcal{G} \rangle$, *where:*

- $\mathcal{S}$ *is a finite set of states;*
- $s_0 \in \mathcal{S}$ *is the initial state;*
- $\mathcal{A}$ *is a finite set of actions;*
- $P : \mathcal{S} \times \mathcal{A} \times \mathcal{S} \to [0,1]$ *is the transition function, such that* $P(s,a,s') = \Pr(s_{t+1} = s' \mid s_t = s, a_t = a)$;
- $c : \mathcal{S} \times \mathcal{A} \to \mathbb{R}_{\geq 0}$ *is the cost function, which assigns a cost* $c(s,a)$ *for taking action* a *at state* s;
- $\mathcal{G} \subset \mathcal{S}$ *is the set of absorbing goal states, i.e.* $c(g,a) = 0$ *and* $P(g,a,g) = 1, \forall g \in \mathcal{G}$ *and* $a \in \mathcal{A}$. *Also,* $c(s,a) > 0$ *for* $s \in \mathcal{S} \setminus \mathcal{G}$.

The objective in an SSP is to find a policy π, a mapping between states to actions, that minimizes the expected cost to goal $V^\pi(s) = \lim_{T \to \infty} \mathbb{E}\left[\sum_{t=0}^{T-1} c_t \mid \pi, s_0 = s\right]$. The probability-to-goal of a policy π from state $s \in \mathcal{S}$ is given by the function $P_G^\pi(s) = \lim_{t \to \infty} \Pr(s_t \in \mathcal{G} \mid \pi, s)$. We also define the maximum probability-to-goal of a state s in an SSP as the function $P_G(s) = \max_{\pi \in \Pi} P_G^\pi(s)$, in which Π is the set of all policies. A policy π is proper if the probability-to-goal when following it is 1, i.e. $P_G^\pi(s_0) = 1$.

When the agent acts for multiple steps $t \in \{0, 1, \ldots, T\}$, she generates a history $h = \{\langle s_1, a_1, c_1 \rangle, \langle s_2, a_2, c_2 \rangle, \ldots, \langle s_{T-1}, a_{T-1}, c_{T-1} \rangle, s_T\}$, where $s_t \in \mathcal{S}$ is the state in which the agent is at step t, $a_t \in \mathcal{A}$ is the action applied at s_t, and $c_t = c(s_t, a_t)$. We define $\mathcal{H} = (\mathcal{S} \times \mathcal{A} \times \mathbb{R}_{>0})^* \times \mathcal{S}$ as the set of all histories.

Among methods to solve SSPs, it is possible to describe an SSP as a linear program and solve it by using this program as an input to a solver [3]. In this LP, the objective is to find the values of $x_{s,a}$, which are variables that represent the expected number of times action $a \in \mathcal{A}$ is executed at state $s \in \mathcal{S}$. This linear program is outlined in LP1.

$$\min_{x} \sum_{s\in\mathcal{S},a\in\mathcal{A}} x_{s,a}c(s,a) \text{ s.t. (C1) – (C6)} \qquad\qquad \text{(LP1)}$$

$$x_{s,a} \geq 0 \qquad\qquad\qquad \forall s \in \mathcal{S}, a \in \mathcal{A} \quad \text{(C1)}$$

$$in(s) = \sum_{s'\in\mathcal{S},a\in\mathcal{A}} x_{s',a}P(s',a,s) \qquad\qquad \forall s \in \mathcal{S} \quad \text{(C2)}$$

$$out(s) = \sum_{a\in\mathcal{A}} x_{s,a} \qquad\qquad\qquad \forall s \in \mathcal{S}\setminus\mathcal{G} \quad \text{(C3)}$$

$$out(s) - in(s) = 0 \qquad\qquad \forall s \in \mathcal{S}\setminus(\mathcal{G}\cup\{s_0\}) \quad \text{(C4)}$$

$$out(s_0) - in(s_0) = 1 \qquad\qquad\qquad\qquad\qquad \text{(C5)}$$

$$\sum_{s_g\in\mathcal{G}} in(s_g) = 1 \qquad\qquad\qquad\qquad\qquad \text{(C6)}$$

$in(s)$ represents the expected flow entering s, while $out(s)$ represents the expected flow leaving s. Constraint (C4) restricts that except for s_0 and goal states, the expected flow entering a state must be equal to the expected flow leaving it. (C5) indicates that the expected flow leaving s_0 must be higher than the one entering s_0 by a difference of 1. In other words, in expectation, the agent must leave the initial state one more time than she enters it (this does not consider the first step of the process). Finally, the total expected flow reaching goal states must equal 1.

The objective function then minimizes the expected total cost of reaching the goal given that the agent starts at s_0, given the constraints specified. Because the optimal policy π^* that can be generated after solving LP1 is guaranteed to be deterministic, it can be defined as $\pi^*(s) = \arg\max_{a\in\mathcal{A}} x_{s,a}$.

2.2 SSPs with Dead Ends and Alternate Criteria

When the policy is improper, V^π is not well-defined and diverges. Thus, when there does not exist a proper policy, the standard criterion for solving SSPs is not well-defined either. This happens when dead ends, states from which no goal state can be reached, are unavoidable. Figure 1 contains an example of an SSP with an unavoidable dead end. This problem has a state s_0, an unavoidable dead-end state s_{de} and a goal state s_g. At the state s_0, action a leads to s_g deterministically with cost c_a. Action b has cost c_b and leads to s_g with probability P, and to s_{de} with probability $1 - P$.

For SSPs with unavoidable dead ends, different criteria have to be used to solve these problems. One can maximize probability without optimizing cost, which is the case of the MAXPROB criterion [7]. Instead of just maximizing probability, several criteria take the dual approach of minimizing a well-defined cost measure in the case of unavoidable dead ends, but only considering policies that maximize probability-to-goal. Throughout this work, we will refer to this class of criteria as lexicographic. S³P [14] and MCMP [16] are examples of such criteria. When considering the example outlined in Fig. 1, lexicographic criteria would choose a as the optimal action at s_0, because it maximizes probability-to-goal.

Other criteria can make trade-offs between probability-to-goal and some cost measures. fSSPUDEs [6], for example, use a finite penalty to pay when a dead end is reached. A discount factor [15] can also be used for this objective. Additionally,

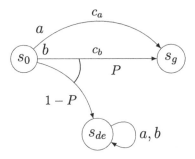

Fig. 1. Example of an SSP with an unavoidable dead-end.

the GUBS criterion [4] has parameters that combine Expected Utility Theory with goal prioritization to make such trade-offs. In the example in Fig. 1, these criteria could either choose between actions a or b, depending on which values for their parameters are selected.

Sections 2.4 and 2.5 will detail the MCMP and GUBS criteria, respectively. Before that, the next subsection will formally define the α-strong probability-to-goal priority property.

2.3 Making Trade-Offs Between Probability-to-Goal and Cost Measures

When analyzing different criteria for solving SSPs with unavoidable dead-ends, we might ask how to evaluate the decisions made by these criteria.

One method for doing so is to take into account how each criterion maintains the α-strong probability-to-goal priority property [2]. The definition for this property is given as follows:

Definition 2 (α-strong probability-to-goal priority [2]). *Consider $0 \leq \alpha \leq 1$, we say that a decision criterion has α-strong probability-to-goal priority if for all SSPs $\mathcal{M}$ and all pairs of policies $\pi, \pi' \in \Pi$, the following condition is true:*

$$\pi \succeq \pi' \implies \frac{P_G^\pi(s_0)}{P_G^{\pi'}(s_0)} \geq \alpha,$$

where $P_G^{\pi'}(s_0) > 0$. If π^* is an optimal policy then for all $\pi' \in \Pi$ the following equation holds:

$$\pi^* \succeq \pi' \implies \frac{P_G^{\pi^*}(s_0)}{P_G^{\pi'}(s_0)} \geq \alpha.$$

Note that $\frac{P_G^\pi(s_0)}{P_G^{\pi'}(s_0)} \geq \alpha$ is a necessary condition for $\pi \succeq \pi'$ to be true.

Lexicographic criteria have 1-strong probability-to-goal priority, as it is guaranteed for these criteria that every policy preferred over another has a larger value of probability-to-goal [2]. The GUBS criterion guarantees α-strong priority for $0 \leq \alpha \leq 1$, while the fSSPUDE and discounted cost criteria only have such guarantees for $\alpha = 0$ (only 0-strong) [2].

2.4 Min-Cost Given Max-Prob (MCMP)

The Min-Cost given Max-Prob (MCMP) criterion [16] shrinks trajectories by pruning histories in the first state that a dead-end state is reached, if that ever happens. This is formulated by the function $\psi : \mathcal{H} \rightarrow \mathcal{H}$:

$$
\psi(h) = \begin{cases} \{s_i\}, & \text{if } |h| = 1 \text{ or } P_G(s_i) = 0 \\ \{\langle s_i, a_i, c_i \rangle\} \cup \psi(\{\langle s_{i+1}, a_{i+1}, c_{i+1} \rangle, \dots \}), & \text{otherwise,} \end{cases}
$$
(1)

such that $h = \{\langle s_i, a_i, c_i \rangle, \langle s_{i+1}, a_{i+1}, c_{i+1} \rangle, \dots \}$.

Based on that, the value of a policy under the MCMP criterion is given by the following function:

$$
\bar{C}^{\pi}_{MCMP}(s) = \mathbb{E} \left[\sum_{t=0}^{|\psi(h)|} c_t \mid \pi, s_0 = s \right].
$$

Finally, a policy π^*_{MCMP} is optimal under MCMP if it minimizes $\bar{C}_{MCMP}$ given that it maximizes $P^{\pi}_G(s_0)$:

$$
\pi^*_{MCMP} = \min_{\pi \in \Pi_{MP}} \bar{C}^{\pi}_{MCMP}(s_0),
$$

such that Π_{MP} is the set that maximizes $P^{\pi}_G(s_0)$, i.e. $\Pi_{MP} = \{\pi \mid P^{\pi}_G(s_0) = \max_{\pi' \in \Pi} P^{\pi'}_G(s_0)\}$.

Solving MCMP is equivalent to solving the following LP 2, a modified version of LP1. The differences between the original LP for SSPs and LP2 are that the constraints (C4) and (C5) were replaced respectively by (C7) and (C8), in which equalities are replaced by inequalities, and (C6) was replaced by (C9), in which 1 is substituted by $P_G(s_0)$. The inequality introduced in (C7) means that the expected flow when reaching a state $(in(s))$ is higher than the expected flow when leaving it $(out(s))$. This can be interpreted as an implicit give-up action with no cost that can be taken by the agent, which represents the pruning of histories that contain dead ends defined in the function ψ. On a similar note, the inequality represented by (C8) means that the expected flow reaching the initial state s_0 is higher than the expected flow leaving this same state. In other words, the agent always enters s_0, since it is the initial state (thus $in(s_0) = 1$), but not always leaves without giving up (which means that $out(s_0) >= 0$).

Also note that a probabilistic Markovian policy $\pi^*(s, a) = x_{s,a}/out(s)$ can be generated after solving LP2.

$$\min_{x} \sum_{s \in \mathcal{S}, a \in \mathcal{A}} x_{s,a} c(s,a) \text{ s.t. (C1) – (C3), (C7) – (C9)} \tag{LP2}$$

$$out(s) - in(s) \leq 0 \qquad\qquad\qquad \forall s \in \mathcal{S} \setminus (\mathcal{G} \cup \{s_0\}) \text{ (C7)}$$

$$out(s_0) - in(s_0) \leq 1 \tag{C8}$$

$$\sum_{s_g \in \mathcal{G}} in(s_g) = P_G(s_0) \tag{C9}$$

To find $P_G(s_0)$, another linear program, referred to as LP3, can be used [16]. In this LP, the objective function maximizes the total expected flow of reaching goal states, which is equivalent to the probability-to-goal from the initial state s_0.

$$\max_{x} \sum_{s_g \in \mathcal{G}} in(s_g) \text{ s.t. (C1) – (C5)} \tag{LP3}$$

Since MCMP is a lexicographic criterion, it follows that it maintains 1-strong probability-to-goal priority [2]. As an example, consider the SSP displayed in the Fig. 1. As mentioned before, MCMP will choose as optimal the policy that takes action a at s_0. Let π be this policy. The cost value under the MCMP criterion for π is $C_{MCMP}^{\pi}(s_0) = c_a$.

Additionally, we can modify MCMP to make trade-offs between probability and cost by simply substituting $P_G(s_0)$ in (C6) to any probability value p, and LP2 will then generate a policy π such that $P_G^{\pi}(s_0) = p$ [8].

In the example of Fig. 1, if we replace $P_G(s_0)$ in (C9) with the value P, for instance, LP2 could return a policy that always takes b at s_0, or some linear combination of actions a and b that would yield probability-to-goal of P, if the MCMP cost of doing so is less than always taking b. To illustrate this more concretely, consider the case of $P = 0.95$, $c_a = 2$, and $c_b = 1$. If we replace $P_G(s_0)$ in (C9) with $P = 0.95$, the optimal policy returned by LP2 always takes b, yielding an expected cost of 1. However, if we replace $P_G(s_0)$ with 0.98 in (C6) instead, the optimal policy returned in this case takes a with probability 0.6 and b with probability 0.4, leading to an expected cost of $0.6 \times 2 + 0.4 \times 1 = 1.6$ and probability-to-goal $0.6 \times 1 + 0.4 \times 0.95 = 0.98$. Note that, in this case, LP2 allows for a policy that, for example, takes a with a probability of 0.98 and takes the implicit give-up action with a probability of 0.02. This also yields a probability-to-goal of 0.98, thus fulfilling (C9), but it is not the optimal policy because its resulting expected cost is $0.98 \times 2 = 1.96$, which is higher than 1.6.

2.5   Goals with Utility-Based Semantics (GUBS)

The Goals with Utility-based Semantics (GUBS) criterion [4] was proposed as an alternative method for solving SSPs with unavoidable dead ends by combining goal prioritization with Expected Utility Theory. It combines these two characteristics by its two parameters: the constant goal utility K_g, and the utility function over cost u, respectively.

Definition 3 (GUBS criterion [4]). *The GUBS criterion evaluates a history by the utility function U:*

$$U(C_T, \beta_T) = u(C_T) + K_g \beta_T. \tag{2}$$

An agent follows the GUBS if she evaluates a policy π *under the following value function:*

$$V_{GUBS}^{\pi}(s) = \lim_{T \to \infty} \mathbb{E}[U(C_T, \beta_T)|\pi, s_0 = s].   (3)$$

A policy π^* is optimal under the GUBS criterion if it maximizes the function V_{GUBS}^{π}, i.e.:

$$V_{GUBS}^{\pi^*}(s) \geq V_{GUBS}^{\pi}(s)   \forall \pi \in \Pi \text{ e } \forall s \in \mathcal{S}.$$

The GUBS criterion is the only criterion among all other mentioned criteria that guarantees α-strong probability-to-goal priority for any value $0 \leq \alpha \leq 1$.

Corollary 1 (α-strong probability-to-goal priority for GUBS [2]). *For an arbitrary $0 \leq \alpha \leq 1$, the GUBS criterion has α-strong probability-to-goal priority.*

For $0 \leq \alpha < 1$, the property is maintained if the following condition is true:

$$K_g \geq \frac{(U_{max} - U_{min})\alpha}{1 - \alpha},$$

such that the values returned by u are in the set $[U_{min}, U_{max}]$.
For $\alpha = 1$, this property is true when $U_{max} = U_{min}$ and $K_g > 0$.

The eGUBS criterion [5] is a specialization of GUBS, in which the function u is an exponential function taken from risk-sensitive SSPs [10].

Definition 4 (eGUBS criterion [2,5]). *The eGUBS criterion considers the GUBS criterion where the utility function u is defined as follows:*

$$u(C_T) = \begin{cases} 0, \text{ if } C_T = \infty \\ e^{\lambda C_T}, \text{ otherwise,} \end{cases}$$

over some accumulated cost C_T and a risk factor $\lambda < 0$.

In the GUBS criterion, generated policies are non-Markovian, since they need to have the state space augmented with the accumulated cost. From the properties of eGUBS, however, it is possible to use algorithms that guarantee that these non-Markovian policies are finite, based on a maximum cost obtained from the problem structure [5]. eGUBS-VI [5] is an algorithm that does this by computing an optimal policy under eGUBS completely via value iteration, while eGUBS-AO* [2] is another algorithm that can compute an optimal policy by using a mix of value iteration and heuristic search.

Also, since for eGUBS $U_{min} = 0$ and $U_{max} = 1$, the lower bound obtained in Corollary 1 can be restricted to $K_g \geq \frac{\alpha}{1-\alpha}$ [2].

3 α-MCMP Criterion

In this section, we will show how to define a new criterion called α-MCMP, which maintains the α-strong probability-to-goal priority by leveraging the MCMP definition.

MCMP is 1-strong since it is a criterion that maximizes probability-to-goal. However, we can relax the LP definition of MCMP to ensure it maintains the α-strong probability-to-goal priority for $0 \leq \alpha \leq 1$. This can be done by replacing the constraint (C9) with a new constraint (C10), which instead of making the total expected flow entering goal states to be equal to $P_G(s_0)$ (the maximum probability-to-goal), it considers this flow as $\alpha P_G(s_0)$.

LP4 contains the modified version of LP2:

$$\min_x \sum_{s \in \mathcal{S}, a \in \mathcal{A}} x_{s,a} c(s, a) \text{ s.t. (C1) – (C3), (C7) – (C8), (C10)} \tag{LP4}$$

$$\sum_{s_g \in \mathcal{G}} in(s_g) = \alpha P_G(s_0) \tag{C10}$$

This generates a new criterion that minimizes a similar yet different cost function than MCMP. Throughout this work, we refer to this modified version of MCMP as α-MCMP. It can be formally defined as the following:

Definition 5. *Given a value of $0 \leq \alpha \leq 1$, a policy $\pi^*_{\alpha MCMP}$ is optimal under the α-MCMP criterion if it minimizes $\bar{C}_{MCMP}$ given policies with probability-to-goal equal to $\alpha P_G(s_0)$:*

$$\pi^*_{\alpha MCMP} = \min_{\pi \in \Pi_{\alpha MP}} \bar{C}^\pi_{MCMP}(s_0),$$

for $\Pi_{\alpha MP} = \{\pi \mid P^\pi_G(s_0) = \alpha P_G(s_0)\}$.

The following theorem demonstrates that α-MCMP thus maintains α-strong priority for any $0 \leq \alpha \leq 1$.

Theorem 1 (α-strong probability-to-goal priority for α-MCMP). *The α-MCMP criterion has α-strong probability-to-goal priority for any value $0 \leq \alpha \leq 1$.*

Proof. By replacing (C9) with the constraint $\sum_{s_g \in \mathcal{G}} in(s_g) \geq \alpha P_G(s_0)$, the condition $\pi^* \succeq \pi \implies \frac{P^{\pi^*}_G(s_0)}{P^\pi_G(s_0)} \geq \alpha$ from Definition 2 holds, in which π^* is the optimal policy generated by LP2 and π is any other policy.

Consider an arbitrary policy π' and let π be defined as the policy that follows π' with probability $\frac{\alpha P_G(s_0)}{P^{\pi'}_G(s_0)}$ and gives up with probability $1 - \frac{\alpha P_G(s_0)}{P^{\pi'}_G(s_0)}$. The probability-to-goal value of π is $P^\pi_G(s_0) = \frac{\alpha P_G(s_0)}{P^{\pi'}_G(s_0)} P^{\pi'}_G(s_0) + (1 - \frac{\alpha P_G(s_0)}{P^{\pi'}_G(s_0)}) \times 0 = \alpha P_G(s_0)$. It then follows that π is at least as good as π', because $\bar{C}^\pi_{MCMP}(s_0) = \alpha P_G(s_0) \bar{C}^{\pi'}_{MCMP}(s_0) \leq \bar{C}^{\pi'}_{MCMP}(s_0)$.

Thus, a policy with probability-to-goal $\alpha P_G(s_0)$ will always be as good as any policy with probability-to-goal higher than $\alpha P_G(s_0)$ when the constraint $\sum_{s_g \in \mathcal{G}} in(s_g) \geq \alpha P_G(s_0)$ is used. This means that using the constraint $\sum_{s_g \in \mathcal{G}} in(s_g) = \alpha P_G(s_0)$ (C10) is equivalent to using $\sum_{s_g \in \mathcal{G}} in(s_g) \geq \alpha P_G(s_0)$. Since the restriction $\sum_{s_g \in \mathcal{G}} in(s_g) \geq \alpha P_G(s_0)$ maintains the α-strong probability-to-goal priority property for $0 \leq \alpha \leq 1$ and (C10) is equivalent to using it, the α-MCMP criterion maintains the α-strong probability-to-goal priority property for $0 \leq \alpha \leq 1$.   □

3.1   Relationship Between α-MCMP and GUBS

By defining α-MCMP, we now have a criterion that maintains the α-strong probability-to-goal priority property for $0 \leq \alpha \leq 1$ like GUBS, but with the advantages of generating a Markovian policy and having efficient solutions inherited from MCMP.

Although both criteria maintain the α-strong priority for $0 \leq \alpha \leq 1$, the probability-to-goal of optimal policies in α-MCMP will always equal $\alpha P_G(s_0)$. In the GUBS criterion, when choosing a value of K_g by following Corollary 1, its resulting policy will not necessarily have a fixed probability-to-goal value of $\alpha P_G(s_0)$. This value can instead be higher than $\alpha P_G(s_0)$, if a deterministic policy with such a value exists and if the chosen utility function allows for that.

For example, consider the eGUBS criterion and a value of $\alpha = 0.95$. From Corollary 1, a value of $K_g \geq \frac{0.95}{0.05} = 19$ guarantees that eGUBS is 0.95-strong. Also, consider the example in Fig. 1, such that $P = 0.95$ and the cost c_a is larger than c_b by a margin of δ, i.e., $c_a = c_b + \delta$, for $\delta > 0$. Under this setting, eGUBS can choose a as the optimal action at s_0 if the expected utility when following it is greater than the one when following b, i.e., $u(c_a) + K_g > 0.95(u(c_b) + K_g) + 0.05u(\infty)$. Considering $K_g = 20 \geq 19$:

$$u(c_a) + 20 > 0.95(u(c_b) + 20) + 0.05u(\infty)$$
$$e^{\lambda c_a} + 20 > 0.95 e^{\lambda c_b} + 0.95 \times 20$$
$$e^{\lambda c_a} - 0.95 e^{\lambda c_b} > 0.95 \times 20 - 20$$
$$e^{\lambda c_a} - 0.95 e^{\lambda c_b} > -1$$

Thus, for values of λ, c_a, and c_b such that $e^{\lambda c_a} - 0.95 e^{\lambda c_b} > -1$, action a will be optimal. For instance, if $\delta = 1$, $\lambda = -0.1$, $c_b = 100$, and $c_a = 100 + 1 = 101$, the utility of taking action a is $e^{-0.1 \times 101} + 20 \approx 20$, which is higher than the utility of taking b that is $0.95(e^{-0.1 \times 100} + 20) \approx 19$. For this same setting, the optimal policy under α-MCMP would be the one that chooses a with probability 0.95, and gives up with probability 0.05. This policy has an expected cost of $0.95 \times 101 = 95.95$.

In summary, for a value of $\alpha = 0.95$ and depending on the trade-off incurred by the chosen value of λ, the GUBS criterion can choose an optimal policy π

such that $P_G^\pi(s_0) \geq \alpha P_G(s_0)$. α-MCMP, on the other hand, will always select a policy that has a probability-to-goal value **equal** to $\alpha P_G(s_0)$.

Table 1 summarizes some of the key differences between α-MCMP and GUBS.

4   Experiments

We have performed experiments in the Navigation [12], River [4], and Triangle Tireworld [9] domains (Figs. 2a, 2b, and 2c illustrate these domains, respectively) to evaluate α-MCMP empirically and compare it to the GUBS criterion.

Table 1. Key differences between α-MCMP and GUBS.

Property	α-MCMP	GUBS
α-strong for $0 \leq \alpha \leq 1$	Yes	Yes
Policy	Markovian and Probabilistic	Non-Markovian and Deterministic
Has give-up action	Yes	No
$P_G^{\pi*}(s_0)$	$= \alpha P_G(s_0)$	$\geq \alpha P_G(s_0)$

The Navigation domain is a grid world, where the agent starts at the rightmost column in the last row, and the goal is in the same column but in the first row. In the middle rows, the actions have a probability of making the agent disappear, thus going to a dead-end state. These probabilities are lower on the left and higher on the right. The River domain is similar, also being a grid world, but representing an agent at a river bank having to cross the river to get to the goal that is on the other side. Actions in the river have a positive probability of making the agent fall down one row, while actions in the bank and the bridge, located in the first row, are deterministic. If the agent gets to the first row after falling down in the river, the waterfall is reached, in which no actions to other states are available, thus being a dead end. Finally, in the Triangle Tireworld domain, the agent is represented by a car that needs to go to a goal location, always with a positive probability of getting a flat tire. Certain states have available tires to change, but these are in general further away from the goal than states with no tires available. In the instances depicted in Fig. 2c, for example, the black circles represent locations that have available tires. If the car gets a flat tire after getting to a state with no tire to replace, then a dead end is reached.

We aimed to evaluate the different probability-to-goal values obtained from optimal policies generated from both criteria when varying their parameters. For that, we varied α between the values of $\{10^{-5}, 10^{-4}, 10^{-3}, 10^{-2}, 0.1, 0.5, 0.999\}$ and computed LP4 to obtain the optimal policy of α-MCMP. For GUBS, we used the value of K_g as in Corollary 1 as a parameter for the given values of α and then computed the eGUBS criterion for this value of

K_g and different values of λ using the eGUBS-VI algorithm. The values of λ were $\{-0.05, -0.06, -0.07, -0.08, -0.09, -0.1, -0.2\}$ for the Navigation domain, $\{-0.01, -0.1, -0.2, -0.3, -0.4, -0.5\}$ for the River domain, and $\{-0.1, -0.2, -0.25, -0.3, -0.35, -0.4\}$ for the Triangle Tireworld domain. A single instance was used for each domain. The number of states for each of these instances is 101, 200 and 19562 for the Navigation, River, and Triangle Tireworld domains, respectively. Note that the numbers of processed states can be considerably higher for eGUBS because it might need to process a high number of augmented states.

The domains were defined as PDDLGym [13] environments, and the code of the implementation is available at https://github.com/GCrispino/ssp-deadends/tree/bracis2023-paper.

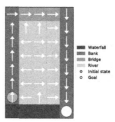

(a) Navigation domain (image extracted from [2]).

(b) River domain (image extracted from [2]).

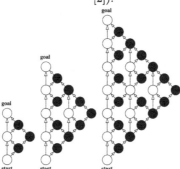

(c) Triangle Tireworld domain (image extracted from [9]).

Fig. 2. Illustration of the domains used in the experiments.

Figures 3, 4, and 5 contain graphs that display the probability-to-goal values for each policy obtained for GUBS and α-MCMP given different values of α (displayed in $\log_{10}$ scale in the x-axis), respectively for domains Navigation, River and Triangle Tireworld. From them, it can be observed that, as mentioned before, while the probability-to-goal obtained by α-MCMP always equals $\alpha P_G(s_0)$, the values obtained from eGUBS' optimal policies are different. In the different lines[1] reflecting different values of the risk factor λ, the probability-to-goal values are always higher than $\alpha P_G(s_0)$. How higher these probabilities are, though, depends on the chosen values of λ. For example, in the River domain, we can observe several different types of probability-to-goal curves when varying λ. The one that results in the widest difference between the minimum and the maximum probability-to-goal values is when $\lambda = -0.5$, for which the minimum probability-to-goal obtained was about 0.07 for $\alpha = 10^{-5}$, and the maximum value was 1, obtained when $\alpha \in \{10^{-3}, 10^{-2}, 0.1, 0.5, 0.999\}$.

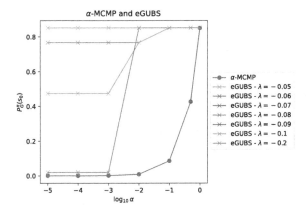

Fig. 3. Results for the Navigation domain.

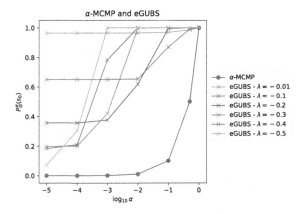

Fig. 4. Results for the River domain.

---

[1] Note that not every line representing a value of λ can be seen in the figures, because the values in these lines might be very close to the values in others, which can make them get covered by these other lines.

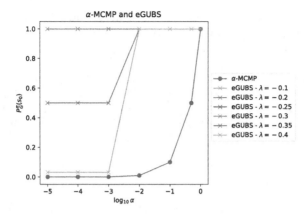

Fig. 5. Results for the Triangle Tireworld domain.

In fact, for all domains, we observed that the higher the absolute value of λ is, the higher the difference between the minimum and maximum probability-to-goal values that were obtained for this value of λ are. It is also interesting to note that intermediate values of λ can cover intermediate values of probability-to-goal that higher values of λ used cannot. For example, in the River domain, when λ is equal to -0.1 and -0.2, values of probability-to-goal close to 0.6 were obtained, which was not the case when the absolute value of λ was higher than -0.2 (i.e. when λ is equal to $-0.3, -0.4$, and -0.5). As another example, in the Navigation domain when $\lambda = -0.1$, the value of about 0.47 was obtained, while for different values of λ such as -0.2 and for absolute values lower than 0.09 (i.e. when λ is equal to $-0.05, -0.06, -0.07$, and -0.08), either a value close to 0 or close to the maximum probability-to-goal was obtained.

5   Conclusion

In this paper, we define α-MCMP, a criterion to solve SSPs with unavoidable dead ends, that is obtained from MCMP. We also show how the modifications that were done to MCMP to generate α-MCMP make it guarantee the α-strong probability-to-goal priority property. Besides this, α-MCMP also provides good advantages inherited from MCMP, such as Markovian policies and efficient state-of-the-art methods to solve it. It also has some differences when compared to GUBS, such as when preferring policies that might give up in the middle of the process instead of policies that do not do it.

Finally, we performed experiments to evaluate policies obtained by α-MCMP and compared them to policies generated by the eGUBS criterion. The results indicate that we can make trade-offs in both criteria by choosing values of α a priori. Nonetheless, the way that these criteria make this trade-off is different. α-MCMP always has probability-to-goal equal to $\alpha P_G(s_0)$, while the compromise made by eGUBS will depend on the value of λ used in its utility function.

This paper attempts to contribute in the general understanding of sequential decision making in the presence of unavoidable dead-end states, such as several other works in this area [2, 4–7, 14–16].

Acknowledgments. This study was supported in part by the *Coordenação de Aperfeiçoamento de Pessoal de Nível Superior* (CAPES) - Finance Code 001, by the São Paulo Research Foundation (FAPESP) grant #2018/11236-9 and the Center for Artificial Intelligence (C4AI-USP), with support by FAPESP (grant #2019/07665-4) and by the IBM Corporation.

References

1. Bertsekas, D.: Dynamic Programming and Optimal Control. Athena Scientific, Belmont, Mass (1995)
2. Crispino, G.N., Freire, V., Delgado, K.V.: GUBS criterion: arbitrary trade-offs between cost and probability-to-goal in stochastic planning based on expected utility theory. Artif. Intell. **316**, 103848 (2023)
3. d'Epenoux, F.: A probabilistic production and inventory problem. Manage. Sci. **10**(1), 98–108 (1963)
4. Freire, V., Delgado, K.V.: GUBS: a utility-based semantic for goal-directed Markov decision processes. In: Proceedings of the 16th Conference on Autonomous Agents and Multiagent Systems, pp. 741–749 (2017)
5. Freire, V., Delgado, K.V., Reis, W.A.S.: An exact algorithm to make a trade-off between cost and probability in SSPs. In: Proceedings of the International Conference on Automated Planning and Scheduling, vol. 29, pp. 146–154 (2019)
6. Kolobov, A., Weld, D., et al.: A theory of goal-oriented MDPs with dead ends. In: Uncertainty in artificial intelligence: proceedings of the Twenty-eighth Conference [on uncertainty in artificial intelligence] (2012), pp. 438–447 (2012)
7. Kolobov, A., Weld, D.S., Geffner, H.: Heuristic search for generalized stochastic shortest path MDPs. In: Proceedings of the Twenty-First International Conference on International Conference on Automated Planning and Scheduling, pp. 130–137 (2011)
8. Kuo, I., Freire, V.: Probability-to-goal and expected cost trade-off in stochastic shortest path. In: Gervasi, O., Murgante, B., Misra, S., Garau, C., Blečić, I., Taniar, D., Apduhan, B.O., Rocha, A.M.A.C., Tarantino, E., Torre, C.M. (eds.) ICCSA 2021. LNCS, vol. 12951, pp. 111–125. Springer, Cham (2021). https://doi.org/10.1007/978-3-030-86970-0_9
9. Little, I., Thiebaux, S., et al.: Probabilistic planning vs. replanning. In: ICAPS Workshop on IPC: Past, Present and Future, pp. 1–10 (2007)
10. Patek, S.D.: On terminating Markov decision processes with a risk-averse objective function. Automatica **37**(9), 1379–1386 (2001)
11. Puterman, M.: Markov decision processes: discrete stochastic dynamic programming. Wiley, New York (1994)
12. Sanner, S., Yoon, S.: IPPC results presentation. In: International Conference on Automated Planning and Scheduling (2011). http://users.cecs.anu.edu.au/ssanner/IPPC_2011/IPPC_2011_Presentation.pdf
13. Silver, T., Chitnis, R.: PDDLGym: Gym environments from PDDL problems. In: International Conference on Automated Planning and Scheduling (ICAPS) PRL Workshop, pp. 1–6 (2020). https://github.com/tomsilver/pddlgym

14. Teichteil-Königsbuch, F.: Stochastic safest and shortest path problems. In: Proceedings of the Twenty-Sixth AAAI Conference on Artificial Intelligence, pp. 1825–1831 (2012)
15. Teichteil-Königsbuch, F., Vidal, V., Infantes, G.: Extending classical planning heuristics to probabilistic planning with dead-ends. In: Proceedings of the Twenty-Fifth AAAI Conference on Artificial Intelligence, pp. 1017–1022 (2011)
16. Trevizan, F.W., Teichteil-Königsbuch, F., Thiébaux, S.: Efficient solutions for stochastic shortest path problems with dead ends. In: Proceedings of the Thirty-Third Conference on Uncertainty in Artificial Intelligence (UAI) (2017), pp. 1–10 (2017)

Specifying Preferences over Policies Using Branching Time Temporal Logic

Warlles Carlos Costa Machado[1(✉)], Viviane Bonadia dos Santos[2],
Leliane Nunes de Barros[2], and Maria Viviane de Menezes[3]

[1] Instituto Federal do Piauí, Teresina, Brazil
`warllescarlos@ifpi.edu.br`
[2] Universidade de São Paulo, São Paulo, Brazil
`{vbonadia,leliane}@ime.usp.br`
[3] Universidade Federal do Ceará - Campus Quixadá, Quixadá, Brazil
`vivianemenezes@ufc.br`

Abstract. Automated Planning is the subarea of AI devoted to developing algorithms that can solve sequential decision making problems. By taking a formal description of the environment, a planning algorithm generates a plan of actions (also called *policy*) that can guide an agent to accomplish a certain task. Classical planning assumes the environment is fully-observed and evolves in a deterministic way considering only simple reachability goals (e.g. a set of states to be reached by a plan or policy). In this work, we approach fully-observed non-deterministic planning (FOND) tasks which allow the specification of complex goals such as the preference over policy quality (weak, strong or strong-cyclic) and preferences over states in the paths generated by a policy. To solve this problem we propose formulae in α-CTL (branching time) temporal logic and use planning as model checking algorithms based on α-CTL to generate a solution that captures both, agent's preferences and the desired policy quality. To evaluate the effectiveness of the proposed formulae and algorithms, we run experiments in the Rovers benchmark domain. Up to our knowledge, this is the first work to solve non-deterministic planning problems with preferences using a CTL temporal logic.

Keywords: Non-Deterministic Planning · Preferences · Temporal Logic

1 Introduction

Automated Planning [9] is the subarea of Artificial Intelligence that studies the process of automatically generating a plan of actions that, when executed, enables an agent to accomplish a task. A *state* refers to a particular configuration of the agent's environment and can be described by a set of properties. The *initial state* is the environment configuration at the beginning of the plan execution. A *planning problem* is defined by means of a formal description of the environment (called *planning domain*) plus an initial state and a *goal description*.

© The Author(s), under exclusive license to Springer Nature Switzerland AG 2023
M. C. Naldi and R. A. C. Bianchi (Eds.): BRACIS 2023, LNAI 14195, pp. 128–143, 2023.
https://doi.org/10.1007/978-3-031-45368-7_9

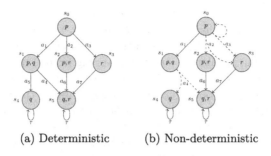

(a) Deterministic   (b) Non-deterministic

Fig. 1. Examples of state transition systems representing planning domains.

The *planning domain* models the environment based on the *set of actions* the agent can perform. Thus an *action* can be seen as a manifestation of the agent's intent to make the environment to change from a state to another. Finally, the *goal description* can vary from a simple reachability goal formula (e.g. a propositional formula that must be satisfied by a set of goal states) to complex temporal goal formulae [22,26], capable to represent the agent's preferences over the plan solution and its quality as well.

Figure 1a shows a planning domain represented by a *state transition system* given by a graph with vertices representing the states - labeled by propositional atoms and directed edges representing the actions - labeled by action names. To represent states we make the *closed world assumption*, i.e. a state is described only by the propositions that are true and any proposition that is false is not included in the state label. For example, in the state s_0 (Fig. 1a) p is *true*, q is *false* and r is *false*. Finding a *solution* to a *planning problem* corresponds to select the best action at each state, which is a computationally hard problem [11]. Therefore many research has been devoted to find efficient solutions for planning problems mainly focusing on making different assumptions about the environment and the capability to express goal formulae.

Deterministic Planning. Research in this area assumes the world is fully-observed and there are no uncertainties about the effects of the actions [9]. Thus, when an action is executed in a state, there is only one possible successor state. The solution of a deterministic planning problem is a sequence of actions that takes the environment from the initial state to a state that satisfies a simple reachability goal description. Examples of efficient algorithms for deterministic planning are based on heuristic search [10] and boolean satisfiability [13].

Example 1 (Deterministic Planning). Suppose the environment (system) depicted in Fig. 1a is currently in the initial state s_0 and the agent goal is to reach the state where property $\varphi = \neg p \wedge q \wedge r$ holds. The sequence of actions $\langle a_1, a_4 \rangle$ is a plan solution since it deterministically takes the agent from state s_0 to state s_5, where φ is satisfied. Other possible plans are: $\langle a_2, a_6 \rangle$ e $\langle a_3, a_7 \rangle$.

Non-deterministic Planning. Also called FOND planning - Fully Observed Non-Deterministic planning, where we make the assumption the environment evolves in a non-deterministic way, due to the occurrence of natural events without the agent's control (called *exogenous events*). In this case, the execution of an action can lead to a set of possible future states [22]. A solution for a non-deterministic planning problem is *policy*: a mapping from states to actions [9]. A policy can be classified into three quality categories: (i) a *weak policy* is a solution that may achieve the goal but offers no guarantee due to the non-determinism of the actions; (ii) a *strong policy* is a solution that guarantees the goal achievement, regardless of non-determinism of the actions; and (iii) a *strong-cyclic policy* is a solution that ensures the goal achievement, but its execution may involve infinite cycles (assuming that the agent will eventually exit all cycles) [6]. Efficient algorithms for non-deterministic planning are based on Model Checking [6,18,21,28] and determinization of the actions to obtain relevant policies [19].

Example 2 (Non-deterministic planning). Consider the non-deterministic planning domain depicted in Fig. 1b, which is similar to the domain shown in Fig. 1a except for the fact that the actions a_2 and a_4 are non-deterministic. These actions can eventually take the agent, respectively, to the states $\{s_2, s_3\}$ and $\{s_4, s_5\}$. Suppose the agent is in state s_0 and its goal is to reach state where the goal formula holds: $\varphi = \neg p \wedge q \wedge r$. A solution is a policy: $\{(s_0, a_1), (s_1, a_4)\}$. This policy indicates that if the agent starts at s_0 and performs the deterministic action a_1, it will reach state s_1. Subsequently, if it takes the non-deterministic action a_4 in s_1, the agent will end up in either s_4 or s_5 (*weak policy*). Other solutions are strong policies $\{(s_0, a_2), (s_2, a_6), (s_3, a_7)\}$ and $\{(s_0, a_3), (s_3, a_7)\}$.

Planning with Preferences. This form of planning allows the specification of complex goal formulae to represent properties that should be preserved or avoided *along the path to finally reach a goal state* (also called *extended reachability goal*). There are two types of preferences: *quantitative preferences* and *qualitative preferences*. In the first type, preferences represent a subset of desirable goals when it is not possible to achieve all goals [25]. In the second type, preferences deal with additional properties that must hold along the path to achieve a goal state. Qualitative preferences can be used to express *safety* and *liveness* properties [16]. Informally, a safety property indicates that "something (bad) should not happen" during the execution of a plan, while a liveness property expresses that eventually "something (good) should happen" during the plan execution [15]. Qualitative preferences can be expressed by operators in PDDL such as: `always`, `sometime`, `sometime-before` and `at-most-once` [3,8].

Example 3 (Planning with Qualitative Preferences in Deterministic Environments). Consider the planning domain depicted in Fig. 1a and a problem where the initial state is s_0 and the goal is to reach a state where the propositional formula $\varphi = \neg p \wedge q \wedge r$ holds. Suppose that the agent may prefer to visit, on the path to reach the goal, *only states where p is true* (*liveness property*). In this case, possible solutions are $\langle a_1, a_4 \rangle$, $\langle a_2, a_6 \rangle$ and $\langle a_3, a_7 \rangle$. This work focuses on the qualitative preference "`always (p)`", which aims to find a plan where the property p is true in *all states along the path to reach the goal*.

Example 4 (Planning with Qualitative Preferences in Non-Deterministic Environments). Consider the planning domain depicted in Fig. 1b and the agent is at state s_0 and its goal is to reach a state where the property $\varphi = \neg p \wedge q \wedge r$ is satisfied, but with the *additional requirement* that the agent *must pass only through states where property p is true before reaching the goal.* The solution $\pi = \{(s_0, a_1), (s_1, a_4)\}$ is a weak police that satisfies also the user preference, once it corresponds to a path that possibly reaches the goal φ and passes only through states where p is satisfied. The solutions $\pi = \{(s_0, a_2), (s_2, a_6), (s_3, a_7)\}$ and $\pi = \{(s_0, a_3), (s_3, a_7)\}$ are strong polices that satisfies the user preference, once it reaches the goal φ for sure and passes only through states where p is satisfied.

Most algorithms for planning with qualitative preferences work on deterministic domains [2,12,14,20]; and the few works on FOND planning do not include preferences over the policy quality. This is because they are based on LTL - Linear Time Logic - [23] which can not have path quantifiers. **To the best of our knowledge, this is the first step toward extending the notion of qualitative preferences in non-deterministic environments to also include preferences over policy quality.** In order to do so, we apply the notion of extended reachability goals in Model Checking based on the branching time temporal logica called α-CTL [27] to express qualitative preferences as extended reachability goals and solve existing algorithms for α-CTL model checking.

In this paper, we propose solutions for planning problems that address two aspects present in real-world environments: (i) we consider that the actions can have *non-deterministic effects* and; (ii) we consider that the user wants to express its *preferences over the policy paths*, and over the policy quality as well. This paper is organized as follows: Sect. 2 presents the related works; Sect. 3 describes the foundations; Sect. 4 details how policy preferences can be expressed using the α-CTL temporal logic [22]; Sect. 5 presents our experimental analysis and, Sect. 6 presents the conclusions and future work.

2   Related Work

The work in [5] presents an approach to **non-deterministic planning** in which the **qualitative preferences** are expressed as extended (reachability) goals in Linear Time Logic (LTL) [23]. Let's call this a problem P and to solve it, the authors proposed the following extra-logical approach: (i) build a Buchi automaton for the extended goal expressed as an LTL formula; (ii) perform the *determinization* of non-deterministic actions (iii) build a new problem P' without extended goals; (iv) use a non-deterministic planner to produce a policy for P' and; (v) convert the resulting policy into a solution for P. In our work, we rely on a solution based on CTL (a branching time propositional logic) able to solve non-deterministic planning without the need to perform determinizations.

The solution for planning with **qualitative preferences** in **deterministic domains**, presented in [20], involves transforming a problem with preferences

P into a problem without preferences P'. For each preference in P, a new goal, which is false in the initial state, is added in P'. The new goal can be achieved by an action with zero cost, but requires the preference to be satisfied, or an action with a cost equal to the utility of the preference that can only be performed if the preference is false. Due to this transformation method, any classical planner can be used to solve problems with preferences.

The work in [4] introduces a method for solving generalized planning problems, where the same plan can be used for multiple problem instances. It generates policies by transforming multiple concrete problems into a single abstract (lifted) problem that captures their common structure. The global structure of the problems can be captured through **qualitative preferences** expressed as formulas in LTL logic. In addition, the authors demonstrate that for a wide class of problems, path constraints can be compiled, which reduces generalized planning into **non-deterministic planning**.

The work in [27] presents a planning algorithm that aims to solve non-deterministic planning problems with temporally extended goals (complex goals), while also considering the quality of the policy (weak, strong, or strong-cyclic). The proposed planner utilizes the α–CTL model checking framework to tackle these kind of problems. Our work is an application of such planner for problems with complex goals involving qualitative preferences.

3   Foundations

3.1   Non-deterministic Planning

In the real-world situations, the nature can cause unpredictability in the effects the actions. As a result, when an agent is in a state s_i and chooses an action a_i with the intention of reaching a particular state s_j, the interference of nature can cause the agent to end up in a different state s_k instead.

A non-deterministic planning domain can be characterized as a state transition system, as showed in Definition 1. In this kind of domain, when an action is performed in one state, it can lead to more than one successor state.

Definition 1. *[Non-Deterministic Planning Domain] Given a set of propositional atoms $\mathbb{P}$ and a set of actions $\mathbb{A}$, a non-deterministic planning domain is defined by a tuple $\mathcal{D} = \langle S, L, T \rangle$ where states are labeled by elements of $\mathbb{P}$ and actions are labeled by elements of $\mathbb{A}$ [9]:*

- *S is a finite set of states;*
- *$L : S \to 2^{\mathbb{P}}$ is a state label function; and*
- *$T : S \times \mathbb{A} \to 2^{S}$ is a non-deterministic state transition function.*

We assume that the set $\mathbb{A}$ contains the trivial action τ and that $T(s, \tau) = \{s\}$, for every final state $s \in S$. Intuitively, this action represents that the agent may choose to do nothing. As the number of propositions increases, it becomes unfeasible to explicitly represent planning domains. Therefore, the planning community uses action languages (such as the Planning Domain Description Language -

PDDL) to concisely describe domains. In this representation, actions are defined by their *preconditions* and *effects*. Preconditions are propositions that must be true in a state for an action to be executed, while effects are the literals that are modified in a state to produce a set of possible next states. After defining the domain, the next step is to formalize the *planning problem*.

Definition 2 (Non-Deterministic Planning Problem). *Given a non deterministic planning domain $\mathcal{D}$, a non-deterministic planning problem is defined by a tuple $P=(\mathcal{D}, s_0, \varphi)$ where: $\mathcal{D}$ is a non-deterministic planning domain; s_0 is the initial state, and φ is a propositional formula representing the goal.*

Due to uncertainties about the effects of the actions, a solution to a non-deterministic problem is a mapping from states to actions, is called *policy* [9].

Definition 3 (Policy). *Let $P = \langle \mathcal{D}, s_0, \varphi \rangle$ be a planning problem in the domain $\mathcal{D} = \langle S, L, T \rangle$ with non-deterministic actions. A policy π is a partial function $\pi : S \rightarrow A$ which maps states to actions; such that, for all state $s \in S$ if π is set to s then $\pi(s) \in \{a \in \mathbb{A} : T(s, a) \neq \varnothing\}$ [21].*

The quality of a policy can be: *weak* are policies that can reach a goal state, but there is no guarantee due to the non-determinism of actions; *strong* are policies that reach a goal state, independently of the non-determinism and, *strong-cyclic* are policy that guarantees to reach a goal state, under the assumption that the agent will eventually exit cycles [6].

Definition 4 (Set of states reachable by a policy). *Let $P = \langle \mathcal{D}, s_0, \varphi \rangle$ be a non-deterministic planning problem and π a policy to a planning problem P. The set of states reachable by a policy π, denoted by $S_{reach[\pi]}$ is defined by $\{s : (s, a) \in \pi\} \cup \{s' : (s, a) \in \pi \ e \ s' \in T(s, a)\}$ [21].*

Definition 5 (Execution structure of a policy). *Let $P = \langle \mathcal{D}, s_0, \varphi \rangle$ be a non-deterministic planning problem and π be a policy for a planning domain $\mathcal{D}$. The execution structure induced by π from $s_0 \in S$ is a tuple $\langle S_{reach[\pi]}, T \rangle$ ($S_{reach[\pi]} \subseteq S$ and $T \subseteq S \times \mathbb{A} \times S$) which contains all the states and transitions that can be reached when executing policy π.*

3.2 Planning as Model Checking

Model checking [7] is a formal technique that explores all possible states of a transition system to verify whether a given property holds. Applying model checking involves: *modeling* the system; *specifying* the property to be verified via a logical formula; and *verifying* the property automatically in the model.

Definition 6 (Action labelled transition system). *Let $\mathbb{P}$ be a not-empty set of atomic proposition and $\mathbb{A}$ a set of action names, an action labelled transition system is a tuple $\mathcal{M} = \langle S, L, T \rangle$ where:*

- *S is a finite nonempty set of states;*

- $L : S \to 2^{\mathbb{P}}$ *is the state labeling function and;*
- $T \subseteq S \times \mathbb{A} \times S$ *is the state transition relation.*

Definition 7 (Path in an action labelled transition system). *A path ρ in an action labelled transition system $\mathcal{M}$ is a sequence of states $s_0, s_1, s_2, s_3, \dots$, such that $s_i \in S$ and $(s_i, a, s_{i+1}) \in T$, for all $i \geq 0$.*

We can combine the model checking and planning approaches to find solutions to planning problems. In this context, the domain represents the model to be checked, and the temporal logic formula expresses the planning goal to be satisfied.

3.3   The Temporal Logic α-CTL

The branching time temporal logic α-CTL [21] is an extension of CTL logic able to consider actions by labeling state transitions. With such extension it is possible to solve FOND planning problems without to appeal to extra-logical procedures. In this logic, temporal operators are represented by "dotted" symbols (Definition 8). The formulae of α-CTL are composed by atomic propositions, logical connectives($\neg$, $\wedge \vee$), path qUantifiers ($\exists$ and $\forall$) and temporal operators: $\odot$ (*next*), $\boxdot$ (*invariantly*), $\diamondsuit$ (*finally*) and $\sqcup$ (*until*).

Definition 8 (α-CTL's syntax). *Let $p \in \mathbb{P}$ be an atomic proposition. The syntax of α-CTL [21] is inductively defined as:*

$$\varphi = p \mid \neg p \mid \varphi_1 \vee \varphi_2 \mid \varphi_1 \wedge \varphi_2 \mid \forall \odot \varphi_1 \mid \exists \odot \varphi_1 \mid \forall \boxdot \varphi_1 \mid \exists \boxdot \varphi_1 \mid \forall [\varphi_1 \sqcup \varphi_2] \mid \exists [\varphi_1 \sqcup \varphi_2]$$

The temporal operators derived from $\diamondsuit$ are $\forall \diamondsuit \varphi_1 = \forall (\top \sqcup \varphi_1)$ and $\exists \diamondsuit \varphi_1 = \exists (\top \sqcup \varphi_1)$. In α-CTL, the temporal model $\mathcal{M}$ with signature $(\mathbb{P}, \mathbb{A})$ is labeled transition system, i.e., a states transition system whose states are labeled by propositions and whose the transitions are labeled by actions (Definition 9):

Definition 9 (α-CTL's model). *A temporal model $\mathcal{M}$ with signature $(\mathbb{P}, \mathbb{A})$ in the logic α-CTL is state transition system $\mathcal{D} = \langle S, L, T \rangle$, where:*

- S : *is a non-empty finite set of states;*
- $L : S \to 2^{\mathbb{P}}$ *is a state labeling function;*
- $T : S \times \mathbb{A} \to 2^S$ *is states transition function labeling by actions.*

The semantics of the local temporal operators ($\exists \odot$ and $\forall \odot$) is given by preimage functions, while the semantics of the global temporal operators ($\exists \boxdot$, $\forall \boxdot$, $\exists \sqcup$ and $\forall \sqcup$) is derived from the semantics of the local temporal operators, by using least (ν) and greatest (μ) fixpoint operations.

Definition 10 (Weak Preimage in α-CTL formula). *Let $Y \subseteq S$ be a set of states. The weak preimage of Y, denoted by $T_{\exists}^{-}(Y)$ is the set $\{s \in S : \exists a \in \mathbb{A}$ and $T(s, a) \cup Y \neq \varnothing\}$.*

Definition 11 (Strong Preimage in α-CTL formula). *Let $Y \subseteq S$ be a set of states. The weak preimage of Y, denoted by $\mathcal{T}_\forall^-(Y)$ is the set $\{s \in S : \exists a \in \mathbb{A} \text{ and } \varnothing \neq T(s,a) \subseteq Y\}$.*

Definition 12 (Intension of an α-CTL formula). *Let $\mathcal{D} = \langle S, L, T \rangle$ be a temporal model (or a non-deterministic planning domain) with signature $(\mathbb{P}, \mathbb{A})$. The intension of an α-CTL formula φ in $\mathcal{D}$ (or the set of states satisfying φ in $\mathcal{D}$), denoted by $[\![\varphi]\!]\mathcal{D}$, is defined as:*

- $[\![p]\!]_\mathcal{D} = \{s : p \in L(s)\}$ (by definition, $[\![\top]\!]_\mathcal{D} = S$ and $[\![\bot]\!]_\mathcal{D} = \varnothing$)
- $[\![\neg p]\!]_\mathcal{D} = S \setminus [\![p]\!]_\mathcal{D}$
- $[\![\varphi_1 \wedge \varphi_2]\!]_\mathcal{D} = [\![\varphi_1]\!]_\mathcal{D} \cap [\![\varphi_2]\!]_\mathcal{D}$
- $[\![\varphi_1 \vee \varphi_2]\!]_\mathcal{D} = [\![\varphi_1]\!]_\mathcal{D} \cup [\![\varphi_2]\!]_\mathcal{D}$
- $[\![\exists \odot \varphi]\!]_\mathcal{D} = \mathcal{T}_\exists^-([\![\varphi]\!]_\mathcal{D})$
- $[\![\forall \odot \varphi]\!]_\mathcal{D} = \mathcal{T}_\forall^-([\![\varphi]\!]_\mathcal{D})$
- $[\![\exists \square \varphi]\!]_\mathcal{D} = \nu Y.([\![\varphi]\!]_\mathcal{D} \cap \mathcal{T}_\exists^-([\![Y]\!]))$
- $[\![\forall \square \varphi]\!]_\mathcal{D} = \nu Y.([\![\varphi]\!]_\mathcal{D} \cap \mathcal{T}_\forall^-([\![Y]\!]))$
- $[\![\exists(\varphi_1 \sqcup \varphi_2)]\!]_\mathcal{D} = \mu Y.([\![\varphi_2]\!]_\mathcal{D} \cup ([\![\varphi_1]\!]_\mathcal{D} \cap \mathcal{T}_\exists^-(Y)))$
- $[\![\forall(\varphi_1 \sqcup \varphi_2)]\!]_\mathcal{D} = \mu Y.([\![\varphi_2]\!]_\mathcal{D} \cup ([\![\varphi_1]\!]_\mathcal{D} \cap \mathcal{T}_\forall^-(Y)))$

Figure 2 shows the semantics of operators used in this work: (a) considering *each effect* of action a when applied in s_0, p is *globally* true; (b) considering *each effect* of action a when applied in s_0, p is true *until* q is true and; (c) considering *some effect* of action a when applied in s_0, p is true *until* q is true.

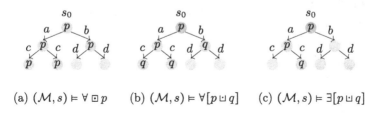

(a) $(\mathcal{M}, s) \vDash \forall \square p$ (b) $(\mathcal{M}, s) \vDash \forall [p \sqcup q]$ (c) $(\mathcal{M}, s) \vDash \exists [p \sqcup q]$

Fig. 2. Semantics of the temporal operators of the logic α-CLT.

3.4   Reasoning About Non-deterministic Actions

The representation of a planning domains through of a *action language*, e.g. PDDL, allow a compact representation of the state space. Thus, we can start from a initial state and progressively advance, determining a sequence of actions and states that lead achievement of the goal state *progressive search* [10] or from goal state and regressively determining the sequence of the actions and state that lead achievement of the initial state *regressive search* [24].

From the representation of the actions through of the preconditions an effect, is possible compute the set of states Y that precede a set X through of the regression operations of a set of states. The regression of a set of states X by action a leads to a set Y of predecessor states. However, due to the non-determinism of actions, to the applying an action a on Y, the states on X can be **required** (strong regression) or **possibly** (weak regression) achieved [18].

Using the representation of states and actions as propositional formulas [18], is possible to compute the set of predecessor states (weak and strong regression operations). These operations was implemented using *Binary Decision Diagram* [1] and incorporated to a planner capable of performing symbolic model checking using directly the actions representation with preconditions and effects.

3.5   Planning as α–CTL Model Checking

In this section, we briefly describe the planning as α–CTL model checking algorithms used in this work. The algorithms receive the planning problem, whose domain is given by actions with preconditions and effects; an initial state s_0; a goal formula φ; and a propositional atom p, representing the qualitative preference to be satisfied in all states along the path to achieve the goal.

The algorithm SAT-AU computes the *submodel* of the domain that satisfies the α-CTL formula $\forall(\varphi_1 \sqcup \varphi_2)$. It performs *strong regression* [17], computing a path that reaches a goal φ_2 while each state along the path towards s_0 also satisfies the formula φ_1. This operation computes the predecessors states directly using actions defined in terms of preconditions and effects (Fig. 3).

Algorithm 1: SAT-AU(φ_1, φ_2)

---

1 W := SAT(φ_1);
2 X:=S;
3 Y := SAT(φ_2);
4 **while** $X \mathrel{!=} Y$ **do**
5 $\quad$ X := Y;
6 $\quad$ Y := Y $\cup$ (W $\cap$ strong_regression(Y));
7 $\quad$ **if** $s_0 \in Y$ **then**
8 $\quad\quad$ | **return** Y;
9 **return** *fail*;

Fig. 3. SAT-AU computes the set of states satisfying $\forall[\varphi_1 \sqcup \varphi_2]$.

Figure 4 provides an example of how to compute the set of states that satisfies $\forall[p \sqcup \varphi]$, where φ is the goal formula $\neg p \wedge q \wedge r$. The algorithm starts by computing the set of pairs state-action whose states satisfies φ ($Y = \{(s_5, \tau)\}$), as showed in Fig. 4(a). In the first iteration (Fig. 4b), the algorithm computes the *strong regression* of the set of states satisfying φ and performs the intersection with the set of states satisfying p, obtaining the set of pairs state-action $Y = \{(s_5, \tau), (s_2, a_6), (s_3, a_7)\}$. In the second iteration (Fig. 4c), the *strong regression* of the set of states Y is computed, the intersection with the set of states satisfying p is done, resulting in the set of pairs state-action $Y = \{(s_5, \tau), (s_2, a_6), (s_3, a_7), (s_0, a_2), (s_0, a_3)\}$. The fixed point is reached after the second iteration, once no new pair state-action is obtained with weak regression operation in the Y set.

Similarly, the algorithm SAT-EU compute the set of states that satisfy the formula $\exists(\varphi_1 \cup \varphi_2)$, but using weak regression operations [17]. In addition, the algorithm SAT-AG computes a submodel satisfying the α-CTL formula $\forall \Box \varphi$. It performs regressive search from goal states (states that satisfies φ_1) towards the initial state s_0 by using strong regression operations [17].

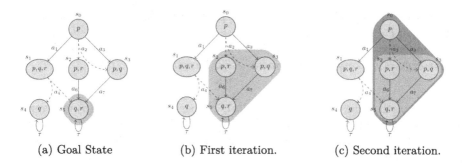

(a) Goal State   (b) First iteration.   (c) Second iteration.

Fig. 4. Computing the `always` preference for strong solutions.

4 Specifying Preferences over Policies in α-CTL

In this section we show how to specify preferences about policies in α–CTL temporal logic. **To the best of our knowledge, this work is the first step toward extending the notion of qualitative preferences in non-deterministic environments to also include the quality of policy** (*weak, strong* or *strong-cyclic*). In order to do so, we apply the notion of extended goals in α-CTL [27] to express qualitative preferences.

In this work, we will specify the qualitative preference `always(p)`, which expresses that the property p must occur in all states in the path to achieve the goal. Consider the planning problem $P = \langle \mathcal{D}, s_0, \varphi \rangle$ where $\mathcal{D}$ is the domain in Figure 4a, s_0 is the initial state and the goal $\varphi = \neg p \wedge q \wedge r$ is satisfied in state s_5. The non-deterministic actions are a_2 and a_4. Notice that:

- The solution $\{(s_0, a_1), (s_1, a_4)\}$ is a weak policy that satisfies the preference *always*(p) on the path to reach the goal state;
- The solution $(s_0, a_2), (s_2, a_5), (s_3, a_7)$ is a strong solution that satisfies the preference *always*(p) on the path to reach the goal state;
- The solution $(s_0, a_3), (s_3, a_7)$ is also a strong solution that satisfies the preference *always*(p).

4.1   Specifying the *always* Preference for Weak Policies

Weak policies can reach a goal state, but there is no guarantee due to the non-determinism of actions. In this section, we define the qualitative preference always for weak policies and present an α-CTL formula for it.

Definition 13 (Preference *always* for weak policies). *Let $P = \langle \mathcal{D}, s_0, \varphi \rangle$ be a non-deterministic planning problem, π a weak policy for P, $S_{\mathcal{D}[\pi]}$ the execution structure induced by weak policy π. The preference always(p) ($p \in \mathbb{P}$) for a weak policy is satisfied in $S_{\mathcal{D}[\pi]}$ iff there is some execution path $\mathcal{P}_\pi$ where $\forall k, 0 \leq k \leq i - 1$, $s_k \in \mathcal{P}_\pi$ and $s_k \vDash p$.*

Definition 14 (Specifying the *always* preference for weak policies in α-CTL). *Let $P = \langle \mathcal{D}, s_0, \varphi \rangle$ be a non-deterministic planning problem with signature $(\mathbb{P}, \mathbb{A})$ and always(p) ($p \in \mathbb{P}$) a preference over P. This preference for a weak policy can be expressed using the following α-CTL formula:*
$$\exists [p \sqcup \varphi].$$

To prove that the α-CTL formula $\exists [p \sqcup \varphi]$ specifies the $always(p)$ preference for a weak solution π for a planning problem $P = \langle \mathcal{D}, s_0, \varphi \rangle$ we must show that it is possible to obtain a submodel $S_{\mathcal{D}[\pi]} \subseteq \mathcal{D}$ induced by π, where: (1) $s_0 \in S_{\mathcal{D}[\pi]}$; (2) there is an execution path $\mathcal{P}_\pi \in S_{\mathcal{D}[\pi]}$ that reaches the goal state $s_i \vDash \varphi$, from the initial state s_0, and $\forall k, 0 \leq k \leq i - 1$, $s_k \in \mathcal{P}_\pi$ and $s_k \vDash p$.

Proof. Consider the non-deterministic planning problem $P = \langle \mathcal{D}, s_0, \varphi \rangle$. Assume that $(\mathcal{D}, s_0) \vDash \exists (p \sqcup \varphi)$. According to α-CTL semantics, there is an action $a \in \mathbb{A}$ such that, for a path $\mathcal{P}_\pi$ started at s_0 there is a state s_i (for $s_i \geq 0$) along this path such that $s_i \vDash \varphi$ and, for every $0 \leq k < i$, we have $(\mathcal{D}, s_k) \vDash p$. Then there is an execution path for a policy π, i.e., that lead the goal φ. Consequently, there is an execution structure $S_{\mathcal{D}[\pi]}$ induced by the policy π (Definition 5) such that $\mathcal{P}_\pi \in S_{\mathcal{D}[\pi]}$. Thus, we have $s_0 \in S_{\mathcal{D}[\pi]}$, satisfying condition 1. Since there is a $\mathcal{P}_\pi \in S_{\mathcal{D}[\pi]}$, which reaches the goal state $s_i \vDash \varphi$, from the initial state s_0 and $\forall k, 0 \leq k < i$, $s_k \in \mathcal{P}_\pi$ we have $s_k \vDash p$, the condition 2 is satisfied.

4.2   Specifying the *Always* Preference for Strong Policies

Strong policies guarantees that all sequences of states, obtained after execution of the policy, reach the goal. In this section we define how one can obtain a strong solution for a non-deterministic planning problem with always preference.

Definition 15 (Preference *always* for strong policies). *Let be $P = \langle \mathcal{D}, s_0, \varphi \rangle$ a non-deterministic planning problem with signature $(\mathbb{P}, \mathbb{A})$, π a strong policy for P, $S_{\mathcal{D}[\pi]}$ the execution structure induced by π. The always(p) ($p \in \mathbb{P}$) preference for a strong policy is satisfied in $S_{\mathcal{D}[\pi]}$ iff for all execution path $\mathcal{P}_\pi$, we have $s_i \vDash \varphi$ and $\forall k, 0 \leq k \leq i - 1, s_k \in \mathcal{P}_\pi$ with $s_k \vDash p$.*

Definition 16 (Specifying the *always* preference for strong policies in α-CTL). *Let be $P = \langle \mathcal{D}, s_0, \varphi \rangle$ a non-deterministic planning problem with signature $(\mathbb{P}, \mathbb{A})$ and always(p) $(p \in \mathbb{P})$ a qualitative preference over P. This preference for a strong policy can be expressed by the following $\alpha-CTL$ formula:*

$$\forall [p \sqcup \varphi].$$

In order to prove that α-CTL formula $\forall [p \sqcup \varphi]$ specifies the *always*(p) preference for a strong policy for a non-deterministic planning problem $P = \langle \mathcal{D}, s_0, \varphi \rangle$ we have to show that is possible to obtain a submodel $S_{\mathcal{D}[\pi]} \subseteq \mathcal{D}$ induced by the strong policy π that satisfies the following conditions: (1) $s_0 \in S_{\mathcal{D}[\pi]}$; (2) for all execution paths $\mathcal{P}_\pi \in S_{\mathcal{D}[\pi]}$, we have $\forall k, 0 \leq k \leq i - 1$, $s_k \in \mathcal{P}_\pi$ and $s_k \vDash p$.

Proof. Consider a non-deterministic planning problem $P = <\mathcal{D}, s_0, \varphi>$. Assume that $(\mathcal{D}, s_0) \vDash \forall (p \sqcup \varphi)$. According α-CTL semantics, there is an action $a \in \mathbb{A}$ such that, for all execution path $\mathcal{P}_\pi$, starting in s_0, there is a state s_i $(s_i \geq 0)$ in each path such that $s_i \vDash \varphi$ and, for each $0 \leq k < i$, we have $(\mathcal{D}, s_k) \vDash p$. Thus, we can affirm that $\mathcal{P}_\pi$ is a execution path. Consequently, there is a execution structure $S_{\mathcal{D}[\pi]}$ induced by the policy π (Definition 5) such that all $\mathcal{P}_\pi \in S_{\mathcal{D}[\pi]}$. Thus, $s_0 \in S_{\mathcal{D}[\pi]}$, satisfying condition 1. Since all $\mathcal{P}_\pi \in S_{\mathcal{D}[\pi]}$ reach a goal state $s_i \vDash \varphi$, from initial state s_0, and $\forall k, 0 \leq k < i$, $s_k \in \mathcal{P}_\pi$ e $s_k \vDash p$, we have that the condition 2 é satisfied.

4.3   Specifying the *Always* Preference for Strong-Cyclic Policies

Strong cyclic policies guarantees that all sequence of states, obtained after policy execution, reach the goal under the assumption that its execution will eventually exit all existing cycles. In this section, we define how to specify the always preference for a strong-cyclic policy.

Definition 17 (Preference *always* for strong-cyclic policies). *Let be $P = \langle \mathcal{D}, s_0, \varphi \rangle$ a non-deterministic planning problem, where $\mathcal{D}$ has signature $(\mathbb{P}, \mathbb{A})$, π a strong cyclic policy for P, $S_{\mathcal{D}[\pi]}$ a execution structure induced by π. The always(p) preference $(p \in \mathbb{P})$ for a strong cyclic policy is satisfied in $S_{\mathcal{D}[\pi]}$ iff for all execution paths $\mathcal{P}_\pi$, we have $\forall k, 0 \leq k \leq i-1, s_k \in \mathcal{P}_{\pi_i}$ e $s_k \vDash p$. Furthermore, it is possible to reach a goal state from each $s_k \in \mathcal{P}_\pi$.*

Definition 18 (Specifying the *always* preference for strong cyclic policies in α-CTL). *Let be $P = \langle \mathcal{D}, s_0, \varphi \rangle$ a non-deterministic planning problem, where $\mathcal{D}$ has signature $(\mathbb{P}, \mathbb{A})$ and always(p) $(p \in \mathbb{P})$ a qualitative preference over P. This preference for a strong cyclic police can be expressed by the following $\alpha-CTL$ formula:*

$$\forall \boxdot \exists [p \sqcup \varphi].$$

In oder to show that α-CTL formula $\forall \boxdot \exists (p \sqcup \varphi)$ specify the *always*(p) preference for a strong solution for a non-deterministic planning problem $P = \langle \mathcal{D}, s_0, \varphi \rangle$ we have to show that it is possible to obtain a submodel $S_{\mathcal{D}[\pi]} \subseteq \mathcal{D}$, such that: (1) $s_0 \in S_{\mathcal{D}[\pi]}$; (2) in all execution path $\mathcal{P}_\pi \in S_{\mathcal{D}[\pi]}$ we have $\forall k, 0 \leq k \leq i-1$, $s_k \in \mathcal{P}_\pi$ and $s_k \vDash p$ and; (3) it is possible to exit the all cycles and reach the goal state from any state in execution path.

Proof. Consider a non-deterministic planning problem $P = \langle \mathcal{D}, s_0, \varphi \rangle$. Assume that $(\mathcal{D}, s_0) \vDash \forall \Box \exists (p \sqcup \varphi)$. According α-CTL semantic, there is an action $a \in \mathbb{A}$ such that for some path $\mathcal{P}_\pi$, starting in s_0 there is a state s_i $(s_i \geq 0)$ in this path such that $s_i \vDash \varphi$ and, for each $0 \leq k < i$, we have $(\mathcal{D}, s_k) \vDash p$. Thus, each path $\mathcal{P}_\pi$ is a execution path. Consequently, there is a execution structure $S_{\mathcal{D}[\pi]}$, induced by π, such that $\mathcal{P}_\pi \in S_{\mathcal{D}[\pi]}$. Thus, we have $s_0 \in S_{\mathcal{D}[\pi]}$ satisfying the condition 1. Since there is a $\mathcal{P}_\pi \in S_{\mathcal{D}[\pi]}$, reaching a goal state $s_i \vDash \varphi$, from s_0 and $\forall k, 0 \leq k < i$, $s_k \in \mathcal{P}_\pi$ and $s_k \vDash p$, attending condition 2. Furthermore, according α-CTL semantic, there is an action $a \in \mathbb{A}$ such that, in all states s_i $(i \geq 0)$ in all paths, starting in s_0, we can reach a goal state by visiting only states where property p is true. Thus, the condition 3 is also satisfied.

5   Experimental Analysis

Our preliminary experimental analysis aims to verify the feasibility of finding policies that satisfy the qualitative preferences specification. For this, we use the deterministic *Rover* domain with preferences from 5th *International Planning Competition*, modified to include non-deterministic actions (to the best of our knowledge, there are currently no benchmark domain with these characteristics). *Rovers* models the mission of planetary exploration on Mars, where rovers are equipped with devices including cameras, sample collectors, and data transmitters. The goal of the rovers is to explore the planet surface and send the collected data to a space station. The qualitative preference `always` entails preventing the rover from sampling soil and rock at a specific *waypoint*. We use the planner PACTL [28] to obtain the set of states satisfying the α-CTL formulas specifying the qualitative preference `always` and also the quality of the policy. The experiments were performed on a Xeon Quad Core Server with 16GB of RAM. A timeout of 45 min was set for solving each instance.

Table 1 presents, for each instance and qualitative preference specified, the execution time and number of steps to obtain a policy. The symbol "-" indicates that timeout was reached. The instances have different levels of difficulty, which are characterized by factors such as the number of regions that need to be traversed, the number of goals to be achieved, and the number of rovers involved. The planner found strong, weak, and strong-cyclic policies for all instances from 1 to 8 (except for instance 6). Starting with instance 9, the planner was able to find solutions to problems 14 and 19. In cases where the planner fails to find a solution, it is difficult to distinguish either there is no solution to the problem, or the planner was unable to find the solution without exceed the timeout.

Table 1. Time and number of steps to compute policies with preferences.

Instance	Preference	Policy					
		Weak		S. Cyclic		Strong	
		T(ms)	Steps	T(ms)	Steps	T(ms)	Steps
p01	at-soil-sample-waypoint0	1813	22	2127	26	1469	14
p02	at-soil-sample-waypoint9	5399	20	6597	25	5031	18
p03	at-soil-sample-waypoint3	10752	24	12131	27	8095	24
p04	at-soil-sample-waypoint0	3324	16	4857	22	2960	16
p05	at-soil-sample-waypoint3	3264	23	3655	31	3163	23
p06	at-soil-sample-waypoint10	2517596	35	-	-	-	-
p07	at-soil-sample-waypoint11	28712	21	39680	26	24313	21
p08	at-soil-sample-waypoint5	9352	36	9834	40	7153	34
p09	at-soil-sample-waypoint1	-	-	-	-	-	-
p10	at-soil-sample-waypoint5	-	-	-	-	-	-
p11	at-soil-sample-waypoint3	-	-	-	-	-	-
p12	at-soil-sample-waypoint8	-	-	-	-	-	-
p13	at-soil-sample-waypoint12	-	-	-	-	-	-
p14	at-soil-sample-waypoint8	7299	23	8176	31	6841	23
p15	at-soil-sample-waypoint5	-	-	-	-	-	-
p16	at-soil-sample-waypoint13	-	-	-	-	-	-
p17	at-soil-sample-waypoint12	-	-	-	-	-	-
p18	at-soil-sample-waypoint10	-	-	-	-	-	-
p19	at-soil-sample-waypoint12	23624	31	25324	38	19351	31
p20	at-soil-sample-waypoint14	-	-	-	-	-	-

6   Conclusion and Future Work

In this work, we propose using the α-CTL temporal logic to specify the **always** qualitative preference in non-deterministic planning problems. In order to obtain weak, strong, or strong cyclic policies for non-deterministic problems with qualitative preferences, we employ planning as α-CTL model checking algorithms. To demonstrate the suitability of these algorithms for solving non-deterministic problems with qualitative preferences, we modify the existing deterministic Rover domain with preferences by incorporating non-deterministic effects on the actions. The instances of this modified domain were used in our experiments, which aimed to show that planning as model checking algorithms can effectively provide solutions for non-deterministic problems with qualitative preferences while considering the policy quality. To the best of our knowledge, this is the first work that addresses non-deterministic problems with preferences while incorporating a preference over the policy quality.

As future work, we aim to investigate the suitability of the α-CTL logic in expressing others types of qualitative preferences and, consequently, if planning as model checking algorithms are appropriated to solve non-deterministic planning problems with these others kinds of preferences. Furthermore, we aim to conduct experiments using different benchmark domains.

References

1. Akers, S.B.: Binary decision diagrams. IEEE Trans. Comput. **27**(06), 509–516 (1978)
2. Baier, J.A., Bacchus, F., McIlraith, S.A.: A heuristic search approach to planning with temporally extended preferences. In: Proceedings of the 20th International Joint Conference on Artifical Intelligence, pp. 1808–1815 (2007)
3. Baier, S., McIlraith, S.A.: Htn planning with preferences. In: 21st Int. Joint Conf. on Artificial Intelligence, pp. 1790–1797 (2009)
4. Bonet, B., De Giacomo, G., Geffner, H., Rubin, S.: Generalized planning: Non-deterministic abstractions and trajectory constraints. arXiv preprint arXiv:1909.12135 [S.l] (2019)
5. Camacho, A., Triantafillou, E., Muise, C., Baier, J.A., McIlraith, S.A.: Non-deterministic planning with temporally extended goals: Ltl over finite and infinite traces. In: Thirty-First AAAI Conference on Artificial Intelligence (2017)
6. Cimatti, A., Pistore, M., Roveri, M., Traverso, P.: Weak, strong, and strong cyclic planning via symbolic model checking. Artif. Intell. **147**(1–2), 35–84 (2003)
7. Edmund, M., Clarke, J., Grumberg, O., Peled, D.A.: Model checking. MIT Press. [S.l] p. 314 (1999)
8. Gerevini, A.E., Haslum, P., Long, D., Saetti, A., Dimopoulos, Y.: Deterministic planning in the fifth international planning competition: Pddl3 and experimental evaluation of the planners. Artif. Intell. **173**(5–6), 619–668 (2009)
9. Ghallab, M., Nau, D., Traverso, P.: Automated Planning: theory and practice. Elsevier (2004)
10. Hoffmann, J.: FF: the fast-forward planning system. AI Mag. **22**(3), 57–57 (2001)
11. Hoffmann, J.: Everything you always wanted to know about planning. In: Annual Conference on Artificial Intelligence, pp. 1–13. Springer (2011)
12. Hsu, C.W., Wah, B.W., Huang, R., Chen, Y.: Constraint partitioning for solving planning problems with trajectory constraints and goal preferences. In: IJCAI, pp. 1924–1929 (2007)
13. Kautz, H., Selman, B.: Unifying sat-based and graph-based planning. In: IJCAI, vol. 99, pp. 318–325 (1999)
14. Kim, J., Banks, C.J., Shah, J.A.: Collaborative planning with encoding of users' high-level strategies. In: Thirty-First AAAI Conference on Artificial Intelligence (2017)
15. Kindler, E.: Safety and liveness properties: a survey. Bull. Eur. Assoc. Theor. Comput. Sci. **53**(268–272), 30 (1994)
16. Lamport, L.: Proving the correctness of multiprocess programs. IEEE Trans. Software Eng. **2**, 125–143 (1977)
17. de Menezes, M.V., de Barros, L.N., do Lago Pereira, S.: Symbolic regression for non-deterministic actions (2014)
18. Menezes, Maria Viviane de: Mudanças em Problemas de Planejamento sem Solução. 2014. 127 f. Ph.D. thesis, Universidade de São Paulo, São Paulo (2014)

19. Muise, C.J., McIlraith, S.A., Beck, C.: Improved non-deterministic planning by exploiting state relevance. In: Twenty-Second International Conference on Automated Planning and Scheduling (2012)
20. Percassi, F., Gerevini, A.E.: On compiling away pddl3 soft trajectory constraints without using automata. In: Proceedings of the International Conference on Automated Planning and Scheduling, vol. 29, pp. 320–328 (2019)
21. Pereira, S.d.L.: Planejamento sob incerteza para metas de alcancabilidade estendidas. Ph.D. thesis, Universidade de São Paulo, São Paulo (2007)
22. Pereira, S.L., de Barros, L.N.: A logic-based agent that plans for extended reachability goals. Auton. Agent. Multi-Agent Syst. **16**(3), 327–344 (2008)
23. Piterman, N., Pnueli, A.: Temporal logic and fair discrete systems. In: Handbook of Model Checking, pp. 27–73 (2018)
24. Rintanen, J.: Regression for classical and nondeterministic planning. In: ECAI 2008, pp. 568–572. IOS Press (2008)
25. Santhanam, G.R., Basu, S., Honavar, V.: Representing and reasoning with qualitative preferences: tools and applications. Synthesis Lectures Artif. Intell. Mach. Learn. **10**(1), 1–154 (2016)
26. Santos, R.M.d.: Especificação de preferências de planos usando metas estendidas na lógica alpha-ctl (2019)
27. Santos, V.B.d., Barros, L.N.d., Pereira, S.d.L., Menezes, M.V.d.: Symbolic fond planning for temporally extended goals. In: Workshop on Knowledge Engineering for Planning and Scheduling (2022)
28. dos Santos, V.M.B., de Barros, L.N., de Menezes, M.V.: Symbolic planning for strong-cyclic policies. In: 2019 8th Brazilian Conference on Intelligent Systems (BRACIS), pp. 168–173. IEEE (2019)

Logic-Based Explanations for Linear Support Vector Classifiers with Reject Option

Francisco Mateus Rocha$^{(\boxtimes)}$, Thiago Alves Rocha ,
Reginaldo Pereira Fernandes Ribeiro , and Ajalmar Rêgo Rocha

Instituto Federal de Educação, Ciência e Tecnologia do Ceará (IFCE), Fortaleza,
Brazil
francisco.mateus.rocha06@aluno.ifce.edu.br

Abstract. Support Vector Classifier (SVC) is a well-known Machine Learning (ML) model for linear classification problems. It can be used in conjunction with a reject option strategy to reject instances that are hard to correctly classify and delegate them to a specialist. This further increases the confidence of the model. Given this, obtaining an explanation of the cause of rejection is important to not blindly trust the obtained results. While most of the related work has developed means to give such explanations for machine learning models, to the best of our knowledge none have done so for when reject option is present. We propose a logic-based approach with formal guarantees on the correctness and minimality of explanations for linear SVCs with reject option. We evaluate our approach by comparing it to Anchors, which is a heuristic algorithm for generating explanations. Obtained results show that our proposed method gives shorter explanations with reduced time cost. Furthermore, although our approach is demonstrated with linear SVCs, it can be easily adapted to other classifiers with reject option, such as neural networks and random forests.

Keywords: Logic-based explainable AI · Support vector machines · Classification with reject option

1 Introduction

It is undeniable that Artificial Intelligence is increasingly being inserted into the daily lives of people [14], influencing the most complex decision taking tasks [22]. Consequently, the most varied classification models in machine learning may come across instances that are difficult to correctly classify, be it due to lack of good data, to feature bias (color, size, gender) [7] or to noise present on the given instances [22].

The SVC is one of the most well-known ML models. The linear Support Vector Machine (SVM), specifically, has been used in a variety of classification problems [2,20,23]. However, this model can also fall into the same previously said pitfalls, failing to give a correct classification. As such, the reject option (RO) strategy depicted in [3] can be used to remedy such cases. Classification with RO is a paradigm that aims to improve reliability in decision support systems by handling the more complex cases and avoiding a higher error rate. It involves withholding and rejecting a sufficiently ambiguous classification result, i.e. when the instance in question is situated too close to the decision boundary of the classifier. These cases are separated to be dealt with in another specialized way, be it through other models or with the assistance of a human specialist [17]. Since classification with RO comprises a set of techniques, in this work we consider the strategy where the classifier is trained as usual and the rejected cases are determined after the training phase. Usually, this process requires finding a trade-off between the costs of misclassifications and rejections.

The linear SVC can already be globally interpreted to a certain extent based on the decision function, where the most important features are often associated with the highest weights [6]. Such analysis, however, is not capable of giving decisive answers on more specific cases [12]. An example of this is depicted and more thoroughly explored in Sect. 3. This gives a margin for questionable explanations, especially when RO is present due to the added level of complexity. Therefore the need for a more thorough, instance-based explanation method.

Then, in this work, we consider instance-based explanations [7]. Specifically, the objective of instance-based explanations is to provide interpretable insights by highlighting relevant features that influenced the output of an ML model on a given instance. By linking the output to specific instances and their features, users can gain a better understanding of the reasons behind the predictions of ML models.

The popularization of concepts of Explainable Artificial Intelligence (XAI) increased efforts to explain most complex models [10]. These have mostly been done through heuristic methods, the more prevalent approach, such as LIME [18], SHAP [15], and Anchors [19]. These tend to be model-agnostic and not able to guarantee a trustworthy explanation with regard to the characteristics of the model or the correctness of the answers [10]. Then, their explanations can be often proven wrong through counterexamples that expose their contradictions, leading to even more doubts regarding how much the ML model can be trusted.

Thus, the importance of trustworthy ML models increases the need for logic-based approaches to explain the decisions made by these models [9,10]. The computation of explanations in these approaches rigorously explores the entire feature space. Due to this, such explanations are provably correct and hold for any point in the space, which therefore makes them trustworthy [7]. Moreover, logic-based approaches can guarantee minimality of explanations. Minimality is important since succinct explanations seem easier to be interpreted by humans. Recent years have seen a surge in research dedicated to investigating logic-based XAI for ML models, such as neural networks, naive Bayes, random forests, decision trees, and boosted trees [1,8,9,11,16].

Due to the importance of reject option and logic-based explainability for trustworthy ML models, this work proposes a logic-based approach to explain linear SVCs with reject option. Given that rejected instances may be further analyzed by specialists, explanations regarding the causes of rejections can reduce the workload of these professionals. Our proposal builds on work from the literature of logic-based explainability for traditional ML models, i.e. without reject option [9]. We compute explanations for linear SVCs with RO by solving a set of logical constraints, specifically, boolean combinations of linear constraints. Despite the fact that we consider a linear SVC with RO in this work, our approach can be easily adapted for other ML models with RO, such as neural networks.

We conducted experiments to compare our approach against Anchors, a heuristic method that generates explanations by locally exploring the feature subspace close to a given instance [19], through six different datasets. The explanations generated by Anchors are expressed as rules designed to highlight important features. Therefore, the resulting explanations of Anchors are similar to the ones computed by our approach. The results of our experiments show that our approach is capable of generating succinct explanations up to 286 times faster than Anchors, in scenarios with and without the presence of reject option.

2   Background

2.1   Machine Learning and Binary Classification Problems

In machine learning, binary classification problems are defined over a set of features $\mathcal{F} = \{f_1, ..., f_n\}$ and a set of two classes $\mathcal{K} = \{-1, +1\}$. In this paper, we consider that each feature $f_i \in \mathcal{F}$ takes its values x_i from the domain of real numbers. Moreover, each feature f_i has an upper bound u_i and a lower bound l_i such that $l_i \leq x_i \leq u_i$. Then, each feature f_i has domain $[l_i, u_i]$. We represent this as a set of domain constraints $D = \{l_1 \leq f_1 \leq u_1, l_2 \leq f_2 \leq u_2, ..., l_n \leq f_n \leq u_n\}$. Besides, the notation $\mathbf{x} = \{f_1 = x_1, f_2 = x_2, ..., f_n = x_n\}$ represents a specific point or instance such that each x_i is in the domain of f_i.

A binary classifier $\mathcal{C}$ is a function that maps elements in the feature space into the set of classes $\mathcal{K}$. For example, $\mathcal{C}$ can map instance $\{f_1 = x_1, f_2 = x_2, ..., f_n = x_n\}$ to class $+1$. Usually, the classifier is obtained by a training process given as input a training set $\{\mathbf{x}_i, y_i\}_{i=1}^{l}$, where $\mathbf{x}_i \in \mathbb{R}^n$ is an input vector or instance and $y_i \in \{-1, +1\}$ is the respective class label. Then, for each input vector $\mathbf{x}_i$, its input values $x_{i,1}, x_{i,2}, ..., x_{i,n}$ are in the domain of corresponding features $f_1, ..., f_n$. A well-known classifier and its training process are presented in Subsect. 2.2.

2.2   Support Vector Machine

The SVM [4] is a supervised machine learning model often used for classification problems. It uses the concept of an optimal separating hyperplane [4] to separate the data. On a $\mathbb{R}^n$ space, such a hyperplane is defined by a set of points $\mathbf{x}$ that satisfies $\mathbf{w}_o \cdot \mathbf{x} + b = 0$, where $\mathbf{w}_o \in \mathbb{R}^n$ is the optimal weight vector, $\mathbf{x} \in \mathbb{R}^n$ is a feature vector with n features and an intercept (bias) $b \in \mathbb{R}$.

A given training set $\{\mathbf{x}_i, y_i\}_{i=1}^k$ is said to be linearly separable if there is $\mathbf{w}_o \in \mathbb{R}^n$ and $b \in \mathbb{R}$ that guarantees the separation between positive and negative class instances without error. In other words, the following inequalities

$$\text{for } i \in \{1, ..., k\}, \begin{cases} \mathbf{w}_o \cdot \mathbf{x}_i + b \geq +1, & \text{if } y_i = +1, \\ \mathbf{w}_o \cdot \mathbf{x}_i + b \leq -1, & \text{if } y_i = -1, \end{cases} \tag{1}$$

must be satisfied to obtain the optimal hyperplane $h_o = \{\mathbf{x} \mid \mathbf{w}_o \cdot \mathbf{x} + b = 0\}$.

A Hard Margin SVM (SVM-HM) [21] can be used when the data is linearly separable and misclassifications are not allowed, maximizing the margin between two hyperplanes, h_+ and h_-, parallel to h_o. There can be no training instances between h_+ and h_-. Once maximizing the margin is similar to minimizing $\frac{1}{2}||\mathbf{w}||^2$, the optimization problem for obtaining the optimal parameters for the SVM can be described as follows:

$$\min_{\mathbf{w},b} \quad \frac{1}{2}||\mathbf{w}||^2 \tag{2}$$
$$\text{s.t.} \quad y_i(\mathbf{w} \cdot \mathbf{x}_i + b) \geq 1, \quad \text{for } i \in \{1, ..., k\}.$$

Thus, the decision function used for classifying input instances $\mathbf{x}$ is defined as $d(\mathbf{x}) = \mathbf{w} \cdot \mathbf{x} + b$, while the predicted label $\hat{y} \in \mathcal{K}$ of an input $\mathbf{x}$ is given by

$$\hat{y} = \begin{cases} +1, & \text{if } d(\mathbf{x}) \geq 0, \\ -1, & \text{if } d(\mathbf{x}) < 0. \end{cases} \tag{3}$$

However, in real-world problems, the training instances from the two classes can not be linearly separated by a hyperplane due to data overlapping. In order to overcome this situation, one must use Soft-Margin SVMs (SVM-SM) in which misclassifications are allowed to happen. To do so, slack variables can be used to relax the constraints of the SVM-HM [4,21]. Given this, the optimization problem for Soft-Margin SVMs is described as

$$\min_{\mathbf{w},b,\xi} \quad \frac{1}{2}||\mathbf{w}||^2 - C\sum_{i=1}^k \xi_i \tag{4}$$
$$\text{s.t.} \quad y_i(\mathbf{w} \cdot \mathbf{x}_i + b) \geq 1 - \xi_i, \quad \text{for } i \in \{1, ..., k\}$$
$$\xi_i \geq 0, \quad \text{for } i \in \{1, ..., k\}$$

where C is a trade-off between $\frac{1}{2}||\mathbf{w}||^2$ and $\sum_i^n \xi_i$. Thus, for a high enough value of C, minimizing the sum of errors while maximizing the separation margin leads toward the optimal hyperplane. This can be the same as the one found through SVM-HM if the data is linearly separable. Moreover, the decision function and predicted label for SVM-HM and SVM-SM are the same.

2.3   Reject Option Classification

Reject option for classification problems, as depicted in [3], is a set of techniques that aim to improve reliability in decision support systems. In the case of a binary problem, it consists in withholding and rejecting a classification result that is ambiguous enough, i.e. when the instance is too close to the decision boundary of the classifier. For the linear SVC, it is when the instance is too close to the separating hyperplane. Then, these rejected instances are analyzed through another classification method or even by a human specialist [17].

In applications where a high degree of reliability is needed and misclassifications can be too costly, rejecting to classify an instance can be more beneficial than the risk of a higher error rate due to wrong classifications [5]. According to [3], the optimal classifiers that best handle such a relation can be achieved by the minimization of $\hat{R}$ through

$$\hat{R} = E + w_r R \tag{5}$$

where R is the ratio of the number of rejected training instances to the number of instances in the entire training dataset; E is the ratio of the number of misclassified instances to the number of all the training instances without including those ones rejected; and w_r is a weight denoting the cost of rejection. A lower w_r gives room for a decreasing error rate at the cost of a higher quantity of rejected instances, with the opposite happening for a higher w_r.

A method is presented in [17] for single, standard binary classifiers that do not provide probabilistic outputs. For SVCs, the proposed rejection techniques are based on the distance of instances to the optimal separating hyperplane. If the distance value is lower than a predefined threshold, then the instance is rejected. As such, a rejection region is determined after the training step of the classifier, with a threshold containing appropriate values being applied to the output of the classifier. Therefore, applying this strategy to a standard binary SVC leads to the following prediction cases:

$$\hat{y} = \begin{cases} +1, & \text{if } f(\mathbf{x}) > t_+, \\ -1, & \text{if } f(\mathbf{x}) < t_-, \\ 0, & \text{otherwise}, \end{cases} \tag{6}$$

where t_+ and t_- are the thresholds for the positive class and negative class, respectively, and 0 is the rejection class. Furthermore, these thresholds are chosen to generate the optimal reject region, which corresponds to the region that minimizes $\hat{R}$ by producing both the ratio of misclassified instances E and the ratio of rejected instances R.

2.4   First-Order Logic

In order to give explanations with guarantees of correctness, we use first-order logic (FOL) [13]. We use quantifier-free first-order formulas over the theory of linear real arithmetic. Then, first-order variables are allowed to take values from the real numbers. Therefore, we consider formulas as defined below:

$$\varphi, \psi := s \mid (\varphi \wedge \psi) \mid (\varphi \vee \psi) \mid (\neg \varphi) \mid (\varphi \rightarrow \psi),$$

$$s := \sum_{i=1}^{n} a_i z_i \le e \mid \sum_{i=1}^{n} a_i z_i < e, \tag{7}$$

such that φ and ψ are quantifier-free first-order formulas over the theory of linear real arithmetic. Moreover, s represents the atomic formulas such that $n \geq 1$, each a_i and e are concrete real numbers, and each z_i is a first-order variable. For example, $(2.5z_1 + 3.1z_2 \geq 6) \wedge (z_1 = 1 \vee z_1 = 2) \wedge (z_1 = 2 \rightarrow z_2 \leq 1.1)$ is a formula by this definition. Observe that we allow standard abbreviations as $\neg(2.5z_1 + 3.1z_2 < 6)$ for $2.5z_1 + 3.1z_2 \geq 6$.

Since we are assuming the semantics of formulas over the domain of real numbers, an *assignment* $\mathcal{A}$ for a formula φ is a mapping from the first-order variables of φ to elements in the domain of real numbers. For instance, $\{z_1 \mapsto 2.3, z_2 \mapsto 1\}$ is an assignment for $(2.5z_1 + 3.1z_2 \geq 6) \wedge (z_1 = 1 \vee z_1 = 2) \wedge (z_1 = 2 \rightarrow z_2 \leq 1.1)$. An assignment $\mathcal{A}$ *satisfies* a formula φ if φ is true under this assignment. For example, $\{z_1 \mapsto 2, z_2 \mapsto 1.05\}$ satisfies the formula in the above example, whereas $\{z_1 \mapsto 2.3, z_2 \mapsto 1\}$ does not satisfy it.

A formula φ is *satisfiable* if there exists a satisfying assignment for φ. Then, the formula in the above example is satisfiable since $\{z_1 \mapsto 2, z_2 \mapsto 1.05\}$ satisfies it. As another example, the formula $(z_1 \geq 2) \wedge (z_1 < 1)$ is unsatisfiable since no assignment satisfies it. The notion of satisfiability can be extended to sets of formulas Γ. A set of first-order formulas is satisfiable if there exists an assignment of values to the variables that makes all the formulas in Γ true simultaneously. For example, $\{(2.5z_1 + 3.1z_2 \geq 6), (z_1 = 1 \vee z_1 = 2), (z_1 = 2 \rightarrow z_2 \leq 1.1)\}$ is satisfiable given that $\{z_1 \mapsto 2, z_2 \mapsto 1.05\}$ jointly satisfies each one of the formulas in the set. It is well known that, for all sets of formulas Γ and all formulas φ and ψ,

$$\Gamma \cup \{\varphi \vee \psi\} \text{ is unsatisfiable iff } \Gamma \cup \{\varphi\} \text{ is unsatisfiable and} \\ \Gamma \cup \{\psi\} \text{ is unsatisfiable.} \tag{8}$$

Given a set Γ of formulas and a formula φ, the notation $\Gamma \models \varphi$ is used to denote *logical consequence*, i.e., each assignment that satisfies all formulas in Γ also satisfies φ. As an illustrative example, let Γ be $\{z_1 = 2, z_2 \geq 1\}$ and φ be $(2.5z_1 + z_2 \geq 5) \wedge (z_1 = 1 \vee z_1 = 2)$. Then, $\Gamma \models \varphi$ since each satisfying assignment for all formulas in Γ is also a satisfying assignment for φ. Moreover, it is widely known that, for all sets of formulas Γ and all formulas φ,

$$\Gamma \models \varphi \text{ iff } \Gamma \cup \{\neg\varphi\} \text{ is unsatisfiable.} \tag{9}$$

For instance, $\{z_1 = 2, z_2 \geq 1, \neg((2.5z_1 + z_2 \geq 5) \wedge (z_1 = 1 \vee z_1 = 2))\}$ has no satisfying assignment since an assignment that satisfies $(z_1 = 2 \wedge z_2 \geq 1)$ also satisfies $(2.5z_1 + z_2 \geq 5) \wedge (z_1 = 1 \vee z_1 = 2)$ and, therefore, does not satisfy $\neg((2.5z_1 + z_2 \geq 5) \wedge (z_1 = 1 \vee z_1 = 2))$.

Finally, we say that two first-order formulas φ and ψ are equivalent if, for each assignment $\mathcal{A}$, both φ and ψ are true under $\mathcal{A}$ or both are false under $\mathcal{A}$. We use the notation $\varphi \equiv \psi$ to represent that φ and ψ are equivalent. For example, $\neg((z_1 + z_2 \leq 2) \wedge z_1 \geq 1)$ is equivalent to $(\neg(z_1 + z_2 \leq 2) \vee \neg(z_1 \geq 1))$. Besides, these formulas are equivalent to $((z_1 + z_2 > 2) \wedge z_1 < 1)$.

2.5   XAI and Related Work

Some ML models are able to be interpreted by nature, such as decision trees [18]. Others, such as neural networks and boosted trees, have harder-to-explain outputs, leading to the use of specific methods to get some degree of explanation [10]. One of the most predominant ways to achieve this is through the use of heuristic methods for generating instance-dependent explanations, which can be defined as a local approach. These analyze and explore the sub-space close to the given instance [7,10].

Some of the most well-known heuristic methods are LIME, SHAP, and Anchors. These approaches are model-agnostic, generating local explanations while not taking into account the instance space as a whole [15,18,19]. This, in turn, allows the explanations to fail when applied, since they can be consistent with instances predicted in different classes. Moreover, they can include irrelevant elements which could be otherwise removed while still maintaining the correctness of the answer [10]. Explanations with irrelevant elements may be harder to understand.

Anchors have been shown as a superior version to LIME, having a better accuracy with the resultant explanations [19]. The explanations obtained by Anchors are decision rules designed to highlight which parts (features) of a given instance are sufficient for a classifier to make a certain prediction while being intuitive and easy to understand. For example, consider an instance $I = \{sepal\_length = 5.0, sepal\_width = 2.3, petal\_length = 3.3, petal\_width = 1.0\}$ of the well known Iris dataset predicted as class *versicolor* by a classifier C. An example of an explanation obtained by Anchors is the decision rule:

IF $sepal\_width \leq 2.8$ **AND** $petal\_length > 1.6$ **THEN** *versicolor*.

However, due to the heuristic characteristic of Anchors, there is still room for wrong explanations, which can lead to cases where, for the same set of given rules, different classes are predicted [10]. Therefore, both explanation correctness and size can be set as of questionable utility if they can not be fully relied upon.

Since heuristic-based explanations are unreliable, logic-based approaches have been proposed recently for a variety of ML models [1,8,9,11,16]. Logic-based approaches can guarantee the correctness of explanations, which makes them trustworthy. Although most of the related work on logic-based explanations focused on traditional ML models, there appears to be a gap in the literature regarding the provision of such explanations specifically in the presence of RO.

Then, in this work we are considering instance-based explanations for linear SVCs with RO, i.e. given a trained linear SVC with RO and an instance $I = \{f_1 = x_1, f_2 = x_2, ..., f_n = x_n\}$ classified as class $c \in \{-1, +1, 0\}$, an *explanation* E is a subset of the instance $E \subseteq I$ sufficient for the prediction c. Therefore, by fixing the values of the features in E, the prediction is guaranteed to be c, regardless of the values of the other features. In other words, features in $I \setminus E$ may take any value in their respective domains without changing the prediction. Moreover, since succinct explanations seem easier to understand, one would like explanations E to be *minimal*, that is, for all subset $E' \subset E$, it follows that E' does not guarantee the same prediction. Informally, a minimal explanation provides an irreducible explanation that is sufficient for guaranteeing the prediction.

3   Linear SVCs May Not Be Instance-Based Interpretable

Efforts have been made to bring explanations to linear SVCs prediction outputs. These have often been done through the analysis of the weights that compose the decision function, where the most important features are associated with the highest weights [6]. However, this may not be enough to enable correct interpretations. Although feature weights can give a rough indication of the overall relevance of features, they do not offer insight into the local decision-making process for a specific set of instances [12].

Assume a binary classification problem where $\mathcal{F} = \{f_1, f_2\}$, and a linear SVC where $\mathbf{w} = \{w_1 = -0.8, w_2 = 2\}$ and $b = 0.05$. The features f_i can take values in range $[0, 1]$. A visual representation is depicted in Fig. 1. Analyzing solely through the values of the weights, it could be assumed that the feature f_2 is determinant and f_1 is not, since $|w_2| > |w_1|$. This would mean that, for any instance, feature f_2 is more important than feature f_1. However, for the instance $\{f_1 = 0.0526, f_2 = 0.3\}$ predicted as class $+1$, feature f_2 is not necessary for the prediction, since for any value of f_2 the class will not change. Therefore, feature f_1 is sufficient for the prediction of this instance. Moreover, feature f_2 is not sufficient for the prediction of this instance, since for $f_1 = 1.0$ and $f_2 = 0.3$ the prediction would change to -1. Therefore, for instance $\{f_1 = 0.0526, f_2 = 0.3\}$, feature f_1 would be determinant and f_2 would not.

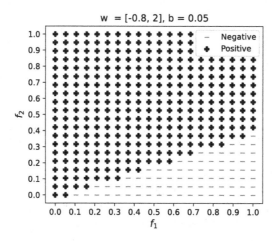

Fig. 1. Classification through features f_1 and f_2.

While it is trivial to observe if feature f_1 and f_2 can be determinant, due to the simplicity of the given decision function, it becomes harder to do so as the number of features increases. Furthermore, the presence of RO adds another layer of complexity to the problem, turning the interpretation task very non-trivial. Hence, explanations given by weights evaluation may give room for incorrect interpretations. This raises the need for an approach to compute explanations with guarantees of correctness for linear SVCs with reject option.

4   Explanations for Linear SVMs with Reject Option

As depicted before, most methods that generate explanations for ML models are heuristic, which can bring issues about how correct and trustworthy the generated answers are. Added that they tend not to take into account whether RO is present, these problems are further aggravated. As such, an approach that guarantees the correctness of explanations for ML models with RO is due.

Our approach is achieved by encoding the linear SVC with RO as a first-order formula, following the ideas of [9] for traditional neural networks, i.e. without RO. Here we need to take into account the reject option and its encoding as a first-order formula. For example, given a trained linear SVC with RO defined by w, b, t_+ and t_- as in Eq. 6, and an instance I classified as class $c = 0$, i.e. in the rejection class, we define the first-order formula P:

$$P = (\sum_{i=1}^{n} w_i f_i + b \leq t_+) \wedge (\sum_{i=1}^{n} w_i f_i + b \geq t_-). \tag{10}$$

Therefore, given an instance $I = \{f_1 = x_1, f_2 = x_2, ..., f_n = x_n\}$ such that its features are defined by $D = \{l_1 \leq f_1 \leq u_1, l_2 \leq f_2 \leq u_2, ..., l_n \leq f_n \leq u_n\}$, an explanation is a subset $E \subseteq I$ such that $E \cup D \models P$. Therefore, each assignment that satisfies all the formulas in $E \cup D$ must also satisfy P. In other words, prediction is guaranteed to be in the rejection class. Moreover, since our goal is to obtain minimal explanations, E must be such that, for all subset $E' \subset E$, it follows that $E' \cup D \not\models P$.

Instances predicted in other classes are treated in a similar way. For example, if instance I is classified as class $c = +1$, P is defined as $\sum_{i=1}^{n} w_i f_i + b > t_+$. It is worth noting that, in these formulas, each f_i is a first-order variable, and b, t_+, t_- and each w_i are concrete real numbers. Note that each first-order variable f_i is analogous to the z_i in Sect. 2.4. Moreover, each first-order variable f_i represents a feature.

Earlier work [9] outlined Algorithm 1 for computing minimal explanations for traditional neural networks, i.e. without RO. By leveraging the insights from this earlier work, we can effectively employ Algorithm 1 to find a minimal explanation for linear SVCs with RO given I, D and P. If instance I is predicted as a class c by the linear SVC with RO, then P is defined accordingly to reflect c. The minimal explanation E of I is calculated by setting E to I and then removing feature by feature from E. For example, given a feature $f_i = x_i$ in E, if $E \setminus \{f_i = x_i\}, D \models P$, then the value x_i of feature f_i is not necessary to ensure the prediction, and then it is removed from E. Otherwise, if $E \setminus \{f_i = x_i\}, D \not\models P$, then $f_i = x_i$ is kept in E since it is necessary for the prediction. This process is performed for all features as described in Algorithm 1. Then, at the end of Algorithm 1, for the values of features in E, the prediction is the same and invariant with respect to the values of the remaining features.

By the result in 9, verifying entailments of the form $(E \setminus \{f_i = x_i\}) \cup D \models P$ can be done by testing whether the set of first-order formulas $(E \setminus \{f_i = x_i\}) \cup D \cup \{\neg P\}$ is unsatisfiable. If P is a conjunction $P_1 \wedge P_2$ as in 10, then $\neg P$ is equivalent to a disjunction $\neg P_1 \vee \neg P_2$. Therefore, we must check whether $(E \setminus \{f_i = x_i\}) \cup D \cup \{\neg P_1\}$ is unsatisfiable and $(E \setminus \{f_i = x_i\}) \cup D \cup \{\neg P_2\}$ is unsatisfiable, by the result in 8.

Algorithm 1. Computing a minimal explanation

Input: instance I predicted as c, domain constraints D, formula P according to c
Output: minimal explanation E
$E \leftarrow I$
for $f_i = x_i \in E$ **do**
    if $(E \setminus \{f_i = x_i\}) \cup D \models P$ **then**
        $E \leftarrow E \setminus \{f_i = x_i\}$
    end if
end for
return E

Moreover, since $(E \setminus \{f_i = x_i\}) \cup D$ is a set of linear constraints and $\neg P_1$ and $\neg P_2$ are linear constraints, the unsatisfiability checkings can be achieved by two queries answered by a linear programming (LP) solver. Therefore, if, for example, the set of linear constraints $(E \setminus \{f_i = x_i\}) \cup D \cup \{\neg P_1\}$ has a solution, then $(E \setminus \{f_i = x_i\}) \cup D \not\models P$. From a computational complexity viewpoint, the linear programming problem is solvable in polynomial time. Then, our approach for computing minimal explanations for linear SVCs with reject option can also be solved in polynomial time. This is achieved by a linear number of calls to an LP solver, which further contributes to the efficiency and feasibility of our approach.

5 Experiments

In this paper, a total of 6 well known datasets are used. The Vertebral Column and the Sonar datasets are available on the UCI machine learning repository[1]. The Pima Indians Diabetes dataset is available on Kaggle[2]. The Iris, the Breast Cancer Wisconsin, and the Wine datasets are available through the scikit-learn package[3]. As a standard procedure, all features were scaled to range $[0, 1]$.

---

[1] https://archive.ics.uci.edu/ml/datasets.php.
[2] https://www.kaggle.com/datasets/uciml/pima-indians-diabetes-database.
[3] https://github.com/scikit-learn/scikit-learn/tree/main/sklearn/datasets/data.

To maintain the $\{-1, +1\}$ class domain in our context of binary classification, the Iris and the Wine dataset (each with three classes) were binarized, changed to *setosa-versus-all* and *class_0-versus-all*, respectively. The other datasets are already binary. The classes were changed to follow values in $\{-1, +1\}$. A summary of the datasets is presented in Table 1.

Table 1. Datasets details.

Dataset	Acronym	$|\mathcal{F}|$	Negative Instances	Positive Instances
Iris	IRIS	4	50	100
Vertebral Column	VRTC	6	100	210
Pima	PIMA	8	500	268
Wine	WINE	13	59	119
Breast Cancer Wisconsin	BRCW	30	212	357
Sonar	SONR	60	111	97

The Classifiers. For each dataset, a linear SVC was trained based on 70% of the original data, to simulate how a model would be trained for usage in real-world problems, with a regularization parameter $C = 1$ and stratified sampling. For finding the rejection region, a value of $w_r = 0.24$ was used together with the decision function outputs based on training data. Although the value itself is not too important for our purposes, it was chosen to ensure that rejections are neither penalized too softly nor too harshly, resulting in some rejected outputs. The selected rejection thresholds were obtained by minimizing $\hat{R}$, as described in Subsect. 2.3.

For defining the values of t_+ and t_-, we determine a range of thresholds $T = \{(t_+^1, t_-^1), ..., (t_+^p, t_-^p)\}$ containing the respective possible candidates. The maximum absolute value for t_+ and t_- in T, respectively, is the highest and lowest output value of the decision function, i.e. the *upper_limit* and the *lower_limit*, based on the training instances. We use decision function output values to achieve a reasonable reject region, that is able to properly reject ambiguous classification results for both known and unknown data. Thus, the attainable thresholds are achieved through $T = \{(i \cdot 0.01 \cdot upper\_limit, i \cdot 0.01 \cdot lower\_limit) \mid i \in \{1, ..., 100\}\}$. Hence, the selected values of t_+ and t_- are the pair that minimizes $\hat{R}$. For each dataset, we obtained the best reject region defined by t_+ and t_-, the test accuracy of the classifier with RO, and the rejection ratio based on test data. The test accuracy of a classifier with RO is the standard test accuracy applied only to non-rejected instances. Afterward, we determine the number of instances per class from both training and test data, i.e. the entire datasets. Table 2 details these results and also the test accuracy of the classifier without reject option.

Observe that the reject region for the IRIS dataset did not return any rejected instances in the dataset. This is likely due to the problem being linearly separable. Therefore, since our experiments rely on explaining the instances present

Table 2. Reject region thresholds using training data. Classifier accuracy without reject option, accuracy with reject option, and rejection ratio for test data. Instances by class for each entire dataset.

Dataset	t_-	t_+	Accuracy w/o RO	Accuracy w/ RO	Rejection	Negative	Rejected	Positive
IRIS	−0.0157	0.0352	100.0%	100.00%	00.00%	50	0	100
VRTC	−0.3334	0.8396	76.34%	89.65%	47.92%	22	139	149
PIMA	−1.1585	0.8312	76.62%	92.20%	59.59%	232	474	62
WINE	−0.0259	0.0243	96.29%	96.29%	00.56%	56	1	121
BRCW	−0.4914	0.2370	97.66%	98.76%	04.02%	190	25	354
SONR	−0.3290	0.2039	74.60%	79.59%	20.00%	76	43	89

in the dataset, this case of the IRIS dataset is treated as if reject option is not present.

The reject region obtained for the WINE dataset did not reject any of the test instances, therefore having no change in accuracy, likely due to the fact that the value chosen for w_r penalizes rejections too harshly. For the other datasets, the presence of reject option lead to a higher accuracy in all cases. This is expected since the rejected instances are not taken into account for evaluation. In addition, there was a substantial increase for both the VRTC and the PIMA dataset at the cost of rejecting roughly 48% and 60% of all the test instances, respectively. Different values for w_r can be used to achieve desirable results, depending on how much a rejection must be penalized in each dataset.

Anchors. We compared our approach against the heuristic method Anchors for computing explanations. Anchors was designed to work with traditional multi-class classification problems, i.e. classifiers without reject option. Since it was proposed as a model-agnostic method for explaining complex models, it should be more than enough for simpler models such as linear SVCs. Moreover, since we are explaining predictions of a linear SVC with RO, we used the Anchors explanation algorithm to treat the classifier with reject option as if it were a traditional classifier with classes in $\{-1, 0, +1\}$.

Our Approach. The prototype implementation of our approach[4] is written in Python and follows Algorithm 1. As an LP solver to check the unsatisfiability of sets of first-order sentences, we used Coin-or Branch-and-Cut (CBC)[5]. Moreover, the solver is accessed via the Python API PuLP[6].

Evaluation. All instances were used for generating explanations, including both the ones used for training and testing. This was done to simulate scenarios closer to real-world problems, where explanations for both previously known (training) and unknown (test) data may prove valuable. Moreover, the main focus is on explaining classification results rather than model evaluation.

---

[4] https://github.com/franciscomateus0119/Logic-based-Explanations-for-Linear-Support-Vector-Classifiers-with-Reject-Option.

[5] https://github.com/coin-or/Cbc.

[6] https://github.com/coin-or/pulp.

Since a per-instance explanation is done for each dataset, both mean elapsed time (in seconds) and the mean of the number of features in explanations are used as the basis for comparison. The results were separated by class to better evaluate and distinguish the rejected class from the others. Moreover, time to compute and size of explanations were chosen as metrics to better reflect the importance of fast and concise explanations when used to assist specialists in better analyzing rejected instances.

5.1 Results

The results based on the running time and the size of explanations are presented in Table 3 and Table 4, respectively. Following the IRIS dataset description in Table 2, there are no results for the rejected class. Our approach has shown to be up to, surprisingly, roughly 286 times faster than Anchors. This happened for the SONR dataset, where the mean elapsed time of our method is 0.86 seconds against 246.14 seconds for Anchors. These results show how much harder it can be for Anchors to give an explanation as the number of features increases. Furthermore, it is worth noting that our approach needs much less time to explain positive, negative, and rejected classes overall.

Table 3. Time comparison between our approach and Anchors (seconds). Less is better.

Dataset	Negative		Rejected		Positive	
	Anchors	Ours	Anchors	Ours	Anchors	Ours
IRIS	0.13 ± 0.045	0.04 ± 0.004	-	-	0.05 ± 0.024	0.04 ± 0.005
VRTC	0.27 ± 0.049	0.07 ± 0.004	0.33 ± 0.147	0.12 ± 0.018	0.28 ± 0.156	0.07 ± 0.002
PIMA	0.85 ± 0.162	0.10 ± 0.001	0.27 ± 0.302	0.16 ± 0.021	0.61 ± 0.108	0.10 ± 0.002
WINE	3.01 ± 0.335	0.17 ± 0.002	1.60 ± 0.000	0.21 ± 0.000	0.38 ± 0.532	0.17 ± 0.002
BRCW	26.58 ± 4.268	0.42 ± 0.015	20.66 ± 3.506	0.55 ± 0.072	5.84 ± 9.686	0.41 ± 0.013
SONR	233.26 ± 70.550	0.86 ± 0.011	257.51 ± 16.267	1.27 ± 0.258	246.14 ± 51.82	0.86 ± 0.019

Table 4. Explanation size comparison between our approach and Anchors. Less is better.

Dataset	Negative		Rejected		Positive	
	Anchors	Ours	Anchors	Ours	Anchors	Ours
IRIS	3.52 ± 0.854	3.00 ± 0.000	-	-	1.89 ± 0.733	2.2 ± 0.400
VRTC	4.95 ± 1.580	5.18 ± 0.574	4.56 ± 1.941	4.99 ± 0.515	4.24 ± 2.077	4.20 ± 1.118
PIMA	7.50 ± 1.534	5.13 ± 0.944	3.02 ± 2.249	4.71 ± 0.897	6.37 ± 2.541	4.85 ± 1.119
WINE	12.98 ± 0.132	7.44 ± 1.252	13.00 ± 0.000	12.00 ± 0.000	2.80 ± 2.803	7.04 ± 1.605
BRCW	29.17 ± 4.372	22.29 ± 3.739	29.64 ± 1.763	27.08 ± 1.016	7.70 ± 11.024	23.08 ± 2.208
SONR	51.10 ± 20.075	50.39 ± 4.283	57.95 ± 9.233	56.16 ± 1.413	54.86 ± 14.625	51.78 ± 1.413

While our approach obtained more extensive explanations in certain cases, it also demonstrated the ability to be more succinct, generating explanations that were up to 43% smaller in other cases. In addition, our method is able to maintain the correctness of explanations. This highlights an expressive advantage of our approach over the heuristic nature of Anchors. While Anchors may provide explanations with fewer features in some cases, this is achieved due to the lack of formal guarantees on correctness.

It is important to note that cases similar to what we discussed in Sect. 3 occurred for the IRIS dataset. Through weights $\mathbf{w} = \{w_1 = 0.8664049, w_2 = -1.42027753, w_3 = 2.18870793, w_4 = 1.7984087\}$ and $b = -0.77064659$, obtained from the trained SVC, it could be assumed that feature f_1 might possibly be not determinant for the classification. However, for the instance $\{f_1 = 0.05555556, f_2 = 0.05833333, f_3 = 0.05084746, f_4 = 0.08333333\}$ in class -1 for example, feature f_1 was in the explanation, while feature f_2 was not present. Furthermore, similar cases happened where f_2 was in the explanation while f_4 was not, even though $|w_4| > |w_2|$.

Similar occurrences are present in other datasets, reinforcing that the points discussed in Sect. 3 are not uncommon. As one more illustrative example, consider the linear SVC with reject option trained on the VRTC dataset such that $\mathbf{w} = \{w_1 = 0.72863148, w_2 = 1.97781269, w_3 = 0.85680605, w_4 = -0.32466632, w_5 = -3.42937211, w_6 = 2.43522629\}$, $b = 1.10008469$, $t_- = -0.3334$ and $t_+ = 0.8396$. For the instance $\{f_1 = 0.25125386, f_2 = 0.4244373, f_3 = 0.7214483, f_4 = 0.20007403, f_5 = 0.71932466, f_6 = 0.15363128\}$ in the reject class, feature f_3 was not in the explanation, while f_4 is present in the explanation, despite the fact that $|w_3| > |w_4|$. This is substantial for our approach since it demonstrates its capability of finding such cases.

6   Conclusions

In this paper, we propose a logic-based approach to generate minimal explanations for a classifier with RO while guaranteeing correctness. A trained linear SVC with RO is used as a target model for explainability. Our approach is rooted in earlier work on computing minimal explanations for standard machine learning models without RO. Therefore, we encode the task of computing explanations for linear SVCs with RO as a logical entailment problem. Moreover, we use an LP solver for checking the entailment since all first-order sentences are linear constraints with real variables, and at most one disjunction occurs.

Our method is compared against Anchors, one of the most well-known heuristic methods, through six different datasets. Anchors was originally proposed as a model-agnostic method, and therefore expected to perform well for simpler models as the linear SVC. Nonetheless, we found that not only our approach takes considerably less time than Anchors, but it also has reduced size of explanations for many instances. Our approach achieved astonishing results in terms of efficiency, surpassing Anchors by an impressive factor of up to approximately 286 times.

However, our approach has limitations. Since we guarantee correctness, the size of explanations may considerably increase, as observed for some cases in Table 4. Moreover, since we only assure minimal explanations, i.e. irreducible explanations that are sufficient for guaranteeing the prediction, it is possible that there are explanations with fewer features that our method does not guarantee to find out.

Our approach can be further improved in future work. For example, it can be easily adapted to other classifiers with RO, such as neural networks and random forests. Another improvement is to provide more general explanations. For example, its possible that some features in an explanation may change values and still lead to the same prediction. Then, a more general explanation may consider a range of values for each feature, such that the prediction does not change for any instances whose feature values fall within such ranges. This may enable a more comprehensive understanding of the model.

Acknowledgments. The authors thank FUNCAP and CNPq for partially supporting our research work.

References

1. Audemard, G., Bellart, S., Bounia, L., Koriche, F., Lagniez, J.M., Marquis, P.: On preferred abductive explanations for decision trees and random forests. In: 31st IJCAI, pp. 643–650 (2022)
2. Chlaoua, R., Meraoumia, A., Aiadi, K.E., Korichi, M.: Deep learning for finger-knuckle-print identification system based on PCANet and SVM classifier. Evol. Syst. **10**(2), 261–272 (2019)
3. Chow, C.: On optimum recognition error and reject tradeoff. IEEE Trans. Inf. Theory **16**(1), 41–46 (1970)
4. Cortes, C., Vapnik, V.: Support-vector networks. Mach. Learn. **20**, 273–297 (1995)
5. De Oliveira, A.C., Gomes, J.P.P., Neto, A.R.R., de Souza, A.H.: Efficient minimal learning machines with reject option. In: 5th BRACIS, pp. 397–402. IEEE (2016)
6. Guyon, I., Weston, J., Barnhill, S., Vapnik, V.: Gene selection for cancer classification using support vector machines. Mach. Learn. **46**, 389–422 (2002)
7. Ignatiev, A.: Towards trustable explainable AI. In: IJCAI, pp. 5154–5158 (2020)
8. Ignatiev, A., Izza, Y., Stuckey, P.J., Marques-Silva, J.: Using MaxSAT for efficient explanations of tree ensembles. In: AAAI, vol. 36, pp. 3776–3785 (2022)
9. Ignatiev, A., Narodytska, N., Marques-Silva, J.: Abduction-based explanations for machine learning models. In: AAAI, vol. 33, pp. 1511–1519 (2019)
10. Ignatiev, A., Narodytska, N., Marques-Silva, J.: On formal reasoning about explanations. In: RCRA (2020)
11. Izza, Y., Ignatiev, A., Marques-Silva, J.: On explaining decision trees. arXiv preprint arXiv:2010.11034 (2020)
12. Krause, J., Dasgupta, A., Swartz, J., Aphinyanaphongs, Y., Bertini, E.: A workflow for visual diagnostics of binary classifiers using instance-level explanations (2017)
13. Kroening, D., Strichman, O.: Decision Procedures. Springer, Heidelberg (2016). https://doi.org/10.1007/978-3-662-50497-0
14. LeCun, Y., Bengio, Y., Hinton, G.: Deep learning. Nature **521**(7553), 436–444 (2015)

15. Lundberg, S.M., Lee, S.I.: A unified approach to interpreting model predictions. In: Advances in Neural Information Processing Systems, vol. 30 (2017)
16. Marques-Silva, J., Gerspacher, T., Cooper, M., Ignatiev, A., Narodytska, N.: Explaining Naive Bayes and other linear classifiers with polynomial time and delay. NeurIPS **33**, 20590–20600 (2020)
17. Mesquita, D.P., Rocha, L.S., Gomes, J.P.P., Rocha Neto, A.R.: Classification with reject option for software defect prediction. Appl. Soft Comput. **49**, 1085–1093 (2016)
18. Ribeiro, M.T., Singh, S., Guestrin, C.: "Why should I trust you?" Explaining the predictions of any classifier. In: 22nd ACM SIGKDD, pp. 1135–1144 (2016)
19. Ribeiro, M.T., Singh, S., Guestrin, C.: Anchors: high-precision model-agnostic explanations. In: AAAI, vol. 32 (2018)
20. Richhariya, B., Tanveer, M., Rashid, A.H., Initiative, A.D.N., et al.: Diagnosis of Alzheimer's disease using universum support vector machine based recursive feature elimination (USVM-RFE). Biomed. Signal Process. Control **59**, 101903 (2020)
21. Vapnik, V.: Statistical Learning Theory. Wiley (1998)
22. Weld, D.S., Bansal, G.: Intelligible artificial intelligence. CoRR abs/1803.04263 (2018)
23. Yang, W., Si, Y., Wang, D., Guo, B.: Automatic recognition of arrhythmia based on principal component analysis network and linear support vector machine. Comput. Biol. Med. **101**, 22–32 (2018)

Resource Allocation and Planning

The Multi-attribute Fairer Cover Problem

Ana Paula S. Dantas[(✉)] [ID], Gabriel Bianchin de Oliveira[ID], Helio Pedrini[ID],
Cid C. de Souza[ID], and Zanoni Dias[ID]

Institute of Computing, University of Campinas, Campinas 13083-852, Brazil
{anadantas,gabriel.oliveira,helio,cid,zanoni}@ic.unicamp.br

Abstract. Alongside the increased use of algorithms as decision making tools, there have been an increase of cases where minority classes have been harmed. This gives rise to study of algorithmic fairness that deals with how to include fairness aspects in the design of algorithms. With this in mind, we define a new problem of fair coverage called Multi-Attribute Fairer Cover, that deals with the task of selecting a subset for training that is as fair as possible. We applied our method to the age regression model using instances from the UTKFace dataset. We also present computational experiments for an Integer Linear Programming model and for the age regression model. The experiments showed significant reduction on the error of the regression model when compared to a random selection.

Keywords: Algorithmic Fairness · Fair Coverage · Integer Linear Programming

1 Introduction

In recent years, there have been a surge of cases where algorithms applications have been linked to some sort of discrimination towards a minority group of people. These cases are known as algorithmic injustice, or even algorithmic racism, when the discrimination is based on race. Silva [12] has compiled a timeline showing that illustrates the increasing frequency of news reports on such cases. Algorithms have been extensively used in the decision making process. The level of human participation varies in these decisions. For example, O'Neil [10] reports on the case where American teachers are evaluated by a proprietary algorithms and are required to keep a certain score. Although these scores can affect their careers, the teachers have no idea how the scores are assigned by the algorithm. In this example, the algorithms are helping on the decision making process, but one extreme case can be found in the report by Soper [13]. The author showcases how some human-resources operations from an online retailer have supposedly been transferred to algorithms. The report discusses how the decisions made by these algorithms are affecting the lives of their collaborators, culminating even on wrongful terminations.

With the increased use of these tools, the scientific community has been reacting and giving attention to the potential problems that the indiscriminate

M. C. Naldi and R. A. C. Bianchi (Eds.): BRACIS 2023, LNAI 14195, pp. 163–177, 2023.
https://doi.org/10.1007/978-3-031-45368-7_11

use of these tools may cause. For instance, Chung [2] has presented an extensive research of cases of algorithmic racism. They pointed out among others things that we need to address the issues of fairness during the design of the algorithm, as well as explicitly including protected attributes in auditing processes in order to properly identify bias. It has been shown that including these attributes even on the training phase can improve accuracy [6,8].

There are several approaches to insert fairness into models and mitigate the problems a biased model may cause. For example, Galhotra et al. [4] investigated the selection of features in order to ensure a fairer model, paying special attention to sensitive characteristics. Roh et al. [11] presented a greedy algorithm to select a fair subset during the sampling step in the batch selection. Asudeh et al. [1] proposed that fairness be considered in a the coverage problem. They defined a combinatorial optimization problem that, given a universe set of colored elements and a family of subsets, aims to select k subsets such that the number of covered elements is the same for each color and the sum of the weight of these elements is maximum, showing that this problem is NP-hard. Based on this interpretation of the problem, Dantas et al. [3] proposed the Fairer Cover problem, that removes restriction of having the same number of covered elements from each class and puts this in the objective function. Therefore, the goal is to find a cover that is as fair as possible. Both of these works consider fairness as a part of the algorithm design, instead of just optimizing models regardless of the impact in the different classes.

In this work, we conjecture that a fair training subset will improve the quality of a machine learning model, specifically of a regression model. We present a new variation and application of a fair coverage problem, and we define a generalization of the Fairer Cover problem introduced by Dantas et al. [3], which we call as Multi-Attribute Fairer Cover. We propose a new Integer Linear Programming (ILP) Model to solve instances of this problem and derive instances from the facial dataset UTKFaces. We also present computational experiments for the ILP model and attest the improvement of the age regression model when using a fair subset in the training. The experiments showed significant reduction on the error of the regression model.

The remainder of this paper is organized as follows. In Sect. 2, we explain our methodology, starting with a few notations, followed by a description of the dataset used in the computational experiments and a detailing of our method to solve the fair coverage problem. In Sect. 3, we present and discuss the results of our computational experiments. Lastly, in Sect. 4, we outline our conclusions and point out future work.

2   Methodology

In order to address the task of selecting a fairer subset as a combinatorial optimization problem, it is essential to begin by introducing a few definitions. Let $\mathcal{U}$ be the universe set of u_j elements. Let $\mathcal{S}$ be a family of subsets $\mathcal{U}$. Lastly, let $\mathcal{C}$ represent a set of χ colors and let $C : \mathcal{U} \to \mathcal{C}$ be a function that assigns one color to each u_j. We call a k-cover, a selection of k subsets from $\mathcal{S}$, and we say that an element u_j is covered if there is a subset $\mathcal{S}_\ell$ from a k-cover that contains u_j. A k-cover is considered *fair* if the number of covered elements is the same for every color.

We refer to as *tabular* a dataset that is represented as a matrix of where the rows are the samples and the columns are the attributes. We consider that the cell from the i-th row and the j-th column of the dataset contains exactly one of the possible values for the attribute described by the column j. That is, if the column j represents eye color and the possible values for eye color are either `blue` or `brown`, the cell from the i-th row and the j-th column has exactly one of these two values. We can use this type of dataset to create a combinatorial optimization problem by translating it into a family of subsets and treating the task of selecting a fair subset of size k for training as the task of finding a fair k-cover.

To translate a tabular dataset into a family of subsets, we start by representing each cell of the dataset by a distinct element u_j and the universe set $\mathcal{U}$ is the set containing every cell. We create subsets $\mathcal{S}_\ell$ of universe $\mathcal{U}$ based on the rows of the dataset, such that each row is represented by one subset $\mathcal{S}_\ell$. For this paper, let us consider that every attribute is discrete. This way, each possible value for an attribute of the dataset can be mapped to a finite set of colors. We color the elements of $\mathcal{U}$ based on the class of the attribute contained in the cell that it represents. Note that, since a subset $\mathcal{S}_\ell$ corresponds to a row of the dataset, every subset $\mathcal{S}_\ell$ contains one element for each attribute (column) and they are all of different colors. We assume that each row of a tabular dataset is complete and there is no missing data. Under this assumption, we have that every row of the dataset has the same number of valid attributes, as do the subsets representing them.

Asudeh et al. [1] defined an optimization problem called Fair Maximum Coverage – FMC. In this problem, each element has a weight and there is a family $\mathcal{S}$ of subsets $\mathcal{S}_\ell$ of $\mathcal{U}$. The goal of the FMC is to select k subsets $\mathcal{S}_\ell$ such that the k-cover formed by these subsets is fair and has a maximum sum of the weights. As pointed out by Dantas et al. [3], the interpretation of Asudeh et al. may result in a limited solution pool, often leading to infeasible instances.

Dantas et al. [3] defined a new problem called Fairer Cover – FC, based on the FMC. This problem relaxes the fairness aspect, allowing for solutions to have a different number of covered elements for each color. This was achieved by removing the need to have a fair coverage as prerequisite and tackling this aspect in the objective function. The FC problem is given in Definition 2.1, and its goal is to find a k-cover that is as fair as possible and covers at least s elements, with k and s being given in the instance. The FC was applied to the Human Protein Atlas dataset obtained from the "Human Protein Atlas - Single Cell Classification". This dataset has a very distinct characteristic: there is a single set of attributes, which are the labels of the identified proteins. We expand the work presented in Dantas et al. [3] to deal with multiple attributes using as an example the UTKFace dataset, which will be presented in the following subsection.

Definition 2.1. Fairer Cover – FC
Input: A universe set $\mathcal{U}$, a family $\mathcal{S}$ of subsets of $\mathcal{U}$, a coloring function C, and positive integers s and k.
Objective: Find a k-cover that is as fair as possible and covers at least s elements.

By dealing with multiple classes of attributes, it becomes necessary to divide the colors that represent the values of the attributes into groups. This happens because each class of attribute may have different sizes and meanings. For example, suppose we have a tabular dataset containing the attributes eye color and age group, where the eye color can be either blue or brown and age group can be either child, teenager or adult. In addition, suppose that each row is made up of a pair of eye color and age group. Each of the values blue, brown, child, teenager and adult will be represented by one of five distinct colors. To obtain a fair coverage of this dataset we would need a subset of rows where each of the five colors has the exact same number of elements. We should not aim to have the same number of covered elements representing adult and blue because they are of different types, therefore FC and FMC are not suitable for this kinds of instances. Also, it is worth noting that in the case of the FMC, where the equality of the covered classes is a characteristic of the solution, the instance may become infeasible.

Given a list of color groups $\mathcal{G} = \{G_1, G_2, \ldots, G_h\}$ that puts each color in exactly one group, that is, $\bigcup_{a=1}^{h} G_a = C$ and $G_a \cap G_b = \emptyset$ for all $a, b \in \{1, 2, \ldots, h\}$ with $a \neq b$. We propose a generalized version of the FC called Multi-Attribute Fairer Cover – MAFC as shown in Definition 2.2. This version of the problem considers the fairness among the colors of the group, that is, our goal is to have a selection that is as fair as possible when analyzed in each group of colors individually. Note that, if $h = 1$, we have a problem equivalent to FC.

Definition 2.2. Multi-Attribute Fairer Cover – MAFC
Input: A universe set $\mathcal{U}$, a family $\mathcal{S}$ of subsets of $\mathcal{U}$, a coloring function C, a list of h of color groups $\mathcal{G}$, and a positive integer k.
Objective: Find the k-cover that is as fair as possible in each group of $\mathcal{G}$.

The definition of the MAFC differs from that of FC in two main points, one is the already discussed need to consider the fairness among the groups of elements and the other is the removal of parameter s that specifies the minimum number of elements to be covered. In an instance where every subset $\mathcal{S}_\ell$ has the same size t, a k-cover always covers exactly $s \times t$ elements. Therefore, there is no need to specify a minimum number of covered elements.

In the remainder of this section, we describe the dataset used in our experiments, as well as the proposed method for selecting a fair coverage and the model

Table 1. Number of images in training, validation, and test sets.

Set	Training	Validation	Test
Number of Images	15,172	3,794	4,742

employed for the age regression task. We also present the evaluation metrics and the setup used for our experiments.

2.1 UTKFace Dataset

To evaluate our proposed method for a fair selection of samples, we use UTKFace dataset [14] in our experiments. This database consists of face images of people with different characteristics, which includes age (integer value ranging from 1 to 116), gender (Male or Female), and race (White, Black, Asian, Indian, and Other, that is, Hispanic, Latino, and Middle Eastern). It is worth noting that even though the age ranges from 1 to 116, the dataset does not contain samples for every value. In fact, there are only 104 distinct values.

From the original UTKFace dataset, we split it into 64%, 16%, and 20% for training, validation, and test sets, respectively. Table 1 presents the number of samples of each set of our split from UTKFace.

The frequencies of gender and race of training, validation, and test sets are presented in Figs. 1a and 1b. Concerning gender, there are more images of males than females in all three sets, although this difference is less accentuated. Considering race, the White class is more frequent, followed by Black, Indian, and Asian, and Other (Hispanic, Latino, and Middle Eastern) have the lowest representation in the sets. It is worth noting that the most frequent class is roughly double the second most frequent class, which showcases how unbalanced the dataset is.

2.2 Proposed Method

In this subsection, we describe the proposed method for fair age regression, divided into ILP and age regression models.

The pipeline of the proposed method, with ILP and age regression models, is illustrated in Fig. 2. From the original UTKFace images and annotations, we created instances for the MAFC combinatorial problem. We then solved each instance of the MAFC using an ILP model. From the ILP solution, we obtained a fair training subset considering age, gender, and race. Lastly, we trained and evaluated the predictions of the age regression model.

ILP Model. In this section we present an ILP formulation that models the MAFC problem. The model is described by Eqs. (1a) - (1g).

The model uses three decision variables: x_j, y_ℓ and Q_c. The first two are binary and the last one is a positive integer. The variable x_j indicates whether

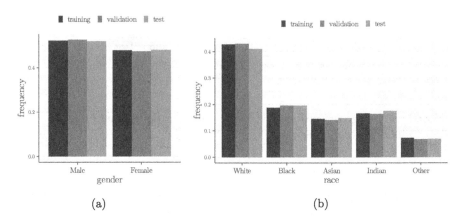

Fig. 1. Gender and race distribution on training, validation, and test sets.

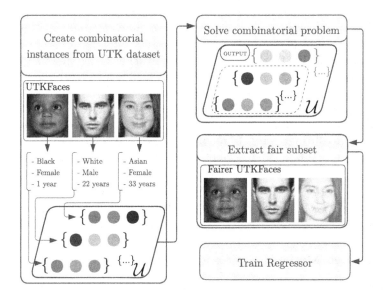

Fig. 2. Pipeline of the proposed method.

the element u_j is covered and the variable y_ℓ indicates whether the subset $\mathcal{S}_\ell$ was selected to be a part of the k-cover. The variable Q_c indicates the difference in the number of covered elements that have color c and the ideal number of covered elements that have color c. The ideal number of each color is given by the constants I_c, the estimation for this value can be adjusted depending on the application.

The model also uses the constants h to represent the number of color groups, m to indicate the number of elements in $\mathcal{U}$, k to indicate the number of subsets to be selected, n to define total number of subsets, and χ to specify the total number of colors used in the instance. We also define subsets C_c, which are the subsets of the elements u_j that are colored with the color c.

Considering the concept of fairness as the equality among the classes in the cover and the objective of the MAFC as it is stated in Definition 2.2, we can consider the task of finding a k-cover that is as fair as possible as the minimization of the differences among the number of covered elements in each class. Dantas et $al.$ [3] modeled this directly in the objective function as the minimization of the difference in the number of color of each pair of colors. This approach might not be viable when the number of color increases, if we have for example 100 colors it would be necessary to minimize the sum of $\frac{100 \times 99}{2} = 4950$ pairs of colors. Therefore, we choose to minimize the difference in the number covered elements for a given color c to its ideal value I_c. The ideal value I_c can be easily estimated, since we know the number of colors in each group of colors and the number of selected subsets. In the case of a tabular dataset we can find the exact number I_c by simply dividing k by the number of colors in the group G_a to which c belongs. Going back to the example given in Sect. 2, the ideal number of elements covered representing the attribute blue is $\frac{k}{2}$, since the class in which blue is contained has size two. With these definitions, the ILP model reads:

$$\min \quad \sum_{c \in G_1} Q_c + \sum_{c \in G_2} Q_c + \cdots + \sum_{c \in G_h} Q_c \tag{1a}$$

$$\text{s. t.} \quad x_j \leq \sum_{\ell \mid u_j \in S_\ell} y_\ell \quad \forall j \in \{1, 2, \ldots, m\} \tag{1b}$$

$$y_\ell \leq x_j \quad \forall u_j \in S_\ell, \forall S_\ell \in S \tag{1c}$$

$$\sum_{\ell=1}^{n} y_\ell = k \tag{1d}$$

$$\sum_{u_j \in C_c} x_j - I_c \leq Q_c \quad \forall c \in C \tag{1e}$$

$$I_c - \sum_{u_j \in C_c} x_j \leq Q_c \quad \forall c \in C \tag{1f}$$

$$x \in \mathbb{B}^m, \ y \in \mathbb{B}^n, \ Q \in \mathbb{Z}_+^\chi \tag{1g}$$

The objective function that reflects this modeling decision is given in Eq. (1a). The remaining equations describe the restrictions of the model, through which we specify the characteristics of a solution.

Equation (1b) indicates that an element u_j can only be considered covered ($x_j = 1$), if there is at least one subset S_ℓ such that $u_j \in S_\ell$ and this subset is selected to be in the cover (i.e., $y_\ell = 1$). Equation (1c) specifies that if a subset S_ℓ is selected to be in the cover ($y_\ell = 1$), then every element u_j of u_ℓ is considered to be covered, along with the previous set of restrictions we guarantee the basic characteristics of a k-cover. Equation (1d) restricts that exactly k subsets must be selected to form a k-cover.

Equations (1e) and (1f) are complementary and together they are used to define the value of the decision variable Q_c. Note that the sums of x_j are used to indicate the number of covered elements with color c. Suppose that this sum is smaller than the ideal value of I_c. Then, Equation (1e) states that Q_c must be larger than a negative value and, since Q_c must be positive, this restriction is satisfied by any possible value of Q_c. Now in the same scenario, Equation (1f) establishes that Q_c must be larger than a positive value. Although there is no positive upper bound on the variable Q_c, we know that in an optimal solution this value will not be greater than the value of the difference. This is true because the objective function is a minimization on the value of Q_c. Lastly, Equation (1g) indicates the domain of the decision variables.

Age Regression Model. For evaluation of ILP model and comparison with random selection of sub-samples of the original UTKFace dataset, we fine-tune ResNet50 architecture [5] for regression task about age prediction of each image. To do so, we remove the original output layer from ImageNet database [7] and we include the regression output layer, which consists of a single neuron, in order to predict a single float value corresponding to the age of each sample.

As each image from the dataset can have different lengths of pixel dimensions, that is, width and height sizes, we resize each image to have the same shape. With this preprocessing step, all images are transformed to 128 pixels of width and 128 pixels of height.

2.3 Evaluation Metrics

To assess our proposed method and comparison with a random selection of images, we apply Balanced MAE during our experiments. Balanced MAE is derived from Mean Absolute Error (MAE), which is presented in Eq. 2, where n indicates the number of samples, y_i and $\hat{y}_i$ represent the ground-truth and predicted values for sample i, respectively.

$$\text{MAE} = \frac{1}{n} \times \sum_{i=1}^{n} |y_i - \hat{y}_i| \tag{2}$$

Based on the MAE score for each age, that is, the value calculated using only the images from a specific age, we calculate Balanced MAE (BMAE), as shown in Eq. 3. In this equation, c represents the number of ages evaluated at the current stage and MAE_j indicates the MAE score for a specific age j.

$$\text{BMAE} = \frac{1}{c} \times \sum_{j=1}^{c} \text{MAE}_j \tag{3}$$

It is important to note that BMAE gives the same importance for each age, that is, independently of the number of samples of a specific age. For example, the age with the majority number of samples, it has the same impact as any other age (for example, the age with the minority number of samples). Since we are aiming for a model that has a fair training in several aspects, including the coverage of the ages, this score is more appropriate when compared to a simple MAE, where we would have that errors in ages that are underrepresented would have a lesser impact on the overall metric.

3   Computational Results

In this section, we present and discuss the results of our computational experiments. Our experiments are divided into two phases: experimentation with the ILP model and experimentation with the age regression model. In both phases, we analyze the concepts of fairness and how they are manifested in each case.

3.1   Computational Setups

Before we present our computational results, let us first present the setups used in each phase of our experiments.

ILP Setup: We used the IBM CPLEX Studio (version 12.8) to solve the combinatorial optimization instances described in the previous sections. We used the programming language C++ to implement the model described in Sect. 2 with compiler g++ (version 11.3.0), flags C++19 and -O3. The experiments reported in Sect. 3.2 were run in a Ubuntu 22.04.2 LTS desktop with Intel® Core™ i7-7700K and 32 GB of RAM.

Age Regression Model Setup: The age regression model was implemented using the programming language Python and TensorFlow library. To train and evaluate our model, we used Google Colaboratory virtual environment.

During the training step, the model was trained during 200 epochs, with early stopping technique with patience of 20 epochs and reduction of learning rate by a factor of 10^{-1} if the model did not improved the results on validation set after 10 epochs. For the optimization process, we employed MAE as loss function and AdamW optimizer [9] with initial learning rate equal to 10^{-4}. It is worth noting that we tuned these parameter with experiments on the full training dataset, without any fairness adjustments. We followed this configuration for both fair and random selection of images for age regression models.

3.2   ILP Model Results

With the model implemented, we started the experiments with the different instances created from the UTKFace dataset. To create the instances, we considered the annotations of gender, age and race provided in the dataset to be a tabular set with three attributes and 15172 samples, that is, we considered only the images from training split of the original dataset.

For each possible value of the three attributes, we assigned a distinct color and created the sets corresponding to the each sample. Note that each subset of elements has size three, since it contains one element for each attribute. We specified four arbitrary values for k to use in the instances: $1000, 2000, 5000, 10000$. These numbers will specify the sizes of the cover.

After defining the elements, the sets, the coloring and the values for the parameter k, we needed to estimate the ideal values for each color. In a perfect scenario, the ideal value I_c for a given color c would be defined as $\frac{k}{|G_a|}$, where $c \in G_a$, but from analyzing the dataset, we saw that this cannot always be achieved since there were sub-represented colors in the dataset. For these cases we devised a simple procedure to compute the value I_c for each color group G_a.

We start the procedure by defining a temporary T_i ideal value as $\frac{k}{|G_a|}$. For every color in $c \in G_a$ that has a frequency smaller than T_i, we set I_c as the value of the frequency of color c, next we update $T_i = \frac{k-k'}{|G_a|-n'}$, where k' is the sum of frequency of the colors that just had their ideal value I_c set and n' is the number of said colors. We repeat this until every color has a frequency of at least T_i. The remaining colors have their I_c value set to T_i. By doing this, we set realistic ideal values for each class. That is, the classes that have a frequency lower than the ideal, have ideal value set to their respective frequencies because the solution has to respect this limitation. Then, the remaining classes have a new ideal value that accounts for the elements that are underrepresented.

In our initial experiments, we noticed a tendency of the model's solutions to have a discrepancy in the covering when considering the combinations of the attributes. In Fig. 3, we show the distribution of gender (Fig. 3a) and race (Fig. 3b) for each of the ages in the training split of the dataset. This figure considers that the parameter k is set to 1000. Since our model is considering only the equality of each attribute separately, we had several cases where all the covered images of an given age were just from the class Male or just the class Female, as shown in Fig. 3a. In Fig. 3b, most ages are predominantly from a single race, in special the race Other is predominant in the ages 20 through 40.

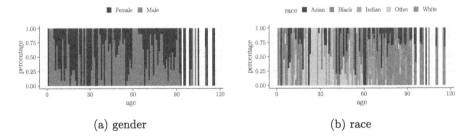

(a) gender   (b) race

Fig. 3. Percentage of gender and race in each of age values in the solution given by the ILP model for the instance with $k = 1000$.

This showed a possible problem for our model, that points towards a necessity to consider also the pairings of the attributes. This aspect could be considered either in the model for the problem or in the instance. For example, we could

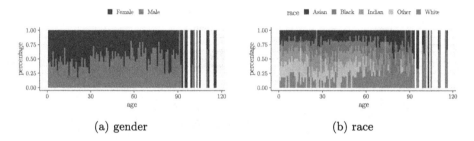

(a) gender   (b) race

Fig. 4. Percentage of gender and race in each of age values in the solution given by the
ILP model for the instance with $k = 1000$ after adding the fourth element.

add to model restrictions that decrease the difference of each pair formed by
one color from group $\mathcal{G}_a$ and one color from group $\mathcal{G}_b$. We choose to contour
this problem by inserting a new element to each set, and color it with colors
from a fourth group. This new element was colored based on the three original attributes combined. That is, if the sample corresponded was annotated
as {male, black, 3 years old}, the new element received the color c and every
other sample that had this exact three characteristics also received color c in
the fourth element. Note that if any of the three characteristics is different, the
color of the fourth element would also be different.

This approach increases the size of the instance, since the new element added
to each set has the potential to add 2 genders ×5 races ×104 ages = 1040 new
colors. This was not the case because the dataset does not have a sample for
each of possible triples formed by the attributes. The increase on the instance
size (more colors and more elements) is an aspect that cannot be avoided, either
we increase the instance or increase the number of restrictions in the modeling.
By creating this new element we avoid modifying the model and the problem
definition. We note that we could also have considered pairs of attributes to color
new elements in the set, but this would result in a much larger instance.

With these modifications, we executed the model with the new instances
and we can see the impact of these changes in Fig. 4. We now have a more even
distribution of the both gender and races in each age. It is worth noting that
the dataset itself, has some gaps that will reflect in the possibility of equality
in our problem. For example, the samples for ages larger than 100 years are all
of female subjects. Over the age of 80 years, we do not have samples of the race
Other, which tends to skew the selection of the classes that are present in these
ages. That is, despite the race White being the most frequent in every age, its
presence becomes more dominant on the larger ages because there are no other
options to select the sets and also maintain the aspect of fairness in the colors
referent to the classes of ages.

The distribution of each attribute in the solutions given by the model can be
seen in Fig. 5. The first two graphs show the frequency of the elements of each
class that were covered in the solution. In Fig. 5a, the values for the gender are
very close to the ideal for each of the values of k, meaning that for this attribute

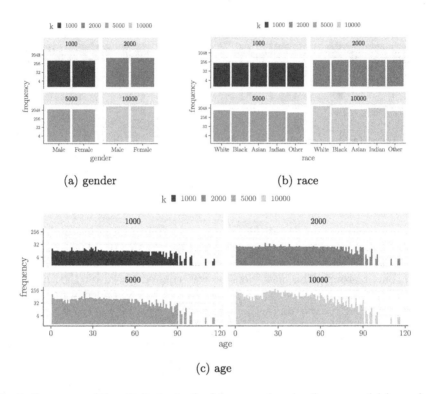

Fig. 5. Frequency of the attributes in the fair cover given by the ILP model for each of the sizes of the parameter k.

the model found a near perfectly fair solution. Now, in Figs. 5b and 5c, we start to see small variations on the number of covered elements of each age and class of race. Such variations are accentuated when k is larger, as expected, because the least common attributes will reach their limits.

It is worth noting that the solutions obtained by the ILP model are optimal solutions. There was a time limit of 30 min for the executions of these experiments, but this limit was never reached. All instances finished execution within a few seconds, even after increasing the size of the instance with the addition of the fourth element to each set.

3.3 Age Regression Results

Based on the fair subsets generated by ILP model, we ran each experiment for the regression task. We also ran experiments for a random selection of the images, considering the same size of each fair set. For each experiment, we executed five runs and compared the mean value of each configuration.

In our first analysis, we evaluated fair and random sets for each size considering BMAE metric. Table 2 presents the results, showing that fair achieved the

Table 2. Comparison of random and fair selections considering BMAE for each training set size.

	k = 1000	k = 2000	k = 5000	k = 10000
Random	10.71	9.41	8.35	7.62
Fair	**8.34**	**8.23**	**7.35**	**7.21**

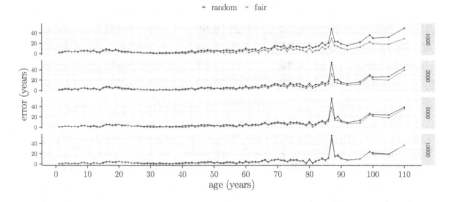

Fig. 6. Comparison of random and fair selections considering BMAE per age.

best outcomes of each size. We can highlight that fair selection outperformed random selection on smallest sets, with the difference of more than 2 years on 1000 set and more than 1 year on 2000 and 5000 sets on average for each age.

The average error for each age of fair selection based on ILP model is lower than random selection, as shown in Fig. 6. The fair selection achieved the lowest errors on ages with more than 80 years on sets with a limited number of images, which are the ages with less number of samples on the dataset.

Then, we assessed ILP fair and random sets considering gender and race classes. To do so, we filtered images from the condition in analysis, e.g., images of gender male, and evaluated using BMAE. Table 3 presents the outcomes of each selection approaches. Concerning gender, the fair set obtained the lowest errors on all the four training sizes against random selection, with the highest difference on training set with 1000 images.

Considering race, fair selection also presented best outcomes compared to random selection in all the classes for each training set size for k greater than 2000, and the best results for most of the classes on k equal to 1000. For White, Black, Asian, and Indian classes, which are the four most frequent ones, the highest difference of fair and random selections occurred on the smallest training set, that is, set with size equal to 1000, and the difference decreases considering the size of the training size. For Other, which is the *least* frequent class, the difference tends to increase considering the size of the training set. We can conclude that this occured due to the frequency of Other class, which represents less than 10% of the original training set, as well as the validation and test sets.

Table 3. Comparison of random and fair selections for gender and race considering BMAE for each training set size.

Attibute class	k = 1000		k = 2000		k = 5000		k = 10000	
	Random	Fair	Random	Fair	Random	Fair	Random	Fair
Male	8.79	**7.17**	7.62	**7.08**	6.74	**6.32**	6.24	**6.02**
Female	11.56	**9.22**	10.28	**8.80**	9.08	**7.68**	8.17	**7.60**
White	9.34	**7.79**	8.00	**7.45**	6.97	**6.56**	6.29	**6.27**
Black	10.79	**8.76**	10.27	**8.90**	9.27	**7.94**	8.59	**8.00**
Asian	8.09	**6.98**	7.19	**6.32**	6.16	**5.75**	5.90	**5.39**
Indian	9.14	**7.33**	7.89	**6.91**	7.04	**5.97**	6.38	**5.92**
Other	**6.41**	6.42	5.84	**5.52**	5.06	**4.73**	4.74	**4.25**

It is worth noting that, although the classes representing race and gender are very balanced, as shown in Fig. 5, indicating that we have a fair cover on these characteristics, we still have a noticeable disparity in the BMAE when considering the different classes of race and gender. For example, the error for the regression of Female is always greater than that of Male. We also have the race Black has a greater error consistently in all cases. This behavior is present in both the training with a fair subset and with a random subset. Although the training with the fair subset has decreased this difference in every case.

4   Conclusions and Future Work

In this paper, we defined a generalized version of the Fairer cover, called Multi-Attribute Fairer Cover. In this version of the problem, we were able to handle the selection of a training subset on a tabular dataset that has multiple attributes and applies the fairness condition to each attribute individually. We present an Integer Linear Programming (ILP) formulation that models this problem and also detail a procedure to create a combinatorial instance from a tabular dataset. We presented experiments with our ILP model and instances derived from a real application dataset. The model achieved a good performance, returning the optimal solution to every instance tested in just a few seconds. We also analyzed the solutions given by the model, and showed how to adjust the instance in order to satisfy the fairness condition through the pairing of the attributes.

We developed an age regression model to test the training with the different sizes of fair subsets obtained with the ILP model and compared these models with their random counterparts. We found that error was reduced when using the fair subsets, specially when dealing with more advanced ages. We also reduced the disparity in error between the samples of the different genders and races.

As future work, we intend to investigate how the fair training subset affects a different regression models specifically tuned for fairness. Since the UTKFace dataset has annotations on both race and genders, we also intend to experiment with our model on other tasks, such as race and gender classification, and investigate how to decrease the error disparity between gender and race classes.

Acknowledgements. The authors would like to thank the São Paulo Research Foundation [grants #2017/12646-3, #2020/16439-5]; the National Council for Scientific and Technological Development [grants #306454/2018-1, #161015/2021-2, #302530/2022-3, #304836/2022-2]; the Coordination for the Improvement of Higher Education Personnel; and Santander Bank, Brazil.

References

1. Asudeh, A., Berger-Wolf, T., DasGupta, B., Sidiropoulos, A.: Maximizing coverage while ensuring fairness: a tale of conflicting objectives. Algorithmica **85**(5), 1287–1331 (2023)
2. Chung, J.: Racism In. Public Citizen, Racism Out - A Primer on Algorithmic Racism (2022)
3. Dantas., A.P.S., de Oliveira., G.B., de Oliveira., D.M., Pedrini., H., de Souza., C.C., Dias., Z.: Algorithmic fairness applied to the multi-label classification problem. In: 18th International Joint Conference on Computer Vision, Imaging and Computer Graphics Theory and Applications - Volume 5: VISAPP, pp. 737–744. SciTePress (2023)
4. Galhotra, S., Shanmugam, K., Sattigeri, P., Varshney, K.R.: Causal feature selection for algorithmic fairness. In: International Conference on Management of Data, SIGMOD2022, pp. 276–285. Association for Computing Machinery (2022)
5. He, K., Zhang, X., Ren, S., Sun, J.: Deep residual learning for image recognition. In: Conference on Computer Vision and Pattern Recognition, pp. 770–778. IEEE (2016)
6. Kleinberg, J., Ludwig, J., Mullainathan, S., Rambachan, A.: Algorithmic fairness. AEA Papers Proc. **108**, 22–27 (2018)
7. Krizhevsky, A., Sutskever, I., Hinton, G.E.: ImageNet classification with deep convolutional neural networks. In: Advances in Neural Information Processing Systems (NIPS), pp. 1097–1105. Curran Associates, Inc. (2012)
8. Lin, Y., Guan, Y., Asudeh, A., Jagadish, H.V.J.: Identifying insufficient data coverage in databases with multiple relations. VLDB Endowment **13**(12), 2229–2242 (2020)
9. Loshchilov, I., Hutter, F.: Decoupled weight decay regularization, pp. 1–11 (2019). arXiv:1711.05101
10. O'Neil, C.: Weapons of math destruction: how big data increases inequality and threatens democracy. In: Crown (2017)
11. Roh, Y., Lee, K., Whang, S., Suh, C.: Sample selection for fair and robust training. Adv. Neural. Inf. Process. Syst. **34**, 815–827 (2021)
12. Silva, T.: Algorithmic racism timeline (2020). https://bit.ly/3O6RJYC
13. Soper, S.: Fired by Bot at Amazon: 'It's You Against the Machine' (2021). https://www.bloomberg.com/news/features/2021-06-28/fired-by-bot-amazon-turns-to-machine-managers-and-workers-are-losing-out, https://bit.ly/3IvhUXy
14. Zhang, Z., Song, Y., Qi, H.: Age progression/regression by conditional adversarial autoencoder. In: Conference on Computer Vision and Pattern Recognition, pp. 5810–5818. IEEE (2017)

A Custom Bio-Inspired Algorithm for the Molecular Distance Geometry Problem

Sarah Ribeiro Lisboa Carneiro[1] , Michael Ferreira de Souza[2] ,
Douglas O. Cardoso[3] , Luís Tarrataca[1] , and Laura S. Assis[1(✉)]

[1] Celso Suckow da Fonseca Federal Center of Technological Education,
Rio de Janeiro, RJ, Brazil
`sarah.carneiro@aluno.cefet-rj.br`,
`{luis.tarrataca,laura.assis}@cefet-rj.br`
[2] Department of Statistics and Applied Mathematics, Federal University of Ceará,
Fortaleza, Ceará, Brazil
`michael@ufc.br`
[3] Smart Cities Research Center, Polytechnic Institute of Tomar, Tomar, Portugal
`douglas.cardoso@ipt.pt`

Abstract. Protein structure allows for an understanding of its function
and enables the evaluation of possible interactions with other proteins.
The molecular distance geometry problem (MDGP) regards determining
a molecule's three-dimensional (3D) structure based on the known distances between some pairs of atoms. An important application consists
in finding 3D protein arrangements through data obtained by nuclear
magnetic resonance (NMR). This work presents a study concerning the
discretized version of the MDGP and the viability of employing genetic
algorithms (GAs) to look for optimal solutions. We present computational results for input instances whose sizes varied from 10 to 10^3 atoms.
The results obtained show that approaches to solving the discrete version
of the MDGP based on GAs are promising.

Keywords: Combinatorial Optimization · Genetic Algorithm ·
Distance Geometry · Molecular Structure

1 Introduction

An important application in molecular biology is to find 3D arrangements of
proteins using the interatomic distances obtained through nuclear magnetic resonance (NMR) experiments [24]. The objective of the Molecular Distance Geometry Problem (MDGP) is to determine molecular structures from a set of inter-
atomic distances. Proteins are essential parts of living organisms. The interest in
the 3D structure of proteins results from the intimate relationship between form
and protein functions. The function of a molecule is determined by its chemical and geometrical structure [4]. Calculating the 3D structure of proteins is an
essential problem for the pharmaceutical industry [16].

M. C. Naldi and R. A. C. Bianchi (Eds.): BRACIS 2023, LNAI 14195, pp. 178–192, 2023.
https://doi.org/10.1007/978-3-031-45368-7_12

The MDGP was studied in [11] where a discretizable version of the MDGP (DMDGP) with a finite search space is also presented. The work proposes a Branch-and-Prune algorithm by searching a binary tree (representing the search space of the DMDGP) for a feasible solution to the problem. The MDGP can be formulated as an optimization problem whose objective is to determine, according to some criteria, the best alternative in a specified universe. In general, combinatorial optimization problems can be defined as given a set of candidate solutions S, a set of constraints Ω, and an objective function f responsible for associating a cost $f(s)$ to each candidate solution $s \in S$. The objective is to find $s^* \in S$ that is globally optimal and with the best cost amongst the candidate solutions [22]. Optimization methods can be divided into two classes: *(i)* deterministic, and *(ii)* stochastic. In the former, it is possible to predict the trajectory according to the input, *i.e.*, the same input always leads to the same exit. Stochastic methods rely on randomness, thus the same input data can produce distinct outputs, as is the case with bio-inspired algorithms [9].

This work presents an evolutionary approach for solving the DMDGP and studies its respective viability. The DMDGP belongs to the class of NP-Hard problems through a reduction of the subset-sum problem [14]. Our aim is to answer the following questions: *(i)* how effective is an evolutionary heuristic for solving the DMDGP, evaluated on real 3D instances of proteins?; *(ii)* for which instance sizes is it possible to find optimal solutions? In order to tackle these issues we present two stochastic approaches based on genetic algorithms (GA) for solving the DMDGP.

The work is organized as follows. Related works are presented in Sect. 2. Section 3 formalizes the problem. The solution approaches proposals are characterized in Sect. 4. The computational experiments and approach comparisons are presented in Sect. 5. The concluding remarks are presented in Sect. 6.

2   Related Work

There are two ways to model the MDGP, namely, a continuous and a discrete formulation. There also exist disparities concerning the instances employed. Some works make use of geometric instances that were randomly generated based on a model [17]. Others employ synthetic instances produced in a random fashion based on a physical model [13]. There are also works that use dense real instances [18] whilst others employ sparse real instances [11] that were taken from the Protein Data Bank (PDB) [1].

Regarding the continuous version of the MDGP, in [6] the authors considered the case where the exact distances between all pairs of atoms are known. The algorithm is based on simple geometrical relationships between atom coordinates and the distances between them. The method was tested with the protein set HIV-1 RT that is composed of 4200 atoms. The algorithm is executed in linear time. In [12] a Divide and Conquer (D&C) algorithm was proposed that makes use of SDP alongside gradient descent refining. A structure consisting of 13000 atoms was solved in an hour. A geometric algorithm for approximate solutions with instances ranging from 402 to 7398 atoms was presented in [25].

In [5] the 3D-As-Synchronized-As-Possible (3D-ASAP) D&C approach was proposed. Protein lengths varied between 200 to 1009 atoms. The work also proposes a more agile version of 3D-ASAP, 3D-SP-ASAP, that uses a spectral partitioning algorithm to preprocess the graph into subgraphs. The results obtained showed that both are robust algorithms for sparse instances. The continuous approach most resembling this work was presented in [23] and describes the MemHPG memetic algorithm (MA). The method combines swarm intelligence alongside evolutionary computation associated with a local search with the objective of optimizing the crossover operator. MemHPG was tested with proteins whose lengths varied between 402 and 1003. The algorithm was able to obtain the solution to the problem. The result obtained was comparable with the then state of the art described in [25].

Regarding the discrete approaches, in [10], a D&C approach (ABBIE) was presented that explores combinatorial characteristics of the MDGP. The idea is to reduce the global optimization problem into subproblems. ABBIE was able to find approximate solutions for the MDGP. In [2] a distributed algorithm was described that does not make use of "known" positions to determine the successor ones. The method divides the problem graph into subgraphs using sampling methods: *(i)* SDP relaxation vying to minimize the sum of errors for specified and estimated distances; and *(ii)* gradient descent method of error reduction.

In the discrete formulations, the works most closely related to the one here presented are [21]. The authors define a discretizable variation of the Geometry Distance Problem known as the DMDGP and propose the Branch-and-Prune algorithm. In [8] a method is described to explore the symmetries that are present in the search space of the DMDGP and that can be represented by a binary tree. The method was able to obtain a relevant speedup for sparse instances against the Branch-and-Prune. Work [26] discusses a sequence of nested and overlapping DMDGP sub-problems. The results exhibited superior performance for randomly generated sparse instances.

3   Presentation and Modelling

3.1   DGP

The Distance Geometry Problem (DGP) consists in finding a set of points in a given geometric space of dimension k by knowing only some distances between them [15]. The DGP can be solved in linear time when all the distances are known [6]. Formally, given a simple, connected and weighted graph, $G = (V, E, d)$, with a set V of n vertices, a set E of m edges, and a weight function $d : E \rightarrow \mathbb{R}_+$ we wish to determine a function $x : V \rightarrow \mathbb{R}^k$ such that $||x(u) - x(v)|| = d(u, v)$ for every $\{u, v\} \in E$. The DGP can be formulated as a global optimization problem whose objective function is described in Eq. (1).

$$\min_{x} g(x) = \sum_{\{u,v\} \in E} (||x(u) - x(v)||^2 - d(u, v)^2)^2, \tag{1}$$

where x solves the problem if and only if $g(x) = 0$. This approach is challenging since an elevated number of local minima exist due to the non-convergence of the associated optimization problem.

3.2   MDGP

The 3D structure of a molecule is intrinsic to its function. Data provided by NMR allows for the calculation of the 3D structure of a protein. This procedure only provides the distances between close atoms. The problem consists in employing this information to obtain the position of every atom in the molecule. The MDGP is a particular case of DGP where the set of vertices V corresponds to the set of atoms of a molecule, the edges E is the set of known distances between atoms and the weight function d is the distance value.

Figure 1(a) shows a solution in $\mathbb{R}^2$ for a graph representing a problem instance. Atoms a, b, c can be fixed in $\mathbb{R}^2$ to eliminate solutions from rotations and translations. Point d can be positioned in any of the countless points of the circle of center c and radius represented by the distance between c and d (Fig. 1(b)). The represented problem has a non-enumerable quantity of solutions. If the distance between b and d is also known (in addition to the one between c and d) then the problem presents a finite number of solutions. In this case, two solutions can be found at the intersection of the circumferences of center c and b with respective radius represented by the distances between: (i) c and d; and (ii) b and d. This case is exemplified in Fig. 1(c).

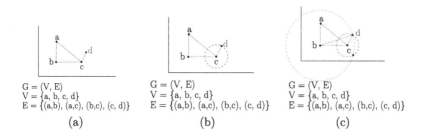

Fig. 1. Solutions may be non-enumerable or finite.

3.3   Discrete Version of the MDGP

As seen, the MDGP is a constraint satisfaction problem, which is generally reformulated as a global optimization one, where the objective is to minimize a penalty function capable of measuring how much the constraints are being violated (Eq. (1)). However, the penalty function of the optimization problem is strongly non-smooth. Optimization methods risk getting stuck at local minima with an objective value very close to the optimal one.

The DMDGP [19] is a particular case of MDGP when the geometric space is $\mathbb{R}^3$, the set of vertices V is finite, and the set of edges E satisfies the following constraints: *(i)* every subset $V_i = \{v_i, v_{i-1}, v_{i-2}, v_{i-3}\} \subset V$ induces a 4−clique in G; *(ii)* the distances in $\{d(v_i, v_j) : i, j \in V_i\}$ define non-null volume tetrahedrons. Therefore, when the discretization assumptions are satisfied, the domain of the penalty function can be reduced to a tree. In [11] the Branch-and-Prune method was proposed where the DMDGP search space is represented by a binary tree and the procedure attempts to fix an atom with each iteration. Assume that for each atom v_i, for $i > 3$, the distances between v_i and the tree immediately predecessors atoms $v_{i-1}, v_{i-2}, v_{i-3}$ are known. Assuming that these three atoms possess realizations x_{i-1}, x_{i-2} e x_{i-3} in $\mathbb{R}^3$, then it is possible to define tree spheres centered in these points and whose radius is, respectively, the distances $d(v_i, v_{i-1})$, $d(v_i, v_{i-2})$ and $d(v_i, v_{i-3})$. Under reasonable assumptions, the intersections of these spheres are the set $\{x_i, x_i'\}$ formed by the coordinates (*branch*) satisfying those three distances. One of these realizations can be chosen for vertex v_i and it is then possible to proceed to fix the next vertex v_{i+1}. It may be possible that other edges (distances) exist that use vertex v_i that invalidate these coordinates. If this occurs then it becomes necessary to change the choices previously made (prune). This methodology defines a binary tree of atomic coordinates that contains every possible position for each vertex v_i, $i > 3$, in the ith tree layer. Each DMDGP solution is associated with a path from the root to a leaf node. This behaviour is exemplified in Fig. 2 where vector $[0, 1, 1]$ represents a solution.

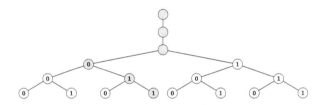

Fig. 2. Solution example - binary tree path.

The representation of the DMDGP as a binary tree evidences several symmetry properties in the search space [20]. Namely, the invariance of the solution set concerning the total reflection of a solution is commonly referred to as the first symmetry. For example, if $s = [0, 1, 1]$ in Fig. 2 is a feasible solution to the problem, then the inverse of all the bits in s ($\bar{s} = [1, 0, 0]$) is also a feasible solution. It is important to emphasize that, in the binary representation, the first three atoms are fixed in $\mathbb{R}^3$. These characteristics led to the approach choice proposed in this work.

4 Solution Methodology

The solution approach developed to solve the DMDGP is based on the concepts of GAs. These are optimization algorithms imbued with a certain oriented randomness that is based on an analogy with the Darwinian principles of evolution and species genetics [7]. GAs have applications in discrete and continuous systems [27]. In this work, we propose two methods based on GAs, referred to as (i) Simplified Genetic Algorithm (SGA); (ii) Feasible Index Genetic Algorithm (FIGA). The main difference resides in the genetic operator employed. Each stage of the GAs proposed and their respective particularities are presented in the following subsections.

4.1 Representation and Initial Population

GAs are population algorithms where each individual represents a candidate solution for a problem. The population can be defined as a set of candidate solutions in the search space. The DMDGP search space is a binary tree, where each of the two possibilities in a given level of the tree symbolizes a position in $\mathbb{R}^3$. A solution can be represented by a binary uni-dimensional chromosome of size n (number of atoms). The first three atoms have their position fixed. As a result, the chromosome can be implemented with size $n-3$. The values assumed for each position, 0 or 1, are known as alleles and indicate the path chosen in the tree. The set of genes is known as the genotype. Figure 1 illustrates the coding implemented for the DMDGP. A constructive random heuristic generates the initial population. Namely, due to the coding of the problem for each position a choice is made (with 50% of probability) if the allele will have value 1 or 0.

Table 1. Chromosomal representation for the MDGP.

Index	0 1 2 3	$\cdots$	$n-1$
Chromosome	1 1 0 1	$\cdots$	0

4.2 Feasibility

The set of distances of the DMDGP can be divided into two distinct subsets, namely: (1) discretization distances; and (2) pruning distances. Discretization distances are the ones utilized in the construction of the binary tree, and pruning distances are the remaining ones. When only the subset of discretization distances is considered then all tree paths lead to feasible solutions. However, a vertex may exist in the tree that does not obey the pruning distances constraints. This results in the solution becoming unfeasible for that tree branch.

In the proposed FIGA algorithm, there is a method responsible for going through the tree and verifying until which chromosome position the solution is feasible, *i.e.*, $||x(u) - x(v)|| = d(u, v)$ for each $\{u, v\} \in E$. This position, which will be referred to as the feasible index, is used to implement crossover and mutation operations. The SGA approach represents the basic implementation of a GA and therefore does not consider the feasible index in its genetic operators. This is the main difference between the previously proposed methods.

4.3   Fitness Function

At each generation, the individuals of a population are evaluated based on their fitness values. As is typical in optimization problems, the fitness function adopted in this work is the inverse of the objective function (OF) with 1 added in the denominator, as can be seen in Eq. (2).

$$fitness = \left(1 + \sum_{\{u,v\} \in E} (||x(u) - x(v)||^2 - d(u, v)^2)^2\right)^{-1}. \tag{2}$$

The OF is equal to 0 when the known distances are equal to the calculated distances. In this case, x is a feasible solution for the DMDGP and the fitness will be equal to 1. The OF value is always greater or equal to zero. The fitness value tends to zero the bigger the OF value.

4.4   Selection

At each generation, the selection process elects the pairs of individuals that will participate in the crossover. These individuals are chosen based on a roulette selection that guarantees that all the individuals have a certain probability of being chosen that is proportional to their aptitude.

4.5   Crossover

Crossover is the process in which an offspring individual is created from the genetic information of the parents. This work employs a one-point crossover operator. In the SGA approach, the crossover consists in generating a random position between $[1, n-1]$, where n is the chromosome size, and the randomly selected position is the cut point of the SGA crossover. In the FIGA strategy, the crossover point is determined based on the feasible index (f) (Subsect. 4.2). This mechanism allows for the generated offspring to always inherit the feasible part of the parent that presents that largest feasible index. The FIGA crossover behavior is illustrated in Fig. 3. The feasible index for each parent is emphasized with the color red. The cut point is determined by the largest feasible index (dotted line). The interval position $[0, c[$ of the son, in which c is the cut point, will have the genetic information of the parent with the largest feasible index. The remaining values of the son (*i.e.*, $]c+1, n[$) are copied from the other parent.

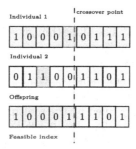

Fig. 3. Crossover operator.

4.6 Mutation

The main objective of this operator is to introduce genetic diversity in the population of the current generation. Namely, each offspring generated can be mutated according to probability p_m. If an offspring is selected for mutation, then the operator considers each gene individually and allows the allele to be inverted in accordance with the same probability. In the FIGA approach, the mutation is performed only on the infeasible part of the solution. Let f represent the feasible index (chromosome position in red - Fig. 4) and n the size of the chromosome, then the FIGA mutation can occur in the position of the interval $[f, n-1]$. In the SGA strategy, the mutation can happen in all the chromosome. Figure 4 illustrates the behavior of the FIGA mutation, with the feasible index being emphasized in red color.

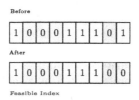

Fig. 4. Mutation Operator. (Color figure online)

4.7 Reset

The reset operation aims to avoid the solution getting stuck in the local minimum. Namely, if the fitness value of the best individual does not improve over a certain number of generations then the reset is activated. The best individual is maintained and the remaining population is discarded. A new population of individuals is generated with the random constructive heuristic.

4.8 Pseudocode

The pseudocode for the FIGA is presented in Algorithm 1. Both the SGA and the FIGA have as a stopping criterion a maximum quantity of generations ($genMax$)

and search for a feasible solution (line 6). Line 1 initializes the variable that stores the largest fitness, $bestFitness$. The algorithm generates in a random fashion an initial population, calculates the fitness of each individual, and their respective feasible indexes (Lines 3, 5 and 5, respectively). The SGA does not consider the feasible index. Therefore, Line 5 is only present in the FIGA algorithm.

The second cycle (Line 8) considers the number of offspring formed with each generation. This value is defined by the multiplication of the crossover rate (t_c) and the population size (p_s). The line 9 selects a pair of individuals for the crossover operation. Line 10 is responsible for performing the crossover of the previously selected individuals. The newly generated offspring individual is stored in the variable $offspring$ and can incur a mutation with probability p_m. The mutation occurs in line 13. The line 15 (only available in FIGA) calculates the feasible index of the new individuals. As previously stated, the genetic operators (crossover and mutation) are different in the proposed algorithms. The fitness of the offspring is then compared against its respective parents. If an improvement is observed, the parent with worse fitness is replaced by its offspring (Lines 16 to 24). Both methods use the elitism scheme to prevent the loss of the fittest member of the population. Therefore, the method "getBestFiness" returns the largest fitness of the current population, this value is part of the stop criteria, and that individual will be part of the population for the next generation.

5   Results

5.1   Parameter Configuration

Due to the high-dimension search space, performing a full grid search of the optimization parameters would have a prohibitive cost. As a consequence, a smaller test was performed for synthetic instances[1] in $\mathbb{R}^2$. in order to define quality values for the GA parameters. The full set of results obtained is detailed in [3]. The evaluation procedure consisted of a set of instances with size $n = [10, 30, 50, 100]$ was randomly generated taking into account the MDGP. These instances are composed of d lines, where d is the number of known distances. Each line represents a graph edge, $i.e.$, a known distance between a pair of atoms. Each edge identifies the connecting atoms as well as the distance between them. For example, [0 1 2.5] represents that atom v_0 is at a distance of 2.5 measurement units from atom v_1.

In these first tests, both methods were executed 10 times each in an algorithm independent run for each instance with each parameter combination. The parameters and values evaluated are presented in Table 2. There were 840 combinations of parameters totaling 8400 executions of the proposed algorithms for the four instance sizes considered in the study.

The tests allowed for a comparison between SGA and FIGA. For instances with only 10 atoms, it is not possible to observe which GA presents the best performance. However, as the number of atoms increases the balance shifts in

---

[1] Instance set is publicly available in: https://bit.ly/3d0ezzo.

Algorithm 1 : FIGA Pseudocode

```
1: bestFitness ← 0
2: gen ← 0
3: population ← GenerateInitialPopulation()
4: ComputeFitness( population )
5: ComputeFeasibleIndex( population )
6: while (gen < genMax) and (bestFitness ≠ 1) do
7:     j ← 0
8:     while j ≤ tc * ps do
9:         (Ind1, Ind2) ← Select a pair ( population )
10:        offspring ← crossover (Ind1,Ind2)
11:        R ← Random(0, 1)
12:        if pm ≥ R then
13:            offspring ← mutation( offspring )
14:        end if
15:        ComputeFeasibleIndex( offspring )
16:        if Fitness( Ind1 ) > Fitness( Ind2 )  then
17:            if Fitness( offspring ) > Fitness( Ind2 ) then
18:                Ind2 ← offspring
19:            end if
20:        else if  Fitness( Ind2 ) > Fitness( Ind1 ) then
21:            if  Fitness( offspring ) > Fitness( Ind1 ) then
22:                Ind1 ← offspring
23:            end if
24:        end if
25:        j ← j + 1
26:    end while
27:    bestFitness← getBestFitness()
28:    gen ← gen + 1
29: end while
```

Table 2. Genetic Algorithm parameter variation in $\mathbb{R}^2$.

Parameter	Description	Set of values
p_m	Mutation probability	$\{0.05,\ 0.1,\ 0.2,\ 0.3,\ 0.4,\ 0.5\}$
t_c	Crossover rate	$\{0.2,\ 0.3,\ 0.4,\ 0.5,\ 0.6,\ 0.7,\ 0.8\}$
p_s	Population size	$\{30, 50, 100, 150, 200\}$
$genMax$	Max. number of generations	$\{30, 50, 100, 150\}$

favor of the FIGA. Namely, SGA is able to find a feasible solution for instances of small size (maximum 30 atoms). On the other hand, the FIGA is able to find the feasible solution for every instance size evaluated. By analyzing the results obtained the value were fixed in $t_c = 0.7$, $p_m = 0.3$, $p_s = 200$ e $genMax = 30$ values.

5.2 Experiments in 3D Space

This section presents the tests performed with real instances in $\mathbb{R}^3$.

Instances. The computational experiments described in this section employed real instances taken from the PDB[2] [1]. The instances selection was made randomly ensuring a linear increase in the number of atoms. The set of instances used is listed in Table 3 and illustrated in Fig. 5.

Table 3. PDB Instances.

	Instance	Number of atoms		Instance	Number of atoms
(a)	1ABZ	114	(f)	1IAD	600
(b)	1CSQ	201	(g)	1DUG	702
(c)	1EJS	300	(h)	1DY6	801
(d)	1B4M	402	(i)	1C80	900
(e)	1FXL	501	(j)	1FOB	1002

Presentation and Result Analysis. This section presents the FIGA results for the above mentioned $\mathbb{R}^3$ instances. The elevated computational time made it unfeasible to perform a parameter study specific to the $\mathbb{R}^3$ instances. As a consequence, the parameter values were empirically chosen based on the tests performed for $\mathbb{R}^2$ and some tests for $\mathbb{R}^3$ instances. After conducting empirical tests with various values for the reset, including not performing it, the value of 25 was selected for this parameter. Additional empirical tests were conducted considering values around the best parameters for $\mathbb{R}^2$, as shown in Table 2. However, since the instances in $\mathbb{R}^3$ are much larger than the synthetic instances in $\mathbb{R}^2$, we had to increase the population size and the number of generations. Consequently, the parameter values for $\mathbb{R}^3$ were defined as indicated in Table 4.

Table 4. Configuration of Genetic Algorithm parameter in $\mathbb{R}^3$.

Parameter	Description	Values
p_m	Mutation probability	0.5
t_c	Crossover rate	0.7
p_s	Population size	1000
$genMax$	Maximum number of generations	800

---

[2] The set of instances is available in https://www.rcsb.org/.

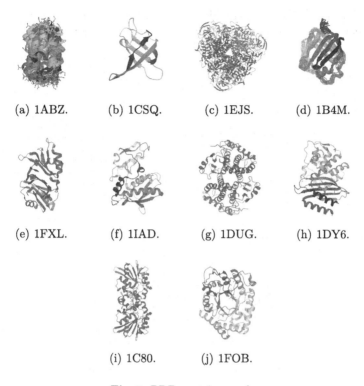

(a) 1ABZ.   (b) 1CSQ.   (c) 1EJS.   (d) 1B4M.

(e) 1FXL.   (f) 1IAD.   (g) 1DUG.   (h) 1DY6.

(i) 1C80.   (j) 1FOB.

Fig. 5. PDB proteins used.

In the graphs that follow (Fig. 6), the green line depicts the largest fitness of the population, the blue line represents the average fitness and the red line corresponds to the smallest fitness.

Figure 6 shows the relationship between the feasible index and the number of generations for the instances with 114, 201, 300, 402, 501, and 702 atoms. When the algorithm finds a feasible solution the feasible index is equal to the chromosome size. The algorithm was able to find a feasible solution for all of the above mentioned instance sizes. However, the algorithm was unable to find a feasible solution for the remaining instances (Table 3 - 600, 801, 900, and 1002 atoms) using the set of stipulated parameters. By analyzing the results presented in the plots it is possible to gain a better understanding of the FIGA evolution until a feasible solution is found. For the smallest instance employed (Fig. 6(a)) the algorithm only required 6 generations to find a feasible solution. In the case of the instance using 702 atoms the algorithm was able to find a feasible solution after 350 generations. Table 5 indicates for each instance what was the iteration where a feasible solution was found or indicates with a character "–" when a solution was not found.

Future work would try to improve the GA in order to explore the symmetry properties of the DMDGP in the crossover and mutation operators. Potential improvement strategies: *i)* using a hierarchically structured population, *ii)* developing a memetic algorithm, and *iii)* evaluating other genetic operator types.

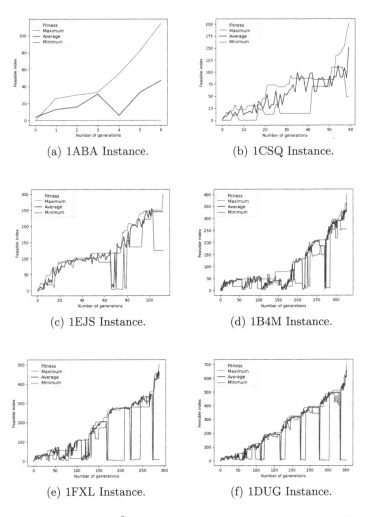

Fig. 6. FIGA performance in $\mathbb{R}^3$ - feasible index × number of generations. (Color figure online)

6 Final Considerations

This work presented a viability study regarding the application of GA for solving the DMDGP. The difference between the proposed algorithms, SGA and FIGA, resides in the partial or total feasibility of the solution and how this information affects the corresponding genetic operators. The FIGA evaluates a tree and

Table 5. FIGA results in $\mathbb{R}^3$.

Instance	Convergence Iteration	Instance	Convergence Iteration
1ABZ	6	1IAD	–
1CSQ	59	1DUG	352
1EJS	112	1DY6	–
1B4M	328	1C80	–
1FXL	288	1FOB	–

verifies up until which chromosome position the solution is feasible. This index is used to perform crossover and mutation operations. The SGA does not consider this information in its genetic operators. Though subtle, this difference was responsible for significant improvements in the tests performed. For instances of size $n = [10, 30, 50, 100]$ in $\mathbb{R}^2$ the FIGA presented better or equal performance when compared against SGA. This performance was measured in terms of fitness and also the generation in which convergence occurred. The FIGA was able to solve real PDB instances for sizes 114, 201, 300, 402, 501, and 700.

Acknowledgements. Douglas O. Cardoso acknowledges the financial support by the Foundation for Science and Technology (Fundação para a Ciência e a Tecnologia, FCT) through grant UIDB/05567/2020, and by the European Social Fund and programs Centro 2020 and Portugal 2020 through project CENTRO-04-3559-FSE-000158.

References

1. Berman, H.M., et al.: The protein data bank. Nucleic Acids Res. **28**(1), 235–242 (2000)
2. Biswas, P., Toh, K.C., Ye, Y.: A distributed SDP approach for large-scale noisy anchor-free graph realization with applications to molecular conformation. SIAM J. Sci. Comput. **30**(3), 1251–1277 (2008)
3. Carneiro, S., Souza, M., Filho, N., Tarrataca, L., Rosa, J., Assis, L.: Algoritmo genético aplicado ao problema da geometria de distâncias moleculares. In: Anais do LII Simpósio Brasileiro de Pesquisa Operacional (2020)
4. Creighton, T.E.: Proteins: Structures and Molecular Properties. Macmillan (1993)
5. Cucuringu, M., Singer, A., Cowburn, D.: Eigenvector synchronization, graph rigidity and the molecule problem. Inf. Infer. J. IMA **1**(1), 21–67 (2012)
6. Dong, Q., Wu, Z.: A linear-time algorithm for solving the molecular distance geometry problem with exact inter-atomic distances. J. Global Optim. **22**(1), 365–375 (2002)
7. Goldberg, D.E.: Genetic Algorithms in Search. Optimization, and Machine Learning (1989)
8. Goncalves, D.S., Lavor, C., Liberti, L., Souza, M.: A new algorithm for the ^{k}dmdgp subclass of distance geometry problems (2020)
9. Gong, Y.J., et al.: Distributed evolutionary algorithms and their models: a survey of the state-of-the-art. Appl. Soft Comput. **34**, 286–300 (2015)

10. Hendrickson, B.: The molecule problem: exploiting structure in global optimization. SIAM J. Optim. **5**(4), 835–857 (1995)
11. Lavor, C., Liberti, L., Maculan, N., Mucherino, A.: The discretizable molecular distance geometry problem. Comput. Optim. Appl. **52**(1), 115–146 (2012)
12. Leung, N.H.Z., Toh, K.C.: An SDP-based divide-and-conquer algorithm for large-scale noisy anchor-free graph realization. SIAM J. Sci. Comput. **31**(6), 4351–4372 (2010)
13. Liberti, L., Lavor, C., Maculan, N., Marinelli, F.: Double variable neighbourhood search with smoothing for the molecular distance geometry problem. J. Global Optim. **43**(2–3), 207–218 (2009)
14. Liberti, L., Lavor, C., Mucherino, A.: The discretizable molecular distance geometry problem seems easier on proteins. In: Mucherino, A., Lavor, C., Liberti, L., Maculan, N. (eds.) Distance Geometry, pp. 47–60. Springer, New York (2013). https://doi.org/10.1007/978-1-4614-5128-0_3
15. Liberti, L., Lavor, C., Mucherino, A., Maculan, N.: Molecular distance geometry methods: from continuous to discrete. Int. Trans. Oper. Res. **18**(1), 33–51 (2011)
16. Maculan Filho, N., Lavor, C.C., de Souza, M.F., Alves, R.: Álgebra e Geometria no Cálculo de Estrutura Molecular. Colóquio Brasileiro de Matemática, IMPA (2017)
17. Moré, J.J., Wu, Z.: Global continuation for distance geometry problems. SIAM J. Optim. **7**(3), 814–836 (1997)
18. Moré, J.J., Wu, Z.: Distance geometry optimization for protein structures. J. Global Optim. **15**(3), 219–234 (1999)
19. Mucherino, A., Lavor, C., Liberti, L.: The discretizable distance geometry problem. Optim. Lett. **6**(8), 1671–1686 (2012)
20. Mucherino, A., Lavor, C., Liberti, L.: Exploiting symmetry properties of the discretizable molecular distance geometry problem. J. Bioinform. Comput. Biol. **10**(03), 1242009 (2012)
21. Mucherino, A., Liberti, L., Lavor, C.: MD-jeep: an implementation of a branch and prune algorithm for distance geometry problems. In: Fukuda, K., Hoeven, J., Joswig, M., Takayama, N. (eds.) ICMS 2010. LNCS, vol. 6327, pp. 186–197. Springer, Heidelberg (2010). https://doi.org/10.1007/978-3-642-15582-6_34
22. Mulati, M.H., Constantino, A.A., da Silva, A.F.: Otimização por colônia de formigas. In: Lopes, H.S., de Abreu Rodrigues, L.C., Steiner, M.T.A. (eds.) Meta-Heurísticas em Pesquisa Operacional, 1 edn., Chap. 4, pp. 53–68. Omnipax, Curitiba, PR (2013)
23. Nobile, M.S., Citrolo, A.G., Cazzaniga, P., Besozzi, D., Mauri, G.: A memetic hybrid method for the molecular distance geometry problem with incomplete information. In: 2014 IEEE Congress on Evolutionary Computation (CEC), pp. 1014–1021. IEEE (2014)
24. Schlick, T.: Molecular Modeling and Simulation: An Interdisciplinary Guide: An Interdisciplinary Guide, vol. 21. Springer, New York (2010). https://doi.org/10.1007/978-1-4419-6351-2
25. Sit, A., Wu, Z.: Solving a generalized distance geometry problem for protein structure determination. Bull. Math. Biol. **73**(12), 2809–2836 (2011)
26. Souza, M., Gonçalves, D.S., Carvalho, L.M., Lavor, C., Liberti, L.: A new algorithm for a class of distance geometry problems. In: 18th Cologne-Twente Workshop on Graphs and Combinatorial Optimization (2020)
27. Yang, X.S.: Engineering Optimization: An Introduction with Metaheuristic Applications. Wiley (2010)

Allocating Dynamic and Finite Resources to a Set of Known Tasks

João da Silva$^{(\boxtimes)}$, Sarajane Peres , Daniel Cordeiro , and Valdinei Freire

School of Arts, Sciences and Humanities, University of São Paulo, São Paulo, Brazil
{jvctr.dasilva,sarajane,daniel.cordeiro,valdinei.freire}@usp.br

Abstract. We consider a generalization of the task allocation problem. A finite number of human resources are dynamically available to try to accomplish tasks. For each assigned task, the resource can fail or complete it correctly. Each task must be completed a number of times, and each resource is available for an independent number of tasks. Resources, tasks, and the probability of a correct response are modeled using Item Response Theory. The task parameters are known, while the ability of the resources must be learned through the interaction between resources and tasks. We formalize such a problem and propose an algorithm combining shadow test replanning to plan under uncertain knowledge, aiming to allocate resources optimally to tasks while maximizing the number of completed tasks. In our simulations, we consider three scenarios that depend on knowledge of the ability of the resources to solve the tasks. Results are presented using real data from the Mathematics and its Technologies test of the Brazilian Baccalaureate Examination (ENEM).

Keywords: Dynamic resource allocation · Adaptive allocation · Shadow test

1 Introduction

The allocation of human resources to solve tasks is part of the daily routine of the most diverse institutions. Going from the allocation of employees in a company to the fulfillment of complex tasks [23], to the assignment of public lawyers to defendants by a court [1]. In the previous examples, if the decision-maker, which allocates resources to perform each set of tasks, is well aware of the abilities of each resource to solve the tasks, one can use the Hungarian Algorithm to obtain the optimal solution [15]. However, this is not always the case. There are numerous scenarios where the abilities of the resources are not known beforehand, for example, in a non-governmental organization allocating tasks to sporadic volunteers [5,14]. Another example is *crowdsourcing*, the recent tendency in certain companies to allocate some tasks to non-specialized outsource workers [26,28].

This study was partially supported by the *Coordenação de Aperfeiçoamento de Pessoal de Nível Superior* (CAPES) - Finance Code 001, by the São Paulo Research Foundation (FAPESP) grant #2021/06867-2 and the Center for Artificial Intelligence (C4AI-USP), with support by FAPESP (grant #2019/07665-4) and by the IBM Corporation.

M. C. Naldi and R. A. C. Bianchi (Eds.): BRACIS 2023, LNAI 14195, pp. 193–208, 2023.
https://doi.org/10.1007/978-3-031-45368-7_13

Some works use previous information about the resource to estimate their ability to solve the tasks [8,11]. To guarantee the quality of the solutions given by unknown resources, it is necessary to estimate their ability to accomplish such tasks. The ability estimation can be done by the conduction of a test before the allocation of the tasks [25]; or by dynamically estimating the abilities during the allocation process, depending on the results of previous assignments [31]. Here, the resources' abilities are dynamically estimated during the process, being updated after each response. Many works consider a threshold so that weaker resources do not receive any task [9,25,31]; instead, we consider the resources scarce and try to use the maximum of their availability. We allocate the dynamic and finite resources that arrive sequentially to try to solve a set of known tasks.

We consider a finite number of human resources arriving sequentially; tasks are assigned to resources dynamically, and the resources can either complete them correctly or fail. Each task must be completed a number of times, and each resource can receive an independent number of tasks. No task is submitted twice to the same resource; each task can be submitted as many times as needed to different resources. Resources, tasks, and the probability of a correct response are modeled using Item Response Theory [6]. The task parameters are known, while the ability of the resources must be learned through the interaction between resources and tasks. We propose an algorithm that combines shadow test replanning [17] to replan under uncertain knowledge, with the goal of optimally allocating resources to tasks while maximizing the number of completed tasks. Each replanning consists in solving an optimization problem through linear programming. The results are compared with two baselines, the random selection and the *easier-first* selection that selects the easier task available as the resources arrive sequentially. In summary, the contributions of this paper are: (*i*) a novel allocation of dynamic finite resources to a set of finite binary tasks problem is formalized; and (*ii*) algorithms are proposed to solve the problem in different scenarios by using most of the resources' availability.

The rest of this paper is structured as follows: the proposed allocation problem is defined in Sect. 2. The proposed framework, with the definition of three algorithms, is presented in Sect. 3. Section 4 gives an overview of related works on the state of the art of task allocation. We present and discuss the results in Sect. 5, before concluding in Sect. 6.

2   The Allocation of Dynamic and Finite Resources to a Set of Binary Tasks Problem

We consider the problem of repeatedly allocating resources to fulfill one of a set of tasks when finite resources are available dynamically. First, the decision-maker knows a set of tasks, in which each task must be solved an arbitrary number of times. Second, resources are available dynamically to receive a number of tasks to try to solve. Third, the decision-maker allocates the current resource to tasks, one after the other, until the resource's availability ends. Fourth, for each allocated task, the resource may fail or may fulfill it. Finally, the decision-maker must allocate tasks to fulfill the largest number of tasks. We consider the following process:

1. while there is time, wait for resource:
   (a) a resource arrives. While the resource is available:
       i chooses and presents a task to the resource under two constraints:
         (i) the task was solved successfully less than a desired level; and
         (ii) the task has not been presented to the current resource;
       ii the resource tries to solve the task; and it is observed whether the task was solved or not.

This process is formalized mathematically by the following definition:

Definition 1 (Dynamic Allocation of Resources to solve Tasks Multiple Times - DART-MT). *Let T be a set of tasks and R a set of human resources. Consider the function $n : T \to \mathbb{N}$, $n(t)$ indicates the number of times the task t must be solved. Also, consider the function $m : R \to \mathbb{N}$, $m(r)$ indicates the number of tasks that can be assigned to resource r. ζ_t is a vector of parameters for task $t \in T$. θ_r is a vector of parameters for resource $r \in R$. $f(\zeta_t, \theta_r)$ is a function that represents the probability of the task t be fulfilled by the resource r. There is an order of resources arriving $Q = (r_1, r_2, \dots, r_{|R|})$, where $r_i \in R$ and $r_j \neq r_i$ for all $i \neq j$. The problem of dynamic resource allocation to solve tasks multiple times is defined by the tuple $(T, R, n, m, \{\zeta_t\}, \{\theta_r\}, f, Q)$ and the objective is to maximize the expected number of solved tasks.*

Because the result of a resource r trying to solve a task t is stochastic, an optimal solution is clearly contingent on past results; in the Markov Decision Process jargon, policies. Here, we considered different scenarios of knowledge about the problem DART-MT, where optimal solutions are contingent on information collected during the process. We consider three different scenarios. In every scenario probability function (f), tasks parameters and demands (T, n, and $\{\zeta_t\}$) are known beforehand; they may be obtained from calibration in another set of resources. Information about the resources ($R, m, \{\theta_r\}$ and Q) may be revealed directly or indirectly during the process. We differentiate knowledge of the current resource and future resources to define the following scenarios:

- **(KK) known-known**: fully-revealed, i.e., resources R, availability function m, parameters $\{\theta_r\}$ and arrival order Q are known before hand;
- **(KU) known-unknown**: the number of resources $|R|$ and the current resource r is fully-revealed (availability $m(r)$ and parameter θ_r); while future resources are known to be drawn from an *a priori* distribution on parameters Θ and availability M; and
- **(UU) unknown-unknown**: the number of resources $|R|$ and the *a priori* distribution on parameters Θ and availability M are revealed; the current resource r reveals its availability $m(r)$, while its parameters must be learnt from results after submitting tasks to the resource.

We note that the objective of managing to solve a task gets more difficult when knowledge is hidden from the decision maker, i.e., a decision maker should solve the largest number of tasks in scenario KK while solving the smallest number of tasks in scenario UU. To obtain an upper bound to the DART-MT problem, we relax the problem regarding the allocation and results of submitting a task to a resource. First, we allow a resource r to be partially allocated to a task t (decision $s_{r,t}$). Second, a resource r solves a task deterministic and partially, given by $P_{r,t} = f(\zeta_t, \theta_r)$. Third, $x_{r,t} = s_{r,t}P_{r,t}$ indicates the amount of task t to be solved by resource r. Then, the following Linear Programming obtains an upper bound to the DART-MT problem:

$$\max \quad \sum_{t \in \mathcal{T}} \sum_{r \in \mathcal{R}} x_{r,t} \tag{1}$$

$$\text{subject to} \quad s_{r,t} \in [0,1] \qquad \qquad \forall r \in \mathcal{R}, t \in \mathcal{T} \tag{2}$$

$$x_{r,t} = s_{r,t}P_{r,t} \qquad \qquad \forall r \in \text{subject to} \mathcal{R}, t \in \mathcal{T} \tag{3}$$

$$\sum_{t \in \mathcal{T}} s_{r,t} \leq m(r) \qquad \qquad \forall r \in \mathcal{R} \tag{4}$$

$$\sum_{r \in \mathcal{R}} x_{r,t} \leq n(t) \qquad \qquad \forall t \in \mathcal{T} \tag{5}$$

3   [STA]²O: A Shadow Test Approach to Skill-Based Task Allocation Optimization

To solve the DART-MT problem in each of the three scenarios (KK, KU, and UU), we consider three levels of time abstractions: episodes, rounds, and steps. An episode comprehends the whole process of DART-MT problems when each resource $r \in \mathcal{R}$ tries to solve $m(r)$ tasks. A round comprehends the whole interaction with a resource r when the resource r tries to solve $m(r)$ tasks. A step comprehends the interaction of a resource r and a task t. For each of the three scenarios, we define three different algorithms; all of them are based on the solution of an LP similar to LP in Eqs. 1–5. Then, at any step j of a round i consider: the current resource r_i and the set of tasks $T_{i,j}$ that was presented to resource r_i before the step j or has already been completely solved as demanded by the function $n(j)$; then, at step j a task t is drawn with probability:

$$P_t = \frac{s_{r_i,t}}{\sum_{t' \in \mathcal{T} \setminus T_{i,j}} s_{r_i,t'}}, \tag{6}$$

where $s_{r,t}$ for any $r \in \mathcal{R}, t \in \mathcal{T}$ is the solution of the LP.

To solve the DART-MT problem in scenario KK, an LP is solved for each episode. To solve the DART-MT problem in scenario KU, an LP is solved for each round. To solve the DART-MT problem in scenario UU, an LP is solved for each step. In scenarios KU and UU, where knowledge is not full-revealed *a priori*, LPs are defined based on population information, and when new information is revealed, per round or per step, a new LP is solved. This strategy of fully planning based on population information and replanning when information is renewed is used in the Shadow Test Approach (STA) [18,19,30]. The Algorithm 1 shows a general solution to DART-MT problems conditioned in one of the three scenarios.

Algorithm 1. The General Solution to DART-MT conditioned for any scenario

---

1: **if KK scenario:** solve the LP described in Equations 1-5
2: **for** for each iteration $i \in \{1, \ldots, |\mathcal{R}|\}$ **do**
3:     draw a resource r_i of $\mathcal{R}$ without replacement
4:     **if KU scenario:** solve the LP described in equations 7-11
5:     **for** $j \in \{1, \ldots, m(r_i)\}$ **do**
6:         **if UU scenario:** solve the LP problem described in equations 12-20
7:         draw a task t_j of $\mathcal{T}$ with probability: P_{t_j} (see equation 6)
8:         submit task t_j to resource r_i
9:         observe the result $x_{i,j}$, if the task was solved $(x_{i,j} = 1)$ or not $(x_{i,j} = 0)$

KK Scenario and per Episode Algorithm - LPPE. Because the scenario KK knowledge is fully revealed *a priori*, the LP to be solved is the one described in Eqs. 1–5. The solution to the LP problem provides us with a set of (partial) tasks to be presented for each resource. Note that tasks are selected stochastically, but guarantee that a task t cannot be submitted repeatedly to the same resource and cannot be submitted beyond its demand $n(t)$. Since planning occurs only once, the only adaptation occurs through set $T_{i,j}$ to avoid violating the constraints of the DART-MT problem.

KU Scenario and the per Round Algorithm - LPPR. Let $\hat{\mathcal{R}}$ be the set of resources that have already been selected from $\mathcal{R}$, $\tilde{\mathcal{R}}$ be the set of $|\mathcal{R}| - |\hat{\mathcal{R}}| - 1$ resources drawn from distributions Θ and M, and r_i be the current resource. An adaptation of the previous LP problem is modeled as follows:

$$\max \quad \sum_{t \in \mathcal{T}} \sum_{r \in \tilde{\mathcal{R}} \cup \{r_i\}} x_{r,t} \tag{7}$$

$$\text{subject to} \quad s_{r,t} \in [0,1] \qquad\qquad \forall r \in \tilde{\mathcal{R}} \cup \{r_i\}, t \in \mathcal{T} \tag{8}$$

$$x_{r,t} = s_{r,t} P_{r,t} \qquad\qquad \forall r \in \tilde{\mathcal{R}} \cup \{r_i\}, t \in \mathcal{T} \tag{9}$$

$$\sum_{t \in \mathcal{T}} s_{r,t} \leq m(r) \qquad\qquad \forall r \in \tilde{\mathcal{R}} \cup \{r_i\} \tag{10}$$

$$\sum_{r \in \tilde{\mathcal{R}} \cup \{r_i\}} x_{r,t} + \sum_{r \in \hat{\mathcal{R}}} \hat{x}_{r,t} \leq n(t) \qquad\qquad \forall t \in \mathcal{T} \tag{11}$$

with $\hat{x}_{r,t} = 1$ if the resource r succeeded on the already presented task t and 0 otherwise. Instead of just solving an LP problem once for the episode, the

LPPR algorithm solves an LP problem after arriving and being revealed the next resource. Replanning is done considering previous results on tasks submitted to previous resources and the knowledge regarding the current resource, whereas future resources are drawn from *a priori* distributions.

UU Scenario and the Per Step Algorithm - LPPS. Let $\hat{\mathcal{R}}$ be the set of resources that have already been selected from $\mathcal{R}$, $\tilde{\mathcal{R}}$ be the set of $|\mathcal{R}| - |\hat{\mathcal{R}}| - 1$ resources drawn from distributions Θ and M, r_i be the current resource, T_i the set of tasks already submitted to resource r_i, and $\hat{P}_{r_i,t}$ be the estimation probability of success given current resource r_i and task t.

An adaptation of the previous LP problem is modeled as follows:

$$\max \quad \sum_{t \in \mathcal{T}} \sum_{r \in \tilde{\mathcal{R}}} x_{r,t} + \sum_{t \in \mathcal{T} \setminus T_i} x_{r_i,t} \tag{12}$$

$$\text{subject to} \quad s_{r,t} \in [0,1] \qquad \forall r \in \tilde{\mathcal{R}}, t \in \mathcal{T} \tag{13}$$

$$s_{r_i,t} \in [0,1] \qquad \forall t \in \mathcal{T} \setminus T_i \tag{14}$$

$$x_{r,t} = s_{r,t} P_{r,t} \qquad \forall r \in \tilde{\mathcal{R}}, t \in \mathcal{T} \tag{15}$$

$$x_{r_i,t} = s_{r_i,t} \hat{P}_{r_i,t} \qquad \forall t \in \mathcal{T} \setminus T_i \tag{16}$$

$$\sum_{t \in \mathcal{T}} s_{r,t} \leq m(r) \qquad \forall r \in \tilde{\mathcal{R}} \tag{17}$$

$$|T_i| + \sum_{t \in \mathcal{T} \setminus T_i} s_{r_i,t} \leq m(r_i) \tag{18}$$

$$x_{r_i,t} + \sum_{r \in \tilde{\mathcal{R}}} x_{r,t} + \sum_{r \in \hat{\mathcal{R}}} \hat{x}_{r,t} \leq n(t) \qquad \forall t \in \mathcal{T} \setminus T_i \tag{19}$$

$$\hat{x}_{r_i,t} + \sum_{r \in \tilde{\mathcal{R}}} x_{r,t} + \sum_{r \in \hat{\mathcal{R}}} \hat{x}_{r,t} \leq n(t) \qquad \forall t \in T_i \tag{20}$$

with $\hat{x}_{r,t} = 1$ if the resource r succeeded on the already presented task t and 0 otherwise.

Here, the LPPS algorithm solves an LP problem on every step (after each task submission). Again, replanning is done considering previous results on tasks submitted to previous resources and the knowledge regarding the current resource, whereas future resources are drawn from *a priori* distributions. However, knowledge regarding the current resource is updated after every trial of solving a task, depending on the parameters of the task already submitted and the given results, success or failure. We postpone the discussion of estimating parameters of the current resource to Sect. 5.

4   Related Work

In this work, a new problem (DART-MT) is defined with specific settings. Therefore, there is not an obvious state-of-the art method to be compared to ours. We used as baselines a random selection and the heuristic *easier first*, even though methods applicable to similar problems are briefly compared to our approach in the following paragraphs.

Many works make use of resource skills to improve the task allocation problem [4,10,14,20,27,29]. Some of them use this approach to improve allocation in project management or in crowdsourcing context, where the tasks are allocated to more than one resource to receive a number of solutions. However, these works assume the ability of the resources are given or can be estimated from a content-based approach, which is not always the case. Our work stands on a cold start, i.e., we do not have any previous information about the resources. In recent works, social search engines have become very useful as it uses information about the resources so as to improve task allocation [2,3,8,11]. These studies use techniques like ranking function and information retrieval in the context of crowdsourcing or the Q&A Problem. Although similar to our work in the sense of optimizing task allocation, we do not use previous information about the resources and estimate their skills from the interaction with the tasks. Also, we do not use a threshold to reject weaker resources.

The STA is largely discussed in the context of adaptive tests [18,19,30]. In this context, many works address the problem of task allocation by learning information about the resources [12,24]. They repeatedly administer tasks to resources; however, the objective of these works is restricted to better estimating the resources' abilities. We use the same approach in a different context and with a different purpose since we use resource characterization to optimize the allocation of tasks by maximizing the number of solutions, not having resource characterization as an end in itself. In some works [9,25,31], resources and tasks are dynamically characterized. The tasks are allocated to a resource if the probability of a correct answer is above a threshold. Our study does not limit the allocation of tasks by a threshold. Also, the purpose is different, in their case, each resource receives only one task, and each task needs only one solution.

5 Experiments and Results

To evaluate the proposed algorithms, we adapt real data from an exam of mathematics to emulate tasks (questions) and resources (students). Within this data, we construct different DART-MT problems to identify the difference in the performance of our algorithms in the three scenarios KK, KU, and UU. Finally, we present the improvement of our algorithms against two baselines: uniformly random and easier first.

5.1 Real World Database

Evaluating in Real World Data. Real data can be obtained, basically, in two ways: online or offline. The online test consists of presenting tasks to human resources while the process is running, whereas the offline test consists of using a database that was previously collected. The advantage of the online test is to apply the method in the real world with resources interacting directly with the allocation process. The downside is that access to such resources is very costly, so getting a reasonable amount of human resources available to receive tasks is very complicated. The advantage of the offline test is precisely that it can take advantage of a database, but the interaction of the process with the real world is lost.

To use an offline test, the database need to meet certain criteria: (i) the independence of task solution conditioned on the resource, and (ii) all tasks have been submitted to all resources. The success or failure of each task submission is obtained from the history taken from the database. We can choose any task for a resource and check in the history if he has solved the task correctly or not. One can arbitrarily choose resources, their arrival order, and their availability without loss of generality. Availability must always be at most equal to the total amount of existing tasks in the database. It is equally possible to choose tasks and the number of solutions for them arbitrarily. The number of solutions demanded by each task must be equal to or less than the total amount of resources in the database. With historical data in hand, it remains to model the parameters ζ, θ, and the success probability function f.

ENEM Database. A database of the 2012 Brazilian baccalaureate examination (ENEM) containing responses from ten thousand people (resources) and 185 questions (tasks) was used. The data are public and can be downloaded from the transparency portal [13]. ENEM uses Item Response Theory to model tasks (ζ), resources (θ), and the success probability function (f). This exam is used to assess the level of knowledge (skills) of students in four areas of knowledge: human sciences; natural sciences; languages; and mathematics. In this work, only the exam of mathematics was considered.

We calibrate task parameters on data from 10,000 students and make use of the three-parameter logistic model, where parameter b indicates the difficulty of a question. We make use of such parameters to construct a baseline policy where the current resource always receives the easiest allowed task, the heuristic *easier first*.

5.2   Defining DART-MT Problems

From the database with the answers of ten thousand ENEM students, the average chance of the task t to be correctly solved is estimated by:

$$k(t) = \frac{\text{number of resources that solved the task } t \text{ correctly}}{\text{number of resources that tried to solve the task } t} \tag{21}$$

The responses of all students in the database were used to calculate $k(t)$ and $k_{mean} = \frac{\sum_{t \in T} k(t)}{|T|}$. However only one hundred of them are sampled to evaluate the method at each episode. So, from now on we consider $|\mathcal{R}| = 100$ and the 45 math items from the ENEM's mathematics test as the tasks, so $|T| = 45$. Depending on the number of solutions for each task, $n(t)$, and the availability of each resource, $m(r)$, solving all the required tasks can be more or less difficult. In this way, we define the following difficulty levels:

– **Level 1:**

$$n(t) = \lceil |\mathcal{R}| \times k_{mean} \rceil \quad \forall t \in \mathcal{T} \tag{22}$$

$$m(r) = \lceil |\mathcal{T}| \times k_{mean} \rceil \quad \forall r \in \mathcal{R} \tag{23}$$

– **Level 2:**

$$n(t) = \lceil |\mathcal{R}| \times k(t) \rceil \quad \forall t \in \mathcal{T} \tag{24}$$

$$m(r) = \lceil |\mathcal{T}| \times k_{mean} \rceil \quad \forall r \in \mathcal{R} \tag{25}$$

– **Level 3:** $n(t)$ is defined as the number of correct solutions for the task t and $m(r)$ as the number of tasks the resource r solved correctly.

In levels 1, 2, and 3, we have that $\sum_{t \in \mathcal{T}} n(t) \approx \sum_{r \in \mathcal{R}} m(r)$. Note that we can only get all the desired solutions for difficulty levels 1 and 2 if all resources have the same ability or/and if all tasks have the same difficulty. To obtain all the solutions for level 3, we have to present for each resource exactly the tasks that user can solve correctly.

One way to relax the levels is to decrease the number of solutions desired for each task, for example, $n'(t) = n(t) \times 0.5$. We apply such relaxed version to level 3 to obtain the **Level 4**.

Another more-relaxed level, let's call **Level 0**, is: given k_{mean}, we compute $n(t)$ and $m(r)$ as constant values such that

$$|\mathcal{R}| \times m(r) \times k_{mean} = |\mathcal{T}| \times n(t) \tag{26}$$

For this level, we choose $m(r) = 16 \approx \sum_{t \in \mathcal{T}} k(t)$ and we get $n(t) = 12$ from the Eq. 26 with $k_{mean} = 0.34$.

Empirical Experiment. For each difficulty level and algorithm, we run 100 episodes. For each episode, one hundred resources are used from a previous random sample among the ten thousand available students. All simulations were performed in a *Google Colaboratory* environment. All codes were written in Python 3 language. To solve LP problems, we used the *Pulp* module and the open-source *solver Coin-or Branch and Cut* (CBC) [22].

5.3 Results Using Real World Data

We use the five levels defined before to evaluate our proposed algorithms. The results obtained for each level and in each scenario are shown in the Table 1. The proposed approaches are compared against two baselines: the uniform random selection and the heuristic *easier first*. The comparison is by the number of tasks solved correctly. Besides solving each scenario with the corresponding time abstraction, we also apply per round algorithm to the scenario KK to be contingent on trial results.

Table 1. Results for all scenarios and algorithms.

Scenario	Algorithm	# Solutions	STD	# Steps
Results for level 0				
All	Random	493,70	15,75	1581,03
All	Easier first	487,56	13,19	1520,94
KK	Per episode	526,85	8,10	1540,06
KK	Per round	530,46	7,16	1556,22
KU	Per round	528,29	9,73	1557,46
UU	Per step	508,94	16,96	1589,55
Results for level 1				
All	Random	568,09	31,43	1600
All	Easier first	703,68	29,84	1600
KK	Per episode	805,03	35,04	1600
KK	Per round	813,97	36,68	1600
KU	Per round	810,44	35,49	1600
UU	Per step	765,34	33,77	1600
Results for level 2				
All	Random	571,53	32,36	1600
All	Easier first	833,6	44,54	1600
KK	Per episode	888,74	43,89	1600
KK	Per round	888,86	43,86	1600
KU	Per round	888,26	44,13	1600
UU	Per step	854,49	45,03	1600
Results for level 3				
All	Random	695,50	68,78	1607,28
All	Easier first	969,57	82,05	1607,28
KK	Per episode	1011,84	82,16	1607,28
KK	Per round	1011,74	82,10	1607,28
KU	Per round	1011,69	81,71	1607,28
UU	Per step	977,05	80,44	1607,28
Results for level 4				
All	Random	701,29	69,97	1606,21
All	Easier first	741,82	54,19	1556,10
KK	Per episode	797,04	49,55	1562,45
KK	Per round	806,12	47,18	1587,62
KU	Per round	800,56	48,23	1574,03
UU	Per step	781,24	54,94	1595,96

Discussion. Regardless of the knowledge of the resource's ability to solve the tasks, the algorithms LPPE, LPPR, and LPPS obtained quantities of solutions greater than the baselines (random and easier first) for all levels. In the case resources' abilities are well known, the algorithms LPPE and LPPR are better than the baselines and can be computed in a feasible time (less than three minutes per episode). If we do not know the ability of the resources to whom the tasks should be submitted, we use the algorithm LPPS, which furnished a better allocation than the baselines for all five levels. The algorithm LPPS takes about forty minutes per episode, and the resources need to wait no more than 1.7 s to receive a task to try to solve (it was the case of the level 0, with the results shown on Table 1).

Note that the algorithms only make sense to be applied in certain scenarios. For example, in scenario UU, it doesn't make sense to use the per episode algorithm since we will allocate the tasks based on the solution of an LP problem whose resources' abilities were unknown. Also, if we are in the scenario KK, it is not necessary to solve an LP problem at each step if the resources' abilities are well known from the beginning. To better visualize the results, box plot graphics are shown for all scenarios, levels, and algorithms in Fig. 1. The name of the algorithms are abbreviated: random (rand), easier first (E-F), per episode (Ep), per round (Res), and per step (Sub).

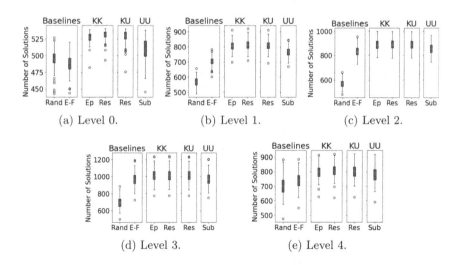

(a) Level 0.   (b) Level 1.   (c) Level 2.

(d) Level 3.   (e) Level 4.

Fig. 1. Box plots results for all scenarios and levels

To better analyze the results, we used statistical tests to verify which algorithm produced the higher number of solutions. The first step was to use the Shapiro-Wilk test to determine the data normality. With 95% of significance, the data do not deviate from a normal distribution, except for the results for some algorithms of level 0. As we have a lot of samples (100 per episode), we used the

Analysis of Variance (ANOVA) test to compare the algorithms for each scenario in all levels, including level 0. The results of the ANOVA tests determined that for all scenarios in all levels, there are statistically significant differences, with 95% of significance. From the ANOVA test results, the number of solutions is not statistically equivalent between the algorithms used in each scenario and level. Then, we used the Tukey pairwise multiple comparisons (*post hoc*) test to verify which means differ, with 95% of significance. Specific algorithms are used in each scenario:

- KK: random; easier first; per episode; and per round.
- KU: random; easier first; and per round.
- UU: random; easier first; and per step.

From the results of the Tukey *post hoc* test, the proposed algorithms per episode, per round and per step are better than the baselines (random and easier first) for all scenarios in all levels, except for the algorithm LPPS and the baseline easier first in the scenario UU at level 3. In this comparison, the Tukey test resulted in no statistical difference in the number of solutions. Figure 2 shows (for the first episode) the number of submissions and solutions, the mean of the difficulties of allocated tasks, and the ability of the resources for these two algorithms for scenario UU at level 3. From this figure, resources with higher abilities are available to receive more tasks at level 3. This is a very appropriate case for the easier first algorithm since the more tasks the resources receive following the easier first heuristic, the more difficult tasks received.

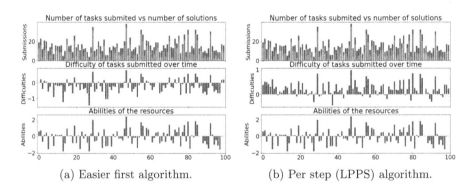

(a) Easier first algorithm.   (b) Per step (LPPS) algorithm.

Fig. 2. Algorithms at level 3, on the first episode.

The algorithms per episode and per round in the scenario KK are statistically equivalent. It is expected since in the scenario KK, all the resources' abilities are well known beforehand, then the gain in doing replannings after each round is not significantly better than planning once at the beginning of the episode.

The heuristic easier first is better than the random selection at most levels, except at level 0. Figure 3 shows (for the mean of all episodes) the number of submissions and solutions, the mean of the difficulties of allocated tasks, and the ability of the resources for these two algorithms at level 0. The mean of all episodes is used so that the influence of the ability of the resources is diminished. We can see that the mean of the abilities goes to zero. In the easier first allocation, it is clear the difficulty of tasks increases over the steps, whereas the number of solutions decreases. While in the random allocation, the difficulty of the tasks is approximately uniform. The last resources receive tasks slightly more difficult because the easier already received the needed number of solutions. These characteristics from both algorithms explain why the random made more allocations (approximately 4%) than the easier first, which led to receiving more solutions (approximately 1%).

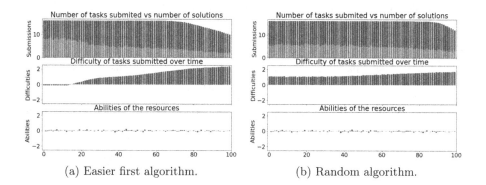

(a) Easier first algorithm.   (b) Random algorithm.

Fig. 3. Algorithms at level 0, mean of all episodes.

6   Conclusion and Future Work

We can compare how far is the algorithm LPPS from the baselines and from the *optimal case*, that is, when we have the most information: scenario KK and the solution of the LP problems is updated per round. We have an increase in the number of solutions compared with the baselines on all levels. Compared with the baseline easier first, we have up to almost 9% more solutions; compared with the random, we have up to 49% more solutions. When we compare with the *optimal case*, the UU scenario using the LPPS algorithm reached up to 96% of the number of solutions of the optimal case.

These processes could be used to submit fewer tasks to resources without losing important information. For example, when asking people to answer polls or to generate databases to be used in important applications such as Q&A, intelligent tutoring systems, or natural language processing. In addition to the abilities of the resources being unknown, there is also the case in which the complexity of solving the tasks is unknown. Therefore, estimating parameters that characterize such tasks become necessary [7,21]. The task parameters can be obtained from the solutions of the tasks or from a description of them [16]. We plan on considering scenarios where the resource is not continuously available to receive tasks, but he arrives and goes. Also, we plan on developing a real experiment with human resources receiving tasks and not only using real data from other applications.

References

1. Abrams, D.S., Yoon, A.H.: The luck of the draw: using random case assignment to investigate attorney ability. Univ. Chic. Law Rev. **74**(4), 1145–1177 (2007). http://www.jstor.org/stable/20141859
2. Ali, I., Chang, R.Y., Hsu, C.H.: SOQAS: distributively finding high-quality answers in dynamic social networks. IEEE Access **6**, 55074–55089 (2018)
3. Aydin, B.I., Yilmaz, Y.S., Demirbas, M.: A crowdsourced "who wants to be a millionaire?" player. Concurr. Comput. Pract. Exp. **33**(8), e4168 (2017)
4. Ben Rjab, A., Kharoune, M., Miklos, Z., Martin, A.: Characterization of experts in crowdsourcing platforms. In: Vejnarová, J., Kratochvíl, V. (eds.) BELIEF 2016. LNCS (LNAI), vol. 9861, pp. 97–104. Springer, Cham (2016). https://doi.org/10.1007/978-3-319-45559-4_10
5. Bezerra, C.M., Araújo, D.R., Macario, V.: Allocation of volunteers in non-governmental organizations aided by non-supervised learning. In: 2016 5th Brazilian Conference on Intelligent Systems (BRACIS), pp. 223–228 (2016)
6. Birnbaum, A.: Some latent trait models and their use in inferring an examinee's ability. In: Lord, F.M., Novick, M.R. (eds.) Statistical Theories of Mental Test Scores, Reading, Charlotte, NC, pp. 397–479. Addison-Wesley (1968)
7. Bock, R.D., Aitkin, M.: Marginal maximum likelihood estimation of item parameters: application of an EM algorithm. Psychometrika **46**(4), 443–459 (1981)
8. Difallah, D.E., Demartini, G., Cudré-Mauroux, P.: Pick-a-crowd: tell me what you like, and i'll tell you what to do: A crowdsourcing platform for personalized human intelligence task assignment based on social networks. In: WWW 2013 - Proceedings of the 22nd International Conference on World Wide Web, pp. 367–377 (2013)
9. Ekman, P., Bellevik, S., Dimitrakakis, C., Tossou, A.: Learning to match. In: 1st International Workshop on Value-Aware and Multistakeholder Recommendation (2017)
10. Fan, J., Li, G., Ooi, B.C., Tan, K.l., Feng, J.: iCrowd: an adaptive crowdsourcing framework. In: Proceedings of the 2015 ACM SIGMOD International Conference on Management of Data, pp. 1015–1030. Association for Computing Machinery (2015)
11. Horowitz, D., Kamvar, S.D.: The anatomy of a large-scale social search engine. In: Proceedings of the 19th International Conference on World Wide Web - WWW 2010, Raleigh, North Carolina, USA, p. 431. ACM Press (2010)
12. Huang, Y.M., Lin, Y.T., Cheng, S.C.: An adaptive testing system for supporting versatile educational assessment. Comput. Educ. **52**(1), 53–67 (2009)
13. INEP: Instituto nacional de educação e pesquisas educacionais anísio teixeira - entenda sua nota no enem (2012). http://download.inep.gov.br/educacao_basica/enem/guia_participante/2013/guia_do_participante_notas.pdf. Accessed 19 June 2021

14. Krstikj, A., Esparza, C.R.M.G., Mora-Vargas, J., Escobar, H.L.: Volunteers in lockdowns: decision support tool for allocation of volunteers during a lockdown. In: Regis-Hernández, F., Mora-Vargas, J., Sánchez-Partida, D., Ruiz, A. (eds.) Humanitarian Logistics from the Disaster Risk Reduction Perspective: Theory and Applications, pp. 429–446. Springer, Cham (2022). https://doi.org/10.1007/978-3-030-90877-5_15

15. Kuhn, H.W.: The Hungarian method for the assignment problem. Nav. Res. Logist. (NRL) **2**(1), 83–97 (1955)

16. Benedetto, L., Cappelli, A., Turrin, R., Cremonesi, P.: R2DE: a NLP approach to estimating IRT parameters of newly generated questions. In: Proceedings of the 10th International Conference on Learning Analytics and Knowledge (2020)

17. van der Linden, W.J.: Constrained adaptive testing with shadow tests. In: van der Linden, W.J., Glas, G.A. (eds.) Computerized Adaptive Testing: Theory and Practice, New York, Boston, Dordrecht, London, Moscow, pp. 27–52. Kluwer Academic Publishers (2000)

18. van der Linden, W.J., Jiang, B.: A shadow-test approach to adaptive item calibration. Psychometrika Soc. **85**(2), 301–321 (2020)

19. van der Linden, W.J., Veldkamp, B.P.: Constraining item exposure in computerized adaptive testing with shadow tests. J. Educ. Behav. Stat. **29**(3), 273–291 (2004)

20. Liu, C., Gao, X., Wu, F., Chen, G.: QITA: quality inference based task assignment in mobile crowdsensing. In: Pahl, C., Vukovic, M., Yin, J., Yu, Q. (eds.) ICSOC 2018. LNCS, vol. 11236, pp. 363–370. Springer, Cham (2018). https://doi.org/10.1007/978-3-030-03596-9_26

21. Mislevy, R.J.: Bayes modal estimation in item response models. Psychometric **51**, 177–195 (1986)

22. Mitchell, S., Kean, A., Mason, A., O'Sullivan, M., Phillips, A., Peschiera, F.: Optimization with pulp (2009). https://coin-or.github.io/pulp/index.html. Accessed 20 June 2021

23. Munkres, J.: Algorithms for the assignment and transportation problems. J. Soc. Ind. Appl. Math. **5**(1), 32–38 (1957)

24. Mwamikazi, E., Fournier-Viger, P., Moghrabi, C., Barhoumi, A., Baudouin, R.: An adaptive questionnaire for automatic identification of learning styles. In: Ali, M., Pan, J.-S., Chen, S.-M., Horng, M.-F. (eds.) IEA/AIE 2014. LNCS (LNAI), vol. 8481, pp. 399–409. Springer, Cham (2014). https://doi.org/10.1007/978-3-319-07455-9_42

25. Negishi, K., Ito, H., Matsubara, M., Morishima, A.: A skill-based worksharing approach for microtask assignment. In: 2021 IEEE International Conference on Big Data (Big Data), pp. 3544–3547 (2021)

26. Paschoal, A.F.A., et al.: Pirá: A bilingual portuguese-english dataset for question-answering about the ocean. In: Demartini, G., Zuccon, G., Culpepper, J.S., Huang, Z., Tong, H. (eds.) CIKM 2021: The 30th ACM International Conference on Information and Knowledge Management, Virtual Event, Queensland, Australia, 1–5 November 2021, pp. 4544–4553. ACM (2021). https://doi.org/10.1145/3459637.3482012

27. Shekhar, G., Bodkhe, S., Fernandes, K.: On-demand intelligent resource assessment and allocation system using NLP for project management. In: AMCIS 2020 Proceedings, vol. 8 (2020)

28. Tran-Thanh, L., Stein, S., Rogers, A., Jennings, N.: Efficient crowdsourcing of unknown experts using bounded multi-armed bandits. Artif. Intell. **214**, 89–111 (2014)

29. Tu, J., Cheng, P., Chen, L.: Quality-assured synchronized task assignment in crowdsourcing. IEEE Trans. Knowl. Data Eng. **33**(3), 1156–1168 (2021)
30. Veldkamp, B.P.: Bayesian item selection in constrained adaptive testing using shadow tests. Psicologica **31**(1), 149–169 (2010)
31. Yu, D., Wang, Y., Zhou, Z.: Software crowdsourcing task allocation algorithm based on dynamic utility. IEEE Access **7**, 33094–33106 (2019)

A Multi-algorithm Approach to the Optimization of Thermal Power Plants Operation

Gabriela T. Justino[1(✉)], Gabriela C. Freitas[1], Camilla B. Batista[1],
Kleyton P. Cotta[1], Bruno Deon[1], Flávio L. Loução Jr.[1],
Rodrigo J. S. de Almeida[2], and Carlos A. A. de Araújo Jr.[2]

[1] Research, Development and Innovation Department (RD&I), Radix Engineering
and Software Development, Rio de Janeiro, RJ, Brazil
{gabriela.justino,pdi}@radixeng.com.br
[2] Paraíba Power Plants S.A. - EPASA, João Pessoa, PB, Brazil
https://www.radixeng.com/

Abstract. A new multi-algorithm approach for the daily optimization
of thermal power plants and the Economic Dispatch was tested. For this,
the dispatches were clustered in different groups based on their duration
and the total power output requested. Genetic Algorithm, Differential
Evolution and Simulated Annealing were selected for implementation
and were employed according to the distinct characteristics of each dispatch. A monthly improvement of up to $2{,}45 \times 10^5$ R\$ in the gross profit
of the thermal power plant with the use of the optimization tool was
estimated.

Keywords: Optimization · Machine Learning · Economic Dispatch ·
Thermoelectric Generation · Fuel Consumption

1 Introduction

The real-time operation of systems, especially in the generation of electricity,
requires agility and assertiveness in decision-making. The availability, reliability
and performance of the generating machines of a Thermoelectric Power Plant
(TPP) are critical issues to maximize the economic results of the business and
to guarantee the fulfillment of the demand of the electric sector.

A part of the Brazilian thermoelectric park may remain idle if hydrological
conditions are favorable. In most countries, combined cycle coal or gas power
plants generally do not experience long-term idleness, due to operation as the
backbone of the system, being dispatched almost continuously. In Brazil, thermal
electricity is normally used to support base generation in the electrical system:
for peak generation, with daily activation or for at least a large fraction of
working days [5].

© The Author(s), under exclusive license to Springer Nature Switzerland AG 2023
M. C. Naldi and R. A. C. Bianchi (Eds.): BRACIS 2023, LNAI 14195, pp. 209–223, 2023.
https://doi.org/10.1007/978-3-031-45368-7_14

System planning is centralized at the ONS (National System Operator) and the daily energy operation, in general, is a problem for the plants, which have availability contracts. This type of plant has high pollutant emission rates and, considering the current scenario of search for solutions to reduce greenhouse gas emissions, it is important to guarantee the maximum energy performance of complementary plants, necessary for the balance of the electrical system. In addition, given the current scenario of the electricity sector and the way in which generating agents are remunerated, the reduction of operating costs brings numerous benefits.

In the context of the plant, the large number of generating units and auxiliary equipment makes its operation arduous and the maximization of performance is complex. This makes the actions taken by operators not completely efficient from an energy and economic point of view. Furthermore, the simultaneous start-up of many generating units creates scenarios of operational complexity that can lead to low-performance operational conditions. Currently, the plants do not have a computational tool to predict trends, identify failures and operational deviations in an automated and reliable way for engines and other large generating machines.

1.1 Motivation

The present work belongs to an ANEEL (National Electric Energy Agency) research and development (R&D) project, which consists of developing a tool to assist the operation of a thermoelectric plant, aiming to improve its energy performance, reduce costs and mitigate process uncertainties, through the optimization of the relationship between fuel used versus energy generated. In general, the objective is to enhance the energy produced for a lower fuel consumption through the health index of the generating units. The expected result obtained by this tool comes in the format of an optimal operational configuration of the thermal power plant: which engines should be used and how much of the generating capacity of each of them should be employed.

2 Method Review and Selection

Heuristic methods are groups of practical guidelines coming from experience. Through trial and error, they help in finding good, but not always optimal, solutions. Their use is of particular interest in the resolution of complex problems, without a clear analytical solution.

From the application of heuristic rules, many meta-heuristic techniques were developed. They can be seen as a generic algorithmic framework which can be applied to different optimization problems with only a few specific modifications [13]. Although they do not guarantee the optimal solution, these methods are widely used, because they provide good solutions (close to the optimal) in a reasonable computation time, enabling their implementation in real time. These methods have also been used successfully to solve high-dimensional problems in many fields of study, including the electric sector [17,18].

The Economic Dispatch (ED) optimization problem is non-convex, therefore finding the optimal solution becomes difficult [15]. Nevertheless, modern meta-heuristic algorithms are promising alternatives for the resolution of such complex problems [7]. Genetic Algorithm was used to solve the ED problem with convex and non-convex cost functions and reached satisfactory results when applied on the electrical grid of Crete Island [4]. The Ant Colony technique is also a viable option for solving ED problems. Silva et al. [13] used this algorithm in conjunction with a Sensitivity Matrix based on the information provided by the Lagrange multipliers to improve the biologically inspired search process. Al-Amyal et al. [1] proposed a multistage ant colony algorithm to solve the ED problem for six (6) thermal generation plants with good results. The Grey Wolf Optimizer is another option for solving ED problems that are nonlinear, non-convex and discontinuous in nature, with numerous equality and inequality constraints. Tests with model systems with up to 80 generators showed that this method is effective in solving ED problems, when compared to other similar algorithms [7]. Nascimento et al. [11] used the Differential Evolution algorithm to solve the ED problem. Their proposed methodology reduced the total fuel cost of the plants significantly. An algorithm based on bat behavior and echolocation was also successfully employed by Hanafi and Dalimi [6] to solve the ED problem in six (6) TPPs. Wang et al. [16] compared traditional and adaptative versions of the Particle Swarm Optimization (PSO) algorithm in the resolution of the ED problem. The modified version of the algorithm used an immune mechanism and adaptative weights controlled by the fitness of the particle. When applied in a system with 10 generating units, the adaptative PSO performed better in terms of solution quality and stability, but the increased complexity of the algorithm also increased its execution time.

The Environmental and Economic Dispatch (EED) is also an important daily optimization task in the operation of many energy plants. It presents two conflicting objectives: fuel cost and gas emission. This problem can be formulated as a highly restricted non-linear multiobjective optimization. Thus, algorithms based on concepts such as Pareto dominance and crowding distance are viable tools for solving the EED problem. One such algorithm is the Elitist Non-Dominated Sorting Genetic Algorithm [3,12]. The goal-attainment method is also an option to solve EED problems. Basu [2] used this method in conjunction with the Simulated Annealing technique to solve an EED problem.

Also in the electric sector, the use of neural networks to solve optimization problems has been increasing. This methodology is usually employed for online optimization of the system power flow [9,14].

All these works focused only on the use of one optimization algorithm to solve the economic dispatch problem, without considering the particularities of each dispatch regarding the power requested of the TPP in relation to its overall capacity. However, recent works in other fields started to use a combination of algorithms in order to find better solutions for complex problems. In some cases, an association of meta-heuristic algorithms was employed; e.g., the combined use of Genetic Algorithm and Simulated Annealing to solve an Emergency Location Routing Problem [10]. Other works tested unions of algorithms with distinct fundamentals, such as the combination of Recurrent Neural Network and Particle Swarm Optimization [8].

With the objective of implementing in plant operation an optimization tool capable of finding good solutions without a high computational cost, five different meta-heuristic algorithms were tested. Four of them had a mono-objective approach: adaptative Particle Swarm Optimization (PSO), Genetic Algorithm (GA), Simulated Annealing (SA) and Differential Evolution (ED). With a multi-objective approach, the Elitist Non-Dominated Sorting Genetic Algorithm (NSGA-II) was tested.

3   Problem Formulation

3.1   System Description

In this article we study the power generation process of a thermoelectric complex, which involves the operation of two identical TPPs. Together they have an installed capacity of 342 MW, enough energy to serve a population of about 980 thousand inhabitants. The structure of the TPP is composed of diesel engines as main equipment, generators and subsystems linked to them, guaranteeing their maintenance and operation.

Each TPP has 19 MAN/STX engines of type 18V32/40 with capacity to generate 8.76 MW and 1 MAN/STX engine of type 9L32/40 with capacity to generate 4.38 MW. The engines are supercharged, nonreversible, 4-stroke, with 320 mm cylinder diameter, 400 mm piston stroke and 720 rpm rotation. The 18V32/40 engines contains 18 cylinders V-shaped and the 9L32/40 engines are L-shaped with 9 cylinders.

The auxiliary systems of the TPP are:

- **Intake Air System:** the air is filtered, compressed, cooled and finally admitted to the engine;
- **Fuel Injection System:** the fuel (usually HFO oil) is heated and filtered to be fed into the engine. If diesel-LFO oil is used, no heating is necessary, as its viscosity is low enough;
- **Exhaust System:** the exhaust gases pass through the turbocharger turbines and exchange heat with the water in the recovery boilers to generate steam. Then, they are sent to exhaustion in the chimneys;
- **Boiler System:** steam from the boilers is used to heat the fuel. The recovery boilers can only be activated when there is generation. Otherwise, auxiliary boilers (diesel oil and natural gas burner) are activated;
- **Lubrication System:** the lubrication oil stored in the crankcase of the engines will promote the lubrication of all moving parts of the engine that come into contact with each other;

– **Cooling System:** water is used for the cooling of the engine and its parts (nozzles and intake air system). This water comes from radiators and also cools heat exchangers from other systems.

3.2 Economic Dispatch Problem

The aim of this work is to solve the ED problem for TPPs. The ED consists in minimizing the operational costs while fulfilling the energy demand requested from the plant. Simultaneously, the plant must also obey physical and operational restrictions. The fuel consumption of a thermal power plant is responsible for approximately 60% of its operational costs [6].

The performance of the generating units regarding other factors necessary for efficient operation was also considered. Using data from the air intake/exhaust, lubrication and cooling systems of each engine, a Health Index (HI) is calculated in real time. This parameter refers to the operational condition of each generating unit regarding need for maintenance and probability of failures and can be used to allocate the engines during operation. Therefore, the objective function of the ED problem considered the fuel consumption and the Health Index.

The specification of the dispatch requested by the ONS is the main input of the algorithm. The daily schedule specifies the power demand for 48 intervals of 30 min each. The schedule is always received a day before operation starts, allowing for time to plan and optimize the operation during the following day. The initial condition of the engines is obtained from sensor data regarding active and reactive power and the hourmeter of each generating unit, as well as machine learning models regarding fuel consumption. The optimization tool outputs an optimized configuration of the TPP for the execution of the requested dispatch, in the form of the values for active and reactive power for each engine in each time interval. These are the decision variables for the model and are hereafter represented by $P_{j,t}$ and $P_{reactive,j,t}$ respectively. Figure 1 shows the inputs for the optimization algorithm.

Objective Function: Equation 1 shows the part of the objective function which represents the fuel consumption (kg) of the TPP for the whole dispatch (F_{fuel}).

$$F_{fuel} = \sum_{j=1}^{N_g} \sum_{t=0}^{T} f(P_{j,t}) \tag{1}$$

where $P_{j,t}$ is the vector corresponding to the generated active power (MW) by the j-th generating unit (among the $N_g = 40$ available ones) for each time interval t of 30 min in a dispatch of $T = 48$ intervals. The function $f(P_{j,t})$ represents the relationship between fuel consumption and generated power.

The Health Index (HI) of each engine was considered as a penalty in the objective function, to apply greater penalties for turning off generating units with the best HI and turning on the ones with the worst HI. Considering HI is set between zero (0) and one (1) and these values respectively represent the

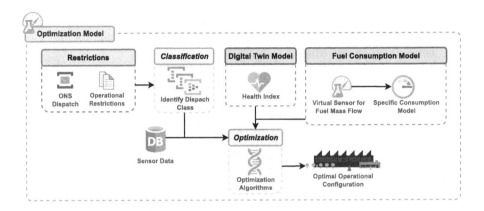

Fig. 1. Structure of the optimization tool.

worst and the best possible operational status for the engine, the penalty F_{HI} was elaborated as described in Eq. 2:

$$F_{HI} = \sum_{j=1}^{N_g} \sum_{t=0}^{T} cos^{-1}(HI_{j,t}M_{j,t})|M_{j,t}| \tag{2}$$

where $HI_{j,t}$ is the Health Index of the j-th generating unit in time interval t and $M_{j,t}$ represents the variation of status of the same engine for the same time interval according to Table 1.

Table 1. Engine status alteration index

Status Alteration	$M_{j,t}$
Turn on the engine	1
Keep the status	0
Turn off the engine	−1

Thus, the sum of these two parts constructs the objective function F_{obj} for the mono-objective algorithms (PSO, GA, SA and DE). The factor λ is used to tune their different orders of magnitude, as described in Eq. 3. For multi-objective NSGA-II, both F_{fuel} and F_{HI} were considered as individual objectives to be minimized.

$$F_{obj} = \sum_{j=1}^{N_g} \sum_{t=0}^{T} f(P_{j,t}) + \lambda \sum_{j=1}^{N_g} \sum_{t=0}^{T} cos^{-1}(HI_{j,t}M_{j,t})|M_{j,t}| \tag{3}$$

Practical Operating Constraints: Physical constraints of the generating units, such as limits for active and reactive power, are some of the main restrictions for the ED problem. Aspects relating to the internal operational rules for the power plant and the energy demand for the requested dispatch are also input for the problem restrictions. Thus, the objective function is subject to:

$$P_{min} \leq P_{j,t} \leq P_{max} \tag{4}$$

$$P_{reactive,min} \leq P_{reactive,j,t} \leq P_{reactive,max} \tag{5}$$

$$PF_{min} \leq PF_{j,t} = \frac{P_{j,t}}{\sqrt{P_{j,t}^2 + P_{reactive,j,t}^2}} \leq PF_{max} \tag{6}$$

$$P_{D,t} - 5MW \leq \sum_{j=1}^{N_g} P_{j,t} \leq P_{D,t} + 5MW \tag{7}$$

$$T_{previous,j,t} = \{\forall t_{previous} \mid M_{j,t_{previous}} \neq 0 \, and \, t_{previous} \leq t\} \tag{8}$$

$$t_{status,j,t} = t - max(T_{previous}) \tag{9}$$

$$\begin{cases} t_{status,j,t} \geq 24, & if \;\; P_{j,t-1} > 0 \, MW \\ t_{status,j,t} \geq 6, & if \;\; P_{j,t-1} = 0 \, MW \end{cases} \tag{10}$$

$$n_{boilers,t} = card(\{\forall j = 2, 7, 12, 14, 22, 27, 32, 37, 39 \mid P_{j,t} > 0 \, MW\}) \tag{11}$$

$$\begin{cases} n_{boilers,t} \geq 4, & if \;\; P_{D,t} < 200 \, MW \\ n_{boilers,t} = 9, & if \;\; P_{D,t} \geq 200 \, MW \end{cases} \tag{12}$$

Equations 4 and 5 represent the physical limits for, respectively, active and reactive power in the j-th generating unit in the time interval t. Equation 6 represents the limits for the power factor ($PF_{j,t}$) of each engine. Equation 7 elucidates that the sum of the power generated in all engines must stay within a 5MW tolerance of the total requested power ($P_{D,t}$) for each time interval t, according to ONS rules. Equation 8 defines the set of time intervals ($T_{previous,j,t}$) before a specific time interval t in which an engine j changed status (turning on or off). Equation 9 defines the difference ($t_{status,j,t}$) between a time interval t and the last change in status for an engine j. Equation 10 shows the minimum operation time of a generating unit and its minimum rest time, after which it can be turned on again. Both variables are represented by $t_{status,j,t}$, depending only on whether the engine is on or off. Equation 11 defines the number of heat recovery boilers active ($n_{boilers,t}$) at a certain time interval t, considering the boilers are associated to engines 2, 7, 12, 14, 22, 27, 32, 37 and 39. Equation 12 elucidates the number of heat recovery boilers that must be operational during a certain time interval t as a function of the scheduled energy demand.

4   Experimental Procedure

4.1   Dispatch Clusterization

The total power requested by ONS is the main input of the optimization algorithm. Evaluating data from January 1st 2018 to February 21st 2021, it was noted that the daily dispatches were diverse in their characteristics. Thus, the sum of all generated power in a daily dispatch and the fraction of hours of scheduled operation in a day were used to cluster the dispatches in five (5) distinct categories employing the Spectral Clustering method. This division allowed the separate evaluation of the optimization algorithms for distinct dispatches.

4.2   Hyperparameter Tuning and Algorithm Selection

A group of dates was randomly selected from each dispatch category to be used as model-dispatches in the estimation of the hyperparameters of the optimization algorithms and the λ factor of the objective function. Due to the great number of possible combinations, the Random Search technique was applied. The following parameters were estimated for each algorithm:

- **Genetic Algorithm:** population size, maximum number of generations, mutation probability and elitism.
- **Elitist Non-Dominated Sorting Genetic Algorithm:** population size, maximum number of generations, mutation probability.
- **Differential Evolution:** population size, maximum number of generations, crossover probability and mutation factor.
- **Simulated Annealing:** initial and minimal temperatures, maximum number of iterations per temperature, cooling factor. During the evolution of the algorithm, the temperature T_i is updated according to the cooling factor κ and the previous temperature T_{i-1}, as described in Eq. 13.

$$T_i = \frac{T_{i-1}}{1 + \kappa T_{i-1}} \tag{13}$$

- **Particle Swarm Optimization:** population size, maximum number of iterations and the inferior and superior limits of the social, cognitive and inertia coefficients. These coefficients are updated according to Eq. 14 for each iteration.

$$C_{i,j} = C_{i,max} - (C_{i,max} - C_{i,min})\frac{j}{J_{max}} \tag{14}$$

where $C_{i,j}$ is the value of coefficient i for iteration j, $C_{i,max}$ and $C_{i,min}$ are respectively the maximum and minimum values possible for coefficient i and J_{max} is the maximum number of iterations.

Afterwards, to compare the algorithms, one model-dispatch was randomly selected from each category. Each algorithm was tested with ten (10) experiments with each model-dispatch, reaching a total of 50 experiments. The feasibility of the algorithms was tested by evaluating their execution time on a computer with Intel Core i7 1.80 GHz processor and 16.0 GB RAM. The effectiveness of each algorithm in solving the ED problem was evaluated through the value of the objective function of the output solution. For multi-objective algorithm NSGA-II, an equivalent objective function was calculated using the final value of both optimized objectives, which correspond to the parts of the objective function described in Eq. 3 and used for the mono-objective algorithms. These metrics were calculated with a 95% confidence interval. This procedure was executed independently for each distinct category of dispatch.

4.3   Testing

After the selection of the algorithms for implementation, the optimization tool was tested with the dispatches scheduled for the month of January 2021. The dispatches were divided into clusters as described in Sect. 4.1 and optimized with the algorithms selected. The results obtained were compared to the performance of the TPP for the same period before the implementation of the optimization tool. The metric used for this comparison was the TPP's estimated gross profit (EGP), calculated according to Eq. 15. During the evaluated period, the price of fuel was R\$ 2,78/kg of fuel and the value received for the generated energy (CVU) was 912,28 R\$/MWh.

$$EGP = CVU * (Generated\ Energy) \\ - (Kg\ of\ Fuel) * (Price\ of\ Fuel) \tag{15}$$

5   Results and Discussion

5.1   Dispatch Clustering

A total of 390 scheduled daily dispatches were analyzed. The sum of the power to be generated throughout the day and the fraction of hours per day in operation were used to clusterize the dispatches. They were separated in five (5) distinct categories, whose characteristics are displayed in Fig. 2. It can be noted that the use of only the generated power is enough to classify the dispatches. Table 2 shows the number of dispatches per category.

Table 2. Distribution of dispatches per category

Category	A	B	C	D	E
Quantity	53	85	92	70	90

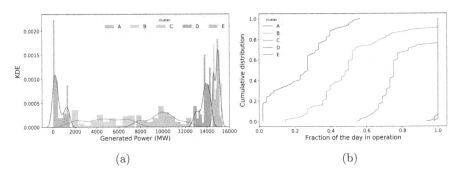

(a)   (b)

Fig. 2. Dispatch categories: (a) Kernel density estimate of dispatches per generated power; (b) Cumulative distribution of dispatches per fraction of the day in operation.

5.2 Hyperparameter Tuning and Algorithm Selection

Using model dispatches selected from each category, the best values were estimated for the λ factor of the objective function and the hyperparameters of each algorithm. Tables 3 and 4, respectively, show these results. It can be noted that the change in the scheduled power demand significantly affects the relationship between the two parts of the objective function and thus distinct values for λ are needed.

Table 3. Factor λ of the objective function

Dispatch Category	A	B	C	D	E
λ	5000	1000	5000	30000	30000

After estimating the hyperparameters, the execution time and the value of the objective function for the optimized solution were used to select the algorithms to be implemented for operation at the TPP. These results are described in Tables 5a to 5e and summarized in Fig. 3. To provide better comparison between the algorithms, for NSGA-II an equivalent objective function was calculated from the two objectives minimized, using the λ factors shown in Table 3 and implemented in the other algorithms.

NSGA-II did not perform well regarding computational cost, with execution times significantly superior to those of the other algorithms, making unfeasible its implementation in production. For dispatches from category A, the best results for objective function were provided by SA, PSO and DE, with this last one also standing out for the low execution time. For category B, GA had the best results for objective function and a reasonable execution time (257 s). Although SA was the fastest algorithm, its performance was much inferior regarding the objective function. For dispatches C, SA and PSO achieved good results for the objective function, with SA performing better in relation to computational

Table 4. Algorithm Hyperparameters

Algorithm	Abbr.	Hyperparameter	Value
Genetic Algorithm	GA	population size	500
		elitism	0.4
		mutation probability	0.4
		max. number of generations	1000
Differential Evolution	DE	population size	400
		crossover probability	0.8
		mutation factor	0.5
		max. number of generations	1500
Simulated Annealing	SA	initial temperature	10000
		min. temperature	50
		cooling factor	1.00×10^{-5}
		max. number of iterations	1500
Particle Swarm Optimization	PSO	population size	400
		max. number of iterations	1500
		min. inertia coefficient	0.01
		max. inertia coefficient	0.8
		min. social coefficient	0.01
		max. social coefficient	1
		min. cognitive coefficient	0.1
		max. cognitive coefficient	0.8
Elitist Non-Dominated Sorting Genetic Algorithm	NSGA-II	population size	400
		mutation probability	0.2
		max. number of generations	1500

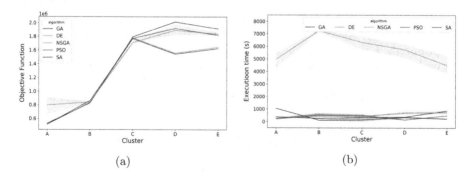

Fig. 3. Objective function (a) and execution time (b) of the algorithms per dispatch cluster.

Table 5. Execution time and objective function for dispatches.

(a) Dispatches A

Algorithm	Exec. time (s)			Fobj (x10^5)		
GA	368	±	5	5.28	±	0.0002
DE	184		0.5	5.13	±	≈ 0
SA	1046	±	14	5.13	±	≈ 0
PSO	246	±	37	5.13	±	0.00003
NSGA-II	4964	±	570	7.95	±	1.28

(b) Dispatches B

Algorithm	Exec. time (s)			Fobj (x10^5)		
GA	257	±	26	8.24	±	0.004
DE	497	±	87	8.34	±	0.005
SA	111	±	22	8.60	±	0.004
PSO	538	±	148	8.29	±	0.01
NSGA-II	7246	±	74	8.40	±	0.10

(c) Dispatches C

Algorithm	Exec. time (s)			Fobj (x10^6)		
GA	338	±	8	1.79	±	0.0001
DE	408	±	22	1.78	±	0.001
SA	140	±	191	1.77	±	0.0009
PSO	491	±	85	1.77	±	0.005
NSGA-II	6316	±	493	1.71	±	0.02

(d) Dispatches D

Algorithm	Exec. time (s)			Fobj (x10^6)		
GA	367	±	1	2.01	±	0.0003
DE	694	±	103	1.56	±	0.0004
SA	339	±	47	1.54	±	0.0001
PSO	143	±	19	1.92	±	0.0002
NSGA-II	5737	±	518	1.89	±	0.07

(e) Dispatches E

Algorithm	Exec. time (s)			Fobj (x10^6)		
GA	868	±	11	1.91	±	0.0002
DE	742	±	107	1.64	±	0.00008
SA	237	±	26	1.63	±	0.0002
PSO	483	±	10	1.82	±	0.002
NSGA-II	4500	±	722	1.84	±	0.02

cost. For category D, SA achieved the best value for the objective function and the second best for execution time. The fastest algorithm (PSO) was not as efficient in minimizing the objective function. For dispatches E, SA was the best algorithm for both evaluated metrics. Table 6 shows the algorithms selected for implementation in operation according to the category of the daily dispatch to be optimized. These results show that the commonly used approach to solve this problem (using only one algorithm regardless of differences in the economic dispatch) may not yield good results when the power requested during dispatch changes considerably. The use of different algorithms allows for optimization methods which best fit the particularities of each problem.

Table 6. Algorithms selected for implementation

Dispatch Category	Algorithm	Abbr.
A	Differential Evolution	DE
B	Genetic Algorithm	GA
C	Simulated Annealing	SA
D	Simulated Annealing	SA
E	Simulated Annealing	SA

5.3 Testing

From the results obtained by the optimization model, it was possible to estimate the gross profit of the TPP for the optimized configuration suggested by the algorithm. Figure 4a and Table 7 show these estimates, calculated using the methodology described in Sect. 4.3, for the month of January/2021.

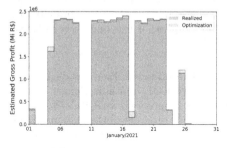
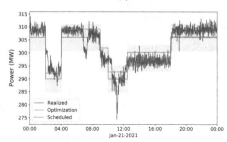

(a) Estimated gross profit.   (b) Dispatch from January 21st, 2021

Fig. 4. Tests for January/2021.

Table 7. Estimated gross profit

Variables	Real Operation	Optimized Configuration
Generated Energy (MWh)	$1{,}19 \times 10^5$	$1{,}20 \times 10^5$
Fuel Consumption (kg)	$2{,}54 \times 10^7$	$2{,}56 \times 10^7$
Estimated Gross Profit (R\$)	$3{,}83 \times 10^7$	$3{,}84 \times 10^7$

Considering an average error of $0{,}344$ kg/MWh for the fuel consumption prediction model in the evaluated period, the use of the operational configuration suggested by the optimization tool should lead to an increase in gross profit between $1{,}44 \times 10^4$ R\$ and $2{,}45 \times 10^5$ R\$. Despite the increased fuel consumption, the increment in the estimated gross profit corroborates the operational and financial benefits that using such an optimization tool can bring to thermoelectric generation.

Figure 4b shows a comparison between the ONS requested power demand, the optimized configuration suggested and the realized operation before the implementation of the algorithms, for the scheduled dispatch on January 21st, 2021. It can be noted that the suggested optimized configuration follows the tendencies of the requested schedule and respects the ONS tolerance limits of 5 MW (shown in the colored area of the graph). However, the optimized configuration is usually above the realized without use of the tool, ratifying the effectiveness of the optimization tool and explaining its superior estimated gross profit.

6 Conclusion

Through the techniques and methodology used and the obtained results, this work provides, during thermoelectric generation, more standardized operational conditions, favoring a more efficient use of the engines. The methodology employing different algorithms for distinct categories of scheduled dispatches helps in adapting the optimization tool to the specificities of these dispatch categories. Moreover, there was a significant financial gain, even if only estimated. The optimized configuration suggested by the algorithms achieved the objective of the ED problem, minimizing the fuel cost for the requested power demand and is ready for implementation in the supervisory system of the thermal power plant. This optimization tool can also be generalized to fit other thermal power plants, regardless of fuel.

Acknowledgments. The authors would like to thank Centrais Elétricas da Paraíba (EPASA) for the financial support to PD-07236-0011-2020 - Optimization of Energy Performance of Combustion Engine Thermal Power Plants with a Digital Twin approach, developed under the Research and Development program of the National Electric Energy Agency (ANEEL R&D), which the engineering company carried out Radix Engenharia e Software S/A, Rio de Janeiro, Brazil.

References

1. Al-Amyal, F., Al-attabi, K.J., Al-khayyat, A.: Multistage ant colony algorithm for economic emission dispatch problem. In: 2019 International IEEE Conference and Workshop in Óbuda on Electrical and Power Engineering (CANDO-EPE), pp. 161–166 (2019). https://doi.org/10.1109/CANDO-EPE47959.2019.9111048
2. Basu, M.: A simulated annealing-based goal-attainment method for economic emission load dispatch of fixed head hydrothermal power systems. Int. J. Electr. Power Energy Syst. **27**, 147–153 (2005). https://doi.org/10.1016/j.ijepes.2004.09.004
3. Basu, M.: Dynamic economic emission dispatch using nondominated sorting genetic algorithm-II. Int. J. Electric. Power Energy Syst. **30**, 140–149 (2008). https://doi.org/10.1016/j.ijepes.2007.06.009
4. Damousis, I., Bakirtzis, A., Dokopoulos, P.: Network-constrained economic dispatch using real-coded genetic algorithm. IEEE Trans. Power Syst. **18**(1), 198–205 (2003). https://doi.org/10.1109/TPWRS.2002.807115
5. Fonseca, M., Bezerra, U.H., Brito, J.D.A., Leite, J.C., Nascimento, M.H.R.: Predispatch of load in thermoelectric power plants considering maintenance management using fuzzy logic. IEEE Access **6**, 41379–41390 (2018). https://doi.org/10.1109/ACCESS.2018.2854612
6. Hanafi, I.F., Dalimi, I.R.: Economic load dispatch optimation of thermal power plant based on merit order and bat algorithm. In: 2019 IEEE International Conference on Innovative Research and Development (ICIRD), pp. 1–5 (2019). https://doi.org/10.1109/ICIRD47319.2019.9074734
7. Jayabarathi, T., Raghunathan, T., Adarsh, B.R., Suganthan, P.N.: Economic dispatch using hybrid grey wolf optimizer. Energy **111**, 630–641 (2016). https://doi.org/10.1016/j.energy.2016.05.105
8. Kumar, R.D., Chakrapani, A., Kannan, S.: Design and analysis on molecular level biomedical event trigger extraction using recurrent neural network-based particle swarm optimisation for covid-19 research. Int. J. Comput. Appl. Technol. **66**(3–4), 334–339 (2021). https://doi.org/10.1504/IJCAT.2021.120459

9. Liu, H., Shen, X., Guo, Q., Sun, H.: A data-driven approach towards fast economic dispatch in electricity-gas coupled systems based on artificial neural network. Appl. Energy **286**, 116480 (2021). https://doi.org/10.1016/j.apenergy.2021.116480

10. Nahavandi, B., Homayounfar, M., Daneshvar, A., Shokouhifar, M.: Hierarchical structure modelling in uncertain emergency location-routing problem using combined genetic algorithm and simulated annealing. Int. J. Comput. Appl. Technol. **68**(2), 150–163 (2022). https://doi.org/10.1504/IJCAT.2022.123466

11. Nascimento, M.H.R., Nunes, M.V.A., Rodríguez, J.L.M., Leite, J.C.: A new solution to the economical load dispatch of power plants and optimization using differential evolution. Electr. Eng. **99**, 561–571 (2017). https://doi.org/10.1007/s00202-016-0385-2

12. e Silva, M.D.A.C., Klein, C.E., Mariani, V.C., Coelho, L.D.S.: Multiobjective scatter search approach with new combination scheme applied to solve environmental/economic dispatch problem. Energy **53**(C), 14–21 (2013). https://doi.org/10.1016/j.energy.2013.02.045

13. Silva, I.C. Jr., do Nascimento, F.R., de Oliveira, E.J., Marcato, A.L., de Oliveira, L.W., Passos Filho, J.A.: Programming of thermoelectric generation systems based on a heuristic composition of ant colonies. Int. J. Electric. Power Energy Syst. **44**, 134–145 (2013). https://doi.org/10.1016/j.ijepes.2012.07.036

14. Sundaram, A.: Multiobjective multi verse optimization algorithm to solve dynamic economic emission dispatch problem with transmission loss prediction by an artificial neural network. Appl. Soft Comput. **124**, 109021 (2022). https://doi.org/10.1016/j.asoc.2022.109021

15. Tian, J., Wei, H., Tan, J.: Global optimization for power dispatch problems based on theory of moments. Int. J. Electric. Power Energy Syst. **71**, 184–194 (2015). https://doi.org/10.1016/j.ijepes.2015.02.018

16. Wang, K., Zhou, C., Jia, R., He, W.: Adaptive variable weights immune particle swarm optimization for economic dispatch of power system. In: 2020 Asia Energy and Electrical Engineering Symposium (AEEES), pp. 886–891 (2020). https://doi.org/10.1109/AEEES48850.2020.9121455

17. Wei, H., Chen, S., Pan, T., Tao, J., Zhu, M.: Capacity configuration optimisation of hybrid renewable energy system using improved grey wolf optimiser. Int. J. Comput. Appl. Technol. **68**(1), 1–11 (2022). https://doi.org/10.1504/IJCAT.2022.123234

18. Zhang, M.J., Long, D.Y., Li, D.D., Wang, X., Qin, T., Yang, J.: A novel chaotic grey wolf optimisation for high-dimensional and numerical optimisation. Int. J. Comput. Appl. Technol. **67**(2–3), 194–203 (2021). https://doi.org/10.1504/IJCAT.2021.121524

Rules and Feature Extraction

An Incremental MaxSAT-Based Model to Learn Interpretable and Balanced Classification Rules

Antônio Carlos Souza Ferreira Júnior[(✉)] and Thiago Alves Rocha

Instituto Federal de Educação, Ciência e Tecnologia do Ceará (IFCE), Fortaleza, Brazil
antonio.carlos.souza60@aluno.ifce.edu.br, thiago.alves@ifce.edu.br

Abstract. The increasing advancements in the field of machine learning have led to the development of numerous applications that effectively address a wide range of problems with accurate predictions. However, in certain cases, accuracy alone may not be sufficient. Many real-world problems also demand explanations and interpretability behind the predictions. One of the most popular interpretable models that are classification rules. This work aims to propose an incremental model for learning interpretable and balanced rules based on MaxSAT, called IMLIB. This new model was based on two other approaches, one based on SAT and the other on MaxSAT. The one based on SAT limits the size of each generated rule, making it possible to balance them. We suggest that such a set of rules seem more natural to be understood compared to a mixture of large and small rules. The approach based on MaxSAT, called IMLI, presents a technique to increase performance that involves learning a set of rules by incrementally applying the model in a dataset. Finally, IMLIB and IMLI are compared using diverse databases. IMLIB obtained results comparable to IMLI in terms of accuracy, generating more balanced rules with smaller sizes.

Keywords: Interpretable Artificial Intelligence · Explainable Artificial Intelligence · Rule Learning · Maximum Satisfiability

1 Introduction

The success of Machine Learning (ML) in recent years has led to a growing advancement in studies in this area [2,8,12]. Several applications have emerged with the aim of circumventing various problems and situations [4,14,20]. One such problem is the lack of explainability of prediction models. This directly affects the reliability of using these applications in critical situations involving, for example, finance, autonomous systems, damage to equipment, the environment, and even lives [1,7,23]. That said, some works seek to develop approaches that bring explainability to their predictions [13,21,22].

Precise predictions with high levels of interpretability are often not a simple task. There are some works that try to solve this problem by balancing the

© The Author(s), under exclusive license to Springer Nature Switzerland AG 2023
M. C. Naldi and R. A. C. Bianchi (Eds.): BRACIS 2023, LNAI 14195, pp. 227–242, 2023.
https://doi.org/10.1007/978-3-031-45368-7_15

accuracy of the prediction with the interpretability [5, 6, 9, 15–17, 24]. It can be seen that some of these works use approaches based on the Boolean Satisfiability Problem (SAT) and the Maximum Boolean Satisfiability Problem (MaxSAT). The choice of these approaches to solve this problem has been increasingly recurrent in recent years. The reasons can be seen in the results obtained by these models.

SAT-based approaches have been proposed recently [18, 19] to learn quantifier-free first-order sentences from a set of classified strings. More specifically, given a set of classified strings, the goal is to find a first-order sentence over strings of minimum size that correctly classifies all the strings. One of the approaches demonstrated is SQFSAT (Synthesis of quantifier-free first-order sentences over strings with SAT). Upon receiving a set of classified strings, this approach generates a quantifier-free first-order sentence over strings in disjunctive normal form (DNF) with a given number of terms. What makes this method stand out is the fact that we can limit both the number of terms and the number of formulas per term in the generated formula. In addition, as the approach generates formulas in DNF, each term of the formula can be seen as a rule. Then, for each rule, its explanation is the conjunction of formulas in the rule, which can be interesting for their interpretability [11, 18]. On the other hand, as the model is based on the SAT problem, in certain situations it may bring results that are not so interesting in terms of interpretability and efficiency, such as in cases where the set of strings is large.

Ghosh, B. et al. created a classification model based on MaxSAT called IMLI [6]. The approach takes a set of classified samples, represented by vectors of numerical and categorical data, and generates a set of rules expressed in DNF or in conjunctive normal form (CNF) that correctly classifies as many samples as possible. In this work, we focus on using IMLI for learning rules in DNF. The number of rules in the set of rules can be defined similarly to SQFSAT, but IMLI does not consider the number of elements per rule. Although IMLI focuses on learning a sparse set of rules, it may obtain a combination of both large and small rules. IMLI also takes into account the option of defining a weighting for correct classifications. As the weighting increases, the accuracy of the model improves, but at the cost of an increase in the size of the generated set of rules. The smaller the weighting, the lower the accuracy of the model, but correspondingly, the generated set of rules tends to be smaller. Furthermore, IMLI uses an incremental approach to achieve better runtime performance. The incremental form consists of dividing the set of samples into partitions in order to generate a set of rules for each partition from the set of rules obtained in the previous partitions.

In this work, we aim to create a new approach for learning interpretable rules based on MaxSAT that unites SQFSAT with the incrementality of IMLI. The motivation for choosing SQFSAT is the possibility of defining the number of literals per clause, allowing us to generate smaller and more balanced rules. The choice of IMLI is motivated by its incrementability technique, which allows the method to train on large sets of samples efficiently. In addition, we propose a technique that reduces the size of the generated rules, removing possible redundancies.

This work is divided into 6 sections. In Sect. 2, we define the general notions and notations. Since all methods presented in this paper use Boolean logic, we also define in Sect. 2 how these methods binarize datasets with numerical and categorical data. In Sect. 3, SQFSAT and IMLI are presented, respectively. We present SQFSAT in the context of our work where samples consist of binary vectors instead of strings, and elements of rules are not first-order sentences over strings. In Sect. 4, our contribution is presented: IMLIB. In Sect. 5, we describe the experiments conducted and the results for the comparison of our approach against IMLI. Finally, in the last section, we present the conclusions and indicate future work.

2   Preliminaries

We consider the binary classification problem where we are given a set of samples and their classifications. The set of samples is represented by a binary matrix of size $n \times m$ and their classifications by a vector of size n. We call the matrix $\mathbf{X}$ and the vector $\mathbf{y}$. Each row of $\mathbf{X}$ is a sample of the set and we will call it $\mathbf{X}_i$ with $i \in \{1, ..., n\}$. To represent a specific value of $\mathbf{X}_i$, we will use $x_{i,j}$ with $j \in \{1, ..., m\}$. Each column of $\mathbf{X}$ has a label representing a feature and the label is symbolized by x^j. To represent a specific value of $\mathbf{y}$, we will use y_i.

To represent the opposite value of y_i, that is, if it is 1 the opposite value is 0 and vice versa, we use $\neg y_i$. Therefore, we will use the symbol $\neg \mathbf{y}$ to represent $\mathbf{y}$ with all opposite values. To represent the opposite value of $x_{i,j}$, we use $\neg x_{i,j}$. Therefore, we will use the symbol $\neg \mathbf{X}_i$ to represent $\mathbf{X}_i$ with all opposite values. Each label also has its opposite label which is symbolized by $\neg x^j$.

A partition of $\mathbf{X}$ is represented by $\mathbf{X}^t$ with $t \in \{1, ..., p\}$, where p is the number of partitions. Therefore, the partitions of vector $\mathbf{y}$ are represented by $\mathbf{y}^t$. Each element of $\mathbf{y}$ is symbolized by y_i and represents the class value of sample $\mathbf{X}_i$. We use $\mathcal{E}^- = \{X_i \mid y_i = 0, 1 \leq i \leq n\}$ and $\mathcal{E}^+ = \{X_i \mid y_i = 1, 1 \leq i \leq n\}$. To represent the size of these sets, that is, the number of samples contained in them, we use the notations: $|\mathcal{E}^-|$ and $|\mathcal{E}^+|$.

Example 1. Let $\mathbf{X}$ be the set of samples

$$\mathbf{X} = \begin{bmatrix} 0 & 0 & 1 \\ 0 & 1 & 0 \\ 0 & 1 & 1 \\ 1 & 0 & 0 \end{bmatrix}$$

and their classifications $\mathbf{y} = [1, 0, 0, 1]$. The samples $\mathbf{X}_i$ are: $\mathbf{X}_1 = [0, 0, 1], ..., \mathbf{X}_4 = [1, 0, 0]$. The values of each sample $x_{i,j}$ are: $x_{1,1} = 0, x_{1,2} = 0, x_{1,3} = 1, x_{2,1} = 0, ..., x_{4,3} = 0$. The class values y_i of each sample are: $y_1 = 1, ..., y_4 = 1$. We can divide $\mathbf{X}$ into two partitions in several different ways, one of which is: $\mathbf{X}^1 = \begin{bmatrix} 0 & 1 & 1 \\ 0 & 1 & 0 \end{bmatrix}$, $\mathbf{y}^1 = [0, 0]$, $\mathbf{X}^2 = \begin{bmatrix} 1 & 0 & 0 \\ 0 & 0 & 1 \end{bmatrix}$ e $\mathbf{y}^2 = [1, 1]$.

Example 2. Let $\mathbf{X}$ be the set of samples from Example 1, then

$$\neg\mathbf{X} = \begin{bmatrix} 1 & 1 & 0 \\ 1 & 0 & 1 \\ 1 & 0 & 0 \\ 0 & 1 & 1 \end{bmatrix}$$

and $\neg\mathbf{y} = [0, 1, 1, 0]$. The samples $\neg\mathbf{X}_i$ are: $\neg\mathbf{X}_1 = [1, 1, 0], ..., \neg\mathbf{X}_4 = [0, 1, 1]$. The values of each sample $\neg x_{i,j}$ are: $\neg x_{1,1} = 1, \neg x_{1,2} = 1, \neg x_{1,3} = 0, \neg x_{2,1} = 1, ..., \neg x_{4,3} = 1$. The class values of each sample $\neg y_i$ are: $\neg y_1 = 0, ..., \neg y_4 = 0$. We can divide $\neg\mathbf{X}$ in partitions as in Example 1: $\neg\mathbf{X}^1 = \begin{bmatrix} 1 & 0 & 0 \\ 1 & 0 & 1 \end{bmatrix}$, $\neg\mathbf{y}^1 = [1, 1]$,

$\neg\mathbf{X}^2 = \begin{bmatrix} 0 & 1 & 1 \\ 1 & 1 & 0 \end{bmatrix}$ e $\neg\mathbf{y}^2 = [0, 0]$.

We define a set of rules being the disjunction of rules and is represented by $\mathbf{R}$. A rule is a conjunction of one or more features. Each rule in $\mathbf{R}$ is represented by R_o with $o \in \{1, ..., k\}$, where k is the number of rules. Moreover, $\mathbf{R}(\mathbf{X}_i)$ represents the application of $\mathbf{R}$ to $\mathbf{X}_i$. The notations $|\mathbf{R}|$ and $|R_o|$ are used to represent the number of features in $\mathbf{R}$ and R_o, respectively.

Example 3. Let $x^1 = $ *Man*, $x^2 = $ *Smoke*, $x^3 = $ *Hike* be labels of features. Let $\mathbf{R}$ be the set of rules $\mathbf{R} = (Man) \vee (Smoke \wedge \neg Hike)$. The rules R_o are: $R_1 = (Man)$ and $R_2 = (Smoke \wedge \neg Hike)$. The application of $\mathbf{R}$ to $\mathbf{X}_i$ is represented as follows: $\mathbf{R}(\mathbf{X}_i) = x_{i,1} \vee (x_{i,2} \wedge \neg x_{i,3})$. For example, Let $\mathbf{X}$ be the set of samples from Example 1, then: $\mathbf{R}(\mathbf{X}_1) = x_{1,1} \vee (x_{1,2} \wedge \neg x_{1,3}) = 0 \vee (0 \wedge 0) = 0$. Moreover, we have that $|\mathbf{R}| = 3$, $|R_1| = 1$ and $|R_2| = 2$.

As we assume a set of binary samples, we need to perform some preprocessing. Preprocessing consists of binarizing a set of samples with numerical or categorical values. The algorithm divides the features into four types: constant, where all samples have the same value; binary, where there are only two distinct variations among all the samples for the same value; categorical, when the feature does not fit in constant and binary and its values are three or more categories; ordinal, when the feature does not fit into constant and binary and has numerical values.

When the feature type is constant, the algorithm discards that feature. This happens due to the fact that a feature common to all samples makes no difference in the generated rules. When the type is binary, one of the feature variations will receive 0 and the other 1 as new values. If the type is categorical, we employ the widely recognized technique of one-hot encoding. Finally, for the ordinal type feature, a quantization is performed, that is, the variations of this feature are divided into quantiles. With this, Boolean values are assigned to each quantile according to the original value.

We use SAT and MaxSAT solvers to implement the methods presented in this work. A solver receives a formula in CNF, for example: $(p \vee q) \wedge (q \vee \neg p)$. Furthermore, a MaxSAT solver receives weights that will be assigned to each clause in the formula. A clause is the disjunction of one or more literals. The weights are represented by $W(Cl) = w$ where Cl is one or more clauses and w represents the weight assigned to each one of them. A SAT solver tries to assign values to the literals in such a way that all clauses are satisfied. A MaxSAT solver tries to assign values to the literals in a way that the sum of the weights of satisfied clauses is maximum. Clauses with numerical weights are considered *soft*. The greater the weight, the greater the priority of the clause to be satisfied. Clauses assigned a weight of ∞ are considered *hard* and must be satisfied.

3   Rule Learning with SAT and MaxSAT

3.1   SQFSAT

SQFSAT is a SAT-based approach that, given $\mathbf{X}, \mathbf{y}, k$ and the number of features per rule l, tries to find a set of rules $\mathbf{R}$ with k rules and at most l features per rule that correctly classify all samples $\mathbf{X}_i$, that is, $\mathbf{R}(\mathbf{X}_i) = y_i$ for all i. In general, the approach takes its parameters $\mathbf{X}, \mathbf{y}, k$ and l and constructs a CNF formula to apply it to a SAT solver, which returns an answer that is used to get $\mathbf{R}$.

The construction of the SAT clauses is defined by propositional variables: $u_{o,d}^j, p_{o,d}, u_{o,d}^*, e_{o,d,i}$ and $z_{o,i}$, for $d \in \{1, ..., l\}$. If the valuation of $u_{o,d}^j$ is true, it means that jth feature label will be the dth feature of the rule R_o. Furthermore, if $p_{o,d}$ is true, it means that the dth feature of the rule R_o will be x^j, in other words, will be positive. Otherwise, it will be negative: $\neg x^j$. If $u_{o,d}^*$ is true, it means that the dth feature is skipped in the rule R_o. In this case, we ignore $p_{o,d}$. If $e_{o,d,i}$ is true, then the dth feature of rule R_o contributes to the correct classification of the ith sample. If $z_{o,i}$ is true, then the rule R_o contributes to the correct classification of the ith sample. That said, below, we will show the constraints formulated in the model for constructing the SAT clauses.

Conjunction of clauses that guarantees that exactly one $u_{o,d}^j$ is true for the dth feature of the rule R_o:

$$A = \bigwedge_{\substack{o \in \{1,...,k\} \\ d \in \{1,...,l\}}} \bigvee_{j \in \{1,...,m,*\}} u_{o,d}^j \tag{1}$$

$$B = \bigwedge_{\substack{o \in \{1,...,k\} \\ d \in \{1,...,l\} \\ j,j' \in \{1,...,m,*\}, j \neq j'}} (\neg u_{o,d}^j \vee \neg u_{o,d}^{j'}) \tag{2}$$

Conjunction of clauses that ensures that each rule has at least one feature:

$$C = \bigwedge_{o \in \{1,...,k\}} \bigvee_{d \in \{1,...,l\}} \neg u_{o,d}^* \tag{3}$$

We will use the symbol $s_{o,d,i}^j$ to represent the value of the ith sample in the jth feature label of $\mathbf{X}$. If this value is 1, it means that if the jth feature label is in the dth position of the rule R_o, then it contributes to the correct classification of the ith sample. Therefore, $s_{o,d,i}^j = e_{o,d,i}$. Otherwise, $s_{o,d,i}^j = \neg e_{o,d,i}$. That said, the following conjunction of formulas guarantees that $e_{o,d,i}$ is true if the jth feature in the oth rule contributes to the correct classification of the sample $\mathbf{X}_i$:

$$D = \bigwedge_{\substack{o\in\{1,\ldots,k\} \\ d\in\{1,\ldots,l\} \\ j\in\{1,\ldots,m\} \\ i\in\{1,\ldots,n\}}} u_{o,d}^j \rightarrow (p_{o,d} \leftrightarrow s_{o,d,i}^j) \tag{4}$$

Conjunction of formulas guaranteeing that if the dth feature of a rule is skipped, then the classification of this rule is not interfered by this feature:

$$E = \bigwedge_{\substack{o\in\{1,\ldots,k\} \\ d\in\{1,\ldots,l\} \\ i\in\{1,\ldots,n\}}} u_{o,d}^* \rightarrow e_{o,d,i} \tag{5}$$

Conjunction of formulas indicating that $z_{o,i}$ will be set to true if all the features of rule R_o contribute to the correct classification of sample $\mathbf{X}_i$:

$$F = \bigwedge_{o\in\{1,\ldots,k\}} \bigwedge_{i\in\{1,\ldots,n\}} z_{o,i} \leftrightarrow \bigwedge_{d\in\{1,\ldots,l\}} e_{o,d,i} \tag{6}$$

Conjunction of clauses that guarantees that $\mathbf{R}$ will correctly classify all samples:

$$G = \bigwedge_{i\in\mathcal{E}^+} \bigvee_{o\in\{1,\ldots,k\}} z_{o,i} \tag{7}$$

$$H = \bigwedge_{i\in\mathcal{E}^-} \bigwedge_{o\in\{1,\ldots,k\}} \neg z_{o,i} \tag{8}$$

Next, the formula Q below is converted to CNF. Then, finally, we have the SAT query that is sent to the solver.

$$Q = A \wedge B \wedge C \wedge D \wedge E \wedge F \wedge G \wedge H \tag{9}$$

3.2   IMLI

IMLI is an incremental approach based on MaxSAT for learning interpretable rules. Given $\mathbf{X}$, $\mathbf{y}$, k, and a weight λ, the model aims to obtain the smallest set of rules $\mathbf{M}$ in CNF that correctly classifies as many samples as possible, penalizing classification errors with λ. In general, the method solves the optimization problem $\min_{\mathbf{M}}\{|\mathbf{M}| + \lambda|\mathcal{E}_M| \mid \mathcal{E}_M = \{\mathbf{X}_i \mid \mathbf{M}(\mathbf{X}_i) \neq y_i\}\}$, where $|\mathbf{M}|$ represents the number of features in $\mathbf{M}$ and $\mathbf{M}(\mathbf{X}_i)$ denotes the application of the set of rules $\mathbf{M}$ to $\mathbf{X}_i$. Therefore, the approach takes its parameters $\mathbf{X}$, $\mathbf{y}$, k and λ and constructs a MaxSAT query to apply it to a MaxSAT solver, which returns an answer that is used to generate $\mathbf{M}$. Note that IMLI generates set of rules in CNF, whereas our objective is to obtain sets of rules in DNF. For that, we will have to use as parameter $\neg\mathbf{y}$ instead of $\mathbf{y}$ and negate the set of rules $\mathbf{M}$ to obtain a set of rules $\mathbf{R}$ in DNF.

The construction of the MaxSAT clauses is defined by propositional variables: b_o^v and η_i, for $v \in \{1, ..., 2m\}$. The v ranges from 1 to $2m$, as it also considers opposite features. If the valuation of b_o^v is true and $v \leq m$, it means that feature x^v will be in the rule M_o, where M_o is the oth rule of $\mathbf{M}$. If the valuation of the b_o^v is true and $v > m$, it means that feature $\neg x^{v-m}$ will be in the rule M_o. If the valuation of η_i is true, it means that sample $\mathbf{X}_i$ is not classified correctly, that is, $\mathbf{M}(\mathbf{X}_i) \neq y_i$. That said, below, we will show the constraints for constructing MaxSAT clauses.

Constraints that represent that the cost of a misclassification is λ:

$$A = \bigwedge_{i \in \{1,...,n\}} \neg \eta_i, W(A) = \lambda \qquad (10)$$

Constraints that represent that the model tries to insert as few features as possible in $\mathbf{M}$, taking into account the weights of all clauses:

$$B = \bigwedge_{\substack{v \in \{1,...,2m\} \\ o \in \{1,...,k\}}} \neg b_o^v, W(B) = 1 \qquad (11)$$

Even though the constraints in 11 prioritize learning sparse rules, they do so by directing attention to the overall set of rules, i.e. in the total number of features in $\mathbf{M}$. Then, IMLI may generate a set of rules that comprises a combination of both large and small rules. In our approach presented in Sect. 4, we address this drawback by limiting the number of features in each rule.

We will use $\mathbf{L}_o$ to represent the set of variables b_o^v of a rule M_o, that is, $\mathbf{L}_o = \{b_o^v | v \in \{1, ..., 2m\}\}$, for $o \in \{1, ..., k\}$. To represent the concatenation of two samples, we will use the symbol $\cup$. We also use the symbol @ to represent an operation between two vectors of the same size. The operation consists of applying a conjunction between the corresponding elements of the vectors. Subsequently, a disjunction between the elements of the result is applied. The following example illustrates how these definitions will be used:

Example 4. Let be $\mathbf{X}_4$ as in Example 1, $\mathbf{X}_4 \cup \neg \mathbf{X}_4 = [1, 0, 0, 0, 1, 1]$ and $\mathbf{L}_o = [b_o^1, b_o^2, b_o^3, b_o^4, b_o^5, b_o^6]$. Therefore, $(\mathbf{X}_4 \cup \neg \mathbf{X}_4)@\mathbf{L}_o = (x_{4,1} \wedge b_o^1) \vee (x_{4,2} \wedge b_o^2) \vee ... \vee (\neg x_{4,6} \wedge b_o^6) = (1 \wedge b_o^1) \vee (0 \wedge b_o^2) \vee ... \vee (1 \wedge b_o^6) = b_o^1 \vee b_o^5 \vee b_o^6$.

The objective of this operation is to generate a disjunction of variables that indicates if any of the features associated with these variables are present in M_o, then sample $\mathbf{X}_i$ will be correctly classified by M_o. Now, we can show the formula that guarantees that if η_i is false, then $\mathbf{M}(\mathbf{X}_i) = y_i$:

$$C = \bigwedge_{i \in \{1,...,n\}} \neg \eta_i \rightarrow (y_i \leftrightarrow \bigwedge_{o \in \{1,...,k\}} ((\mathbf{X}_i \cup \neg \mathbf{X}_i)@\mathbf{L}_o)), W(C) = \infty \qquad (12)$$

We can see that C is not in CNF. Therefore, formula Q below must be converted to CNF. With that, finally, we have the MaxSAT query that is sent to the solver.

$$Q = A \wedge B \wedge C \qquad (13)$$

The set of samples $\mathbf{X}$, in IMLI, can be divided into p partitions: $\mathbf{X}^1$, $\mathbf{X}^2$, ..., $\mathbf{X}^p$. Each partition, but the last one, contains the same values of $|\mathcal{E}^-|$ and $|\mathcal{E}^+|$. Also, the samples are randomly distributed across the partitions. Partitioning aims to make the model perform better in generating the set of rules $\mathbf{M}$. Thus, the conjunction of clauses will be created from each partition $\mathbf{X}^t$ in an incremental way, that is, the set of rules $\mathbf{M}$ obtained by the current partition will be reused for the next partition. In the first partition, constraints (10), (11), (12) are created in exactly the same way as described. From the second onwards, (11) is replaced by the following constraints:

$$B' = \bigwedge_{\substack{v \in \{1,\ldots,2m\} \\ o \in \{1,\ldots,k\}}} \begin{cases} b_o^v, \text{ if } b_o^v \text{ is true in the previous partition;} \\ \neg b_o^v, \text{ otherwise.} \end{cases}, W(B') = 1 \quad (14)$$

The IMLI also has a technique for reducing the size of the generated set of rules. The technique removes possible redundancies in ordinal features as the one in Example 5. In the original implementation of the model, the technique is applied at the end of each partition. In our implementation for the experiments in Sect. 5, this technique is applied only at the end of the last partition. The reason for this is training performance.

Example 5. Let $\mathbf{R}$ be the following set of rules with redundancy in the same rule:

$$(Age > 18 \wedge Age > 20) \vee (Height \leq 2).$$

Then, the technique removes the redundancy and the following set of rules is obtained:

$$(Age > 20) \vee (Height \leq 2).$$

4   IMLIB

In this section, we will present our method IMLIB which is an incremental version of SQFSAT based on MaxSAT. IMLIB also has a technique for reducing the size of the generated set of rules. Therefore, our approach partitions the set of samples $\mathbf{X}$. Moreover, our method has one more constraint and weight on all clauses. With that, our approach receives five input parameters $\mathbf{X}$, $\mathbf{y}$, k, l, λ and tries to obtain the smallest $\mathbf{R}$ that correctly classifies as many samples as possible, penalizing classification errors with λ, that is, $\min_{\mathbf{R}}\{|\mathbf{R}|+\lambda|\mathcal{E}_R| \mid \mathcal{E}_R = \{\mathbf{X}_i \mid \mathbf{R}(\mathbf{X}_i) \neq y_i\}\}$. That said, below, we will show the constraints of our approach for constructing MaxSAT clauses.

Constraints that guarantee that exactly only one $u_{o,d}^j$ is true for the dth feature of the rule R_o:

$$A = \bigwedge_{\substack{o \in \{1,\ldots,k\} \\ d \in \{1,\ldots,l\}}} \bigvee_{j \in \{1,\ldots,m,*\}} u_{o,d}^j, W(A) = \infty \quad (15)$$

$$B = \bigwedge_{\substack{o \in \{1,\ldots,k\} \\ d \in \{1,\ldots,l\} \\ j,j' \in \{1,\ldots,m,*\}, j \neq j'}} \neg u_{o,d}^{j} \vee \neg u_{o,d}^{j'}, W(B) = \infty \tag{16}$$

Constraints representing that the model will try to insert as few features as possible in $\mathbf{R}$:

$$C = \bigwedge_{\substack{o \in \{1,\ldots,k\} \\ d \in \{1,\ldots,l\} \\ j \in \{1,\ldots,m\}}} \neg u_{o,d}^{j} \wedge \bigwedge_{\substack{o \in \{1,\ldots,k\} \\ d \in \{1,\ldots,l\}}} u_{o,d}^{*}, W(C) = 1 \tag{17}$$

Conjunction of clauses that guarantees that each rule has at least one feature:

$$D = \bigwedge_{o \in \{1,\ldots,k\}} \bigvee_{d \in \{1,\ldots,l\}} \neg u_{o,d}^{*}, W(D) = \infty \tag{18}$$

The following conjunction of formulas ensures that $e_{o,d,i}$ is true if the jth feature label in the oth rule contributes to correctly classify sample $\mathbf{X}_i$:

$$E = \bigwedge_{\substack{o \in \{1,\ldots,k\} \\ d \in \{1,\ldots,l\} \\ j \in \{1,\ldots,m\} \\ i \in \{1,\ldots,n\}}} u_{o,d}^{j} \rightarrow (p_{o,d} \leftrightarrow s_{o,d,i}^{j}), W(E) = \infty \tag{19}$$

Conjunction of formulas guaranteeing that the classification of a specific rule will not be interfered by skipped features in the rule:

$$F = \bigwedge_{\substack{o \in \{1,\ldots,k\} \\ d \in \{1,\ldots,l\} \\ i \in \{1,\ldots,n\}}} u_{o,d}^{*} \rightarrow e_{o,d,i}, W(F) = \infty \tag{20}$$

Conjunction of formulas indicating that the model assigns true to $z_{o,i}$ if all the features of rule R_o support the correct classification of sample $\mathbf{X}_i$:

$$G = \bigwedge_{o \in \{1,\ldots,k\}} \bigwedge_{i \in \{1,\ldots,n\}} z_{o,i} \leftrightarrow \bigwedge_{d \in \{1,\ldots,l\}} e_{o,d,i}, W(G) = \infty \tag{21}$$

Conjunction of clauses designed to generate a set of rules $\mathbf{R}$ that correctly classify as many samples as possible:

$$H = \bigwedge_{i \in \mathcal{E}^+} \bigvee_{o \in \{1,\ldots,k\}} z_{o,i}, W(H) = \lambda \tag{22}$$

$$I = \bigwedge_{i \in \mathcal{E}^-} \bigwedge_{o \in \{1,\ldots,k\}} \neg z_{o,i}, W(I) = \lambda \tag{23}$$

Finally, after converting formula Q below to CNF, we have the MaxSAT query that is sent to the solver.

$$Q = A \wedge B \wedge C \wedge D \wedge E \wedge F \wedge G \wedge H \wedge I \tag{24}$$

IMLIB can also partition the set of samples $\mathbf{X}$ in the same way IMLI. Therefore, all constraints described above are applied in the first partition. Starting from the second partition, the constraints in (17) are replaced by the following constraints:

$$C' = \bigwedge_{\substack{o \in \{1,\dots,k\} \\ d \in \{1,\dots,l\} \\ j \in \{1,\dots,m,*\}}} \begin{cases} u_{o,d}^j, \text{ if } u_{o,d}^j \text{ is true in the previous} \\ \text{partition;} \\ \neg u_{o,d}^j, \text{ otherwise.} \end{cases}, W(C') = 1 \qquad (25)$$

IMLIB also has a technique for reducing the size of the generated set of rules demonstrated in Example 5. Moreover, we added two more cases which are described in Example 6 and Example 7.

Example 6. Let $\mathbf{R}$ be the following set of rules with opposite features in the same rule:

$$(Age > 20) \vee (Height \leq 2 \wedge Height > 2) \vee (Hike \wedge Not\ Hike).$$

Therefore, the technique removes rules with opposite features in the same rule obtaining the following set of rules:

$$(Age > 20).$$

Example 7. Let $\mathbf{R}$ be the following set of rules with the same feature occurring twice in a rule:

$$(Hike \wedge Hike) \vee (Age > 20).$$

Accordingly, our technique for removing redundancies eliminates repetitive features, resulting in the following set of rules:

$$(Hike) \vee (Age > 20).$$

5   Experiments

In this section, we present the experiments we conducted to compare our method IMLIB against IMLI. The two models were implemented[1] with Python and MaxSAT solver RC2 [10]. The experiments were carried out on a machine with the following configurations: Intel(R) Core(TM) i5-4460 3.20GHz processor, and 12GB of RAM memory. Ten databases from the UCI repository [3] were used to compare IMLI with IMLIB. Information on the datasets can be seen in Table 1. Databases that have more than two classes were adapted, considering that both models are designed for binary classification. For purposes of comparison, we measure the following metrics: number of rules, size of the set of rules, size of the largest rule, accuracy on test data and training time. The number of rules, size of the set of rules, and size of the largest rule can be used as interpretability metrics.

---

[1] Source code of IMLIB and the implementation of the tests performed can be found at the link: https://github.com/cacajr/decision_set_models.

Table 1. Databases information.

Databases	Samples	$|\mathcal{E}^-|$	$|\mathcal{E}^+|$	Features
lung cancer	59	31	28	6
iris	150	100	50	4
parkinsons	195	48	147	22
ionosphere	351	126	225	33
wdbc	569	357	212	30
transfusion	748	570	178	4
pima	768	500	268	8
titanic	1309	809	500	6
depressed	1429	1191	238	22
mushroom	8112	3916	4208	22

For example, a set of rules with few rules and small rules is more interpretable than one with many large rules.

Each dataset was split into 80% for training and 20% for testing. Both models were trained and evaluated using the same training and test sets, as well as the same random distribution. Then, the way the experiments were conducted ensured that both models had exactly the same set of samples to learn the set of rules.

For both IMLI and IMLIB, we consider parameter configurations obtained by combining values of: $k \in \{1, 2, 3\}$, $\lambda \in \{5, 10\}$ and $lp \in \{8, 16\}$, where lp is the number of samples per partition. Since IMLIB has the maximum number of features per rule l as an extra parameter, for each parameter configuration of IMLI and its corresponding **R**, we considered l ranging from 1 to one less than the size of the largest rule in **R**. Thus, the value of l that resulted in the best test accuracy was chosen to be compared with IMLI. Our objective is to evaluate whether IMLIB can achieve higher test accuracy compared to IMLI by employing smaller and more balanced rules. Furthermore, it should be noted that this does not exclude the possibility of our method generating sets of rules with larger sizes than IMLI.

For each dataset and each parameter configuration of k, λ and lp, we conducted ten independent realizations of this experiment. For each dataset, the parameter configuration with the best average of test accuracy for IMLI was chosen to be inserted in Table 2. For each dataset, the parameter configuration with the best average of test accuracy for IMLIB was chosen to be inserted in Table 3. The results presented in both tables are the average over the ten realizations.

Table 2. Comparison between IMLI and IMLIB in different databases with the IMLI configuration that obtained the best result in terms of accuracy. The column Training time represents the training time in seconds.

Databases	Models	Number of rules	$|\mathbf{R}|$	Largest rule size	Accuracy	Training time
lung cancer	IMLI	**2.00 ± 0.00**	3.60 ± 0.84	2.20 ± 0.63	**0.93 ± 0.07**	**0.0062 ± 0.0016**
	IMLIB	**2.00 ± 0.00**	**2.20 ± 0.63**	**1.10 ± 0.32**	**0.93 ± 0.07**	0.0146 ± 0.0091
iris	IMLI	**2.00 ± 0.00**	7.60 ± 1.35	4.50 ± 1.08	**0.90 ± 0.08**	**0.0051 ± 0.0010**
	IMLIB	**2.00 ± 0.00**	**4.90 ± 1.20**	**2.50 ± 0.71**	0.84 ± 0.12	0.0523 ± 0.0378
parkinsons	IMLI	**2.00 ± 0.00**	5.00 ± 2.05	2.90 ± 1.37	**0.80 ± 0.07**	**0.0223 ± 0.0033**
	IMLIB	**2.00 ± 0.00**	**3.00 ± 1.41**	**1.60 ± 0.84**	0.79 ± 0.06	0.0631 ± 0.0263
ionosphere	IMLI	**2.90 ± 0.32**	12.00 ± 1.63	5.20 ± 0.63	**0.81 ± 0.05**	**0.0781 ± 0.0096**
	IMLIB	3.00 ± 0.00	**7.70 ± 3.02**	**2.70 ± 1.16**	0.79 ± 0.04	0.2797 ± 0.1087
wdbc	IMLI	**2.90 ± 0.32**	8.70 ± 2.50	3.70 ± 1.34	**0.89 ± 0.03**	**0.0894 ± 0.0083**
	IMLIB	3.00 ± 0.00	**5.30 ± 2.36**	**1.80 ± 0.79**	0.86 ± 0.06	0.2172 ± 0.0800
transfusion	IMLI	**1.00 ± 0.00**	3.10 ± 0.88	3.10 ± 0.88	**0.72 ± 0.08**	**0.0291 ± 0.0026**
	IMLIB	**1.00 ± 0.00**	**2.00 ± 0.82**	**2.00 ± 0.82**	0.68 ± 0.08	0.5287 ± 0.3849
pima	IMLI	**1.00 ± 0.00**	5.10 ± 0.74	5.10 ± 0.74	0.68 ± 0.09	**0.0412 ± 0.0032**
	IMLIB	**1.00 ± 0.00**	**1.90 ± 1.10**	**1.90 ± 1.10**	**0.74 ± 0.04**	0.6130 ± 0.5093
titanic	IMLI	**1.00 ± 0.00**	6.90 ± 1.91	6.90 ± 1.91	0.71 ± 0.07	**0.0684 ± 0.0040**
	IMLIB	**1.00 ± 0.00**	**1.70 ± 0.67**	**1.70 ± 0.67**	**0.75 ± 0.06**	1.9630 ± 3.2705
depressed	IMLI	**1.80 ± 0.42**	7.50 ± 2.64	5.30 ± 1.89	0.74 ± 0.08	**0.2041 ± 0.0059**
	IMLIB	2.00 ± 0.00	**6.20 ± 3.36**	**3.30 ± 1.95**	**0.79 ± 0.04**	0.5175 ± 0.2113
mushroom	IMLI	**2.90 ± 0.32**	16.30 ± 2.91	8.20 ± 2.20	**0.99 ± 0.01**	**0.3600 ± 0.0340**
	IMLIB	3.00 ± 0.00	**12.30 ± 7.24**	**4.30 ± 2.54**	0.97 ± 0.03	2.3136 ± 0.6294

In Table 2, when considering parameter configurations that favor IMLI, we can see that IMLIB stands out in the size of the generated set of rules and in the size of the largest rule in datasets. Furthermore, our method achieved equal or higher accuracy compared to IMLI in four out of ten datasets. In datasets where IMLI outperformed IMLIB in terms of accuracy, our method exhibited a modest average performance gap of only three percentage points. Besides, IMLI outperformed our method in terms of training time in all datasets.

In Table 3, when we consider parameter configurations that favor our method, we can see that IMLIB continues to stand out in terms of the size of the generated set of rules and the size of the largest rule in all datasets. Moreover, our method achieved equal or higher accuracy than IMLI in all datasets. Again, IMLI consistently demonstrated better training time performance compared to IMLIB across all datasets.

As an illustrative example of interpretability, we present a comparison of the sizes of rules learned by both methods in the Mushroom dataset. Table 4 shows the sizes of rules obtained in all ten realizations of the experiment. We

Table 3. Comparison between IMLI and IMLIB in different databases with the IMLIB configuration that obtained the best result in terms of accuracy. The column Training time represents the training time in seconds.

Databases	Models	Number of rules	|**R**|	Largest rule size	Accuracy	Training time
lung cancer	IMLIB	**2.00 ± 0.00**	**2.20 ± 0.63**	**1.10 ± 0.32**	**0.93 ± 0.07**	0.0146 ± 0.0091
	IMLI	**2.00 ± 0.00**	3.60 ± 0.84	2.20 ± 0.63	**0.93 ± 0.07**	**0.0062 ± 0.0016**
iris	IMLIB	2.90 ± 0.32	**6.80 ± 1.48**	**2.50 ± 0.53**	**0.90 ± 0.07**	0.0373 ± 0.0095
	IMLI	**2.50 ± 0.53**	9.10 ± 1.91	4.80 ± 0.92	0.86 ± 0.09	**0.0062 ± 0.0011**
parkinsons	IMLIB	**3.00 ± 0.00**	**4.90 ± 1.66**	**1.70 ± 0.67**	**0.82 ± 0.07**	0.0868 ± 0.0510
	IMLI	**3.00 ± 0.00**	8.40 ± 1.90	3.70 ± 1.06	0.79 ± 0.07	**0.0295 ± 0.0064**
ionosphere	IMLIB	**2.00 ± 0.00**	**5.00 ± 1.70**	**2.50 ± 0.85**	**0.82 ± 0.06**	0.2002 ± 0.0725
	IMLI	**2.00 ± 0.00**	7.90 ± 1.79	4.90 ± 1.45	0.80 ± 0.07	**0.0531 ± 0.0106**
wdbc	IMLIB	**1.00 ± 0.00**	**1.20 ± 0.42**	**1.20 ± 0.42**	**0.89 ± 0.04**	0.0532 ± 0.0159
	IMLI	**1.00 ± 0.00**	2.50 ± 0.71	2.50 ± 0.71	0.86 ± 0.09	**0.0357 ± 0.0048**
transfusion	IMLIB	**1.00 ± 0.00**	**1.70 ± 0.67**	**1.70 ± 0.67**	**0.72 ± 0.03**	0.2843 ± 0.1742
	IMLI	**1.00 ± 0.00**	3.10 ± 0.74	3.10 ± 0.74	0.71 ± 0.06	**0.0273 ± 0.0032**
pima	IMLIB	**1.00 ± 0.00**	**1.90 ± 1.10**	**1.90 ± 1.10**	**0.74 ± 0.04**	0.6130 ± 0.5093
	IMLI	**1.00 ± 0.00**	5.10 ± 0.74	5.10 ± 0.74	0.68 ± 0.09	**0.0412 ± 0.0032**
titanic	IMLIB	**1.00 ± 0.00**	**1.40 ± 0.97**	**1.40 ± 0.97**	**0.76 ± 0.08**	0.8523 ± 1.7754
	IMLI	**1.00 ± 0.00**	6.80 ± 1.87	6.80 ± 1.87	0.68 ± 0.12	**0.0649 ± 0.0047**
depressed	IMLIB	3.00 ± 0.00	**13.30 ± 5.12**	**4.70 ± 1.89**	**0.80 ± 0.04**	0.7263 ± 0.1692
	IMLI	**2.90 ± 0.32**	14.80 ± 2.25	6.70 ± 1.70	0.69 ± 0.08	**0.2520 ± 0.0140**
mushroom	IMLIB	**1.00 ± 0.00**	**6.70 ± 0.95**	**6.70 ± 0.95**	**0.99 ± 0.00**	1.2472 ± 0.2250
	IMLI	**1.00 ± 0.00**	8.90 ± 1.10	8.90 ± 1.10	**0.99 ± 0.01**	**0.1214 ± 0.0218**

can observe that IMLIB consistently maintains a smaller and more balanced set of rules across the different realizations. This is interesting because unbalanced rules can affect interpretability. See realization 6, for instance. The largest rule learned by IMLI has a size of 10, nearly double the size of the remaining rules. In contrast, IMLIB learned a set of rules where the size of the largest rule is 6 and the others have similar sizes. Thus, interpreting three rules of size at most 6 is easier than interpreting a rule of size 10. Also as illustrative examples of interpretability, we can see some sets of rules learned by IMLIB in Table 5.

Table 4. Comparison of the size of the rules generated in the ten realizations of the Mushroom base from Table 2. The configuration used was $lp = 16$, $k = 3$ and $\lambda = 10$. As the value of l used in IMLIB varies across the realizations, column l will indicate which was the value used in each realization. In the column Rules sizes, we show the size of each rule in the following format: $(|R_1|, |R_2|, |R_3|)$. We have highlighted in bold the cases where the size of $|R_o|$ is the same or smaller in our model compared to IMLI.

Realizations	l	Models	Rules sizes
1	–	IMLI	(4, 6, 5)
	1	IMLIB	(**1**, **1**, **1**)
2	–	IMLI	(6, 11, **0**)
	8	IMLIB	(**3**, **6**, 6)
3	–	IMLI	(**3**, 8, **5**)
	5	IMLIB	(5, **5**, **5**)
4	–	IMLI	(6, 4, 4)
	2	IMLIB	(**2**, **2**, **2**)
5	–	IMLI	(3, 3, 4)
	2	IMLIB	(**2**, **2**, **2**)
6	–	IMLI	(**5**, 10, **6**)
	6	IMLIB	(6, **5**, **6**)
7	–	IMLI	(**5**, 9, **3**)
	8	IMLIB	(8, **8**, 8)
8	–	IMLI	(**4**, 10, **3**)
	6	IMLIB	(5, **5**, 6)
9	–	IMLI	(9, **5**, **4**)
	6	IMLIB	(**6**, 6, 6)
10	–	IMLI	(4, 9, 5)
	1	IMLIB	(**1**, **1**, **1**)

Table 5. Examples of set of rules generated by IMLIB in some tested databases.

Databases	Sets of rules
lung cancer	$(AreaQ \leq 5.0$ and $Alkhol > 3.0)$
iris	$(petal\ length \leq 1.6$ and $petal\ width > 1.3)$ or $(sepal\ width \leq 3.0$ and $petal\ length \leq 5.1)$
parkinsons	$(Spread2 > 0.18$ and $PPE > 0.19)$ or $(Shimmer{:}APQ3 > 0.008$ and $Spread2 \leq 0.28)$
wdbc	$(Largest\ area > 1039.5)$or $(Area \leq 546.3$ and $Largest\ concave\ points \leq 0.07)$
depressed	$(Age > 41.0$ and $Living\ expenses \leq 26692283.0)$or $(Education\ level > 8.0$ and $Other\ expenses \leq 20083274.0)$

6   Conclusion

In this work, we present a new incremental model for learning interpretable and balanced rules: IMLIB. Our method leverages the strengths of SQFSAT, which effectively constrains the size of rules, while incorporating techniques from IMLI, such as incrementability, cost for classification errors, and minimization of the set of rules. Our experiments demonstrate that the proposed approach generates smaller and more balanced rules than IMLI, while maintaining comparable or even superior accuracy in many cases. We argue that sets of small rules with approximately the same size seem more interpretable when compared to sets with a few large rules. As future work, we plan to develop a version of IMLIB that can classify sets of samples with more than two classes, enabling us to compare this approach with multiclass interpretable rules from the literature [11, 24].

References

1. Biran, O., Cotton, C.: Explanation and justification in machine learning: a survey. In: IJCAI-17 Workshop on Explainable AI (XAI), vol. 8, pp. 8–13 (2017)
2. Carleo, G., et al.: Machine learning and the physical sciences. Rev. Mod. Phys. **91**(4), 045002 (2019)
3. Dua, D., Graff, C.: UCI machine learning repository (2017). http://archive.ics.uci.edu/ml
4. Ghassemi, M., Oakden-Rayner, L., Beam, A.L.: The false hope of current approaches to explainable artificial intelligence in health care. Lancet Digit. Health **3**(11), e745–e750 (2021)
5. Ghosh, B., Malioutov, D., Meel, K.S.: Efficient learning of interpretable classification rules. J. Artif. Intell. Res. **74**, 1823–1863 (2022)
6. Ghosh, B., Meel, K.S.: IMLI: an incremental framework for MaxSAT-based learning of interpretable classification rules. In: Proceedings of the 2019 AAAI/ACM Conference on AI, Ethics, and Society, pp. 203–210 (2019)
7. Gunning, D., Stefik, M., Choi, J., Miller, T., Stumpf, S., Yang, G.Z.: XAI-explainable artificial intelligence. Sci. Robot. **4**(37), eaay7120 (2019)
8. Huang, H.Y., et al.: Power of data in quantum machine learning. Nat. Commun. **12**(1), 2631 (2021)
9. Ignatiev, A., Marques-Silva, J., Narodytska, N., Stuckey, P.J.: Reasoning-based learning of interpretable ML models. In: IJCAI, pp. 4458–4465 (2021)
10. Ignatiev, A., Morgado, A., Marques-Silva, J.: RC2: an efficient MaxSAT solver. J. Satisfiability Boolean Modeling Comput. **11**(1), 53–64 (2019)
11. Ignatiev, A., Pereira, F., Narodytska, N., Marques-Silva, J.: A SAT-based approach to learn explainable decision sets. In: Galmiche, D., Schulz, S., Sebastiani, R. (eds.) IJCAR 2018. LNCS (LNAI), vol. 10900, pp. 627–645. Springer, Cham (2018). https://doi.org/10.1007/978-3-319-94205-6_41
12. Janiesch, C., Zschech, P., Heinrich, K.: Machine learning and deep learning. Electron. Mark. **31**(3), 685–695 (2021)
13. Jiménez-Luna, J., Grisoni, F., Schneider, G.: Drug discovery with explainable artificial intelligence. Nat. Mach. Intell. **2**(10), 573–584 (2020)

14. Kwekha-Rashid, A.S., Abduljabbar, H.N., Alhayani, B.: Coronavirus disease (COVID-19) cases analysis using machine-learning applications. Appl. Nanosci. 1–13 (2021)
15. Lakkaraju, H., Bach, S.H., Leskovec, J.: Interpretable decision sets: a joint framework for description and prediction. In: Proceedings of the 22nd ACM SIGKDD International Conference on Knowledge Discovery and Data Mining, pp. 1675–1684 (2016)
16. Maliotov, D., Meel, K.S.: MLIC: a MaxSAT-based framework for learning interpretable classification rules. In: Hooker, J. (ed.) CP 2018. LNCS, vol. 11008, pp. 312–327. Springer, Cham (2018). https://doi.org/10.1007/978-3-319-98334-9_21
17. Mita, G., Papotti, P., Filippone, M., Michiardi, P.: LIBRE: learning interpretable Boolean rule ensembles. In: AISTATS, pp. 245–255. PMLR (2020)
18. Rocha, T.A., Martins, A.T.: Synthesis of quantifier-free first-order sentences from noisy samples of strings. In: 2019 8th Brazilian Conference on Intelligent Systems (BRACIS), pp. 12–17. IEEE (2019)
19. Rocha, T.A., Martins, A.T., Ferreira, F.M.: Synthesis of a DNF formula from a sample of strings using Ehrenfeucht-Fraïssé games. Theoret. Comput. Sci. **805**, 109–126 (2020)
20. Sharma, A., Jain, A., Gupta, P., Chowdary, V.: Machine learning applications for precision agriculture: a comprehensive review. IEEE Access **9**, 4843–4873 (2020)
21. Tjoa, E., Guan, C.: A survey on explainable artificial intelligence (XAI): toward medical XAI. IEEE Trans. Neural Netw. Learn. Syst. **32**(11), 4793–4813 (2020)
22. Vilone, G., Longo, L.: Notions of explainability and evaluation approaches for explainable artificial intelligence. Inf. Fusion **76**, 89–106 (2021)
23. Yan, L., et al.: An interpretable mortality prediction model for COVID-19 patients. Nat. Mach. Intell. **2**(5), 283–288 (2020)
24. Yu, J., Ignatiev, A., Stuckey, P.J., Le Bodic, P.: Computing optimal decision sets with SAT. In: Simonis, H. (ed.) CP 2020. LNCS, vol. 12333, pp. 952–970. Springer, Cham (2020). https://doi.org/10.1007/978-3-030-58475-7_55

d-CC Integrals: Generalizing CC-Integrals by Restricted Dissimilarity Functions with Applications to Fuzzy-Rule Based Systems

Joelson Sartori[1], Giancarlo Lucca[1(✉)], Tiago Asmus[2], Helida Santos[1],
Eduardo Borges[1], Benjamin Bedregal[3], Humberto Bustince[4],
and Graçaliz Pereira Dimuro[1]

[1] Centro de Ciêcias Computacionais, Univ. Fed. do Rio Grande, Porto Alegre, Brazil
{joelsonsartori,giancarlo.lucca,helida,eduardoborges,
gracalizdimuro}@furg.br
[2] Instituto de Matemática, Estatística e Física, Univ. Fed. do Rio Grande,
Porto Alegre, Brazil
tiagoasmus@furg.br
[3] Departamento de Informática e Matemática Aplicada, Univ. Fed. do Rio Grande
do Norte, Natal, Brazil
bedregal@dimap.ufrn.br
[4] Dept. Estadistica, Informa. y Matem., Univ. Publica de Navarra, Pamplona, Spain
bustince@unavarra.es

Abstract. The discrete Choquet Integral (CI) and its generalizations
have been successfully applied in many different fields, with particularly
good results when considered in Fuzzy Rule-Based Classification Systems
(FRBCSs). One of those functions is the CC-integral, where the product
operations in the expanded form of the CI are generalized by copulas.
Recently, some new Choquet-like operators were developed by generalizing the difference operation by a Restricted Dissimilarity Function (RDF)
in either the usual or the expanded form of the original CI, also providing good results in practical applications. So, motivated by such developments, in this paper we propose the generalization of the CC-integral
by means of RDFs, resulting in a function that we call d-CC-integral.
We study some relevant properties of this new definition, focusing on its
monotonicity-like behavior. Then, we proceed to apply d-CC-integrals
in a classification problem, comparing different d-CC-integrals between
them. The classification acuity of the best d-CC-integral surpasses the
one achieved by the best CC-integral and is statistically equivalent to
the state-of-the-art in FRBCSs.

Keywords: Choquet Integral · CC-integral · X-dC-integral ·
Fuzzy-rule based classification systems

This research was funded by FAPERGS/Brazil (Proc. 19/2551-0001660-3, 23/2551-
0000126-8), CNPq/Brazil (301618/2019-4, 305805/2021-5, 150160/ 2023-2).

M. C. Naldi and R. A. C. Bianchi (Eds.): BRACIS 2023, LNAI 14195, pp. 243–258, 2023.
https://doi.org/10.1007/978-3-031-45368-7_16

1   Introduction

The discrete Choquet Integral (CI) is an interesting aggregation operator that can capture the relationship between the aggregated data by means of a fuzzy measure (FM) [10]. It has been successfully applied in many different fields, such as decision-making [13,29], classification [13,28] and image processing [24]. This yielded the theoretical development of several generalizations [13] of the CI, such as the CC-integral [22] (a generalization by copulas [3]) and $C_{F_1 F_2}$-integral [19] (a generalization by fusion functions F_1 and F_2 under some constraints), all of them providing at least competitive results when applied in practical problems.

In particular, Fuzzy Rule-Based Classification Systems (FRBCSs) [17], which are notable for their high interpretability while still achieving good classification results, seem to benefit from the application of the CI and their generalizations in the aggregation process that occur in the reasoning method. This can be seen in the works of Lucca et al. [19], where the application of such integrals produced results that rival the state of the art in FRBCSs.

Bustince et al. [8] generalized the CI using Restricted Dissimilarity Functions (RDF), in the form of d-integrals. Following that, Wieczynski et al. studied dC_F-integrals [28] and d-XC-integrals [29], obtaining promising results not only in decision making and classification but also in signal processing, based on motor-imagery brain-computer interface. Recently, Boczek and Kaluszka [4] introduced the preliminary notion of the extended Choquet-Sugeno-like (eCS) operator, which generalizes most of the modifications of the CI known in the literature (briefly discussed above), and also generates some new CI-type operators. This inspired us to further research their properties and applications.

Then, the objectives of this paper are: (1) to study some important theoretical properties of an instance of the eCS-operator, introducing a generalization of CC-integrals by RDFs, called d-CC-integrals (Sect. 3); (2) to apply d-CC-integrals in FRBCSs in an experimental study, where we compare the classification accuracy of different d-CC-integrals based on combinations of RDFs and copulas with the best results obtained by other approaches in the literature (Sect. 4 and 5). Additionally, Sect. 2 recalls preliminary concepts and Sect. 6 is the Conclusion.

2   Preliminaries

Consider $N = \{1, \ldots, n\}$, with $n > 0$, and $\mathbf{x} = (x_1, \ldots, x_n)$. A function $A: [0,1]^n \to [0,1]$ is an aggregation function if **(A1)** it is increasing, and **(A2)** $A(0, \ldots, 0) = 0$ and $A(1, \ldots, 1) = 1$. Copulas are a special type of aggregation function that, in the context of the theory of metric spaces, link (2-dimensional) probability distribution functions to their 1-dimensional margins . A bivariate function $C: [0,1]^2 \to [0,1]$ is a copula if, for all $x, x', y, y' \in [0,1]$ with $x \leq x'$ and $y \leq y'$: **(C1)** $C(x,y) + C(x',y') \geq C(x,y') + C(x',y)$; **(C2)** $C(x,0) = C(0,x) = 0$; **(C3)** $C(x,1) = C(1,x) = x$ [3].

Table 1. Example of Copulas used as the basis for the d-CC integrals

T-norms	Overlap functions	Other Copulae
$T_M(x,y) = \min\{x,y\}$	$O_B(x,y) = \min\{x\sqrt{y}, y\sqrt{x}\}$	$C_F(x,y) = xy + x^2 y(1-x)(1-y)$
$T_P(x,y) = xy$	$O_{mM}(x,y) = \min\{x,y\}\max\{x^2, y^2\}$	$C_L(x,y) = \max\{\min\{x, \frac{y}{2}\}, x+y-1\}$
$T_L(x,y) = \max\{0, x+y-1\}$	$O_\alpha(x,y) = xy(1+\alpha(1-x)(1-y))$	
$T_{HP}(x,y) = \begin{cases} 0 & \text{if } x = y = 0 \\ \frac{xy}{x+y-xy} & \text{otherwise} \end{cases}$	$O_{Div}(x,y) = \frac{xy+\min\{x,y\}}{2}$	

Considering that $\alpha \in [-1, 0[\,\cup\,]0, 1]$

Table 2. Examples of RDFs based on Proposition 2

Name	Definition	Name	Definition
δ_0	$\delta_0(x,y) = \|x - y\|$	δ_3	$\delta_3(x,y) = \|\sqrt{x} - \sqrt{y}\|$
δ_1	$\delta_1(x,y) = (x-y)^2$	δ_4	$\delta_4(x,y) = \|x^2 - y^2\|$
δ_2	$\delta_2(x,y) = \sqrt{\|x-y\|}$	δ_5	$\delta_5(x,y) = (\sqrt{x} - \sqrt{y})^2$

Proposition 1 [3]. *For each copula C it holds that: (i) C is increasing; (ii) C satisfies the Lipschitz property with constant 1, that is, for all $x_1, x_2, y_1, y_2 \in [0,1]$, one has that $\mid C(x_1, y_1) - C(x_2, y_2) \mid \leq \mid x_1 - x_2 \mid + \mid y_1 - y_2 \mid$.*

Table 1 shows examples of copulas, which are used in the rest of the paper. We divided them into three groups: t-norms [3], overlap functions [6], and copulas that are neither t-norms nor overlap functions.

A Restricted Dissimilarity Function (RDF) [7] $\delta : [0,1]^2 \to [0,1]$ is a function such that, for all $x, y, z \in [0,1]$: (d1) $\delta(x,y) = \delta(y,x)$; (d2) $\delta(x,y) = 1$ if and only if $\{x,y\} = \{0,1\}$; (d3) $\delta(x,y) = 0$ if and only if $x = y$; (d4) if $x \leq y \leq z$, then $\delta(x,y) \leq \delta(x,z)$ and $\delta(y,z) \leq \delta(x,z)$. One convenient way of constructing RDFs is through automorphisms (strictly increasing bijections) of the unit interval:

Proposition 2 [7]. *Let $\varphi_1, \varphi_2 : [0,1] \to [0,1]$ be two automorphisms. Then, the function $\delta^{\varphi_1, \varphi_2} : [0,1]^2 \to [0,1]$, given, for all $x, y \in [0,1]$, by $\delta^{\varphi_1, \varphi_2}(x,y) = \varphi_1(\mid\varphi_2(x) - \varphi_2(y)\mid)$, is a Restricted Dissimilarity Function.*

All functions from Table 2 are examples of RDFs, constructed via Proposition 2, which are considered in our experiments presented in Sect. 5.

A function $m: 2^N \to [0,1]$ is a fuzzy measure (FM) [10] if, for all $X, Y \subseteq N$, it satisfies the following properties: (m1) Increasing: if $X \subseteq Y$, then $m(X) \leq m(Y)$; (m2) Boundary conditions: $m(\emptyset) = 0$ and $m(N) = 1$. An example of FM is the power measure $m_{PM}: 2^N \to [0,1]$, which is defined, for all $X \subseteq N$, by $m_{PM}(X) = \left(\frac{|X|}{n}\right)^q$, where the exponent $q > 0$ can be learned genetically from the data. It is the only FM considered in this paper, since it provides excellent results in classification problems, as discussed in [21].

The discrete CI [10] was generalized in many forms [4,13], one of them, based on its expanded form, considers copulas instead of the product operation:

Definition 1 [22]. *Let $m: 2^N \to [0,1]$ be a fuzzy measure and $C: [0,1]^2 \to [0,1]$ be a bivariate copula. The CC-integral is defined as a function $\mathfrak{C}_m^C : [0,1]^n \to [0,1]$, given, for all $\boldsymbol{x} \in [0,1]^n$, by*

$$\mathfrak{C}_m^C(\boldsymbol{x}) = \sum_{i=1}^{n} C\left(x_{(i)}, m\left(A_{(i)}\right)\right) - C\left(x_{(i-1)}, m\left(A_{(i)}\right)\right), \tag{1}$$

where $\left(x_{(1)}, \ldots, x_{(n)}\right)$ is an increasing permutation on the input x, that is, $0 \leq x_{(1)} \leq \ldots \leq x_{(n)}$, where $x_{(0)} = 0$ and $A_{(i)} = \{(i), \ldots, (n)\}$ is the subset of indices corresponding to the $n - i + 1$ largest components of $\boldsymbol{x}$.

In another direction, the CI in its expanded form was generalized considering RDFs in the place of the subtraction operation:

Definition 2 [29]. *The generalization of the CI expanded form by RDFs* $\delta\colon [0,1]^2 \to [0,1]$ *with respect to an FM* $m\colon 2^N \to [0,1]$, *named d-XChoquet integral (d-XC), is a mapping* $X\mathfrak{C}_{\delta,m}\colon [0,1]^2 \to [0,n]$, *defined, for all* $\boldsymbol{x} \in [0,1]^n$, *by:*

$$X\mathfrak{C}_{\delta,m}(\boldsymbol{x}) = x_{(1)} + \sum_{i=2}^{n} \delta\left(x_{(i)} \cdot m(A_{(i)}),\ x_{(i-1)} \cdot m(A_{(i)})\right), \tag{2}$$

where $x_{(i)}$ *and* $A_{(i)}$ *are defined according to Definition 1.*

3   d-CC Integrals

In this section, we introduce the definition of d-CC-integrals, by combining the concepts of CC-integral (Definition 1) and d-XC-integral (Definition 2). Following that, we study some aspects of d-CC-integrals that are relevant to our application in classification, such as different forms of monotonicity.

Definition 3. *Let* $C\colon [0,1]^2 \to [0,1]$ *be a copula. The generalization of the CC-integral by RDFs* $\delta\colon [0,1]^2 \to [0,1]$ *with respect to a FM* $m\colon 2^N \to [0,1]$, *named d-CC-integral (d-CC), is a mapping* $\mathfrak{C}\mathfrak{C}_{\delta,m}^C\colon [0,1]^2 \to [0,n]$, *defined by:*

$$\mathfrak{C}\mathfrak{C}_{\delta,m}^C(\boldsymbol{x}) = x_{(1)} + \sum_{i=2}^{n} \delta\left(C(x_{(i)}, m(A_{(i)})), C(x_{(i-1)},\ m(A_{(i)}))\right), \tag{3}$$

for all $\boldsymbol{x} \in [0,1]^n$, *where the ordered* $x_{(i)}$, *and* $m(A_{(i)})$, *with* $0 \leq i \leq n$, *were stated in Definition 1.*

From now on, we denote $m(A_{(i)})$ simply by $m_{(i)}$. Observe that d-CC-integrals can also be obtained by generalizing X-dC-integrals (Definition 2), where the product is replaced by a copula, or a restriction of $C_{F_1 F_2}$ [19], putting $F_1 = F_2 = C$.

Proposition 3. *Under the conditions given in Definition 3,* $\mathfrak{C}\mathfrak{C}_{\delta,m}^C$ *is well defined, for all RDF* δ, *FM* m *and copula* C.

Proof. For $i = 1, \ldots, n$, one has that $0 \leq x_{(i)} \leq 1$, and for $i = 2, \ldots, n$, we have that $0 \leq m_{(i)} \leq 1, 0 \leq C(x_{(i)}, m_{(i)}) \leq 1$ and $0 \leq \delta(C(x_{(i)}, m_{(i)}), C(x_{(i-1)}, m_{(i)})) \leq 1$. It is immediate that, for $\boldsymbol{x} \in [0,1]^n, 0 \leq X\mathfrak{C}_{\delta,m}(\boldsymbol{x}) \leq n$, for any RDF δ, FM m and copula C. Consider an input vector $\boldsymbol{x} \in [0,1]^n$, for which there may be different increasing permutations, meaning that $\boldsymbol{x}$ has repeated elements. For the sake of simplicity, but without loss of generality, consider that there exists $r, s \in \{1, \ldots, n\}$ such that $x_r = x_s = z \in [0, 1]$ and, for all $i \in \{1, \ldots, n\}$, with $i \neq r, s$, it holds that $x_i \neq x_r, x_s$. The only two possible increasing permutations are:

$$(x_{(1)}, \ldots, x_{(k-1)} = x_r, x_{(k)} = x_s, \ldots, x_{(n)}), \tag{4}$$

$$(x_{(1)}, \ldots, x_{(k-1)} = x_s, x_{(k)} = x_r, \ldots, x_{(n)}) \tag{5}$$

Denote by $m_{(i)}^{(1)} = m^{(1)}(A_{(i)})$ and $m_{(i)}^{(2)} = m^{(2)}(A_{(i)})$, with $i \in \{1, \ldots, n\}$, the fuzzy measures of the subsets of $A_{(i)}$ of indices corresponding to the $n - i + 1$ largest components of $\boldsymbol{x}$ with respect to the permutations (4) and (5), respectively. Observe that

$$m_{(i)}^{(1)} = m_{(i)}^{(2)}, \text{ for all } i \neq k, \text{ and} \tag{6}$$

$$m_{(k)}^{(1)} = m(\{s, (k+1), \ldots, (n)\}), \quad m_{(k)}^{(2)} = m(\{r, (k+1), \ldots, (n)\}), \tag{7}$$

hold, meaning it may be the case that $m_{(k)}^{(1)} \neq m_{(k)}^{(2)}$. Now, denote by $\mathfrak{CC}_{\delta,m}^C(1)$ and $\mathfrak{CC}_{\delta,m}^C(2)$ the d-CC integrals with respect to the permutations (4) and (5), respectively, and suppose that

$$\mathfrak{CC}_{\delta,m}^C(1)(\boldsymbol{x}) \neq \mathfrak{CC}_{\delta,m}^C(2)(\boldsymbol{x}). \tag{8}$$

From Eqs. (6) and (7), whenever $k \neq 1$, it follows that:

$\mathfrak{CC}_{\delta,m}^C(1)(\boldsymbol{x}) - \mathfrak{CC}_{\delta,m}^C(2)(\boldsymbol{x})$

$= \delta\left(C(x_{(k)}, m_{(k)}^{(1)}), \ C(x_{(k-1)}, m_{(k)}^{(1)})\right) - \delta\left(C(x_{(k)}, m_{(k)}^{(2)}), \ C(x_{(k-1)}, m_{(k)}^{(2)})\right)$

$= \delta\left(C(x_s, m(\{s, (k+1), \ldots, (n)\})), C(x_r, m(\{s, (k+1), \ldots, (n)\}))\right)$

$\quad - \delta\left(C(x_r, m(\{r, (k+1), \ldots, (n)\})), C(x_s, m(\{r, (k+1), \ldots, (n)\}))\right)$

$= \delta\left(C(z, m(\{s, (k+1), \ldots, (n)\})), C(z, m(\{s, (k+1), \ldots, (n)\}))\right)$

$\quad - \delta\left(C(z, m(\{r, (k+1), \ldots, (n)\})), C(z, m(\{r, (k+1), \ldots, (n)\}))\right) = 0, \quad \text{by (d3)},$

which is in contradiction to Eq. (8). Similarly, there is also a contradiction for $k = 1$. The result can be easily generalized for any subsets of repeated elements in the input $\boldsymbol{x}$. The conclusion is that for any different increasing permutations of the same input $\boldsymbol{x}$, one always obtains the same output value of $\mathfrak{CC}_{\delta,m}^C(\boldsymbol{x})$.

Remark 1. Note that, in $\mathfrak{CC}_{\delta,m}^C$ definition, the summation first element is just $x_{(1)}$ instead of $\delta\left(C(x_{(1)}, m_{(1)}), \ C(x_{(0)}, m_{(1)})\right)$. According to [19,29], this can be used to avoid a discrepant behavior of non-averaging[1] functions in the initial phase of the aggregation process.

Example 1. It is immediate that any choices of RDF δ from Table 2 and copula C from Table 1 can be combined to obtain an example of d-CC-integral $\mathfrak{CC}_{\delta,m}^C$.

In the following, we first study some general properties and then monotonicity-like properties, for the RDFs of Table 2, any fuzzy measure m and copula C.

---

[1] A function F is said to be *averaging* if $min \leq F \leq max$.

3.1 Some Important Properties for Aggregation-Like Processes

The next proposition shows that whenever the adopted RDF is constructed using Proposition 2, then, for some choices of the automorphism φ_2, the result obtained is upper limited in relation to the dissimilarities (by δ_0) of the inputs:

Proposition 4. *Let δ be an RDF constructed by Proposition 2, for any automorphism φ_1 and the identity as φ_2. Let $\mathfrak{CC}^C_{\delta,m} : [0,1]^n \rightarrow [0,1]$ be the derived d-CC integral for any fuzzy measure m and copula C. Then, for all $\boldsymbol{x} \in [0,1]^n$, it holds that:*

$$\mathfrak{CC}^C_{\delta,m}(\boldsymbol{x}) \leq x_{(1)} + \sum_{i=2}^{n} \varphi_1 \left(x_{(i)} - x_{(i-1)} \right) = x_{(1)} + \sum_{i=2}^{n} \varphi_1 \left(\delta_0(x_{(i)}, x_{(i-1)}) \right).$$

Proof. Let δ be an RDF constructed by Proposition 2. Then, since φ_1 is an automorphism, for any $\boldsymbol{x} \in [0,1]$, FM m and copula C, we have that:

$$\mathfrak{CC}^C_{\delta,m}(\boldsymbol{x}) = x_{(1)} + \sum_{i=2}^{n} \delta(C(x_{(i)}, m_{(i)}), C(x_{(i-1)}, m_{(i)}))$$

$$= x_{(1)} + \sum_{i=2}^{n} \varphi_1 \left(C(x_{(i)}, m_{(i)}) - C(x_{(i-1)}, m_{(i)}) \right) \text{ by Props. 2 and 1(i)}$$

$$\leq x_{(1)} + \sum_{i=2}^{n} \varphi_1 \left(x_{(i)} - x_{(i-1)} \right) \text{ by Prop. 1 (ii)}$$

The following three results are immediate:

Proposition 5. *For δ, m and C, $\mathfrak{CC}^C_{\delta,m}(\boldsymbol{x}) \geq \min(\boldsymbol{x})$, for all $\boldsymbol{x} \in [0,1]$.*

Proposition 6. *$\mathfrak{CC}^C_{\delta,m}(\boldsymbol{x}) \leq \max(\boldsymbol{x})$ if and only if, for all $0 \leq a_1 \leq \ldots \leq a_n$ and FM m, the RDF δ satisfies: $\sum_{i=2}^{n} \delta(C(a_i, m_i), C(a_{i-1}, m_i)) \leq a_n - a_1$, where $m_i = m(A_i)$, for $A_i = \{i, \ldots, n\}$.*

Corollary 1. *$\mathfrak{CC}^C_{\delta,m}$ is averaging if and only if it satisfies Proposition 6.*

Proposition 7. *For any RDF δ, FM m and copula C, $\mathfrak{CC}^C_{\delta,m}$ is idempotent.*

Proof. If $\boldsymbol{x} = (x, \ldots, x)$, then, by **(d3)**, one has that:

$$\mathfrak{CC}^C_{\delta,m}(\boldsymbol{x}) = x + \sum_{i=2}^{n} \delta \left(C(x, m_{(i)}), C(x, m_{(i)}) \right) = x + 0 = x.$$

Since the range of the dCC-integral is $[0, n]$, we avoid the term "boundary condition" when referring to condition **(A2)** in this context. Instead, we simply call it 0, 1-condition. The same term was also adopted in [29].

From Proposition 7, the following result is immediate:

Corollary 2. *The d-CC integral satisfies the 0, 1-condition for any m, δ and C.*

3.2 Monotonicity-Like Properties of d-CC Integrals

As explained in [13], one important property of aggregation-like operators is to present some kind of "increasingness property" to guarantee that the more information is provided the higher the aggregated value is in the considered direction. From the discussions in [29], it is immediate that d-CC integrals can not be fully monotonic, in general, since they have to satisfy the following conditions: (i) For $z_1, z_2, z_3, z_4 \in [0,1]$, with $z_1 \leq z_2 \leq z_3 \leq z_4$, $w_1, w_2 \in [0,1]$, with $w_1 \geq w_2$, we have: $\delta(C(z_1, w_1), C(z_3, w_1)) + \delta(C(z_3, w_2), C(z_4, w_2)) \geq \delta(C(z_1, w_1), C(z_2, w_1))$ $+\delta(C(z_2, w_2), C(z_4, w_2))$; (ii) For all $z_1, z_2, z_3 \in [0,1]$, with $z_1 \leq z_2 \leq z_3$, $w \in [0,1]$, we have: $z_2 + \delta(C(z_3, w), C(z_2, w)) \geq z_1 + \delta(C(z_3, w), C(z_1, w))$.

Considering the RDFs of Table 2, for any copula C and FM m, only $\mathfrak{CC}_{\delta_0, m}^C$ is increasing, since it satisfies both conditions. Observe that $\mathfrak{CC}_{\delta_0, m}^C$ is, in fact, a CC-integral [22], which is fully monotonic, for any copula C. When C is the product, both coincide with the d-XC integral (Definition 2) for δ_0, which is the standard CI in the expanded form. Nevertheless, d-CC integrals do satisfy weaker forms of monotonicity, as we discuss in this section.

Directional Monotonicity of d-CC Integrals. Directional monotonicity is one of the most adopted weaker notions of monotonicity, which enlarged the scope of aggregation processes in applications (e.g., [13,18,24,25]).

Definition 4 [5]. *Let $r = (r_1, \ldots, r_n)$ be a real n-dimensional vector such that $r \neq 0 = (0, \ldots, 0)$. A function $F\colon [0,1]^n \to [0,1]$ is said to be r-increasing if, for all $x = (x_1, \ldots, x_n) \in [0,1]^n$ and $c > 0$ such that $x + cr = (x_1 + cr_1, \ldots, x_n + cr_n) \in [0,1]^n$, it holds that $F(x + cr) \geq F(x)$.*

Theorem 1. *Let $m\colon 2^N \to [0,1]$, $\delta\colon [0,1]^2 \to [0,1]$ and $C\colon [0,1]^2 \to [0,1]$ be an FM, an RDF, and a copula, respectively. $\mathfrak{CC}_{\delta,m}^C$ is 1-increasing if and only if one of the following conditions hold: (i) the RDF δ is 1-increasing; (ii) for all $0 \leq z_1 \leq \ldots \leq z_n \leq 1$ and $c > 0$, such that $z_i + c \in [0,1]$, for all $i = 1, \ldots, n$:*

$$\sum_{i=2}^{n} \delta(C(z_i + c, m_i), C(z_{i-1} + c, \cdot m_i)) \geq \sum_{i=2}^{n} \delta(C(z_i \cdot m_i), C(z_{i-1} \cdot m_i)) - c,$$

where $m_i = m(A_i)$, for $A_i = \{i, \ldots, n\}$.

Proof. ($\Leftarrow$)(i) Suppose that δ is 1-increasing and let $c = (c, \ldots, c)$, $c > 0$, such that $x, x + c \in [0,1]^n$. Then:

$$\mathfrak{CC}_{\delta,m}^C(x + c1) = (x_{(1)} + c) + \sum_{i=2}^{n} \delta\left(C(x_{(i)} + c, m_{(i)}), C(x_{(i-1)} + c, m_{(i)})\right)$$

$$> x_{(1)} + \sum_{i=2}^{n} \delta\left(C(x_{(i)}, m_{(i)}), C(x_{(i-1)}, m_{(i)})\right) = \mathfrak{CC}_{\delta,m}^C(x).$$

Now suppose that (ii) holds. Then, for all $x \in [0,1]^n$ and $c = (c, \ldots, c)$, with $c > 0$, such that $x_i + c \in [0,1]$, $\forall i = 1, \ldots, n$, it follows that $\sum_{i=2}^{n} \delta(C(x_{(i)} +$

$c, m_{(i)}), C(x_{(i-1)} + c, m_{(i)})) \geq \sum_{i=2}^{n} \delta(C(x_{(i)}, m_{(i)}), C(x_{(i-1)}, m_{(i)})) - c$. This implies that $(x_{(1)} + c) + \sum_{i=2}^{n} \delta(C(x_{(i)} + c, m_{(i)}), C(x_{(i-1)} + c, m_{(i)})) \geq x_{(1)} + \sum_{i=2}^{n} \delta(C(x_{(i)}, m_{(i)}), C(x_{(i-1)}, m_{(i)}))$. Then, $\mathfrak{CC}_{\delta,m}^{C}(x + c) \geq \mathfrak{CC}_{\delta,m}^{C}(x)$. Therefore, if (i) or (ii) holds, then $\mathfrak{CC}_{\delta,m}^{C}$ is **1**-increasing.

($\Rightarrow$) Suppose that $\mathfrak{CC}_{\delta,m}^{C}$ is **1**-increasing, that is $\mathfrak{CC}_{\delta,m}^{C}(x + c) \geq \mathfrak{CC}_{\delta,m}^{C}(x)$. Then it follows that $(x_{(1)} + c) + \sum_{i=2}^{n} \delta(C(x_{(i)} + c, m_{(i)}), C(x_{(i-1)} + c, m_{(i)})) \geq x_{(1)} + \sum_{i=2}^{n} \delta(C(x_{(i)}, m_{(i)}), C(x_{(i-1)}, m_{(i)}))$ which implies that condition (ii) holds.

Ordered Directional Monotonicity of d-CC Integrals. Any d-CC integral is ordered directionally monotonic [9]. Such functions are monotonic along different directions according to the ordinal size of the coordinates of each input.

Definition 5 [9]. *Consider a function $F: [0,1]^n \to [0,1]$ and let $r = (r_1, \dots, r_n)$ be a real n-dimensional vector, $r \neq 0$. F is said to be ordered directionally (OD) r-increasing if, for each $x \in [0,1]^n$, any permutation $\sigma: \{1, \dots, n\} \to \{1, \dots, n\}$ with $x_{\sigma(1)} \geq \dots \geq x_{\sigma(n)}$, and $c > 0$ such that $1 \geq x_{\sigma(1)} + cr_1 \geq \dots \geq x_{\sigma(n)} + cr_n$, it holds that $F(x + cr_{\sigma^{-1}}) \geq F(x)$, where $r_{\sigma^{-1}} = (r_{\sigma^{-1}(1)}, \dots, r_{\sigma^{-1}(n)})$.*

Theorem 2. *For any FM m, RDF δ, copula C and $k > 0$, the d-CC integral $\mathfrak{CC}_{\delta,m}^{C}$ is an (OD) $(k, 0, \dots, 0)$-increasing function.*

Proof. For all $x \in [0,1]^n$ and permutation $\sigma: \{1, \dots, n\} \to \{1, \dots, n\}$, with $x_{\sigma(1)} \geq \dots \geq x_{\sigma(n)}$, and $c > 0$ s.t. $x_{\sigma(i)} + cr_i \in [0,1]$, for $i \in \{1, \dots, n\}$, and $1 \geq x_{\sigma(1)} + cr_1 \geq \dots \geq x_{\sigma(n)} + cr_n$, for $r_{\sigma^{-1}} = (r_{\sigma^{-1}(1)}, \dots, r_{\sigma^{-1}(n)})$, one has that:

$$X\mathfrak{CC}_{\delta,m}^{C}(x + cr_{\sigma^{-1}}) = x_{(1)} + c \cdot r_{\sigma^{-1}(1)}$$

$$+ \sum_{i=2}^{n-1} \delta(C((x_{(i)} + c \cdot r_{\sigma^{-1}(i)}), m_{(i)}), C((x_{(i-1)} + c \cdot r_{\sigma^{-1}(i-1)}), m_{(i)}))$$

$$+ \delta(C((x_{(n)} + c \cdot r_{\sigma^{-1}(n)}), m_{(n)}), C((x_{(n-1)} + c \cdot r_{\sigma^{-1}(n-1)}), m_{(n)}))$$

$$= x_{(1)} + c \cdot 0 + \sum_{i=2}^{n-1} \delta(C((x_{(i)} + c \cdot 0), m_{(i)}), C((x_{(i-1)} + c \cdot 0), m_{(i)}))$$

$$+ \delta(C((x_{(n)} + c \cdot k), m_{(n)}), C((x_{(n-1)} + c \cdot 0), m_{(n)}))$$

$$= x_{(1)} + \sum_{i=2}^{n-1} \delta(C(x_{(i)}, m_{(i)}), C(x_{(i-1)}, m_{(i)}))$$

$$+ \delta(C((x_{(n)} + c \cdot k), m_{(n)}), C(x_{(n-1)}, m_{(n)}))$$

$$\geq x_{(1)} + \sum_{i=2}^{n-1} \delta(C(x_{(i)}, m_{(i)}), C(x_{(i-1)}, m_{(i)}) + \delta(C(x_{(n)}, m_{(n)}), C(x_{(n-1)}, m_{(n)})))$$

$$= \mathfrak{CC}_{\delta,m}^{C}(x) \qquad\qquad \text{by Prop. 1-(i) and (d4)}$$

4   Application of d-CC Integrals in Classification

A classification problem consists of P training examples $\vec{x}_p = (x_{p1}, \ldots, x_{pn})$, $p \in \{1, \ldots, P\}$ where x_{pi} is the value of the i-th variable of the p-th example and each example belongs to one of M classes in $C = \{C_1, \ldots, C_M\}$. The goal of the learned classifier is to identify the class of new/unknown examples.

A FRBCS [17] is a type of classification system that is based on rules with linguistic labels, modeled by fuzzy sets. The inference process, known as the Fuzzy Reasoning Method (FRM), is determined by four sequential steps:

(1) *Matching degree* (A_j) - It measures, for the example x_p to be classified, the strength of the IF-part of a rule R_j. A_j is calculated by a fuzzy conjunction operator $\mathfrak{c}$, as follows: $A_j(x_p) = \mathfrak{c}(A_{j1}(x_{p1}), \cdots, A_{jn}(x_{pn}))$.

(2) *Association degree* (b_j^k) - It weights the matching degree A_j by the rule weight RW_j, through a product operation:

(3) *Example classification soundness degree for all classes* (Y_k) - For each class $C_k \in C$, we aggregate all the positive association degrees b_j^k that were obtained in the previous step with respect to C_k, through an aggregation function A: $Y_k = A(b_j^k)$, with $j = 1, \ldots, L$, and $b_j^k > 0$.

(4) *Classification* - The final decision is made in this step. For that, a function $F : [0,1]^M \rightarrow C$ is applied over all example classification soundness degrees calculated in the previous step: $F(Y_1, \ldots, Y_M) = \arg \max_{k=1,\ldots,M} (Y_k)$.

In this paper, we will apply d-CC integrals as the aggregation operator A in the third step of the FRM to analyze their effect on the classification process.[2]

4.1   Experimental Framework

The application of our new family of functions is based on a benchmark composed of 33 different public datasets found in KEEL [2] dataset repository[3]. It is important to mention that this benchmark has been used for all the most important generalizations of the CI discussed before. The selection of the same datasets and partitions allows us to directly compare the methods and consequently provide a more complete analysis. In Table 3, we provide the information of the datasets along with their identification (ID), Number of instances (#NoI), Number of Attributes (#NoA), and the Total of Classes (#ToC).

The experimental framework lies in the same context as other works in the literature (See [19,20,23]), i.e. a 5-fold cross validation approach. Consequently, the results provided in Sect. 5 are the Accuracy Mean (AM) related to these folds. Moreover, following the same approach seen in [22], the hyperparameters used by the model are the same originally performed by FARC-HD [1].

---

[2] For more information about this approach see [19,20,23].

[3] https://keel.es.

Table 3. Summary of the considered datasets in the experimental study.

ID.	Dataset	#NoI.	#NoA.	#ToC	ID.	Dataset	#NoI.	#NoA.	#ToC
App	Appendicitis	106	7	2	Pen	Penbased	10,992	16	10
Bal	Balance	625	4	3	Pho	Phoneme	5,404	5	2
Ban	Banana	5300	2	2	Pim	Pima	768	8	2
Bnd	Bands	365	19	2	Rin	Ring	740	20	2
Bup	Bupa	345	6	2	Sah	Saheart	462	9	2
Cle	Cleveland	297	13	5	Sat	Satimage	6,435	36	7
Con	Contraceptive	1473	9	3	Seg	Segment	2,310	19	7
Eco	Ecoli	336	7	8	Shu	Shuttle	58,000	9	7
Gla	Glass	214	9	6	Son	Sonar	208	60	2
Hab	Haberman	306	3	2	Spe	Spectfheart	267	44	2
Hay	Hayes-Roth	160	4	3	Tit	Titanic	2,201	3	2
Ion	Ionosphere	351	33	2	Two	Twonorm	740	20	2
Iri	Iris	150	4	3	Veh	Vehicle	846	18	4
Led	led7digit	500	7	10	Win	Wine	178	13	3
Mag	Magic	1,902	10	2	Wis	Wisconsin	683	11	2
New	Newthyroid	215	5	3	Yea	Yeast	1,484	8	10
Pag	Pageblocks	5,472	10	5					

5   Analysis of the Experimental Results

This section explores the application of d-CC Integrals. So, it is presented the performance of d-CC integrals over the datasets to achieve it. Then, the best function is pointed out, and we compare it against other operators found in the literature to highlight the best-performing method efficiency.

5.1   The Performance of the d-CC Integrals in Classification

This section provides the results obtained by the application of the d-CC Integrals in the FRM. To ease the comprehension, in Table 4, we present the combination of the RDF and d-CC integral per dataset that led to the largest accuracy. Moreover, we show in Table 5 the mean accuracy for our approach considering the 33 different datasets[4] In order to provide a better analysis, we also highlight, per RDF, the largest mean in **boldface**.

The combination of the RDF δ_5 and copula O_{mM} achieves the largest AM in the study. Moreover, considering δ_5, for more than half of the considered d-CC Integrals, the AM is superior to 80%. The RDF δ_2 is the next one to achieve satisfactory performance. In fact, only for two copulas, T_M, and T_L, this RDF presents an AM inferior to 80%. Observe that δ_5 and δ_2 also performed well in [27,28]. Additionally, both δ_0 and δ_3 presented similar AMs, around 79%. On the other hand, the remaining cases (δ_1 and δ_4) presented the worst scenarios in this study, with AMs around 78%.

---

[4] To analyze the particular cases and results per fold please check the following link - https://github.com/Giancarlo-Lucca/d-CC-Integrals-generalizing-CC-integrals-by-restricted-dissimilarity-functions.

Table 4. The best-achieved results in a test for each dataset.

Dataset	RDF	Copula	Accuracy	Dataset	RDF	Copula	Accuracy
App	δ_2	T_L	87.79	Pen	δ_2, δ_5	C_F, T_P	92.45
Bal	δ_2, δ_2	T_M, O_{Div}	89.12	Pho	δ_0	O_B	83.55
Ban	δ_0	O_{mM}	87.00	Pim	δ_2	C_{CL}	77.08
Bnd	δ_3	T_L	72.67	Rin	δ_5	T_{HP}	92.43
Bup	δ_4	O_α	68.41	Sah	δ_5	O_α	71.43
C_{CL}e	δ_5	O_{mM}	60.27	Sat	δ_2	C_{CL}	80.25
Con	δ_5	O_α	54.51	Seg	δ_2	O_{Div}	93.59
Eco	δ_2	T_{HP}	83.04	Shu	δ_3	O_{Div}	99.03
Gla	δ_5	T_{HP}	71.05	Son	δ_2	O_{mM}	84.65
Hab	δ_1	T_P	76.78	Spe	δ_2	O_{Div}	82.40
Hay	δ_0	T_M	81.74	Tit	All	All	78.87
Ion	δ_2	T_{HP}	91.18	Two	δ_2	T_{HP}	92.43
Iri	δ_2	O_{mM}	96.00	Veh	δ_0	T_M	69.86
Led	δ_0	O_{mM}	70.00	Win	δ_0	C_{CL}	98.32
Mag	δ_2	O_{mM}	81.12	Wis	δ_5	T_P	97.36
New	δ_2	T_P	97.67	Yea	δ_5	T_L	58.56
Pag	δ_5	T_M	95.07				

Table 5. Obtained accuracy mean in all datasets.

	RDF	T_M	T_P	T_L	T_{HP}	O_B	O_{mM}	O_α	O_{Div}	C_F	C_{CL}
Mean	δ_0	79.54	79.20	79.00	79.54	78.94	78.90	79.44	79.26	**79.55**	79.36
	δ_1	77.30	77.87	77.71	77.79	77.74	77.60	77.59	**78.01**	77.54	77.63
	δ_2	79.67	80.23	79.84	80.28	80.06	80.29	80.46	**80.49**	80.17	80.38
	δ_3	79.23	79.76	79.53	79.38	79.63	79.71	79.55	79.58	**79.83**	79.31
	δ_4	**78.87**	78.21	78.41	78.47	78.46	78.21	78.34	78.39	78.54	78.45
	δ_5	79.98	80.44	79.88	80.56	79.99	**80.62**	80.35	80.27	80.33	79.95

Up to this point, the superiority of δ_2 and δ_5 among the RDFs can be perceived. However, to complete the analysis we have conducted a set of statistical tests to reinforce this conclusion. Considering that the conditions necessary to perform parametric tests are not fulfilled, we considered non-parametric statistical tests. Hence, the analysis was made by a group comparison using the Align-Friedman Rank test [14], and the results were analyzed in terms of the adjusted p-value (APV) computed by Holm's post hoc test [15].

The analysis was done by performing a statistical test per RDF, with a local statistical analysis among the different functions. After that, we chose the control variables of each group, the ones having the lowest obtained rank, and once again we statistically compared them. This approach allowed us to point out the best d-CC integral, used to compare against other aggregation operators.

The results, in Table 6, sort the methods by ranks per RDF. The last column shows a comparison among the selected control variables. If the obtained APV is smaller than 0.1 (10% of significance level), it means the methods are statistically different and, so, we underline them. Among the groups of RDFs, in general, the number of statistical differences are low. The test rejected the null hypothesis when comparing the control variable against three different approaches for δ_4, two for δ_0 and δ_2, one for δ_5 and none for δ_1 and δ_2.

Table 6. Align Friedman rank test and Holm's post hoc test for the d-CC integrals.

δ_0			δ_1			δ_2			δ_3			δ_4			δ_5			Control Variables			
d-CC	Rank	APV	d-CC	Rank	APV	d-CC	Rank	APV	d-CC	Rank	APV	d-CC	Rank	APV	d-CC	Rank	APV	Diss	d-CC	Rank	APV
C_F	128.35	-	O_{Div}	131.15	-	O_α	135.85	-	C_F	143.62	-	T_M	119.03	-	O_{mM}	134.55	-	δ_5	O_{mM}	62.88	-
T_{HP}	134.09	0.81	T_{HP}	145.64	0.68	O_{Div}	137.42	1.00	O_{mM}	144.77	1.00	O_B	152.80	0.21	O_α	146.23	0.62	δ_2	O_α	67.59	0.73
T_M	157.41	0.76	T_P	157.27	0.68	C_{CL}	147.91	1.00	T_P	146.95	1.00	T_{HP}	160.67	0.21	T_{HP}	147.71	0.62	δ_0	C_F	90.86	0.09
C_{CL}	158.83	0.76	T_L	159.55	0.68	T_P	153.73	1.00	O_B	160.47	1.00	C_F	161.83	0.21	T_P	149.74	0.62	δ_3	C_F	97.36	0.04
O_α	159.11	0.76	O_B	163.88	0.65	T_{HP}	156.68	1.00	O_α	162.65	1.00	T_L	166.42	0.21	O_{Div}	156.65	0.62	δ_4	T_M	128.23	0.00
O_{Div}	167.89	0.46	O_{mM}	176.27	0.35	O_{mM}	158.65	1.00	O_{Div}	168.52	1.00	C_{CL}	166.68	0.21	C_F	161.27	0.62	δ_1	O_{Div}	150.08	0.00
T_P	170.00	0.46	C_{CL}	176.91	0.35	C_F	165.30	1.00	T_L	169.47	1.00	O_α	169.53	0.19	C_{CL}	183.95	0.21				
T_L	183.36	0.13	O_α	177.20	0.35	O_B	186.61	0.21	C_{CL}	180.02	0.88	O_{Div}	178.45	0.08	T_M	186.59	0.19				
O_B	197.27	0.03	C_F	180.20	0.29	T_L	200.68	0.05	T_{HP}	181.14	0.88	O_{mM}	188.26	0.03	T_L	197.80	0.14				
O_{mM}	198.68	0.02	T_M	186.94	0.16	T_M	212.17	0.01	T_M	197.39	0.20	T_P	191.32	0.02	O_B	190.50	0.06				

The last column performs a group comparison among the control variables of each family based on an RDF. See that the d-CC integral with δ_5 and O_{mM}, when compared against almost all approaches, is statically superior. But this does not hold when δ_2 and O_α are considered. Finally, $\delta_5 - O_{mM}$ provides the winning combination in this first analysis and so, it is the representing method of the class of d-CC integrals.

5.2 Comparing the Best d-CC Integral Against Classical Operators

This subsection states the position of the d-CC integrals among the generalizations of the CI as well as known FRMs found in the literature. We have selected averaging and non-averaging operators:

- **Averaging operators:** [WR] (Winning Rule [11], based on the maximum as aggregation function), [Cho] (CI), [CC] (the best CC-integral [22]), [C_T] (the best C_T-integral, a CI generalization by t-norms [20]), [C_{AVG}^F] (the best averaging C_F-integral [23]), [d-CF] (the best averaging d-CF integral [28]).
- **Non-Averaging functions:** [AC] (the Additive Combination [12], based on the normalized sum, FARC-HD algorithm basis), [PS] (the Probabilistic Sum [12]), [C_{F1F2}] (the C_{F1F2}-integral [19], being the state-of-the-art), [C_{N-AVG}^F] (the best on-averaging C_F-integral [23]).

The obtained results are in Table 7, detailed per dataset. We highlight in **boldface** the largest obtained AM among all results.

We observed that the C_{F1F2}-integral is the approach having the largest AM in the study. However, the d-CC integral (our new approach) is the second one, followed by the C_F and the d-CF integrals. It is also noticeable that the d-CC integral achieves a superior mean than the classical FRMs of AC and PS.

Considering the methods that present averaging characteristics and their obtained AM, it is easy to see that their performance is, in general, inferior to the non-averaging ones. The accuracy in these cases is around 79% which is inferior to the best d-CC integral, with the exception of the d-CF integral.

Once again, we performed a set of analyses to statistically compare the approaches. Therefore, we have compared the best d-CC integral (based on OmM and δ_5) against the averaging and non-averaging approaches. The results

Table 7. Comparing the best d-CC integral with other operators in the literature.

Dataset	Averaging						Non-Averaging				d-CC
	WR	Cho	CC	C_T	C^F_{AVG}	d-CF	AC	PS	C_{F1F2}	C^F_{N-AVG}	
App	83.03	80.13	85.84	82.99	82.99	83.98	83.03	85.84	**86.80**	85.84	85.93
Bal	81.92	82.40	81.60	82.72	82.56	87.36	85.92	87.20	**89.12**	88.64	88.80
Ban	83.94	**86.32**	84.30	85.96	86.09	83.55	85.30	84.85	84.79	84.60	84.25
Bnd	69.40	68.56	71.06	**72.13**	69.40	69.42	68.28	68.82	71.30	70.48	71.55
Bup	62.03	66.96	61.45	65.80	**67.83**	65.80	67.25	61.74	66.96	64.64	64.35
Cle	56.91	55.58	54.88	55.58	57.92	**60.94**	56.21	59.25	56.22	56.55	60.27
Con	52.07	51.26	52.61	53.09	52.27	**54.72**	53.16	52.21	54.72	53.16	54.04
Eco	75.62	76.51	77.09	80.07	78.88	81.55	**82.15**	80.95	81.86	80.08	79.79
Gla	64.99	64.02	69.17	63.10	64.51	66.39	65.44	64.04	68.25	66.83	**70.09**
Hab	70.89	72.52	**74.17**	72.21	73.51	71.88	73.18	69.26	72.53	71.87	69.90
Hay	78.69	79.49	**81.74**	79.49	78.72	78.72	77.95	77.95	78.66	79.43	79.43
Ion	90.03	90.04	88.89	89.18	**90.60**	87.76	88.90	88.32	88.33	89.75	90.32
Iri	94.00	91.33	92.67	93.33	93.33	**97.33**	94.00	95.33	94.00	94.00	94.67
Led	69.40	68.20	68.40	68.60	68.60	**70.00**	69.60	69.20	**70.00**	69.80	69.80
Mag	78.60	78.86	79.81	79.76	80.02	80.07	80.76	80.39	**80.86**	79.70	79.44
New	94.88	94.88	93.95	95.35	93.49	94.42	94.88	94.42	**96.74**	96.28	96.28
Pag	94.16	94.16	93.97	94.34	93.97	95.25	95.07	94.52	**95.25**	94.15	94.71
Pen	91.45	90.55	91.27	90.82	91.45	91.36	92.55	**93.27**	92.91	92.91	91.73
Pho	82.29	82.98	82.94	**83.83**	82.86	81.62	81.70	82.51	81.42	81.44	81.18
Pim	74.60	73.95	74.21	74.87	75.64	75.13	74.74	**75.91**	75.38	75.52	75.00
Rin	90.00	90.95	87.97	88.78	90.27	90.68	90.95	90.00	**91.89**	89.86	91.35
Sah	68.61	69.69	70.78	70.77	68.61	**71.65**	68.39	69.69	71.43	70.12	70.77
Sat	79.63	79.47	79.01	80.40	78.54	79.16	79.47	80.40	79.47	**80.41**	79.94
Seg	93.03	**93.46**	92.25	93.33	92.55	93.07	93.12	92.94	93.29	92.42	93.07
Shu	96.00	97.61	**98.16**	97.20	96.78	97.10	95.59	94.85	96.83	97.15	97.06
Son	77.42	77.43	76.95	79.34	78.85	80.82	78.36	82.24	**85.15**	83.21	81.28
Spe	77.90	77.88	78.99	76.02	78.26	**81.25**	77.88	77.90	79.39	79.77	79.41
Tit	78.87	78.87	78.87	78.87	78.87	78.87	78.87	78.87	78.87	78.87	78.87
Two	86.49	84.46	85.14	85.27	83.92	89.59	90.95	90.00	92.30	**92.57**	88.51
Veh	66.67	68.44	**69.86**	68.20	67.97	67.97	68.56	68.09	68.20	68.08	69.03
Win	96.60	93.79	93.83	96.63	96.03	96.62	96.03	94.92	95.48	96.08	**97.17**
Wis	96.34	**97.22**	95.90	96.78	96.34	96.93	96.63	97.22	96.78	96.78	96.78
Yea	55.32	55.73	57.01	56.53	56.40	57.68	58.96	**59.03**	58.56	57.08	55.66
Mean	79.15	79.20	79.54	79.74	79.64	80.57	80.12	80.07	**81.02**	80.55	80.62

Table 8. Statistical analysis comparing the d-CC integral against classical operators.

Averaging			N-Averaging		
Method	Rank	APV	Method	Rank	APV
d-CC	74.21212	–	C_{F1F2}	57.56	–
d-CF	81.51515	0.65	d-CC	76.74	0.10
C_T	111.2424	0.04	C^F_{N-AVG}	81.36	0.08
C^F_{AVG}	123.9545	0.00	AC	98.82	0.00
CC	128.6818	0.00	PS	100.52	0.00
Cho	142.1818	0.00			
WR	150.2121	0.00			

are presented in Table 8, following the same structure used in the previous analysis.

Regarding the results obtained when comparing the d-CC integral against the averaging approaches, we see that it is statistically superior to all other approaches, with the exception of the d-CF integral, considered equivalent. It is noteworthy that the CC-integral is among the surpassed methods, pointing out that its generalization by RFDs resulted in better performances. If the non-averaging approaches are considered, it is noticeable that our new method is the only one that is able to be considered statistically equivalent to the state-of-the-art. For all the other cases, the $C_{F_1F_2}$-integral is still superior.

6   Conclusions

New generalizations of the discrete CI have been developed in recent years, by generalizing the product and/or the difference operations from the original expression or its expanded form. The study of those generalizations has proven to be worthwhile since they can provide promising results in applications. In this paper, we continue this trend by focusing on the generalization of the CC-integral by means of RDFs, calling them d-CC-integrals. Then, we studied some relevant properties for application purposes.

The final part of our work applied the d-CC-integrals in a classification problem, followed by a two-step analysis. First, we compared different d-CC-integrals and identified that the one based on the copula OmM and the RDF δ_5 had the best overall performance. Then, we compared this particular d-CC-integral with other methods from the literature, showing that it is statistically equivalent to the state-of-the-art $C_{F_1F_2}$-integral and surpasses most of the other considered approaches, including the CC-integral.

References

1. Alcala-Fdez, J., Alcala, R., Herrera, F.: A fuzzy association rule-based classification model for high-dimensional problems with genetic rule selection and lateral tuning. IEEE Trans. Fuzzy Syst. **19**(5), 857–872 (2011)
2. Alcalá-Fdez, J., et al.: Keel: a software tool to assess evolutionary algorithms for data mining problems. Soft. Comput. **13**(3), 307–318 (2009)
3. Alsina, C., Frank, M.J., Schweizer, B.: Associative Functions: Triangular Norms and Copulas. World Scientific Publishing Company, Singapore (2006)
4. Boczek, M., Kaluszka, M.: On the extended Choquet-Sugeno-like operator. Int. J. Approx. Reason. **154**, 48–55 (2023)
5. Bustince, H., Fernandez, J., Kolesárová, A., Mesiar, R.: Directional monotonicity of fusion functions. Eur. J. Oper. Res. **244**(1), 300–308 (2015)
6. Bustince, H., Fernandez, J., Mesiar, R., Montero, J., Orduna, R.: Overlap functions. Nonlinear Anal. Theory Methods App. **72**(3–4), 1488–1499 (2010)
7. Bustince, H., Jurio, A., Pradera, A., Mesiar, R., Beliakov, G.: Generalization of the weighted voting method using penalty functions constructed via faithful restricted dissimilarity functions. Eur. J. Oper. Res. **225**(3), 472–478 (2013)

8. Bustince, H., et al.: d-Choquet integrals: choquet integrals based on dissimilarities. Fuzzy Sets Syst. (2020)
9. Bustince, H., et al.: Ordered directionally monotone functions: justification and application. IEEE Trans. Fuzzy Syst. **26**(4), 2237–2250 (2018)
10. Choquet, G.: Theory of capacities. Institut Fourier **5**, 131–295 (1953–1954)
11. Cordón, O., del Jesus, M.J., Herrera, F.: A proposal on reasoning methods in fuzzy rule-based classification systems. Int. J. Approx. Reason. **20**(1), 21–45 (1999)
12. Cordon, O., del Jesus, M.J., Herrera, F.: Analyzing the reasoning mechanisms in fuzzy rule based classification systems. Math. Soft Comput. **5**(2–3), 321–332 (1998)
13. Dimuro, G.P., et al.: The state-of-art of the generalizations of the Choquet integral: from aggregation and pre-aggregation to ordered directionally monotone functions. Inf. Fusion **57**, 27–43 (2020)
14. Hodges, J.L., Lehmann, E.L.: Ranks methods for combination of independent experiments in analysis of variance. Ann. Math. Stat. **33**, 482–497 (1962)
15. Holm, S.: A simple sequentially rejective multiple test procedure. Scand. J. Stat. **6**, 65–70 (1979)
16. Ishibuchi, H., Nakashima, T.: Effect of rule weights in fuzzy rule-based classification systems. Fuzzy Syst. IEEE Trans. **9**(4), 506–515 (2001)
17. Ishibuchi, H., Nakashima, T., Nii, M.: Classification and Modeling with Linguistic Information Granules, Advanced Approaches to Linguistic Data Mining. Advanced Information Processing. Springer, Berlin, Heidelberg (2005). https://doi.org/10.1007/b138232
18. Ko, L., et al.: Multimodal fuzzy fusion for enhancing the motor-imagery-based brain computer interface. IEEE Comput. Intell. Mag. **14**(1), 96–106 (2019)
19. Lucca, G., Dimuro, G.P., Fernandez, J., Bustince, H., Bedregal, B., Sanz, J.A.: Improving the performance of fuzzy rule-based classification systems based on a nonaveraging generalization of CC-integrals named $C_{F_1 F_2}$-integrals. IEEE Trans. Fuzzy Syst. **27**(1), 124–134 (2019)
20. Lucca, G., Sanz, J., Pereira Dimuro, G., Bedregal, B., Mesiar, R., Kolesárová, A., Bustince Sola, H.: Pre-aggregation functions: construction and an application. IEEE Trans. Fuzzy Syst. **24**(2), 260–272 (2016)
21. Lucca, G., Sanz, J.A., Dimuro, G.P., Borges, E.N., Santos, H., Bustince, H.: Analyzing the performance of different fuzzy measures with generalizations of the Choquet integral in classification problems. In: 2019 FUZZ-IEEE, pp. 1–6 (2019)
22. Lucca, G., et al.: CC-integrals: choquet-like copula-based aggregation functions and its application in fuzzy rule-based classification systems. KBS **119**, 32–43 (2017)
23. Lucca, G., Sanz, J.A., Dimuro, G.P., Bedregal, B., Bustince, H., Mesiar, R.: CF-integrals: a new family of pre-aggregation functions with application to fuzzy rule-based classification systems. Inf. Sci. **435**, 94–110 (2018)
24. Marco-Detchart, C., Lucca, G., Lopez-Molina, C., De Miguel, L., Pereira Dimuro, G., Bustince, H.: Neuro-inspired edge feature fusion using Choquet integrals. Inf. Sci. **581**, 740–754 (2021)
25. Mesiar, R., Kolesárová, A., Bustince, H., Dimuro, G., Bedregal, B.: Fusion functions based discrete Choquet-like integrals. EJOR **252**(2), 601–609 (2016)
26. Sanz, J.A., Bernardo, D., Herrera, F., Bustince, H., Hagras, H.: A compact evolutionary interval-valued fuzzy rule-based classification system for the modeling and prediction of real-world financial applications with imbalanced data. IEEE Trans. Fuzzy Syst. **23**(4), 973–990 (2015)
27. Wieczynski, J., et al.: Applying d-XChoquet integrals in classification problems. In: 2022 IEEE International Conference on Fuzzy Systems (FUZZ-IEEE), pp. 1–7 (2022)

28. Wieczynski, J., et al.: dc_F-integrals: generalizing c_F-integrals by means of restricted dissimilarity functions. IEEE TFS **31**(1), 160–173 (2023)
29. Wieczynski, J.C., et al.: d-XC integrals: on the generalization of the expanded form of the Choquet integral by restricted dissimilarity functions and their applications. IEEE TFS **30**(12), 5376–5389 (2022)

FeatGeNN: Improving Model Performance for Tabular Data with Correlation-Based Feature Extraction

Sammuel Ramos Silva[1,2]([✉]) [ID] and Rodrigo Silva[1] [ID]

[1] Universidade Federal de Ouro Preto, Ouro Preto 35402-163, Brazil
`sammuel.silva@aluno.ufop.edu.br`, `rodrigo.silva@ufop.edu.br`
[2] Cloudwalk, Inc., São Paulo, São Paulo 05425-070, Brazil
`sammuel@cloudwalk.io`

Abstract. Automated Feature Engineering (AutoFE) has become an important task for any machine learning project, as it can help improve model performance and gain more information for statistical analysis. However, most current approaches for AutoFE rely on manual feature creation or use methods that can generate a large number of features, which can be computationally intensive and lead to overfitting. To address these challenges, we propose a novel convolutional method called FeatGeNN that extracts and creates new features using correlation as a pooling function. Unlike traditional pooling functions like max-pooling, correlation-based pooling considers the linear relationship between the features in the data matrix, making it more suitable for tabular data. We evaluate our method on various benchmark datasets and demonstrate that FeatGeNN outperforms existing AutoFE approaches regarding model performance. Our results suggest that correlation-based pooling can be a promising alternative to max-pooling for AutoFE in tabular data applications.

Keywords: Automated Feature Engineering · feature creation · correlation-based pooling · tabular data · machine learning

1 Introduction

Creating effective features is a crucial aspect of machine-learning projects. Essentially, it involves deriving new features from existing data to train a model or extract more information for statistical analysis. Discovering novel features from raw datasets is often the key to improving model performance [1].

The author would like to acknowledge FAPEMIG (Fundação de Amparo à Pesquisa do Estado de Minas Gerais), CAPES (Coordenação de Aperfeiçoamento de Pessoal de Nível Superior), UFOP (Universidade Federal de Ouro Preto) and Cloudwalk, Inc, for the financial support which has been instrumental in the successful execution of our research endeavors.

M. C. Naldi and R. A. C. Bianchi (Eds.): BRACIS 2023, LNAI 14195, pp. 259–273, 2023.
https://doi.org/10.1007/978-3-031-45368-7_17

Traditionally, feature creation is a manual process that heavily relies on an analyst's domain knowledge and programming skills. However, this approach can be limiting, as an analyst's intuition and expertise often influence the features created. To overcome these limitations, researchers have been exploring the field of Automated Feature Engineering (AutoFE). AutoFE aims to automate the feature creation process, enabling the discovery of more complex and effective features without relying solely on human input.

Automated feature engineering methods involve applying transformations to raw data to create new features. One commonly used technique is the expansion-reduction method [4], which generates a large number of features and then applies a feature selection algorithm to reduce their dimensionality. During the expansion phase, various transformations, such as logarithmic, max/min, or sum, are applied to the raw data. In the reduction phase, a feature selection method is utilized to identify the most effective set of features, which can significantly enhance a model's performance.

The possible number of transformation operations that can be performed on already-transformed features is practically infinite, which leads to an exponential increase in the feature space. This issue can cause a problem in reducing the number of feature evaluations required. To address this issue, researchers have proposed adaptive methods for AutoFE. For instance, Khurana et al. [5] introduced a Q-learning agent capable of performing feature transformation search, achieving higher performance but still generating a large number of features. In another study [6], a Multi-Layer Perceptron (MLP) was trained to suggest the best transformations for each raw feature, resolving the problem of excessive feature generation. More recently, DIFER [7], a gradient-based method for differentiable AutoFE, has demonstrated superior performance and computational efficiency compared to other approaches, although it still requires significant computation

In recent years, the use of deep neural networks (DNNs) has become increasingly widespread across a range of fields, such as computer vision and natural language processing [14,15]. Typically, these models extract new features by feeding input features into the hidden layers of a DNN. While this approach is effective in capturing complex interactions between implicit and explicit features, it may not always generate useful new features due to a lack of relevant interactions in the dataset [9]. Moreover, most existing works use max-pooling in the pooling layer, which may not be optimal for tabular data because it does not preserve the order and context of features in the data matrix. Additionally, max-pooling is intended to identify the most significant features within an image, which may not always be relevant or effective for tabular data.

To address the limitations of existing AutoFE methods, we propose Feat-GeNN, a convolutional approach that leverages correlation as a pooling function to extract and generate new features. FeatGeNN first applies convolutional filters to the raw data to extract high-level representations. Then, instead of using traditional pooling functions like max or average pooling, it computes the correlation between the extracted features, which helps to identify the most informative features. The selected features are then passed through a multi-layer perceptron (MLP) to create the final set of new features. Preliminary results indicate that FeatGeNN outperforms existing AutoFE methods in both the number of generated features and model performance, demonstrating its potential as a potent tool for creating features in machine learning.

2   Related Work

The main goal of feature engineering is to transform raw data into new features that can better express the problem to be solved. Training a model with the generated features can increase the performance of the model. However, the process of feature engineering can be limited by the expertise, programming skills, and intuition of the person working with the data. For this reason, AutoFE approaches have recently gained attention.

The authors of [7] propose a differentiable AutoML model that efficiently extracts low and high-order features. The model includes three steps: Initialization, Optimizer Training, and Feature Evaluation. In initialization, features are constructed randomly and evaluated using a machine-learning model in the validation set. In training the optimizer, a tree-like structure is created with an encoder, a predictor, and a decoder, called a parse tree. The encoder maps the post-order traversal string to a continuous space, the predictor is a 5-layer MLP that maps the representation to the score computed by a machine learning model, and the decoder maps the embedding to the discrete feature space. In the final step of feature evolution, the best n features are selected and optimized using a gradient-based approach.

In [4], the authors present an algorithm that uses mathematical functions to generate new features for relational databases. The algorithm begins by identifying the entities that make up the database and defines a set of mathematical functions that are applied at both the entity level and the relational level. The proposed approach first enumerates all possible transformations on all features and then directly selects features based on their impact on model performance. However, due to the potentially large number of features generated, it is necessary to perform feature selection and dimensionality reduction to avoid overfitting and improve the interpretability of the model.

In [6], the authors propose a novel model for feature engineering in classification tasks that can generalize the effects of different feature transformations across multiple datasets. The model uses an MLP for each transformation to predict whether it can produce more useful features than the original set. The Quantile Sketch Array (QSA) achieves a fixed-size representation of feature values to handle features and data of different lengths. The QSA uses Quantile Data Sketch to represent feature values associated with a class label.

The authors of [11] have proposed an RNN-based approach to address the feature explosion problem in feature engineering and support higher-order transformations. Their architecture uses an RNN to generate transformation rules with a maximum order for each raw feature within a fixed time limit. For datasets with multiple raw features, the authors use multiple RNNs as controllers to generate transformation rules for each feature. The transformed features are evaluated using a machine learning algorithm and the controller is trained using policy gradients. The model includes two special unary transformations: "delete" and "terminate", which remove a feature and terminate the current transformation, respectively, to determine the most appropriate transformation order.

In [5] they propose a heuristic model for automating feature engineering in supervised learning problems. Their model is based on a tree structure, where the raw dataset is the root, each node is a transformed dataset, and the edges represent the transformation functions. The goal is to find the node with the highest score, reducing the feature construction problem to a search problem.

The authors present three exploration strategies to traverse the tree. The first is "depth-first traversal", in which a random transformation is applied to the root and the algorithm then explores a branch until there is no further improvement. Then it chooses another node with the highest score and starts the process again. The second is the "Global Traversal", where a global search is performed to find the most promising node out of all the nodes explored so far. The third is "Balanced Traversal", in which the algorithm chooses either an exploration or exploitation strategy at each step based on a time or node budget. To handle the explosive growth of columns, feature selection is required as they grow. Cognito allows the selection of features after each transformation to clean up the dataset and ensure a manageable size. In addition, at the end of the model execution, the algorithm performs another feature selection for all columns in the dataset, including the newly created columns.

AutoFeat is a method presented in [10] that generates and selects non-linear input features from raw inputs. The method applies a series of transformations to the raw input and combines pairs of features in an alternating multi-step process to generate new features. However, this leads to an exponential increase in the size of the feature space, so a subsampling procedure is performed before computing new features. The authors have shown that two or three steps of the feature technique are usually sufficient to generate new features.

After feature engineering, the new dataset has a higher number of features than the original dataset. To reduce the dimensionality, the authors developed a feature selection procedure. First, they remove new features that are highly correlated with the original or simpler features. Then they apply a wrapper method with L1-regular linear models to select the most informative and non-redundant features from the dataset. In the end, only a few dozen features are retained and used after the feature creation and selection process.

In [17], autolearn is proposed, a learning model based on regression between pairs of features and aimed at discovering patterns and their variations in the data. The method selects a small number of new features to achieve the desired performance. The proposed method consists of four phases: Pre-processing to reduce dimensionality, where the authors perform feature selection based on information gain (IG); Mining of correlated features to define and search for pairwise correlated features, where the distance correlation [8] is calculated to determine if there is an interesting predictive relationship between a pair of features; Feature generation, where regularized regression algorithms are used to search for associations between features and generate new features; and Feature selection, where features that do not add new information to the dataset are discarded.

The authors of [3] have proposed a novel model that achieves both memorization and generalization by simultaneously training a linear model component and a neural network component. The model consists of two components: The Wide component, which is a generalized linear model of the form yW^txb, where y denotes prediction, x denotes features, w denotes model parameters and b denotes bias. The input features can be either raw or transformed, the most important transformation being the cross-product transformation; and the Deep component, which is a feed-forward neural network. For categorical features, an embedding is created, which is then added to the dataset and fed into the network.

The authors of [2] have proposed a model for predicting CTR that can handle interactions between low and high-order features by introducing a factorization-machine (FM) based neural network. The model consists of two parts: the FM component, which generates low-order features and can generate interactions between 1st and 2nd-order features with low computational cost, and the deep component, a feed-forward neural network that learns interactions between higher-order features. The input to the network is a high-dimensional vector of sparse data containing categorical and continuous variables as well as grouped fields.

The FGCNN is another approach proposed in CTR for prediction [9]. This model consists of two components, namely the Feature Generation and the Deep Classifier. The Feature Generation component uses the mechanisms inherent in the Convolutional Neural Network (CNN) and the Multilayer Perceptron (MLP) to identify relevant local and global patterns in the data and generate new features. The Deep Classifier component then uses the extended feature space to learn and make predictions.

Our work introduces a CNN-based model with correlation-pooling for extracting high-order features and improving model performance. Unlike traditional pooling functions such as max-pooling, which focus on selecting the maximum value within a pooling region, correlation-pooling considers the linear relationships between features in the data matrix. It measures the correlation coefficient between the features and aggregates them based on their correlation values to capture the interdependencies and patterns in the data. By incorporating correlation-based pooling into the feature extraction process, FeatGeNN can effectively extract high-order features that reflect the underlying relationships among input variables. Our proposed method achieves competitive results on a range of problems, suggesting that correlation-based pooling is a promising technique for working with tabular data in neural networks.

3   Proposed Approach

In this section, we describe the proposed Feature Generation with Evolutionary Convolutional Neural Networks (FeatGeNN) model in detail.

3.1   Problem Formulation

Given an dataset $D = \langle F, tg \rangle$, where $F = \{f_1, f_2, ..., f_n\}$ are the raw features and tg the target vector. We denote as $L_E^M(D, tg)$ the performance of the machine learning model M that is learned from D and measured by the evaluation metric E (e.g. accuracy). In addition, we transform a raw set of features D into D_{new} by applying a set of transformation functions $T = \{t_1, t_2, ..., t_n\}$.

Formally, the goal of the AutoFE is to search the optimal transformed feature set D^* where $L_E^M(D^*, tg)$ is maximized.

3.2   FeatGeNN Model

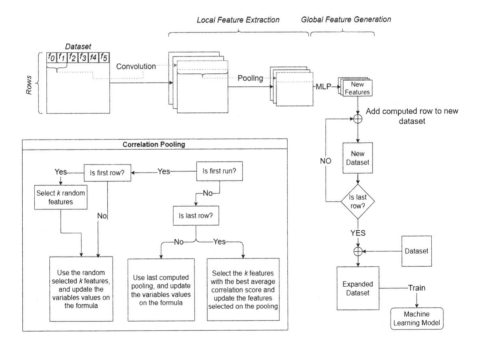

Fig. 1. The FeatGeNN process.

In this study, we use a convolutional neural network to extract features that can improve the performance of a machine learning model (i.e., Random Forest). As explained earlier, using an MLP alone to generate new features would not result in a good set of new features. The reason for this is the relationship between the number of informative interactions between features and the total number of features in the feature space. Also, using a CNN alone might not lead to good performance because a CNN only considers local interactions and does not consider many important global interactions between features [9].

To overcome this problem, we use an architecture that combines the MLP with the CNN. The FeatGeNN model includes two main blocks, namely local feature extraction and global feature generation (Fig. 1). The first block attempts to identify the most informative interactions between local features, while the second block generates new features from the features extracted by the local feature extraction block and combines them globally.

The Local Feature Extraction block includes two main operations, namely Pooling and Convolution. Among these operations, the pooling operation plays a crucial role in reducing dimensionality and preserving the most informative features for subsequent layers. In previous work on feature generation for tabular data with CNN, max-pooling was mainly used. However, we found that using max-pooling for tabular data may not give the desired result because the model may not compare closely related features, thus affecting the features generated by the model. Therefore, we propose the use of correlation-pooling to address this issue.

In correlation pooling (Fig. 2), the variant of pooling used in our Local Feature Extraction block, uses Pearson correlation [19] to group features that are highly correlated. By grouping these features, correlation-pooling can preserve the relationship between closely related features and thus improve the quality of the features extracted by the CNN model. This is in contrast to max-pooling, which preserves only the most dominant feature in a group and may ignore other relevant features that are closely related. Therefore, by incorporating Pearson correlation in the pooling operation, correlation-pooling can effectively circumvent the limitation of max-pooling and help generate more informative features for subsequent layers in the CNN model. The Pearson correlation coefficient can be formulated as follows for our problem:

$$r = \frac{n \sum xy - (\sum x)(\sum y)}{\sqrt{[n \sum x^2 - (\sum x)^2][n \sum y^2 - (\sum y)^2]}} \tag{1}$$

where x and y represent the values of the features X and Y respectively, and $X, Y \in F$, where F is the set of all features. The variable n denotes the number of samples in the dataset D.

To avoid having to run the Pearson algorithm twice, we have introduced an iterative calculation of the Pearson coefficient. This means that at the current stage of model development, we compute the Pearson coefficient r to perform the pooling operation for the subsequent evolutionary generation of the model. To reduce the computations required, we also added a threshold to limit the number of data sent to the correlation calculation, i.e., a model can only use 70% of the data to calculate the correlation value for the features.

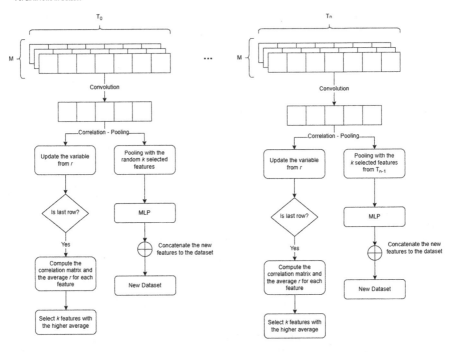

Fig. 2. Correlation-Pooling process.

While the Pearson correlation is a statistical measure that describes the linear relationship between two variables, it is not suitable for analyzing relationships between more than two characteristics. To overcome this limitation, we use the multivariate correlation matrix, which consists of pairwise Pearson correlation coefficients between all pairs of variables. This matrix allows us to analyze relationships between multiple variables and identify the most highly correlated variables. The overall correlation value for the feature f can be formulated as follows:

$$CS_f = \frac{\sum_{k}^{N} r_{fk}}{N} \qquad (2)$$

where CS_f is the correlation score for the feature f, r_{fk} represent the person correlation score for the feature tuple (f,k) and N the total number of feature in the dataset.

In the Global Feature Generation block, an MLP is utilized to merge the features extracted from the Local Feature Extraction block and generate novel features. These novel features are then appended to the original dataset and used in the machine-learning model.

3.3   Evolution Process

In this work, we adopt an evolution process for conducting AutoFE, as depicted in Fig. 3. This process involves three distinct steps: (1) Feature Selection, (2) Population Initialization, and (3) Feature Evolution.

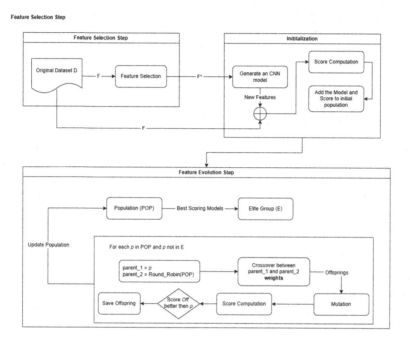

Fig. 3. Feature Evolution process.

The first step of our proposed approach is to reduce the combination of uncorrelated and redundant features using a feature selection method. We used the Maximum Relevance-Minimum Redundancy (MRMR) [12] method for this purpose. By minimizing the combination of such features, we aim to reduce the introduction of noise into the model and improve the quality of the features generated by the CNN model.

During the population initialization step, we generate a population of the CNN model POP that is evolved in the Feature Evolution step. To evaluate this initial population, we use a machine learning model on the dataset resulting from step (1) Specifically, we take the set of features $F*$ from the Feature Selection step and input them into the CNN model p (where $p \in POP$) to generate n new features f. These newly created features are concatenated with the original dataset D to create a new dataset $D * \{F \cup f\}$, which is then evaluated by the machine learning model L^m to obtain a score S_p.

In the trait evolution step, a genetic algorithm [13] is used to evolve the population and identify the most effective traits to improve the performance score obtained by L^m. During each epoch of the genetic algorithm, for each model p that is not part of the elite group E (where $Eis \in POP$), a crossover is performed between its weights and those of a second model p', which is selected using a round-robin tournament [18]. Following the crossover process, the offspring generated by this operation can be subjected to mutation. The features produced by the offspring are then evaluated, as described in the initialization of the population initialization step. If the score obtained by L^m is better than the current score for p or if depreciation is allowed, the offspring replaces the current model p and the score is updated.

4   Results

In this section, we aim to answer the following research questions:

- **RQ1:** How effective is correlation-pooling compared to Max-Pooling?
- **RQ2:** Study of the impact of the number of data on the correlation pooling computation?
- **RQ3:** How effective is the proposed FeatGeNN approach? (Comparison with literature)

4.1   Experimental Setup

To evaluate the performance of the FeatGeNN model, on classification problems, 6 classification datasets from the UCI repository, which were used in the state-of-the-art methods [7,11], were selected. The description of each dataset in terms of the Number of Features and Number of Samples is presented in Table 1.

Table 1. Statistics of the benchmarks used to perform the evaluation of the FeatGeNN features.

Datasets	Samples	Features
SpamBase	4601	57
Megawatt1	253	37
Ionosphere	351	34
SpectF	267	44
Credit_Default	30000	25
German Credit	1001	24

In our experiments, we use the *f1-score* as the evaluation measure, which is also commonly used in the related works [11] and [7]. The threshold for questions RQ1 and RQ2 was set at 80% of the available data in the dataset. To ensure robustness and reliability, we use 5-fold cross-validation, in which the dataset is divided into five subsets or folds and the evaluation is performed five times, with each fold serving once as a test set. This approach helps mitigate the effects of data variability and provides a more comprehensive assessment of the model's performance. As for the chosen algorithm, we use Random Forest as the base method in all our experiments. Random Forest is a popular and widely used ensemble learning method known for its robustness and ability to handle different types of data.

4.2   Effectiveness of Correlation-Pooling vs. Max-Pooling (RQ1)

In this subsection, this experiment aims to answer: *Can our FeatGeNN with Correlation-Pooling achieve competitive results compared to the version with Max-Pooling?* Table 2 shows the comparison results in terms of F1 score. The results show that the FeatGeNN with correlation-pooling outperforms the version with max-pooling in most datasets. The only exceptions are the Megawatt1 and Credit_Default datasets, where the results are very similar. This result can be attributed to the fact that correlation-pooling takes into account the relationships between features when generating new features, which contributes to its relatively better performance.

Table 2. Comparing FeatGeNN performance with Correlation-Pooling and Max-Pooling. The * denotes the version of the FeatGeNN that was executed with Correlation-Pooling. The results are the average score, and the standard deviation, after 30 runs

Dataset	Base	FeatGeNN	FeatGeNN*
SpamBase	0.9102	0.9422 (0.011)	**0.9530** (0.016)
Megawatt1	0.8890	0.9148 (0.002)	0.9151 (0.002)
Ionosphere	0.9233	0.9587 (0.012)	**0.9667** (0.004)
SpectF	0.7750	0.8682 (0.018)	**0.8776** (0.013)
Credit_Default	0.8037	0.8092 (0.003)	0.8095 (0.003)
German Credit	0.7401	0.7775 (0.006)	**0.7814** (0.002)

4.3   Impact of the Number of Data on the Correlation Pooling Computation (RQ2)

In this subsection, our experiment aims to answer the question: *What is the influence of the number of available data on the Correlation-Pooling computation?*. Figure 4 shows the performance of three versions of FeatGeNN: FeatGeNN (using all available data), FeatGeNN (using 60% of the data), and FeatGeNN* (using 30% of the data).

The results show that, as expected, the performance of the model varies with the amount of data used to compute the correlation-pooling. On average, the version with access to the entire dataset achieves a performance improvement of 0.76% and 1.38% compared to the FeatGeNN and FeatGeNN* versions, respectively. Compared to the version that used 80% of the available data, the result after 30 epochs is very similar, although the version with more data performs better in fewer epochs. These results indicate that the performance of FeatGeNN is still competitive with the original version, even though the performance decreases slightly with less available data.

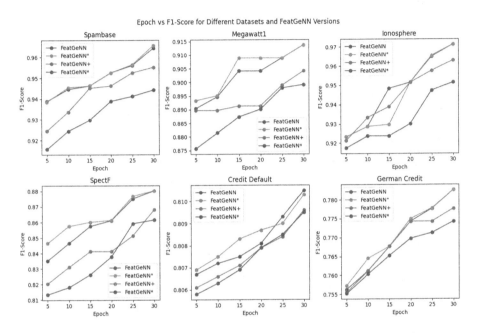

Fig. 4. The performance of the different versions of FeatGeNN is compared in terms of the amount of data used for computation. In the image, the symbol represents the version that used 60% of the available data, the * symbol represents the version that used 30% of the data, and the symbol ° represents the version that used 100% of the data. The FeatGeNN without symbol stands for the version that used 80% of the available data.

4.4   Effectiveness of FeatGeNN (RQ3)

In this subsection, this experiment aims to answer: *Can our FeatGeNN with Correlation-Pooling achieve competitive results when compared to the state-of-the-art models?*. We compare FeatGeNN on 6 datasets with state-of-the-art methods, including (a) Base: Raw dataset without any transformation; (b) Random: randomly apply a transformation to each raw feature; (c) DFS [4]; (d) AutoFeat [10]; (e) LFE [6]; (f) NFS [11]; and (g) DIFER [7].

Table 3. Comparison between FeatGeNN with other methods from the literature, reported on [7]. * reports the average and standard deviation across 30 runs, while the FeatGeNN column reports the maximum value across the same runs.

Dataset	Base	Random	DFS	AutoFeat	NFS	DIFER	FeatGeNN*	FeatGeNN
SpamBase	0.9102	0.9237	0.9102	0.9237	0.9296	0.9339	0.9530 (0.016)	**0.9644**
Megawatt1	0.8890	0.8973	0.8773	0.8893	0.9130	**0.9171**	0.9151 (0.002)	**0.9171**
Ionosphere	0.9233	0.9344	0.9175	0.9117	0.9516	**0.9770**	0.9644 (0.012)	0.9713
SpectF	0.7750	0.8277	0.7906	0.8161	0.8501	0.8612	0.8776 (0.013)	**0.8802**
Credit_Default	0.8037	0.8060	0.8059	0.8060	0.8049	0.8096	0.8095 (0.003)	**0.8102**
German Credit	0.7410	0.7550	0.7490	0.7600	0.7818	0.7770	0.7814 (0.002)	**0.7827**

Table 3 shows the comparative results of FeatGeNN relative to existing methods (results reported in [7]). From Table 3 we can observe that in the classification tasks, the comparison shows that FeatGeNN, performs the best for the SpamBase, Credit_Default, German Credit, and SpectF benchmarks, the second best for the Ionosphere benchmark and achieves the same result as the DIFER method for the Megawatt1 benchmark. Although DIFER achieves the best performance in the Ionosphere benchmark, they only achieve 0.58% more than the best result obtained by our proposed method.

Regarding the number of features, Table 4 shows that FeatGeNN excels in producing fewer features for the Megawatt1, SpectF, and Credit_Default datasets compared to other methods. For the remaining datasets, FeatGeNN achieves comparable results with the same number of features.

Compared to the performances of Base and Random, FeatGeNN achieved an average improvement of 5.89% considering all datasets, which demonstrates the potential of the features generated by our proposed model.

Table 4. Comparison between FeatGeNN, DIFER, AutoFeat, and Random (* the results reported on [7]).

Dataset	Random	AutoFeat*	NFS*	DIFER*	FeatGeNN
SpamBase	1	46	57	1	1
Megawatt1	8	48	37	29	8
Ionosphere	1	52	34	1	1
SpectF	8	37	44	9	8
Credit_Default	4	30	25	5	4
German Credit	1	22	24	1	1

5   Conclusion

In this study, we presented a novel approach for generating new features in tabular data that combines feature selection and feature generation to improve the performance of predictive models. Our proposed method uses a CNN architecture to effectively capture local features during convolution operations (Local Feature Extraction), thereby reducing the number of combinations required in the MLP phase (Global Feature Generation). In addition, we integrated a correlation-pooling operation as a dimensionality reduction step. Our approach demonstrates efficient feature learning and achieves competitive results compared to the architecture used by Max-Pooling and state-of-the-art methods.

As a direction for future research, we intend to explore information theory methods as possible alternatives for pooling operations. This could further increase the effectiveness of our approach to learning new features.

References

1. Domingos, P.M.: A few useful things to know about machine learning. Commun. ACM **55**, 78–87 (2012)
2. Guo, H., Tang, R., Ye, Y., Li, Z., He, X.: DeepFM: a factorization-machine based neural network for CTR prediction. arXiv (2017). https://doi.org/10.48550/arXiv.1703.04247
3. Cheng, H., et al.: Wide & deep learning for recommender systems. arXiv (2016). https://doi.org/10.48550/arXiv.1606.07792
4. Kanter, J.M., Veeramachaneni, K.: Deep feature synthesis: towards automating data science endeavors. In: 2015 IEEE International Conference on Data Science and Advanced Analytics (DSAA), pp. 1–10 (2015)
5. Khurana, U., Turaga, D., Samulowitz, H., Parthasrathy, S.: Cognito: automated feature engineering for supervised learning (2016)
6. Nargesian, F., Samulowitz, H., Khurana, U., Khalil, E.B., Turaga, D.S.: Learning feature engineering for classification. In: IJCAI (2017)
7. Zhu, G., Xu, Z., Guo, X., Yuan, C., Huang, Y.: DIFER: differentiable automated feature engineering. ArXiv abs/2010.08784 (2020)
8. Székely, G.J., Rizzo, M.L., Bakirov, N.K.: Measuring and testing dependence by correlation of distances. Ann. Statist. **35**(6), 2769–2794 (2007). https://doi.org/10.1214/009053607000000505
9. Liu, B., Tang, R., Chen, Y., Yu, J., Guo, H., Zhang, Y.: Feature generation by convolutional neural network for click-through rate prediction. In: The World Wide Web Conference (2019)
10. Horn, F., Pack, R.T., Rieger, M.: The autofeat python library for automatic feature engineering and selection. ArXiv abs/1901.07329 (2019)
11. Chen, X., et al.: Neural feature search: a neural architecture for automated feature engineering. In: 2019 IEEE International Conference on Data Mining (ICDM), pp. 71–80 (2019)
12. Peng, H., Long, F., Ding, C.: Feature selection based on mutual information criteria of max-dependency, max-relevance, and min-redundancy. IEEE Trans. Pattern Anal. Mach. Intell. **27**, 1226–1238 (2005)

13. Katoch, S., Chauhan, S.S., Kumar, V.: A review on genetic algorithm: past, present, and future. Multimed. Tools Appl. **80**, 8091–8126 (2021)

14. Bahdanau, D., Cho, K., Bengio, Y.: Neural machine translation by jointly learning to align and translate. CoRR abs/1409.0473 (2015)

15. Iandola, F.N., Moskewicz, M.W., Ashraf, K., Han, S., Dally, W.J., Keutzer, K.: SqueezeNet: AlexNet-level accuracy with 50x fewer parameters and <1MB model size. ArXiv abs/1602.07360 (2016)

16. Cheng, H.-T., et al.: Wide & deep learning for recommender systems. In: Proceedings of the 1st Workshop on Deep Learning for Recommender Systems (2016)

17. Kaul, A., Maheshwary, S., Pudi, V.: AutoLearn - automated feature generation and selection. In: 2017 IEEE International Conference on Data Mining (ICDM), pp. 217–226 (2017). https://doi.org/10.1109/ICDM.2017.31

18. Goldberg, D.E., Deb, K.: A comparative analysis of selection schemes used in genetic algorithms. In: Rawlins, G.J.E. (ed.) Foundations of Genetic Algorithms, pp. 69–93. Morgan Kaufmann Publishers Inc., San Francisco (1991)

19. Pearson, K.: Note on regression and inheritance in the case of two parents. Proc. R. Soc. Lond. **58**, 240–242 (1895)

Hierarchical Time-Aware Approach for Video Summarization

Leonardo Vilela Cardoso[(✉)][iD], Gustavo Oliveira Rocha Gomes[iD],
Silvio Jamil Ferzoli Guimarães[iD],
and Zenilton Kleber Gonçalves do Patrocínio Júnior[iD]

Image and Multimedia Data Science Laboratory (IMSCIENCE), Pontifícia
Universidade Católica de Minas Gerais (PUC Minas), Belo Horizonte, Minas Gerais,
Brazil
{leonardocardoso,sjamil,zenilton}@pucminas.br, ggomes@sga.pucminas.br
http://www.imscience.icei.pucminas.br/

Abstract. Video summarization consists of generating a concise video representation that captures all its meaningful information. However, conventional summarization techniques often fall short of capturing all the significant events in a video due to their inability to incorporate the hierarchical structure of the video content. This work proposes an unsupervised method, named **Hie**rarchical **T**ime-aware **Summ**arizer–HieTaSumm, that uses a hierarchical approach for that task. In this regard, hierarchical strategies for video summarization have emerged as a promising solution, in which video content is modeled as a graph to identify keyframes that represent the most relevant information. This approach enables the extraction of the frames that convey the central message of the video, resulting in a more effective and precise summary. Experimental results indicate that the proposed approach has great potential. Specifically, it seems to enhance coherence among different video segments, reducing frame redundancy in the generated summaries, and enhancing the diversity of selected keyframes.

Keywords: Video summarization · Hierarchical graph-based clustering · Unsupervised learning

1 Introduction

Video summarization is a challenging task that has gained significant attention in the computer vision and multimedia communities [1,19]. One of the goals of video summarization is to extract essential information from a video and present it in a condensed format [9,10,16,18]. This task is essential for applications such as video captioning, surveillance, synopsis of news videos [4,5,13], and video retrieval, among others [1]. The video summarization task involves

---

Code available at https://github.com/IMScience-PPGINF-PucMinas/HieTaSumm.

M. C. Naldi and R. A. C. Bianchi (Eds.): BRACIS 2023, LNAI 14195, pp. 274–288, 2023.
https://doi.org/10.1007/978-3-031-45368-7_18

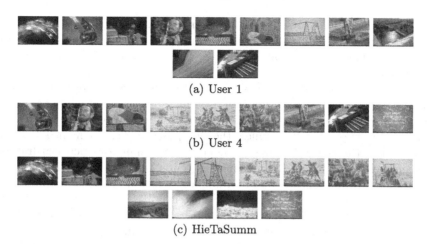

(a) User 1

(b) User 4

(c) HieTaSumm

Fig. 1. Example of the summary generated by the HieTaSumm method compared with the groundtruth. In this case, the HieTaSumm method returns 13 keyframes in contrast to two annotators, the first one (User 1) selects 11 keyframes while the second (User 4) selects 9 keyframes for video $v21$ of the OpenVideo dataset.

several sub-tasks, such as keyframe extraction, object tracking, and summarization itself. The keyframe extraction step selects representative frames that capture the essence of the video, while object tracking aims to track important objects across frames [2]. The summarization step involves selecting a subset of keyframes that provide a comprehensive summary of the video while minimizing redundancy [6]. Video summarization techniques can be categorized into unsupervised and supervised approaches, depending on the availability of training data. While unsupervised techniques aim to identify patterns in the video data without any prior knowledge, supervised techniques require labeled data to train the summarization model [1,19].

Video summarization is particularly useful when dealing with a video collection containing lots of repeated or redundant information spread out over many points in time [6]. In such cases, it becomes a challenge to analyze the entire video and efficiently extract useful information. Video summary techniques can help identify the most important frames in the video that are likely to contain unique and relevant information. By summarizing the video, one can achieve a condensed version that retains the most relevant information while reducing the overall size of the video collection [6]. This allows us to efficiently analyze large video datasets and highlight the most important information, improving the overall effectiveness of video analysis tasks [10,16].

Figure 1 shows that creating a single summary for a video that accurately reflects every user's perception and preferences can be a challenging task. Since groundtruth data is generated by humans, the interpretation of each user of what is essential and relevant may vary. To generate the groundtruth for the video summarization task, annotators must watch the entire video and identify the most crucial moments. However, what one annotator perceives as essential may differ from another, leading to a subjective groundtruth. Hence, the subjectivity of groundtruth generated for each user is a critical aspect to consider in a machine learning method [2,16].

Regardless of the difficulties related to the subjectivity of groundtruth generated by several users, many unsupervised methods have been proposed over the years. In [11], the authors presented a platform for customizing video summaries. Using clustering techniques, they proposed a method named VISTO, which analyzed low-level features to determine the similarity between frames. Keyframe selection is done by selecting the center of each cluster then a post-processing step is responsible for analyzing and removing possible frame redundancies. In [9], the authors presented a clustering-based strategy to solve the video summarization task named VSUMM. First, a sampling process is made to reduce the number of frames under analysis. Then, frames represented by color histograms were grouped into similar sets by a k-means algorithm. VSUMM results tended to group dispersed frames in time that may have a considerable temporal separation. In [17], the authors presented a graph-based approach for video summarization named HSUMM. The proposed approach was hierarchical and comprised keyframe extraction, scene segmentation, and video summarization stages. During the keyframe extraction stage, their method selected representative frames based on image quality and diversity. In the scene segmentation stage, the video was divided into different scenes based on the visual similarity between frames. Finally, keyframes were combined to generate a video summary. The proposed approach employed a hierarchical graph-based clustering that was capable of generating effective video summaries. In [15], the authors presented an unsupervised approach for summarizing a collection of videos. They developed a diversity-aware optimization method for multi-video summarization by exploring the videos' complementarity.

The video summarization landscape has evolved over the last few years, especially after the introduction of deep learning algorithms. The study in [14] focused on egocentric' video summarization and the challenges of this task. In [3], the authors concentrated on summarization methods that are directly applied to the compressed domain. Finally, the authors in [20] presented the relevant bibliography for dynamic video summarization. According to [1], in deep-learning-based video summarization methods, the video content is represented by deep feature vectors extracted by pre-trained neural networks. The extracted features are then utilized by a deep summarizer network and its output can be either a set of keyframes (i.e., a static summary) or a set of video fragments (that form a dynamic summary).

The study conducted by [6] introduces a hierarchical approach for generating a fixed number of keyframes for video captioning. Their approach involves generating a fixed number of keyframes per video, which serves as a representative distribution of the video content. These keyframes are then used as input for a transformer-based method. In contrast to the proposed by [6], our proposed approach takes a different direction. We propose the creation of a dynamic number of keyframes to serve as a comprehensive summary for each video. This approach recognizes the inherent variability in video content and aims to capture the most salient frames specific to each video. Consequently, the number of keyframes in the summary differs across videos within the same dataset. By employing a dynamic keyframe selection process, the proposed approach enables a more discriminating representation of the video content compared to the fixed number approach suggested by [6].

This work proposes an unsupervised method for video summarization that considers changes in video content over time, named **Hie**rarchical **T**ime-aware **Summ**arizer–HieTaSumm. Similarly to recent deep-learning-based approaches, the proposed method uses pre-trained neural networks to generate video frame descriptions. However, it does not adopt a deep summarizer network to avoid the challenges related to its training. Instead, a hierarchical graph-based clustering strategy is adopted. It is worth mentioning that the proposed method assesses frame importance over time for selecting keyframes that comprise the video summary, which is different from other hierarchical approaches. The major contributions of this work are two-fold: (i) a strategy for video summarization that incorporates frame importance over time for selecting keyframes; and (ii) the identification of keyframes through a hierarchical graph-based clustering using deep-learning-based descriptors and a dynamic strategy to define summary sizes.

This work is organized as follows. Section 2 defines many concepts used in this work. Section 3 presents the proposed method, followed by the experimental results in Sect. 4. Finally, Sect. 5 draws some conclusions and future work proposals.

2   Fundamental Concepts

Let $\mathbb{A} \subset \mathbb{N}^2$, $\mathbb{A} = \{0, \ldots, H-1\} \times \{0, \ldots, W-1\}$, where H and W are the width and height of each frame, respectively, and, $\mathbb{T} \subset \mathbb{N}$, $\mathbb{T} = \{0, \ldots, N-1\}$, in which N is the number of frames of a video. A frame f is a function from $\mathbb{A}$ to $\mathbb{R}^3$, where for each spatial position (x, y) in $\mathbb{A}$, $f(x, y)$ represents the color value at pixel location (x, y). A video V_N, in domain $\mathbb{A} \times \mathbb{T}$, can be seen as a sequence of frames f. It can be described by $V_N = (f)_{t \in \mathbb{T}}$, where N is the number of frames contained in the video.

A frame f is usually described in terms of a global descriptor $d(f)$. Let f_{t_1} and f_{t_2} be two video frames at locations t_1 and t_2, respectively. The (dis)similarity between f_{t_1} and f_{t_2} can be evaluated by a distance measure $\mathcal{D}(d(f_{t_1}), d(f_{t_2}))$ between their descriptors. There are several choices for $\mathcal{D}(d(f_{t_1}), d(f_{t_2}))$, i.e., the distance measure between two frames depending on the global descriptor, e.g. histogram/frame difference, histogram intersection, difference of histograms means, and even the L_2 norm.

A time-aware frame similarity graph $G_\delta = (V, E_\delta)$ is a weighted undirected graph. Each node $v_t \in V$ represents a frame $f_t \in V_N$. There is an edge $e \in E_\delta$ with a weight $w(e) = \mathcal{D}(d(f_{t_1}), d(f_{t_2}))$ between two nodes v_{t_1} and v_{t_2} if the difference between their time indexes falls below a specified threshold δ, i.e.,

$$E_\delta = \{ (v_{t_1}, v_{t_2}, \mathcal{D}(d(f_{t_1}), d(f_{t_2}))) \mid v_{t_1}, v_{t_2} \in V, v_{t_1} \neq v_{t_2}, |t_2 - t_1| \leq \delta\}. \quad (1)$$

This constraint over the frames' time indexes limits the connections between distant video frames, effectively allowing the proposed method to consider two frames as similar only if they are not very far in time. This is a noteworthy distinction from many other approaches in the literature, which may consider two frames as similar independently from their time occurrence. Doing that permits the proposed method to assess frame importance over time for selecting it as a keyframe to form the video summary even when it seems to reoccur throughout the video. Figure 2(a) illustrates a time-aware frame similarity graph with $\delta = 4$.

Similar to [17], this work also constructs a hierarchy based on a minimum spanning tree (MST) of the original graph. So, we define an edge-weighted tree of frames $T_{G_\delta} = (V, E_\delta^*)$ is a connected acyclic subgraph of G_δ, i.e., $E_\delta^* \subseteq E_\delta$. The weight of T_{G_δ} is equal to the sum of weights of all edges belonging to E_δ^*, i.e., $w(T_{G_\delta}) = \sum_{e \in E_\delta^*} w(e)$. The minimum spanning tree of frames $T_{G_\delta}^*$ is a tree of frames whose weight is minimal.

Given a finite set V, a *partition* of V is a set $\mathbf{P}$ of nonempty disjoint subsets of V whose union is V. Any element of $\mathbf{P}$, denoted by $\mathbf{R}$, is called a *region* of $\mathbf{P}$. Given two partitions $\mathbf{P}$ and $\mathbf{P}'$ of V, $\mathbf{P}'$ is said to be a (total) refinement of $\mathbf{P}$, denoted by $\mathbf{P}' \preceq \mathbf{P}$, if any region of $\mathbf{P}'$ is included in a region of $\mathbf{P}$. Let $\mathcal{H} = (\mathbf{P}_1, \ldots, \mathbf{P}_\ell)$ be a set of ℓ partitions on V. $\mathcal{H}$ is a hierarchy if $\mathbf{P}_{i-1} \preceq \mathbf{P}_i$, for any $i \in \{2, \ldots, \ell\}$. According to [8], an MST can be utilized to represent a hierarchy, and a weighted MST of a graph can address any connected hierarchy for that graph. Additionally, the work in [12] demonstrated that creating a hierarchical graph segmentation involves reweighting an MST using a dissimilarity measure between regions. Thus, the proposed method utilizes an MST of frame similarity graph $T_{G_\delta}^*$ to obtain a hierarchy $\mathcal{H}$ which is then used to obtain frame clusters.

Finally, a hierarchical segmentation of G_δ into k components is equivalent to the partition of a hierarchy $\mathcal{H}$ into k regions (containing more similar elements) and can be done by removing $k - 1$ edges that present higher weights (representing greater dissimilar) from the $T_{G_\delta}^*$ (since it represents $\mathcal{H}$). This strategy incorporates a similarity measure between clusters while partitioning the graph, providing a more comprehensive approach than traditional methods that only consider the similarity between isolated frames.

3   Hierarchical Time-Aware Video Summarization

Figure 2 illustrates the proposed method steps. The main steps of HieTaSumm method are the following: (i) generation of a time-aware frame similarity graph G_δ to represent a video; (b) computation of a minimum spanning tree $T_{G_\delta}^*$ for that graph; (c) creation of a hierarchy $\mathcal{H}$ based on the $T_{G_\delta}^*$; (d) generation of subsets of frames through cuts on the hierarchy; and (e) selection of keyframes to represent each subset.

The HieTaSumm method (see Algorithm 1) created and uses a frame similarity graph G_δ. Each vertex represents a distinct video frame and there is an edge between two vertices if the difference between their time indexes falls below a

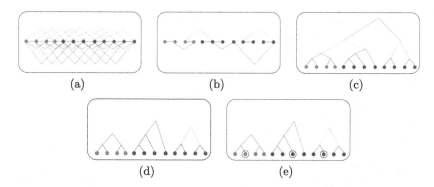

Fig. 2. Illustration of the proposed method steps: (a) generation of a time-aware frame similarity graph G_δ for a video; (b) computation of its minimum spanning tree $T^*_{G_\delta}$; (c) creation of a hierarchy $\mathcal{H}$ based on $T^*_{G_\delta}$; (d) generation of subsets of frames through hierarchy cuts (edge removals); and (e) selection of keyframes to represent each subset. These keyframes are the result of the summarization process.

specified threshold δ_t. Equation 2 represents this constraint and is implemented at line 7 of Algorithm 1.

$$|t_2 - t_1| < \delta_t \tag{2}$$

in which t_f and $t_{f'}$ represent the time indexes of frames f and f', respectively. Additionally, the edge weight represents the (dis)similarity between frames.

The proposed method employs the Kruskal algorithm to obtain the MST $T^*_{G_\delta}$ from G_δ, while the *watershed by area* [7] is used to generate a hierarchy $\mathcal{H}$ from $T^*_{G_\delta}$. Once a hierarchy $\mathcal{H}$ is constructed, a hierarchical segmentation of G_δ generates a video summary of size k. For that, The proposed method needs only to remove the $k - 1$ edges with higher weights from $\mathcal{H}$. Instead of generating a fixed-size video summary, we adopt a strategy for identifying the moment when stability is reached during the edge removal process that is similar (but distinct) to the one used in [17]. Let e' be the edge with the highest weight in the hierarchy $\mathcal{H}$. Thus, the edge e' is removed only when its weight $w(e')$ is greater than or equal to an equilibrium measure function $F(e')$, i.e., $w(e') \geq \mathbf{F}(e)$. In this work, the equilibrium measure function is given by Eq. 3.

$$\mathbf{F}(e) = \gamma \sigma_w(e) \tag{3}$$

in which $\sigma_w(e)$ represents the standard deviation of all edge weights of the connected component that contains edge e, and γ is a parameter related to the allowed variability. During tests, we have set γ empirically.

Finally, after dividing the hierarchy into several connected components, central frames (concerning chronological order) are selected as keyframes for the video summary.

This dynamic choice of the number of components and, consequently, the size of the video summary becomes essential when it comes to videos that contain numerous very similar scenes. In such cases, employing a static number of frames

---

Algorithm 1. Hierarchical time-aware video summarization

Input: A video V_N, threshold value δ
Output: A list of keyframes $\mathcal{K}$
1: Create a graph G_δ with a vertice set $V = \emptyset$ and an edge set $E_\delta = \emptyset$
2: **for all** $f_t \in V_N$ **do**
3:    $V := V \cup \{f\}$ // Insert f in V if f does not belong to it
4:    $d(f_t) := GenerateDescriptor(f_t)$ // Obtain a descriptor for frame f_t
5: **end for**
6: **for all** $f_{t_1} \in V_N$ **do**
7:    **for all** $f_{t_2} \in V_N$ such that $f_{t_2} \neq f_{t_1}$ and $|t_2 - t_1| < \delta$ **do**
8:        $w = \mathcal{D}(d(f_{t_1}), d(f_{t_2}))$
9:        $G.AddEdge(f_{t_1}, f_{t_2}, w)$ // Insert edge (f_{t_1}, f_{t_2}) with (dis)similarity as weight
10:   **end for**
11: **end for**

12: $T^*_{G_\delta} := G_\delta.Obtain\_MST\_from\_Graph()$
13: $\mathcal{H}$ $:= T^*_{G_\delta}.Generate\_Hierarchy\_from\_MST()$

14: $\mathcal{K} := \mathcal{H}.Dynamic\_Selection\_of\_Keyframes()$ // Remove edges from $\mathcal{H}$ to obtain
                                          // frame sets and select the central
                                          // vertice of each set as keyframe

15: Return $\mathcal{K}$;

---

for all videos can result in redundant and repetitive content in the summary. By adopting a dynamic approach, the method can infer an adequate summary size based on the specific video content and characteristics.

4   Experimental Results

This section provides a comprehensive analysis of results obtained by the proposed approach to video summarization with a dynamic selection of video summaries.

We compared HieTaSumm with other unsupervised video summarization methods, namely HSUMM [17], VSUMM1 [9], VSUMM2 [9], VISTO [11] and Open Video summaries (referred to like OVSummary). These comparative assessments allow for a comprehensive review of the performance and effectiveness of HieTaSumm against these established approaches.

4.1   Implementation Details and Dataset

Similar to [11,17], we applied the proposed method to the same collections of videos from the OpenVideo dataset (referred to as the VSUMM dataset in [19]). This dataset contains 50 videos of different genres. All videos are in MPEG-1 format (30 fps, 352 × 240 pixels). The genres are distributed into documentary, educational, ephemeral, historical, and lecture. The time duration of each video varies from 01 to 04 min. The process of creating of user summary consists of the collaboration of 50 different persons. Each user is dealing with the task of choosing the keyframes for 5 videos. Thus, 250 were created for the dataset each video has 05 different user summaries generated manually. And, as a way to pre-process the video dataset we extracted 04 fps from all videos.

For the creation of the frame similarity graph, we use ResNet50 and VGG16 (both pre-trained on ImageNet) to extract frame descriptors. The cosine similarity was used to assess the similarity between two frame descriptors. And, we also set $\delta_t = 32$ (i.e., 08 s with 04 fps) and $\gamma = 0.05$, during the experiments. The parameter δ_t plays a crucial role in enhancing the temporal threshold and restricting vertex connections to avoid the creation of edges that span across all frames of the video. This strategy is used since, if all frames were connected, temporal dependencies may be neglected. Similarly, the parameter γ is employed to regulate the variance amplification in feature differences. Its utilization helps control the level of distinction among features, ensuring a balanced representation of the underlying data.

4.2 Evaluation Metrics

Assessing frame quality in the context of video summarization poses a distinct challenge because of the many ways in which frames can be constructed while conveying similar meanings. These variations can arise from using different analyzes of resources from different informational aspects. Although humans have an intuitive understanding of this process, abstract evaluation remains an open question without a specific framework. As a result, the conventional practice involves adapting similar metrics that have been stretched to accommodate the specific requirements of the video summary task. By re-purposing and customizing these metrics, researchers, and practitioners can assess the effectiveness and fidelity of summaries generated in video summarization, despite the inherent complexities and subjectivity involved in sentence evaluation [9,17].

To compute the improvement of the frame selection, we will evaluate the obtained results following the same approach used by the authors of [9,17]. They reported their results using metrics widely disseminated in the literature such as CUSa, CUSe [9,17], and COV [17], defined by the Eqs. 4–6, respectively, to evaluate the similarity between the frames generated by their summarization method and the GT results.

$$CUSa = \frac{m_A}{n_U} \tag{4}$$

$$CUSe = \frac{\overline{m}_A}{n_U} \tag{5}$$

in which m_A denotes the number of matching keyframes generated from the Automatic Summary (AS), $\overline{m}_A$ represent non-matching keyframes from AS, and n_U are the number of keyframes selected for the user to represent the user summary (U) to each video.

$$COV = \frac{\sum_{U \in US} |M(AS, U)|}{\sum_{U \in US} |U|} \tag{6}$$

in which $M(X,Y)$ and $|.|$ are the maximum matching between two sets of different elements X and Y, and the cardinality of a set, respectively.

While those two first metrics provide valuable insights, they often fail to measure the diversity displayed in user summaries as COV does. Furthermore, the calculation of averages for each user's measurements can introduce distortions and inaccuracies. Specifically, the CUSa, which is commonly employed to assess user opinions, fails to effectively capture the diversity of these opinions. To illustrate, consider two users, A and B, providing summaries for the same video. Let the summary of user A be $U_A = \{X,Y\}$ while the summary of user B is $U_B = \{M,N,O,P,Q,R,S,T,U,V\}$, in which each character denotes a single frame of video. Now suppose that three distinct methods generate summaries: $AS_1 = \{X,Y\}$, $AS_2 = \{M,N,O,P,Q,R,S,T,U,V\}$, and $AS_3 = \{X,M,N,O,P,Q\}$. Despite these summaries being completely different, they provide the same accuracy rate (i.e., CUSa = 0.5). This highlights the limitations of CUSa in accurately assessing divergence of opinion and the need for more comprehensive assessment metrics [9,17].

Unlike CUSa, COV assesses the extent to which an automatic summary covers all user-generated summaries. This measure takes into account both the diversity of opinions expressed by users and the degree of agreement among them. Specifically, the CUSa measure calculates the average ratio between each user's summary and an automatic summary, thus capturing the level of agreement between the two. In contrast, COV assesses the proportion of an automatic summary that aligns with all user summaries, providing a measure of overall covering. We use COV as the first metric to compute the effectiveness of the HieTaSumm. The reader should refer to [9,17] for more information about those metrics.

4.3 Quantitative Analysis

Table 1 presents the HieTaSumm results. We used ResNet50 and VGG16 to extract frame descriptors for the construction of the frame similarity graph. During the evaluation of the results, we also used ResNet50 and VGG16 to extract frame descriptors but the cosine similarity was used to verify the agreement between the groundtruth and automatic summaries. We have also used color histograms (CH) during the assessment of the results.

Table 1 presents the average values of all metrics for the 50 videos belonging to the dataset. The results are presented for different levels of precision (between groundtruth and automatic summaries). It is possible to notice that the use of ResNet50 presents a slight improvement compared to the results with VGG16 (under a greater precision in evaluation), and the VGG16 presented better results (under a lower preciseness in evaluation). Moreover, it is also possible to observe the high values of COV and CUSa achieved by HieTaSumm method, and even under a higher precision in evaluation, the proposed method still presents competitive results.

Table 1. Performance of HieTaSumm method for different levels of precision in evaluation of video summaries. CUSa, CUSe, and COV values were multiplied by 10^2 to improve readability.

Metrics ($\times 10^2$)	Precision of Matches (%)												
	100	99	98	95	90	85	80	75	70	65	60	55	50
ResNet50 + CH													
COV	23.09	35.48	41.81	56.55	68.87	77.87	84.53	87.73	89.55	89.90	90.16	90.33	90.42
CUSa	23.27	35.91	42.42	57.17	69.64	78.90	85.75	88.89	90.68	91.01	91.26	91.46	91.54
CUSe	76.73	64.10	57.58	42.83	30.36	21.11	14.25	11.11	09.32	08.99	08.74	8.54	8.46
VGG16 + CH													
COV	22.85	35.38	42.20	56.74	67.86	77.58	85.65	88.35	89.21	90.08	90.27	90.33	90.42
CUSa	23.01	35.59	42.64	56.34	68.54	78.52	86.85	89.48	90.34	91.22	91.40	91.46	91.54
CUSe	76.99	64.41	57.36	43.66	31.46	21.47	13.15	10.52	09.66	08.78	08.60	08.54	08.46
ResNet50 + ResNet50													
COV	08.29	15.66	20.32	30.67	41.93	49.28	53.66	57.64	60.39	63.16	65.24	67.94	70.60
CUSa	08.45	15.90	20.58	31.12	42.20	49.68	54.17	58.22	60.91	63.70	65.82	68.50	71.30
CUSe	91.55	84.10	79.42	68.88	57.80	50.32	45.83	41.77	39.09	36.30	34.18	31.50	28.70
VGG16 + VGG16													
COV	01.43	12.66	20.90	35.21	49.35	57.92	63.57	69.63	74.92	76.96	78.78	81.24	82.95
CUSa	01.60	12.66	20.93	35.43	50.00	58.32	64.02	70.16	75.38	77.54	79.36	81.89	83.55
CUSe	98.40	87.34	79.07	64.57	50.00	41.68	35.98	29.84	24.62	22.46	20.64	18.11	16.45

4.4 Qualitative Analysis

To provide a better understanding of the results obtained and their improvements, Figs. 3 and 4 present samples of summaries generated by various approaches in the literature, including HSUMM [17], VSUMM1 [9], VSUMM2 [9], VISTO [11] and Open Video summaries (referred to like OVSummary), and the groundtruth (GT) results, alongside those generated by the HieTaSumm method. This comparison enables the evaluation of time awareness, similarity with the GT results, and the rate of the frames selected by the HieTaSumm method and others.

Figure 3 shows the results generated by the HieTaSumm method along with HSUMM [17] results, and the summaries generated by two users. Each frame list created for each user encapsulates a distinct selection of frames, reflecting individual preferences and perspectives. Employing cosine similarity, we can quantify the degree of similarity between the GT of the users and that generated for HieTaSumm method. However, it is essential to recognize that similarity is subjective and may vary among observers. Factors such as the weighting of different frames, the level of granularity in frame selection, and the specific context of the video all influence perceived similarity. Therefore, when evaluating the cosine similarity between two lists of frames, it is crucial to consider the subjective nature of the perception and the different perspectives that individuals bring to the comparison. The result obtained for the HSUMM has a much higher number of frames than the others and, due to this, they present a large number

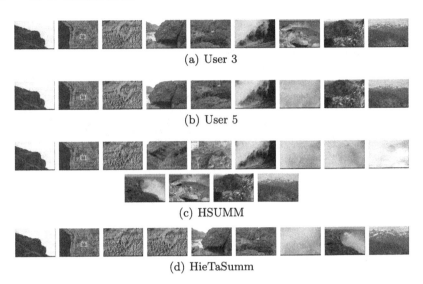

(a) User 3

(b) User 5

(c) HSUMM

(d) HieTaSumm

Fig. 3. Comparative example of HieTaSumm results compared with HSUMM results and with the frames selected by the User 3 and User 5 (both selected 9 frames). The video summary generated by HieTaSumm contains 9 frames.

of frames with high similarity. Furthermore, HSUMM results may not preserve chronological order.

On the other hand, HieTaSumm method presents a fluid and coherent result. Furthermore, the select keyframes are very similar to those frames in GT. For all keyframes selected by HieTaSumm method, only one frame does not have another directly correlated with those selected by the two users. But, in all cases, even with different keyframes selected by the users, the automatic summary generated by HieTaSumm method is very close to theirs (especially for Users 3 and 5 shown in Fig. 3). In addition, the unrelated keyframe preserves temporal order and, when we look at the three keyframes in which the map is present, it is possible to observe that a refinement process takes place to identify the correct highlighted region, starting from a global visualization to an analysis local that identifies the region in focus as the most important point of location on the map of the region presented in the video.

Figure 4 also presents some subjective characteristics for the keyframes selected by users 2 and 3. Considering the number of frames selected, 15 and 17 respectively, it tends to suggest the existence of a larger number of scene modifications. This variation can cause the selection of a greater number of frames returned by automatic methods, but the increase in the number of scenes can cause frames to be repeated by automatic methods. In this way, the returned summaries have a great challenge of maintaining temporal coherence, but without two highly similar frames being selected without the presence of other events. With this difficulty in mind, OVSummary presents a series of repeated frames side by side. Seen displays some repeated frames, but a reduced number of frames

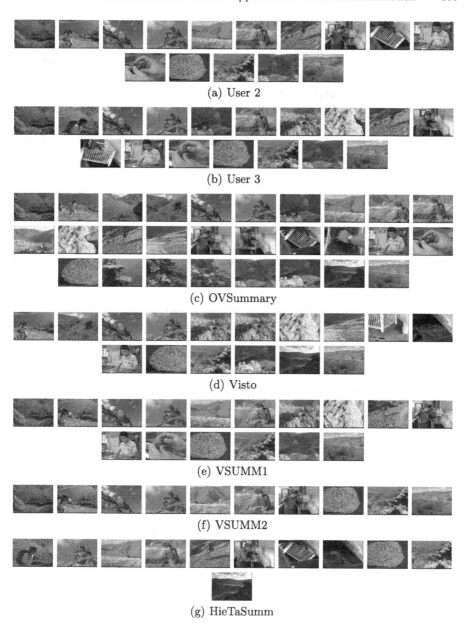

(a) User 2

(b) User 3

(c) OVSummary

(d) Visto

(e) VSUMM1

(f) VSUMM2

(g) HieTaSumm

Fig. 4. Comparative example of HieTaSumm results compared with the results of VSUMM1 [9], VSUMM2 [9], VISTO [11], OVSummary and with the frames selected by the User 2 and User 3.

with more similar information. VSUMM1 observes more scene modification and has some information that tends to be more similar. VSUMM2 tends to keep the results without redundancy but without the presence of some scenes more relevant to the user. Finally, the hierarchical approach used by HieTaSumm tends to reduce the redundancy of information with a lot of similarity. HieTaSumm results has a smaller number of keyframes, but these keyframes are more related to user summaries. Moreover, keyframes selected by HieTaSumm method keep the temporal ordering and shows that the dynamic selection of summary size helps to better capture the changing scenes more smoothly.

5   Conclusion

This work proposes an unsupervised method for video summarization that considers changes in video content over time, named **Hie**rarchical **T**ime-aware **Summ**arizer– HieTaSumm. It uses pre-trained neural networks to generate video frame descriptions with a hierarchical graph-based clustering strategy. The proposed method explores a time-aware frame similarity graph to represent video content considering changes over time. Moreover, a dynamic strategy for defining summary size is adopted. Experimental results indicate that the proposed approach has great potential. Specifically, it seems to enhance coherence among different video segments, reducing frame redundancy in the generated summaries, and enhancing the diversity of selected keyframes.

Future works may explore other strategies for selecting keyframes and different hierarchies. It might also be interesting to investigate the impact of different datasets with little scene modifications. Following these future research directions, we can advance the video summary field and further refine the dynamic frame selection approach to provide more accurate, informative, and user-centric video summaries.

Acknowledgements. The authors would like to thank Conselho Nacional de Desenvolvimento Científico e Tecnológico - CNPq - (Universal 407242/2021-0 and PQ 306573/2022-9), and Fundação de Amparo Pesquisa do Estado de Minas Gerais - FAPEMIG - (Grants PPM- 00006-18). This study was also financed in part by PUC Minas and by the Coordenação de Aperfeiçoamento de Pessoal de Nível Superior - Brasil (CAPES) - Finance Code 001.

References

1. Apostolidis, E., Adamantidou, E., Metsai, A.I., Mezaris, V., Patras, I.: Video summarization using deep neural networks: a survey. Proc. IEEE **109**(11), 1838–1863 (2021)
2. Asha Paul, M.K., Kavitha, J., Jansi Rani, P.A.: Key-frame extraction techniques: a review. Recent Patents Comput. Sci. **11**(1), 3–16 (2018)
3. Basavarajaiah, M., Sharma, P.: Survey of compressed domain video summarization techniques. ACM Comput. Surv. **52**(6), 1–29 (2019)
4. Cardoso, L.V., Guimaraes, S.J.F., Patrocínio, Z.K.G.: Enhanced-memory transformer for coherent paragraph video captioning. In: 2021 IEEE 33rd International Conference on Tools with Artificial Intelligence (ICTAI), pp. 836–840. IEEE (2021)
5. Cardoso, L.V., Guimaraes, S.J.F., Patrocinio, Z.K.G.: Exploring adaptive attention in memory transformer applied to coherent video paragraph captioning. In: 2022 IEEE Eighth International Conference on Multimedia Big Data (BigMM), pp. 37–44. IEEE (2022)
6. Cardoso, L.V., Guimaraes, S.J.F., Patrocinio Junior, Z.K.G.: Hierarchical time-aware summarization with an adaptive transformer for video captioning. Int. J. Semant. Comput. (2023)
7. Cousty, J., Najman, L.: Incremental algorithm for hierarchical minimum spanning forests and saliency of watershed cuts. In: Soille, P., Pesaresi, M., Ouzounis, G.K. (eds.) ISMM 2011. LNCS, vol. 6671, pp. 272–283. Springer, Heidelberg (2011). https://doi.org/10.1007/978-3-642-21569-8_24
8. Cousty, J., Najman, L., Kenmochi, Y., Guimarães, S.: Hierarchical segmentations with graphs: quasi-flat zones, minimum spanning trees, and saliency maps. J. Math. Imaging Vis. **60**(4), 479–502 (2018)
9. De Avila, S.E.F., Lopes, A.P.B., da Luz Jr., A., de Albuquerque Araújo, A.: Vsumm: A mechanism designed to produce static video summaries and a novel evaluation method. Pattern Recognit. Lett. **32**(1), 56–68 (2011)
10. Ejaz, N., Tariq, T.B., Baik, S.W.: Adaptive key frame extraction for video summarization using an aggregation mechanism. J. Vis. Commun. Image Represent. **23**(7), 1031–1040 (2012)
11. Furini, M., Geraci, F., Montangero, M., Pellegrini, M.: VISTO: visual storyboard for web video browsing. In: Proceedings of the 6th ACM International Conference on Image and Video Retrieval, pp. 635–642 (2007)
12. Guimarães, S., Kenmochi, Y., Cousty, J., Patrocinio, Z., Najman, L.: Hierarchizing graph-based image segmentation algorithms relying on region dissimilarity: the case of the felzenszwalb-huttenlocher method. Math. Morphol.-Theory Appl. **2**(1), 55–75 (2017)
13. Lu, G., Zhou, Y., Li, X., Yan, P.: Unsupervised, efficient and scalable key-frame selection for automatic summarization of surveillance videos. Multimed. Tools Appl. **76**, 6309–6331 (2017)
14. del Molino, A.G., Tan, C., Lim, J.H., Tan, A.H.: Summarization of egocentric videos: a comprehensive survey. IEEE Trans. Hum.-Mach. Syst. **47**(1), 65–76 (2017)
15. Panda, R., Mithun, N.C., Roy-Chowdhury, A.K.: Diversity-aware multi-video summarization. IEEE Trans. Image Process. **26**(10), 4712–4724 (2017)
16. Pandey, S., Dwivedy, P., Meena, S., Potnis, A.: A survey on key frame extraction methods of a mpeg video. In: 2017 International Conference on Computing, Communication and Automation (ICCCA), pp. 1192–1196 (2017)

17. dos Santos Belo, L., Caetano Jr., C.A., Patrocínio Jr., Z.K.G., Guimarães, S.J.F.: Summarizing video sequence using a graph-based hierarchical approach. Neurocomputing **173**, 1001–1016 (2016)

18. Song, Y., Vallmitjana, J., Stent, A., Jaimes, A.: TVSum: summarizing web videos using titles. In: Proceedings of the IEEE Conference on Computer Vision and Pattern Recognition (CVPR), pp. 5179–5187 (2015)

19. Tiwari, V., Bhatnagar, C.: A survey of recent work on video summarization: approaches and techniques. Multimed. Tools Appl. **80**(18), 27187–27221 (2021)

20. Vivekraj, V., Debashis, S., Balasubramanian, R.: Video skimming: taxonomy and comprehensive survey. ACM Comput. Surv. **52**(5) (2019)

IA and Education

Analyzing College Student Dropout Risk Prediction in Real Data Using Walk-Forward Validation

Rodolfo Sanches Santos, Moacir Antonelli Ponti,
and Kamila Rios Rodrigues[(✉)]

Instituto De Ciências Matemáticas e de Computação, Universidade de São Paulo,
Av. Trabalhador São Carlense, 400, São Carlos, SP, Brazil
rodolfosanches@usp.br, {moacir,kamila.rios}@icmc.usp.br

Abstract. College dropout is a concern for educational institutions since it directly impacts educational management and academic results, as well as being directly related to social problems. Therefore, there is significant incentive for studies that use data to support decisions by predicting risk of dropout so that institutions can attempt to prevent such cases. Although machine learning techniques were shown to have potential for this task, there are many steps involved when it comes to the use of real data, which comes from scattered systems and present issues such as need for data cleaning and preparation, high dimensionality of the data requiring adequate feature selection, as well as class imbalance. In this paper, we used data from 32.892 students enrolled between 2008 and 2020 from all courses offered by a public high-education institution. A protocol for data preparation is proposed and found to be more important than designing complex classifiers. We present guidelines when modelling a college dropout classification task using a public university data and experiments using Walk-Forward Validation that showed the predictive capacity for the first years.

Keywords: Higher Education · College Dropout · Machine Learning

1 Introduction

Students of higher education are exposed to several positive and negative events in college. The successes often outweigh the efforts from the enrolment until the student effectively earns a degree, however adversities may lead to dropout [36,39]. This phenomenon also related to college attrition, is common to both public and private higher education, directly interfering in their management and in the results of education quality [29,35], which consequently generates a necessary concern, as the students' departure from the study cycle induces several consequences [18,31]. The costs associated with college attrition include hindering of future job prospects with impacts in the countries' economy, the personal and professional costs for the former student, waste of institutional and

M. C. Naldi and R. A. C. Bianchi (Eds.): BRACIS 2023, LNAI 14195, pp. 291–305, 2023.
https://doi.org/10.1007/978-3-031-45368-7_19

federal resources, potential damage to university reputation and demoralization of students [5,13].

Several factors may influence dropout [3], the main ones reported in the literature related to financial and family reasons, unfulfilled expectations and lack of motivation [5]. Xenos, Pierrakeas and Pintelas [40] stated that the identification of such specific factors is essential to provide special assistance to students, and categorizes them as related to: (i) internal or students' perception; (ii) course and professors and; (iii) student demographic characteristics.

Additionally, many studies point out a high prevalence of mental health problems in college students, which are even most common in those exposed to relational stressors and low social support [9,22]. In a survey conducted by Eisenberg et al. [9] in an university in the US, the prevalence of depressive or anxiety disorders was 15.6% for undergraduate students and 13.0% for graduate students, as well as presence of suicidal ideation reported by 2% of students.

According to Leonhardt and Sahil [18], in North American higher education, in 2018, about one in three students who enrol in college never earn a degree. They looked at data from 368 colleges arranged by what they would expect their graduation rates to be, based on the average for colleges with similar student bodies. They found higher success rates in those that remove hurdles for students. Also, such success was correlated with students having better connection to the college community, having less financial issues, and when the university offered good infrastructure and support.

In Brazil, a special committee on dropout studies was established in 1995, from Ministry of Education ordinance, with the purpose of evaluating the performance of Federal Higher Education Institutions (FHEI; IFES, in Portuguese). In 2002, this became a major concern with the significant increase in the number of places offered by IFES in Brazil [39].

In this context, Data Mining and Machine Learning techniques have been explored to investigate this problem. Previous studies often to use data to identify student's risk of dropout [1,7,15,16,21,25,32,34]. Some of those focus on specific patterns such as student's personal profile [8,23], others on behavior, e.g. procrastination [11] and others in social issues [6].

In this paper, we use data gathered from different information systems at the Federal University of São Carlos (UFSCar), in the interior of the State of São Paulo - Brazil, to predict the risk of dropout in groups of students. The objective is to integrate it to the university solutions for decision support. When using real world data is not trivial. Thus, we present guidelines to evaluate and deploy predictive models and the necessary preparation steps, including data cleaning. In addition to evaluate the model using Walk-Forward Validation evaluation, we propose four experiments using different subset of data and tests on data not seen during the training of the learning model. Our results represent a milestone in terms of making it possible, in practice, to identify students in higher risk and direct them to services that address issues related to emotional distress, financial problems, among others.

The paper is structured as follows: Sect. 2 describes related works; Sect. 3 describes the main steps of data exploration, as well as what information was collected and processed, the data preparation and cleaning, experiments applied, and the results obtained and, finally, Sect. 4 points out the final remarks.

2   Related Work

Previous studies found that college attrition depend on the type of course [8, 40], student's year at college [8, 18, 19, 36], social issues such as parental background [2, 8], class-cultural discontinuities [17] and economic profile [28]; as well as quantitative academic data [25].

In particular, Lozano et al. [19] found the first and second years to be crucial when it comes to retention, as it allows for more numerous and more intensive interventions. According to Casanova et al. [8], the first years are particularly difficult for students who enrolled in the course as a second option. The authors suggest improving motivation in students who have to adapt to this situation.

A study carried out in Catalan universities used academic data, face-to-face interviews and telephone interviews, concluding that main factors for dropout risk were: lack of motivation; work-related reasons; unfulfilled expectations; timetable mismatch; family reasons; financial reasons and other opportunities [38]. However, the need for individual interviews at the dropout risk assessment stage can hamper scalability.

Interventions may represent a positive approach to improve retention [30]. This was shown at a Brazilian public university, in which dropout decreased 12.3% in 2018 when compared to 2017 [4]. This was due to institutional interventions that increased the number of registered students, accounting for new admissions and the re-inclusion of students who had taken a time off from college.

Data Mining and Data Science techniques are also relevant in the educational context [12]. In da Silva et al. [33], the authors used an ensemble of regressors to predict dropout rate percentage of higher-education institutions. Pal [26], in turn, used a Bayesian Machine Learning (BML) classification method with high school data, income, family position, type of admission, gender, and other school data. Their strategy was to compare the student profile with the classification provided by the model [26]. Manhaes et al. [20] studied academic data of a Brazilian university and found dropout patterns related to the first year, indicating failure in early courses are present in all dropouts.

A study carried out by Tontini and Walter [37] analyzed the effectiveness of an institutional intervention, which consisted of the institution's contact with the student in order to influence the student to give up dropping out. They observed a decrease of 18% in the semester following the interventions. They also employed descriptive quantitative research with an online questionnaire with the aim of exploring the perception regarding the retention attributes. To predict dropouts using the gathered data, the authors used Radial Basis Function (RBF) neural networks and cluster analysis to group new students by similarity. Although the author had positive results, the techniques and steps for replicating the study were not detailed.

Hippel and Hofflinger [10] conducted a study to identify students at risk of dropping out at 8 Chilean public universities using Logistic Regression (LR). The authors used both personal/family data, e.g. parents' education, high school grades, and entrance exam scores, as well as academic data collected during college. They studied the effect of programs focused on helping students adapt to university life, develop study skills and manage anxiety. The assisting programs dictated a reduction of 30–40% in the chances of dropping out in 2 of 4 universities where such programs existed. They also found the dropout risk to be inferior for older students, and for those receiving scholarships. This makes it evident there is no clear consensus on the causes of dropout, and also the variety of factors protecting against it.

In this paper, we use data as available at most Brazilian higher education institutions, including personal and academic data. We show the main steps to be considered when designing predictive models to provide educational managers a decision-making support, as well as important insight for similar initiatives in other institutions. In particular, we show there is a significant gap between analyzing historical data and future data, and specific features play a major role in this type of problem. As far as we know, this is the first paper to address a complete pipeline for a data-driven dropout risk assessment tool, including data preparation, feature selection and proper evaluation protocol, as well as feature importance analysis. Therefore, we intend this paper to shed light on the application, and serve as guideline for other higher-education institutions.

3   Data-Driven Dropout Risk Predictive Model

This section describes the exploration of data and the assessment of a predictive model which includes data collection from information systems, pre-processing, experiments using different subsets of the data for training and testing models using Walk-Forward Validation, as summarized in Fig. 1.

Formally, let S be a dataset where an instance $\mathbf{s} \in S$ represents a student enrolled in a course, in a given semester, by attributes or variables with γ dimensions, and $Y = \{0, 1\}$ represents the label describing the student's s situation: studying/regular (0 or negative) or avoidant (1 or positive). First, we sought to obtain a subset of the original data to pre-process, transform and obtain X that has γ original characteristics, and $X' \in R^m$ with attributes such that $m \leq \gamma$. From the preprocessed dataset, X' supervised learning methods are used to infer $f : X' \to Y$.

Then, S was evaluated in different subsets grouped by the current semester of the student, with the test data being used only to test the generalization of the models for a group of students of a certain semester not seen during the training.

3.1   University Data Exploration

Data Description. The UFSCar has several computational information systems for managing institutional processes. Those are separated by purpose such

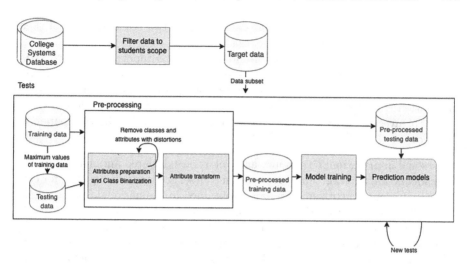

Fig. 1. Overall flow for the design of our dropout prediction model. First, the maximum values of training data variables are computed for scaling testing data. Then the data undergoes preprocessing, always fitting on training and applying on testing. Finally, the training data is used for training the model, while the preprocessed testing data is used for evaluation.

as: undergraduate studies; postgraduate studies; university restaurant; library, among others. The main systems used to manage academic and student data from which we gathered data were:

- *Integrated Academic Management System*: responsible for the academic processes related to the student, lecturers and the main activities, such as course enrolments, registration of grades and frequencies, teaching plans, subjects, courses, among other activities;
- *Integrated University Management Support System*: general management of the university including registration of people, permissions, and applications student cards;
- *Dean of Extension*: management of outreach and extension courses, as well as postgraduate courses *lato sensu* e *stricto sensu*.

In order to collect and analyze the students' academic, economic and personal data, in addition to the approval of the UFSCar ethics committee (CAAE number: 34343920.5.0000.5504), a formal request was made to the department responsible for the database, which was stored in a single database management system. To maintain data secure, the stored files were encrypted and the access to the computer and the database was granted only for authorized users.

Data Gathering. The UFSCar information technology team defined business rules related to undergraduate students in order to define the data to be gathered and integrated into a single dataset S. In this way, the source database schema was simplified by querying the target data into just two schemas, both of them totalling 222 tables and 15 views.

We follow the empirical evidence found by Pal [26], Hippel and Hofflinger [10]; and de Souza [35] for filtering relevant data to be used. Those authors indicate that personal data, academic information prior to college enrolment, academic information collected during college, as well as economic information, are sufficient to investigate college attrition and student dropout. In this paper we included:

– Personal information: city, genre, age, color/race declared and marital status;
– Academic information: undergraduation course, grades and frequency for each course taken, current semester, grade obtained in the national college admission test (acronym: ENEM, in Portuguese).
– Economic and social information: category of admission (which is defined by lower income levels, if previous education was on public schools, ethnic/racial information, special needs); and also whether self-claimed to below income during registration at the college admission test.

Note that, Brazilian federal universities have affirmative actions, so that the category of admission define quotas of student places related to social/economic status of the student as well as race.

Our analysis obtained data from students who entered the UFSCar between 2008 and 2020, from all undergraduate courses offered by the institution during this period. In each year, between 2000 and 3000 students are admitted. In total, information was collected from 32.892 students, separated by course, year and period/semester.

Guidelines: include personal, academic and economic/social data available in the information systems, taking into account opinions from the information technology team, as well as the team of the dean of undergraduate students. In principle, sensible personal data should not be included, and be encoded if needed. Also, the resulting data should be encrypted and not given broad access.

3.2   Data Preparation

As we show later in results, such step significantly improves the quality of dropout prediction. Below, we describe the data preparation/cleaning and standardization in detail.

Some important characteristics of the source data to be observed when designing a Machine Learning model are:

1. students in the first period still have no grades and attendance therefore cannot be used to train the model;
2. there are missing values in particular for personal data, in our case for the attributes: *Marital Status*, *Lower income* and *declared Color/Race*;
3. data for active students includes status from different academic periods (semester) of all courses, while data for those that concluded the degrees are complete, and those that abandoned are partial;

4. class imbalance: around 70% of students are regularly enrolled or earned a degree, and around 30% are considered dropouts – value referring to all data collected (all years and all academic periods, including students who are out of regular period);
5. Academic periods: most courses have a regular curriculum of 4 years and 8 academic periods, except for a few courses that have a curriculum of 5 years.

Data Preparation and Cleaning. Below we list the cleaning procedures carried out only when those showed to improve results in validation sets within a 10-fold cross-validation experiment using the training data only:

- **Instances removed:** students without grade or attendance information recorded;
- **Attributes removed:** *Start year, Registration number; City of birth of the student; Name of the country; Start period; Has lower income* were removed because they are not related to the classification task, possible insertion of bias, because they have missing values and/or have only one value for all students;
- **Obtaining the binary label:** the original database contains different *status* for the student which will represent the positive label. Following the university's guidelines, all status indicating the student left college were defined as *Evader* (positive), i.e.: *Change of area, Cancelled, Loss of vacancy, Retired and External Transfer.* On the other hand: *Studying, Undergraduate Candidate, Internal Transfer, In Appeal, Graduated* and *Graduating* were considered *Studying* (negative), in which the student still maintains a relationship with the university.

In addition to cleaning, we filled the null values for the features: *Civil Status, Declared Color/Race, Type of School in High School, Gender, Type of Admission* were filled using the value *Not informed*, while the features *Course, Course Regime, Course City, Marital Status, Declared Color/Race, Type of High School, Nationality Type, Course Shift, Degree, Admission Type, Admission Form and Sexuality* attributes were one-hot-encoded. The decision for one-hot-encoding transforming was also made based on cross-validation results.

The min-max normalization of the numerical data was applied in order to maintain the range from 0 to 1 of all values and, consequently, to avoid that larger scale variables were considered of greater importance during the training of the algorithms. The following transformations were performed: a) Expected values between 0 and 10 were divided by 10; b) Expected values between 0 and 100 were divided by 100; c) Values of *Profile; Total number of courses enrolled, cancelled, approved and disapproved; Income coefficient; Workload completed and Classification obtained in Enem* were divided by the maximum value found in the database. In this case, the maximum values obtained in the training of the classifiers were stored in memory and used for data normalization in the inference (test) step.

The target variable *Status* was designed as binary. However, the original dataset has 17 distinct status, so we grouped those into class "Dropout" (positive) for the statuses *Area Change, Canceled, Loss of Vacancy, Temporary Academic Leave and External Transfer*). The class "Studying" (negative) included *Studying, Under Graduation Candidate, Internal Transfer, In Appeal and Forming*.

Guidelines: remove instances that do not provide representative information, and study cases of missing values, filling those if those account for a significant number of instances. Convert attributes using one-hot-encoding approaches depending on their nature (categorical). Normalize the numeric data to eliminate larger scales from the values.

4   Experiments and Results

Three classification algorithms were used in the experiments: Support Vector Classification, Decision Tree (CART) and LightGBM.

We do not aim at comparing which is the best classifier, but instead use their results to understand the problem in terms of the available data and the features. The SVM offers a proxy measure for the linear separability of the classes, therefore its performance allows us to draw conclusions about the difficulty of the problem in terms of the given variables [24]. On the other hand, the Decision Tree allows understanding if it is possible to infer a single set of rules based on the features that allow to estimate dropout for all students. Finally, LightGBM [14] would be the attempt to combine several decision trees and, the more its results surpass the other methods, it means different students require different decision trees, needing different ways of shattering the feature space [27].

The following hyperparameters were used:

- *Support Vector Classification:* linear kernel, cost 1 and maximum of 100 iterations;
- *Decision Tree:* gini criterion, best divisor and 2 minimum samples to split;
- *LightGBM:* binary target and 100 iterations.

The experiments consisted of training the models and applying the tests using the techniques of WFV (Walk-Forward Validation), in which the methodology is applied to data that have time intervals with the possibility of making good forecasts in each period. We sought to use this methodology to simulate the real scenario of application of an AM algorithm at the university, in which it will learn temporally according to the years and also, validate the behavior of the algorithms with the training of different time windows. Table 1 shows the number of instances per year and class (1st to 5th year) of the dataset used in the experiments, and Fig. 2 illustrates all the experiments performed.

The first experiment aimed to evaluate whether the use of data from students from all previous years allows for the prediction of dropout, simulating learning from year to year successively according to the results in Table 2. The second experiment verified whether the use of data from students from the immediately

Table 1. Number of instances per year and class – Positive (Evading) and Negative (Studying).

Year/Class	Positive	Negative
1st	4426	3040
2nd	3528	2794
3rd	2616	2164
4th	1904	1084
5th	2239	548

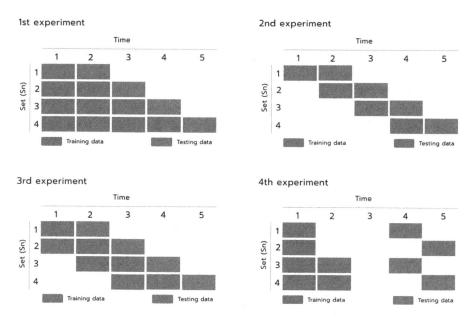

Fig. 2. Experiments carried out using the WFV methodology.

previous year is sufficient to classify those from the current year, without the accumulation of all previous years, with results in Table 3. The third experiment followed the methodology of the second experiment, but with the use of instances with up to 2 years prior to the test, as shown in Table 3. Finally, we sought to assess whether the first years would provide enough signal to allow predicting dropout in instances beyond the 2nd and 3rd year, see Table 5 (Table 4).

Because the second test of the first experiment had the best result in general, a list of the most relevant *features* was obtained according to the LightGBM classifier. Thus, it was possible to observe that the relevant *features* are, for the most part, academic data and have a direct relationship with the student's performance in the subjects and in the course. Table 6 displays the 10 *features*.

Table 2. Classification metrics results from experiment 1, training with all pre-test years in WFV for all test sets (Sn) and classifiers (Cl).

Sn/Cl	Accuracy			Recall			Precision			F1-score		
	DT	SVC	LGBM	DT	SVC	LGBM	DT	SVC	LGBM	DT	SVC	LGBM
S1	0.79	0.56	0.83	0.77	0.96	0.82	0.84	0.56	0.87	0.80	0.71	0.84
S2	0.77	0.75	0.83	0.71	0.84	0.77	0.77	0.69	0.84	0.74	0.76	0.80
S3	0.62	0.71	0.87	0.80	0.64	0.58	0.35	0.40	0.77	0.49	0.49	0.66
S4	0.82	0.88	0.91	0.29	0.01	0.28	0.26	0.62	0.91	0.27	0.62	0.43

Table 3. Classification metrics results from experiment 2, training with the year immediately preceding the test in WFV for all test sets (Sn) and classifiers (Cl).

Sn/Cl	Accuracy			Recall			Precision			F1-score		
	DT	SVC	LGBM	DT	SVC	LGBM	DT	SVC	LGBM	DT	SVC	LGBM
S1	0.77	0.57	0.84	0.75	0.98	0.85	0.82	0.57	0.86	0.78	0.72	0.85
S2	0.77	0.56	0.72	0.62	0.06	0.4	0.82	0.70	0.95	0.71	0.11	0.56
S3	0.74	0.35	0.35	0.66	0.97	0.97	0.44	0.25	0.25	0.53	0.40	0.40
S4	0.86	0.88	0.85	0.24	0.17	0.59	0.35	0.49	0.67	0.28	0.25	0.63

Table 4. Classification metrics results from experiment 3, training with (up to) two years immediately prior to testing in WFV for all test sets (Sn) and classifiers (Cl).

Sn/Cl	Accuracy			Recall			Precision			F1-score		
	DT	SVC	LGBM	DT	SVC	LGBM	DT	SVC	LGBM	DT	SVC	LGBM
S1	0.78	0.57	0.84	0.76	0.98	0.83	0.83	0.57	0.88	0.79	0.72	0.85
S2	0.72	0.75	0.83	0.73	0.84	0.75	0.68	0.69	0.86	0.70	0.76	0.80
S3	0.77	0.40	0.83	0.40	0.86	0.32	0.48	0.25	0.76	0.44	0.39	0.45
S4	0.86	0.89	0.90	0.25	0.01	0.11	0.35	1.00	0.91	0.29	0.02	0.20

Table 5. Classification metrics results from experiment 4, training with data from years 1, 2, 1+2 for test sets (Sn) 4 and 5, and all classifiers (Cl).

Sn/Cl	Accuracy			Recall			Precision			F1-score		
	DT	SVC	LGBM	DT	SVC	LGBM	DT	SVC	LGBM	DT	SVC	LGBM
S1	0.66	0.49	0.82	0.53	0.76	0.56	0.54	0.27	0.93	0.53	0.4	0.7
S2	0.62	0.7	0.9	0.71	0.56	0.51	0.3	0.2	0.95	0.42	0.29	0.66
S3	0.77	0.77	0.76	0.55	0.14	0.35	0.74	0.45	0.94	0.63	0.21	0.51
S4	0.86	0.85	0.85	0.41	0.32	0.27	0.79	0.35	0.93	0.54	0.33	0.42

Table 6. 10 most relevant *features*, in order from most to least important, according to the *LightGBM* classifier modelled during S2 of experiment 1. We show the absolute number of times the *feature* was used by the model in training, its relative percentage against all *features* and the cumulative percentage.

Description	Importance of *features*		
	Absolute	Percentage	Cumulative percentage
Total completed workload of the course	1142	3.8%	47.4%
Minimum credits to complete the course	1232	4.1%	51.5%
Percentage of courses completed	1238	4.1%	55.7%
Age in years	1346	4.4%	60.1%
Grade Point Average	1400	4.6%	64.8%
Standard deviation of frequency registers	1439	4.7%	69.6%
Average attendance in subjects	1456	4.8%	74.5%
Standard deviation of grades	1550	5.1%	79.6%
Classification obtained in ENEM	1843	6.1%	85.8%
Current number of enrolled subjects	2114	7.0%	92.8%
ENEM score	2139	7.1%	100%

4.1 Discussion

- **Data engineering and processing:** as in every real-world problem, and unlike databases *benchmark* prepared and cleaned, knowledge of the data available in the databases is necessary to extract and prepare them. The variables have different ranges, many are categorical, data have missing or inconsistencies values, and is unbalanced. Thus, it is necessary to make decisions to treat the data in order to allow the subsequent modeling;

- **Training regime:** if it is necessary to obtain a model capable of predicting whether a student is a potential dropout, the training regime must consider the temporal aspect, always using past data to predict a current instance. Experiments show that accumulating all past history is beneficial for up to the third year, but for the fourth year, using only the previous year is better;

- **Predictive ability:** The predictive ability is higher for students in the first and second years. From the third stage, the capacity is degraded, making it more difficult to differentiate evaders from non-evaders. For the best *f1-score* result in the third and fourth years, a lower recall than precision was observed, indicating a greater presence of false negatives. Thus, a specific study may be necessary to define the training set for each year;

- **Classification method:** the decision tree *ensemble LightGBM* showed the best results. The results with SVC indicate the difficulty of finding a linear separation with the available *features*, in particular for the third and fourth years, confirming the conclusions about the predictive capacity. A single decision tree is also insufficient to solve the problem, so that there is no single set of rules that is suitable for all students, needing ensembles as to better tackle the problem;

– **Features:** in terms of importance of *features*, considering the result for S2, the ENEM grade and classification is one of the most important, indicating the need for the university to have specific programs for students with grades lower and, possibly, have entered advanced calls and with the initial semester already started. The importance of the dispersion of grades and attendance (measured by the standard deviation) and the mean of attendance, indicates that the student with very different results for some subjects may be less motivated to continue. Next, it was observed the coefficient of performance, higher grade and the *features* that relate to how much of the course is already completed (subjects, workload and percentage), which may indicate that the greater the student's success and the shorter the time to complete, the less likely you are to drop out. Age in years also appears to be important, showing that younger people are less likely to drop out. Finally, it is worth noting that the difficulty in a single (or a few) subjects can induce dropout as indicated by the lower grade, lower frequency and the *features* related to the cancellation of subjects.

About the importance of the variables and factors for dropout, Hippel and Hofflinger [10] mentioned that enrolment variables (high school grades, entrance exam grades) are not sufficient for the prediction. They also suggested the use of variables like family income, high school, first, second or third option course and financial assistance to influence the prediction. In this study, we confirm that actual academic data, directly related to the student experience during undergraduation is needed, including grades, attendance, admission modality and workload. Although we did not obtain some of the attributes related to financial aid and course option (as in [10]), as these data do not exist in the UFScar systems.

5   Conclusions

This paper offered guidelines for the design of a college dropout risk prediction model. We used real data and proposed a protocol for data preparation and a study of classification models by modelling the task as a binary classification problem. It was possible to show the difficulties of processing the data and its challenges. By using the real data available at the college database, the best results for second and third year students (which is still early allowing for interventions), had an accuracy of 0.83 and F1-Score of 0.80 (S2) and 0.83 (S3) in a realistic scenario. This represents a milestone in terms of making it viable to use as a tool in practice.

We can also interpret the results of classifiers in such context: the SVC method is comparable to the decision tree in particular for early years, which indicates there is some degree of linear separability but not sufficient to solve the problem. The LightGBM was in general better than using a single decision tree, showing that some instances might have conflicting patterns and that a single set of rules based on the available features is not sufficient to classify the students according to their risk of dropout.

Because the systems and databases studied in this paper are often available in institutions, our proposal is scalable for other educational institution that wishes to improve its college attrition management. It allows providing statistical reports for university management, as well as defining dropout risk thresholds so that to apply interventions, anticipating and avoiding dropouts.

Future work may investigate the role of fees from academic centers and evasion by type of admission. Also, we believe there is space for improvement by segmenting the data to study it in a more fine-grained level, such as into different course areas and specific degrees.

References

1. Abu-Oda, G.S., El-Halees, A.M.: Data mining in higher education: university student dropout case study. Int. J. Data Min. Knowl. Manage. Process **5**(1), 15 (2015)
2. Aina, C.: Parental background and university dropout in Italy. High. Educ. **65**(4), 437–456 (2013)
3. Aina, C., Baici, E., Casalone, G., Pastore, F.: The determinants of university dropout: a review of the socio-economic literature. Socioecon. Plann. Sci. **79**, 101102 (2022)
4. Advisory at the Federal University of Alagoas, C.: Ufal comemora a redução do índice de evasão de estudantes de graduação. Technical report (2019). https://ufal.br/ufal/noticias/2019/10/ufal-comemora-a-reducao-do-indice-de-evasao-de-estudantes-de-graduacao
5. Ataíde, J., Lima, L., de Oliveira Alves, E.: A repetência e o abandono escolar no curso de licenciatura em física: um estudo de caso. Physicae **6**, 21–32 (2006)
6. Bayer, J., Bydzovská, H., Géryk, J., Obsivac, T., Popelinsky, L.: Predicting dropout from social behaviour of students. Int. Educ. Data Min. Soc. (2012)
7. Burgos, C., Campanario, M.L., de la Peña, D., Lara, J.A., Lizcano, D., Martínez, M.A.: Data mining for modeling students' performance: a tutoring action plan to prevent academic dropout. Comput. Electr. Eng. **66**, 541–556 (2018)
8. Casanova, J.R., Cervero Fernández-Castañón, A., Núñez Pérez, J.C., Almeida, L.S., Bernardo Gutiérrez, A.B., et al.: Factors that determine the persistence and dropout of university students. Psicothema **30** (2018)
9. Eisenberg, D., Gollust, S., Golberstein, E., Hefner, J.: Prevalence and correlates of depression, anxiety, and suicidality among university students. Am. J. Orthopsychiatry **77**, 534–542 (2007)
10. Hippel, P.T.V., Hofflinger, A.: The data revolution comes to higher education: identifying students at risk of dropout in Chile. J. High. Educ. Policy Manage. 1–22 (2020)
11. Hooshyar, D., Pedaste, M., Yang, Y.: Mining educational data to predict students' performance through procrastination behavior. Entropy **22**(1), 12 (2020)
12. Huo, H., et al.: Predicting dropout for nontraditional undergraduate students: a machine learning approach. J. Coll. Student Retent.: Res. Theory Pract. **24**(4), 1054–1077 (2023)
13. Ivankova, N.V., Stick, S.L.: Students' persistence in a distributed doctoral program in educational leadership in higher education: a mixed methods study. Res. High. Educ. **48**(1), 93–135 (2007)
14. Ke, G., et al.: LightGBM: a highly efficient gradient boosting decision tree. Adv. Neural Inf. Process. Syst. **30** (2017)

15. Kelly, J.D.O., Menezes, A.G., de Carvalho, A.B., Montesco, C.A.: Supervised learning in the context of educational data mining to avoid university students dropout. In: 2019 IEEE 19th International Conference on Advanced Learning Technologies (ICALT), vol. 2161, pp. 207–208. IEEE (2019)
16. Kotsiantis, S.: Educational data mining: a case study for predicting dropout-prone students. Int. J. Knowl. Eng. Soft Data Paradigms **1**(2), 101–111 (2009)
17. Lehmann, W.: "I just didn't feel like i fit in": The role of habitus in university dropout decisions. Can. J. High. Educ. **37**(2) (2007)
18. Leonhardt, D., Chinoy, S.: The college dropout crisis. The New York Times (2019). https://www.nytimes.com/interactive/2019/05/23/opinion/sunday/college-graduation-rates-ranking.html
19. Lozano, J.M., Rua Vieites, A., Bilbao-Calabuig, P., Casadesús-Fa, M.: University student retention: best time and data to identify undergraduate students at risk of dropout. Innov. Educ. Teach. Int. **57**, 1–12 (2018)
20. Manhaes, L., Manhães, B., Cruz, S., Costa, M., Zavaleta, J., Silva, G.: Identificação dos fatores que influenciam a evasão em cursos de graduação através de sistemas baseados em mineração de dados: uma abordagem quantitativa. In: VIII Simpósio Brasileiro de Sistemas de Informação (2012)
21. Martins, L.C.B., Carvalho, R.N., Carvalho, R.S., Victorino, M.C., Holanda, M.: Early prediction of college attrition using data mining. In: 2017 16th IEEE International Conference on Machine Learning and Applications (ICMLA), pp. 1075–1078. IEEE (2017)
22. da Matta, K.W.: Evasão Universitária Estudantil: Precursores Psicológicos do Trancamento de Matrícula por Motivo de Saúde Mental. Master's thesis, Universidade de Brasília (2011)
23. Meedech, P., Iam-On, N., Boongoen, T.: Prediction of student dropout using personal profile and data mining approach. In: Lavangnananda, K., Phon-Amnuaisuk, S., Engchuan, W., Chan, J. (eds.) Intelligent and Evolutionary Systems. Proceedings in Adaptation, Learning and Optimization, vol. 5, pp. 143–155. Springer, Cham (2016). https://doi.org/10.1007/978-3-319-27000-5_12
24. Mello, R.F., Ponti, M.A.: Machine Learning: A Practical Approach on the Statistical Learning Theory. Springer, Heidelberg (2018). https://doi.org/10.1007/978-3-319-94989-5
25. Nistor, N., Neubauer, K.: From participation to dropout: quantitative participation patterns in online university courses. Comput. Educ. **55**(2), 663–672 (2010)
26. Pal, S.: Mining educational data using classification to decrease dropout rate of students. Int. J. Multidisc. Sci. Eng. **3**, 35–39 (2012)
27. Ponti, M.: Combining classifiers: from the creation of ensembles to the decision fusion. In: 2011 24th SIBGRAPI Conference on Graphics, Patterns, and Images Tutorials, pp. 1–10. IEEE (2011)
28. Powdthavee, N., Vignoles, A.: The socio-economic gap in university dropout. BE J. Econ. Anal. Policy **9**(1) (2009)
29. Ribeiro, M.: O projeto profissional familiar como determinante da evasão universitária: um estudo preliminar. Rev. Brasileira Orientacao Prof. **6**, 55–70 (2005)
30. Santos, R.S., Ponti, M.A., Rodrigues, K.R.H.: Evasão universitária e estratégias para retenção de alunos com base em intervenções remotas. In: Anais Estendidos do Simpósio Brasileiro de Fatores Humanos em Sistemas Computacionais (IHC) (2022)
31. dos Santos Baggi, C.A., Lopes, D.A.: Evasão e avaliação institucional no ensino superior: uma discussão bibliográfica. Avaliação: Rev. Avaliação Educação Superior (Campinas) **16**, 355–374 (2011)

32. Sarra, A., Fontanella, L., Di Zio, S.: Identifying students at risk of academic failure within the educational data mining framework. Soc. Indic. Res. **146**(1), 41–60 (2019)
33. da Silva, P.M., Lima, M.N., Soares, W.L., Silva, I.R., Roberta, A.D.A., de Souza, F.F.: Ensemble regression models applied to dropout in higher education. In: 2019 8th Brazilian Conference on Intelligent Systems (BRACIS), pp. 120–125. IEEE (2019)
34. Solís, M., Moreira, T., Gonzalez, R., Fernandez, T., Hernandez, M.: Perspectives to predict dropout in university students with machine learning. In: 2018 IEEE International Work Conference on Bioinspired Intelligence (IWOBI), pp. 1–6. IEEE (2018)
35. de Souza, A.M.: Machine learning e a evasão escolar: análise preditiva no suporte à tomada de decisão. Master's thesis, Faculdade de Ciências Empresariais (2020). https://repositorio.fumec.br/xmlui/handle/123456789/420
36. Stein, C.: The push for higher education: college attrition rates. PA Times Org (2018). https://patimes.org/the-push-for-higher-education-college-attrition-rates/
37. Tontini, G., Walter, S.: Pode-se identificar a propensão e reduzir a evasão de alunos? ações estratégicas e resultados táticos para instituições de ensino superior. Avaliação Rev. Avaliação Educação Superior (Campinas) **19**, 89–110 (2014)
38. Triado, X., Sallán, J., Feixas, M., Figuera, P., Chueca, P., Fonseca, M.: Student dropout rates in catalan universities: profile and motives for disengagement. Qual. High. Educ. **20**, 165–182 (2014)
39. Veloso, T.C.M.A., de Almeida, E.P.: Evasão nos cursos de graduação da universidade federal de mato grosso, campus universitário de cuiabá - um processo de exclusão. Série-Estudos - Perioódico Mestrado Educação UCDB (13), 133–148 (2002)
40. Xenos, M., Pierrakeas, C., Pintelas, P.: A survey on student dropout rates and dropout causes concerning the students in the course of informatics of the hellenic open university. Comput. Educ. **39**(4), 361–377 (2002)

The Artificial Intelligence as a Technological Resource in the Application of Tasks for the Development of Joint Attention in Children with Autism

Nathália Assis Valentim[1]([✉]), Fabiano Azevedo Dorça[1], Valéria Peres Asnis[1], and Nassim Chamel Elias[2]

[1] Universidade Federal de Uberlândia, Uberlândia, Minas Gerais, Brazil
{nathaliavalentim,fabianodor,valeria.asnis}@ufu.br
[2] Universidade Federal de São Carlos, São Carlos, São Paulo, Brazil
nassim@ufscar.br

Abstract. People with autism spectrum disorder (ASD) may present, in addition to deficits in communication, social interaction and patterns of restricted and repetitive behaviors, also present a deficit in joint attention (JA), which refers to the response repertoire of following and/or directing an adult's visual attention to objects or events in the environment. By having a strong relationship with the learning process, joint attention deficits can compromise a person's learning process. In this way, the use of technology can help in the development of abilities in people with autism, such as, for example, improving joint attention, communication and social skills. In this context, the general objective of the work proposal was to develop a computational approach for intervention that allows the interaction of the student with autism, with 4 and 5 years old, with deficit in joint attention and social-communicative difficulties. Artificial intelligence (AI) techniques were used to model the most appropriate sequence and level of complexity of exercises for each child. AI resources were used with the intention of providing an intelligent environment to guide the child, dynamically and adaptively, in order to promote stimuli and adequate personalization of the process. In this way, it is intended to contribute significantly to the advancement of the state of the art regarding the production of computational technologies for people with ASD.

Keywords: artificial intelligence · intelligent tutoring systems · autism · learning process · joint attention

Supported by Research Support Foundation of the State of Minas Gerais (FAPEMIG) - UNIVERSAL DEMAND Process: APQ-00837-21.
This research has an opinion embodied by the Research Ethics Committee number 5.273.182, with CAAE 54880921.7.0000.5152, the Proposing Institution being the Faculty of Computing of the Federal University of Uberlândia.

M. C. Naldi and R. A. C. Bianchi (Eds.): BRACIS 2023, LNAI 14195, pp. 306–320, 2023.
https://doi.org/10.1007/978-3-031-45368-7_20

1   Introduction

Autism Spectrum Disorder (ASD) refers to a group of neurodevelopmental disorders that are characterized by the American Psychiatric Association by levels of severity based on two main areas of deficit: communication and social interaction; restricted and repetitive behavior patterns [2]. It is now known that there is not one autism, but many types, caused by different combinations of genetic and environmental influences. The term "spectrum" reflects the wide variation in challenges and strengths possessed by each person with autism [17].

There is a strong interest in developing accurate techniques for early diagnosis and treatment of people with autism [13], considering that the number of diagnosed cases worldwide has increased considerably in recent years and that these diagnoses often depend on judgments correct as to whether each symptom, listed in the diagnostic criteria, is met or not. There are several ways to work with children with autism, one of them is through assistive technologies. In the area of education, assistive technology can be used as a computational approach capable of serving and helping students according to their respective special educational needs [19].

People with autism may present, in addition to deficits in communication, social interaction and restricted and repetitive behavior patterns, also present a deficit in Joint Attention (JA), which refers to the repertoire of responses to follow and/or direct an adult's visual attention to objects or environmental events [5]. By having a strong relationship with the learning process, joint attention deficits can compromise a person's learning process. Thus, a new challenge arose to develop an assistive technology, using the architecture of an Intelligent Tutoring System (ITS) to support the strengthening of joint attention, whose deficit hinders learning, in children with autism in the preschool phase.

This work is organized as follows. In Sect. 2 the fundamental concepts are defined. In Sect. 3, related works are presented. The objective of the STI proposal that was developed in this work was also described and will be presented in detail in Sect. 4, which also presents the proposal of the algorithms that were developed in the work. Section 5 presents the methodology that was used in the work. Section 6 presents how the experiments were carried out and Sect. 7 presents the results obtained and discussion. Finally, in Sect. 8, the final considerations of the work are detailed.

2   Fundamental Concepts

This Section presents the theoretical foundations related to work. Intelligent Tutoring Systems are described in Subsect. 2.1. Subsection 2.2 describes Machine Learning concepts, specifically about Convolutional Neural Networks and Reinforcement Learning. Section 2.3 describes Affective Learning concepts. Finally, the concept of Shared Attention is detailed in Subsect. 2.4.

2.1 Intelligent Tutoring System

The Intelligent Tutoring System (ITS) is a computer system that aims to provide immediate and personalized instruction or feedback to users, usually without the intervention of a human tutor [12]. The ITS aims to enable learning in a meaningful and effective way using a variety of computing technologies.

They also include metacognitive and supportive activities that are relevant. Considering differences in student's abilities, preferences, and metacognitive needs. Considering an ITS an assistive technology is possible because a student diagnosed with ASD, depending on their characteristics, can learn better through individual teaching than through classroom teaching.

2.2 Machine Learning

In Artificial Intelligence, an important concept is machine learning. Machine learning is the research field devoted to the formal study of learning systems, a highly interdisciplinary one that builds on ideas from statistics, computer science, engineering, cognitive science, optimization theory, and many other disciplines of science and mathematics [9]. Within this field, it is possible to distinguish different types of machine learning, among them machine learning. Supervised machine learning and unsupervised machine learning. A combination of these two approaches were used in this work: Convolutional Neural Network and Reinforcement Learning.

Convolutional Neural Network. Supervised learning is done when starting from a previously defined set of labeled data, where the objective is to find a function that is able to predict unknown labels. A Convolutional Neural Network (CNN) is a computational resource inspired by the functioning of the human brain and is commonly used in the processing and analysis of digital images. They are also known to require a small processing time when compared to other algorithms.

The focus of this work was to use CNN for the Detection of Facial Emotions. A commonly used CNN architecture for detecting facial emotions was presented in [4]. CNN's output for a given image is one of eight possible emotion classes: neutral, anger, fear, happy, sadness, surprise, contempt e disgust.

CNN was implemented with the *Caffe* framework considering the *VGG13* architecture and was pre-trained on the *ImageNet* set, one of the most popular image sets for working with CNN because it has millions of images and several classes [11]. Due to the focus of the work, a CNN tuning process is performed on the set of FER+ faces [3]. This dataset is commonly used in works that deal with emotion detection and has more than 30,000 images [8]. The authors provided an *ONNX* file with the training of the model created by CNN, which allowed the use of the model through the *OpenCV* library of the Python language [14], which includes a module of neural networks for loading the *ONNX* file.

Reinforcement Learning. In unsupervised machine learning, the dataset used does not have any type of label and, based on analysis, it is possible to discover similarities between objects. This type of learning has variations and, among them, is reinforcement learning. In reinforcement learning, the model learns by performing actions and evaluating rewards.

The algorithms developed in this work were inspired by a reinforcement learning algorithm, known as *Q-Learning* [10], which learns the quality of actions to tell an agent which action to take and under what circumstances.

2.3 Affective Learning

Affective Learning was defined as learning that relates to the learner's interests, attitudes and motivations. The ITS can deal with personalized learning experiences, containing emotion and affection, approaching successful human tutors such as guiding rational behavior, helping memory retrieval, supporting decision-making, increasing creativity, etc.

In the proposal presented in this work, the use of affective learning happens through characters, with the function of virtual tutors, aiming at a good interaction between the student and the ITS.

2.4 Joint Attention

The Joint Attention is a skill that refers to the response repertoire of following and/or directing an adult's visual attention to objects or events in the environment. The JA is a very influential factor in the development of learning and research indicates that, in the case of children with autism, there are greater JA deficits when compared to children with typical development. Thus, it is suggested that deficits in the JA repertoire are among the early signs of ASD. JA is divided into two phases: responding to joint attention (RJA) and initiating joint attention (IJA) [7].

RJA and IJA are functionally independent components and children with autism can either have difficulties with just one of the two or with both. RJA is the response of following another person's gaze or directing your gaze where the other person points. IJA consists of establishing eye contact or pointing, alternately, at another person and at the object of interest [7].

There are protocols and exercises with the aim of evaluating and reducing JAdeficits in children with autism and in this work, specialists in the field of teaching people with autism guided the exercises presented in the ITS.

3 Related Work

In [6] the study aimed to evaluate how caregivers of children with autism can use assistive technologies to support the child's routine/activities. In the study done in [16] describes requirements for building an accessible user interface for users with autism spectrum disorders (ASD). In [21] proposed an intelligent learning assistant to provide adequate teaching material for students with autism. In [18] proposed a model for the automated detection of ASD using CNN. In [1] 11 analysis patterns were developed, where each one was based on characteristics and problems faced daily by autistic people. Related works presented previously focused on the use of assistive technologies for routine activities in people with autism. However, the aforementioned computational approaches consider people diagnosed with autism, no studies were found that explicitly address ways to strengthen Joint Attention (JA).

The proposal presented in this work promotes innovations in relation to the state of the art as it uses Artificial Intelligence resources such as CNN to identify facial emotions and reinforcement learning, with the intention of providing an intelligent environment to guide the child, in a dynamic, personalized and adaptive way to her. In addition, its target audience is children with autism in the preschool phase, with deficits in JA and, consequently, with impairment in their learning process and sociocommunicative difficulties. The exercises in this intelligent environment were defined by specialists in the field of teaching people with autism. In addition, the children in the sample underwent three assessments, before and after performing the exercises. These assessments were based on the *Early Social Communication Skills* (ESCS) [15] evaluation instrument, which measures the development of different dimensions of non-verbal communication associated with JA.

4   Proposed Approach

The approach proposed in this work was the development of an ITS for intervention that allows the interaction of students with autism reported by professionals, aged between 4 and 5 years, with a deficit in JA. Each student in the sample was submitted to an initial evaluation, carried out by the researcher after a period of training with specialists, to verify their level of JA, based on the *Early Social Communication Skills* evaluation instrument, the ESCS, which measures the development of different dimensions of non-verbal communication associated with JA. From this, the student participated in individual assistance sessions where he solved, with the support of an instructor, exercises with themes of his preference, receiving positive reinforcers, in case of success, and recommendations, in case of errors. The intent of the ITS was to model exercise routes appropriately for each student, based on their preferences and performance.

The modeling of appropriate exercise routes for the student is carried out by combining two Artificial Intelligence approaches: Convolutional Neural Networks (CNN) for detecting facial emotions, and Reinforcement Learning based on the *Q-Learning* algorithm. 24 exercises were developed, distributed in 4 levels of complexity. The advancement of exercises and their respective levels happened following the performance calculation of each exercise. The architecture of the proposed approach is based on the ITS architecture.

The Fig. 1 shows how the proposed ITS works, in which the Control Module is responsible for implementing the integration between the modules according

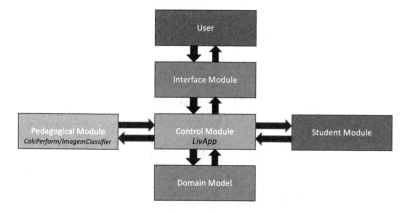

Fig. 1. Functioning of the proposed ITS.

to the events received from the user interface through the *LivApp* algorithm. The Interface Module is responsible for communicating in a friendly and appropriate way with the student with autism. The Student Module stores data related to the student's performance in performing the exercises. It also models a knowledge base about the autistic student as he interacts with the system. The Pedagogical Module models the individual characteristics of each student, presenting the most suitable exercises at each moment. In this module, two algorithms were implemented: the *CalcPerform* algorithm, where CNN infers whether the student is satisfied or dissatisfied, as shown in the *ImageClassifier* algorithm, and the performance calculation is performed. Finally, the Domain Model contains the basis of the ITS exercises.

4.1   Developed Algorithms

The combination of Artificial Intelligence approaches was possible through the implementation of three algorithms: *LivApp*, *CalcPerform* and *ImageClassifier*. The Fig. 2 shows the flowchart of interaction with ITS.

When the session with the student is started and the ITS is accessed, the *LivApp* Algorithm is started, where the final score (PF) is configured, which will be the criterion for stopping the exercise route, as shown in Algorithm 1. Next, the *CalcPerform* Algorithm, responsible for customizing the Session's exercise route, is activated. A third algorithm is activated, *ImageClassifier*, which detects facial emotions. The captured image (I) is classified as "Satisfied", "Neutral" or "Dissatisfied" through a *CNN* and sends this classification as one of the inputs to *CalcPerform*, which will calculate the performance, also considering the response time (T) of each exercise and whether this response was correct or incorrect (Dis). The cumulative score (PA) is generated from the performance calculation (D).

Each exercise (E) had a score range. Thus, for example, if the student had a PA between 0 and 4, the ITS presented the first exercise of the first level until the PA exceeded 4 and so on. While PA is smaller than PF, the *LivApp*

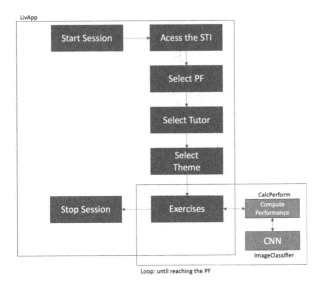

Fig. 2. Operation LivApp Algorithm

Algorithm is executed, presenting the respective exercises to the student. After meeting the stopping criterion (PA greater than or equal to PF), the algorithm ends the exercise route.

5 Research Method

The ITS was developed for desktop and mobile platforms from a web application. The entire process was accompanied by an instructor and the knowledge base, containing the exercises and their respective levels of complexity, was built by specialists in the field of education for people with autism. The process was divided into three phases: initial and final evaluation (phase 1 and phase 3, respectively); interaction with ITS (phase 2) divided into two other phases: preference evaluation and performance evaluation, as shown in the Fig. 3. The sample included two groups of children diagnosed with autism (Group 1 and Group 2), in order to make a comparison between the groups. Group 1 contained students identified as *L1*, *L2* and *L3*. Group 2 contained students identified as *V1*, *V2* and *V3*.

In phases 1 and 3, each student in the sample was submitted to three evaluations (AC1, AC2, AC3) in all, carried out by the researcher, to verify their level of JA, based in part on the ESCS evaluation instrument. For Group 1, phase 1 consisted of AC1, followed by the interaction with the STI (phase 2) and ending with phase 3, composed of AC2 and AC3. For Group 2, phase 1 consisted of AC1 and AC2, followed by interaction with the STI (phase 2) and ending with phase 3, consisting of AC3. All JA evaluations were performed in the same environment at the educational centers, at the same time, using the same objects

---

Algorithm 1: *LivApp*

Data: *PF* : Final score
Result: *E*

```
1 begin
2 │   PA ← 0;
  │   /* Scoring Interval Setup                              */
  │   /* User selects favorite Character                     */
  │   /* User selects favorite Theme                         */
3 │   E ← first exercise;
4 │   while PA < PF do
5 │   │   I, T, Dis ← user solves exercise E;
6 │   │   D ← CalcPerform(I, T, Dis);
7 │   │   PA ← PA + D;
8 │   │   E ← SelectExercise(PA);
9 end
```

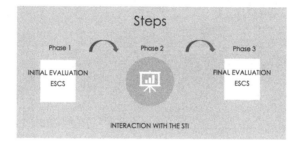

Fig. 3. The three stages of the experiment process.

and following the same evaluation order, and were recorded for later evaluation of the JA level. After quotation, the results were reviewed by the advisors.

In the interaction phase with the ITS, initially, a preference evaluation is carried out with the child, at the beginning of the session, to detect which character and theme is preferred. The Fig. 4 shows procedure for evaluating preference was, initially, to present the child with profile characters, applying the theory of *Affective Learning*, where the character was related to the interests, attitudes and motivations of the student. After that, the child was presented with different themes, so that he could make his choices under the guidance of the instructor. From the choice of theme, the child solves 3 simple level exercises and then returns to the other themes (total of 12 exercises). This stage had five sessions. After the preference evaluation, the performance evaluation takes place, where the exercises become more complex according to the student's performance in each exercise. In this step, the session's exercise route is customized. This stage also had five sessions and the Fig. 5 shows the advance of the complexity levels.

The ITS exercises were accompanied by specialists in the area and that new activities are being included to provide better adaptability of the process. The degree of difficulty/attention required increases as the child advances in the

Fig. 4. Choice of characters for virtual tutors and theme for exercises.

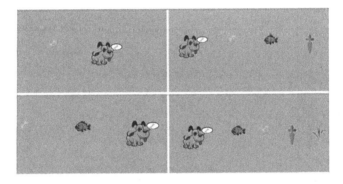

Fig. 5. Exercises at different levels.

exercise route or may decrease if the child demonstrates difficulties in the process, thus being a personalized exercise route in each session for each student, according to Fig. 5.

6 Experiments

The experiments began with Groups 1 and 2, in two different educational centers (EC). Group 1 was composed of three students (one female and two males) from the Educational Center located in the *Laranjeiras* neighborhood, in the city of Uberlândia. Group 2 was composed of three students (all male) from the Educational Center located in the *Vigilato Pereira* neighborhood, also in the city of Uberlândia. Of the six children, two of them, present in Group 2, were children who did not verbalize and needed help from teachers and caregivers to carry out the sessions. Figure 6 shows two children from the sample interacting with the ITS during a session.

The sessions for using the ITS took place on consecutive working days, for 3 weeks in October 2022. In all meetings, the researcher played the role of instructor and wore clothes in sober tones, without accessories and avoided any fragrances such as body moisturizers, eau de colognes and/or perfumes in order

Fig. 6. Children interacting with the ITS.

not to distract or compromise the comfort of the children during the sessions, which lasted an average of 10 min with each student. The calculation of the JA level from quotation tables was carried out using a spreadsheet created based on the ESCS spreadsheet, however considering only IJA and RJA.

The children in Group 1 went through phase 1 where they did the AC1 evaluation. After that, they proceeded to phase 2 where they interacted with the ITS. Then, they went through phase 3: they did the AC2 evaluation, had a 7-day break, and then they did the AC3 evaluation. Children in Group 2 went through phase 1: they took the AC1 evaluation, had a 7-day break, and then took the AC2 evaluation (phase 1). Then they interacted with the STI (phase 2). Then, they went through phase 3 with the AC3 evaluation.

The interaction of groups with ITS (phase 2) was divided into two phases: preference evaluation and performance evaluation. This ITS consisted of a web application that, at the *front-end* level, was developed using *HyperText Markup Language (HTML)*, *Javascript* and resources such as *Cascade Style Sheets (CSS)*. In terms of *back-end* the application was developed in the Python language, using *framework* Django. Data were stored using the *Sqlite3* database.

Detection of facial emotions was performed using *OpenCV*, as described in Sect. 2. The experiments of this work had images analyzed for the proposed classification of "Satisfied", "Dissatisfied" and "Neutral". Images labeled as *anger*, *fear*, *disgust* and *sadness* were analyzed for the proposed classification of "dissatisfied". Images labeled *happy* and *surprise* were analyzed for the proposed classification of "satisfied". Images labeled *neutral* were analyzed for the proposed "neutral" classification.

The Fig. 7 shows the images captured during the interaction with the ITS of three children in the sample and the classification of facial emotions of these respective images. The first child smiled while interacting with the ITS and had his facial emotion classification *"happy"*, that is, classified as "Satisfied". The second child's facial expression did not show any relevant emotion, causing his facial emotion classification to be *"neutral"*, that is, classified as "Neutral". And the third child had a facial expression of sadness, especially with the eyebrows with the final point lower than the initial point and had his facial emotion classification *"sad"*, that is, classified as "Dissatisfied".

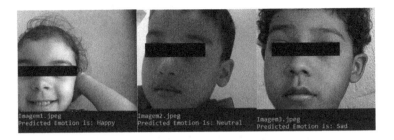

Fig. 7. Classification of Facial Emotions.

7 Discussion of Results

This section presents a discussion of the results obtained related to the work described. The results are presented about preference evaluation, performance evaluation and joint attention level. The authors provide data from the experiments carried out at [20].

7.1 Preference Evaluation

The preference evaluation had the following character options for the tutor: *Girl, Boy, Emoji, Baby, Rabbit* and *Dog*. For themes, the options were: *Car, Candy, Ball* and *Book*. Table 1 and Table 2 show the results of the choices made by each child in the five sessions for characters and themes, respectively.

Table 1. Character Preference Evaluation Result.

Child	Session 1	Session 2	Session 3	Session 4	Session 5
L1	Dog	Girl	Girl	Girl	Rabbit
L2	Boy	Boy	Boy	Boy	Boy
L3	Dog	Dog	Rabbit	Dog	Rabbit
V1	Boy	Girl	Girl	Girl	Girl
V2	Dog	Rabbit	Dog	Dog	Dog
V3	Emoji	Baby	Emoji	Emoji	Emoji

The children did not show patterns among themselves in the choices of tutors and themes, but they did show some patterns of individual behavior. However some children like L1 always selected the same tutor, children V1 and V3 always selected the same theme and L3 and V2 selected tutors related to animals.

7.2 Performance Evaluation

In addition to calculating the performance of each exercise, the performance of the session was calculated from the inverse of the number of exercises that each

Table 2. Theme Preference Evaluation Result.

Child	Session 1	Session 2	Session 3	Session 4	Session 5
L1	Car	Car	Candy	Book	Car
L2	Candy	Ball	Candy	Ball	Ball
L3	Candy	Ball	Ball	Car	Ball
V1	Car	Car	Car	Car	Car
V2	Ball	Ball	Car	Ball	Ball
V3	Ball	Ball	Ball	Ball	Ball

student solved in each session. Because the higher the student's performance in the exercise, the faster he reached the Final Score and, consequently, he solved fewer exercises. About the results of the performance evaluation, Fig. 8 presents a graph that shows that the children showed improvements throughout the sessions.

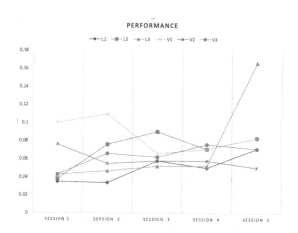

Fig. 8. Joint Attention Level.

Children L2 and L3 had constant improvements. Child L1 had a drop in performance in Session 4, which can be explained by having been without contact with the ITS for 3 consecutive days. Child V1 also showed a drop in his performance in Session 3, which can be explained by having been without contact with the ITS for 4 consecutive days. Child V2, who does not vocalize, had a drop in performance after Session 1. However, this was the only session in which assistance from the advisor was needed. In the other sessions, the child interacted with the ITS without assistance and showed improvements in their performance. The second drop in performance after session 4 can be explained by the fact that the child was absent from school for health reasons for 8 consecutive days, justifying the variation in his performance. Finally, child V3, who also does not vocalize, despite the variation in performance, had assistance during all sessions, however, being gradually smaller at each session.

7.3 Joint Attention Level

About the results of the JA levels, Fig. 9 shows that AC1, the greatest difficulties of the children were related to IJA, even those with lower levels of JA deficits, did not show any object to the evaluator and had difficulties in making eye contact with the instructor, alternating or not. Regarding RJA, children in Group 1 had less difficulties, responding to most notes, proximal touches and reference lines. Children V2 and V3 presented great difficulties both for IJA and for RJA, having the level JA considerably lower than the other children.

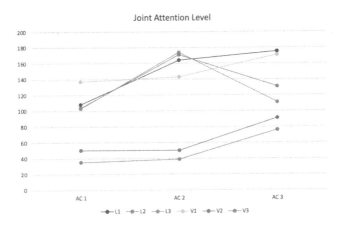

Fig. 9. Joint Attention Level.

In AC2, Group 1 had an improvement compared to Group 2 in JA levels, with considerable improvement related to IJA. In addition, the duration of AC2 with Group 1 had an average of 2 min more with each child in relation to the time in AC1, which indicates an improvement in interaction and social communication with the evaluator. Group 2 had subtle changes in relation to JA, but there was still no contact with the ITS at that time. In AC3, after a period of seven days after the last session, Group 1 showed changes in relation to AC2, with a decrease in JA levels related to both IJA and RJA. Group 2, in turn, showed a significant improvement in JA levels, where AC3 was applied right after all sessions in which there was contact with the ITS.

8 Conclusion

The main contributions expected with the completion of the work proposal presented here are the development of an ITS for strengthening JA in students with autism. The ITS utilized AI capabilities such as facial emotion detection and reinforcement learning to build personalization of the exercise journey for each user session after session. The results showed that the developed ITS is an effective tool as there was an increase in the level of JA and reports of improvements in the children's social interaction throughout the sessions carried out with contact with the ITS. The next expected steps with the completion of the work proposal presented here are the extension/adaptation of the effective ITS to foster the development of underdeveloped abilities in children with autism. It is intended to carry out the addition of more variations/types of activities, in addition to the inclusion of a dashboard for real-time monitoring of the children's evolution and also the improvement of the algorithms already implemented. In addition, the use of new suitable artificial intelligence approaches to help automate the process to foster the development of underdeveloped abilities in children with autism. As future work, it is intended the extension/adaptation of the effective STI to foster the development of underdeveloped abilities in children with autism, to increase the repertoire of STI exercises and evelop a tool so that professionals can monitor the levels of preference and performance evaluations of children in real time. Also, indication/use of new suitable artificial intelligence approaches to help automate the process to foster the development of underdeveloped abilities in children with autism. If they exist, statistical comparisons with other algorithms will be performed. So availability to the community of the STI developed with each new feature update, with host the application to facilitate access.

Acknowledgement. I offer my sincerest gratitude to my right arm Research Support Foundation of the State of Minas Gerais (FAPEMIG) - UNIVERSAL DEMAND Process: APQ-00837-21.

References

1. de Almeida, L.G.S.: Padrões de Projeto de Análise para Desenvolvimento de Software do Domínio do Transtorno do Espectro Autista (TEA). Master's thesis, Universidade Federal Fluminense (2021)
2. American Psychiatric Association: DSM-5: manual diagnóstico e estatístico de transtornos mentais. Artmed Editora (2014)
3. Barsoum: Fer+ (face expression recognition plus dataset) (2017). https://github.com/Microsoft/FERPlus
4. Barsoum, E., Zhang, C., Ferrer, C.C., Zhang, Z.: Training deep networks for facial expression recognition with crowd-sourced label distribution. In: ACM International Conference on Multimodal Interaction, pp. 279–283 (2016). https://doi.org/10.1145/2993148.2993165
5. Bates, E., Benigni, L., Bretherton, I., Camaioni, L., Volterra, V.: The Emergence of Symbols: Cognition and Communication in Infancy. Academic Press, New York (1979)
6. Cardon, T.A., Wilcox, M.J., Campbell, P.H.: Caregiver perspectives about assistive technology use with their young children with autism spectrum disorders. Infants Young Child. **24**(2), 153–173 (2011). https://doi.org/10.1097/IYC.0b013e31820eae40

7. Elias, N.C.: Teorias comportamentais sobre a etiologia do autismo e uma nova proposta. UEL (2019)

8. Gera, D., Balasubramanian, S.: Landmark guidance independent spatio-channel attention and complementary context information based facial expression recognition. Pattern Recogn. Lett. **145**, 58–66 (2021). https://doi.org/10.1016/j.patrec.2021.01.029

9. Ghahramani, Z.: Unsupervised learning. In: Bousquet, O., von Luxburg, U., Rätsch, G. (eds.) ML 2003. LNCS (LNAI), vol. 3176, pp. 72–112. Springer, Heidelberg (2004). https://doi.org/10.1007/978-3-540-28650-9_5

10. Goodfellow, I., Bengio, Y., Courville, A., Bengio, Y.: Deep Learning, vol. 1. MIT Press, Cambridge (2016)

11. Jia, Y., et al.: Caffe: convolutional architecture for fast feature embedding. In: ACM International Conference on Multimedia. Association for Computing Machinery (2014). https://doi.org/10.1145/2647868.2654889

12. Juárez Ramírez, R., Navarro-Almanza, R., Gomez-Tagle, Y., Licea, G., Huertas, C., Quinto, G.: Orchestrating an adaptive intelligent tutoring system: towards integrating the user profile for learning improvement. Procedia. Soc. Behav. Sci. **106**, 1986–1999 (2013). https://doi.org/10.1016/j.sbspro.2013.12.227

13. Chinea Manrique de Lara, A., Jiménez de Espinoza, C., González-Mora, J.: A fast automated diagnosis system for autism spectrum disorders based on eye tracking technology (2016). https://doi.org/10.13140/RG.2.2.32220.28809

14. Mordvintsev, A., Abid, K.: Opencv-python tutorials documentation (2014). https://media.readthedocs.org/pdf/opencv-python-tutroals/latest/opencv-python-tutroals.pdf

15. Mundy, P., Delgado, C., Block, J., Venezia, M., Hogan, A., Seibert, J.: Early Social Communication Scales (ESCS). University of Miami, Coral Gables (2003)

16. Pavlov, N.: User interface for people with autism spectrum disorders. J. Softw. Eng. Appl. (2014). https://doi.org/10.4236/jsea.2014.72014

17. Pimenta, T.: Transtorno do espectro autista ou autismo: causas e tratamento (2018). https://www.vittude.com/blog/transtorno-do-espectro-autista-ou-autismo/. Accessed May 2019

18. Sherkatghanad, Z., et al.: Automated detection of autism spectrum disorder using a convolutional neural network. Front. Neurosci. **13** (2019). https://doi.org/10.3389/fnins.2019.01325

19. Tenório, M., Vasconcelos, N.: Autismo: a tecnologia como ferramenta assistiva ao processo de ensino e aprendizagem de uma criança dentro do espectro. CINTEDI-Práticas pedagógicas direitos humanos e interculturalidade (2015)

20. Valentim, N.A.: Experiment data (2022). https://l1nk.dev/experimentdata. Accessed Oct 2022

21. Vijayan, A., Janmasree, S., Keerthana, C., Syla, L.B.: A framework for intelligent learning assistant platform based on cognitive computing for children with autism spectrum disorder. In: International CET Conference on Control, Communication, and Computing, pp. 361–365 (2018). https://doi.org/10.1109/CETIC4.2018.8530940

Machine Teaching: An Explainable Machine Learning Model for Individualized Education

Eduardo Vargas Ferreira[1,2]([⊠]) [ID] and Ana Carolina Lorena[1] [ID]

[1] Aeronautics Institute of Technology, São José dos Campos, Brazil
aclorena@ita.br
[2] Federal University of Paraná, Curitiba, Brazil
e.ferreira@ufpr.br

Abstract. Education is the single most important investment that people can make in their futures, and since the Universal Declaration of Human Rights in 1948, the goal of achieving universal education has been on the international agenda. In this regard, there is no doubt that the Web has had a profound impact on making education both universally available and more relevant. Nevertheless, online courses are based on "static" learning material (one-size-fits-all). For that reason, it is not straightforward to assess learning with a great number of learners who differ considerably in their educational background, engagement styles, and cognitive skills. In this work, we aim to address these aforementioned challenges by proposing an explainable Machine Learning algorithm for personalizing web-based education systems. The method has a "deep" architecture mimicking the information representation structure in human brains, and it is continuously adapted based on the signals of the students, understanding their performance through micro-steps and maximizing the learning outcome. While our methodology is general and can be applied in numerous scenarios, we demonstrate its performance by a real case study which comprises a non-mandatory, standardized exam, that evaluates high school students in Brazil.

Keywords: Explainable artificial intelligence · Deep Boltzmann Machines · Education

1 Introduction

Education is the single-most important investment that people can make in their futures, and the best way to beat poverty, promoting more inclusive societies [40]. Even in the worst circumstances, it is the basic building block to give people confidence face to future. Nevertheless, the cost of attaining higher education has skyrocketed over the past few decades [39]. Faced with this, the Web has enabled one of the most prominent developments in higher education in recent years, not only taking classes out of the traditional classroom and making them

© The Author(s), under exclusive license to Springer Nature Switzerland AG 2023
M. C. Naldi and R. A. C. Bianchi (Eds.): BRACIS 2023, LNAI 14195, pp. 321–336, 2023.
https://doi.org/10.1007/978-3-031-45368-7_21

open, accessible and dynamic [26], but also minimizing the lack of quality due to poor conditions of schools and the unavailability of adequately trained teachers. In 2011, the deployment of massive open online courses (MOOCs), such as edX consortium [8] and Coursera [19], exploded around the world and rapidly became the current trend for online learning [25], available to all those who wish to learn regardless of educational background and individual needs.

Despite their importance and a high degree of interest, there are significant limitations of moving from open content towards open educational practices, as described by [12]. Since online courses are based on "static" learning material, i.e. one-size-fits-all program and content [29], it is not straightforward to assess learning with those great number of students who differ considerably in their educational background, engagement styles and cognitive skills, which might merely lead to surface learning.

In this spirit, Machine Learning plays an important role to open up the "black box of learning", giving us deeper and fine-grained understandings of how learning actually happens, creating a relevant and transformative approach for making the education both universally available and more relevant. This paper aims to address these aforementioned challenges by proposing an explainable Machine Learning algorithm for personalizing web-based education systems. The key feature of the method is automatically tailoring the content to the proficiency level of the examinees by using an unsupervised Deep Learning algorithm, such that the next item (or set of items) selected to be administered is optimally informative for a given person. The items' personalized recommendations are elicited by using an post hoc explanation approach.

This paper is structured as follows: Sect. 2 presents some related work. Section 3 presents the problem formulation and describes the proposed method to deal with personalized learning. Section 4 explains the scenario of the simulations and results. Section 5 presents some results on the analysis of real data from a national exam for students in Brazil. Section 6 discusses and concludes the paper.

2   Related Work

One of the most important latent variable models, the item response theory (IRT) models [15], have been very successful in educational research [23]. In this context, latent traits refer to unobservable characteristics of individuals that may influence their performance in educational settings. By considering those abilities, it is possible to gain insights into the complex interplay between individual differences and educational outcomes (item responses), allowing for more targeted interventions and instructional approaches tailored to students' unique needs and characteristics. Considering the rich and broad corpus of findings devoted to IRT models [22], the two-parameter logistic (2PL) [7] is one of the most prominent. Let a response of a person j $(j = 1, \ldots, n)$ to the item i $(i = 1, \ldots, I)$ be represented by a stochastic vector $\mathbf{x}_j = (x_{1j}, x_{2j}, \ldots, x_{Ij})$ with elements

$$x_{ij} = \begin{cases} 1, & \text{if person } j \text{ gives a correct response in item } i \\ 0, & \text{otherwise.} \end{cases}$$

Selecting a response alternative is the result of the complex interaction between the characteristics of the test item and the latent trait of the person, y_j, which is commonly known as 'ability' (or proficiency) in the framework of educational and psychological measurement [23]. The probability of a correct response for an item i by individual j ($X_{ij} = 1$), is given by:

$$P(X_{ij} = 1 | y_j, \boldsymbol{\Theta}) = \frac{e^{(W_i y_j - a_i)}}{1 + e^{(W_i y_j - a_i)}}, \tag{1}$$

where $\boldsymbol{\Theta} = (\mathbf{W}, \mathbf{a})$ groups the model parameters. $\mathbf{W} = (W_1, \ldots, W_I)$ is the vector of items discrimination parameters, which reflects the capability of a generic item i to discriminate between individuals with different levels of ability. The difficulty parameters of the items are $d_i = a_i / W_i$. In IRT, each item is assumed to have its own unique item characteristic curve (ICC) and it provides valuable information about the properties of an item. The left panel of Fig. 1 presents three examples of ICCs with varying W-parameters and the right panel presents ICCs with varying a-parameters. Greater values of W represent curves with higher slopes. Therefore, even individuals with almost the same latent traits will be well discriminated, because they will produce different response probabilities. On the other hand, higher values of a indicate a proportionally higher level of difficulty of the item, since greater latent traits are needed to obtain $P(X_{ij} = 1 | y_j, \boldsymbol{\Theta})$.

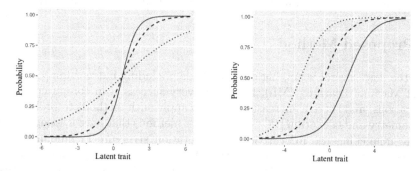

Fig. 1. Item characteristic curves for the 2PL with W-parameters $0.3, 1.0, 1.7$ (left) and a-parameters $-1, 0, 1$ (right), respectively.

While the 2PL is often easy to comprehend, interpret and can have various important measurement properties, the complexities of the brain are even more challenging. Many pieces of knowledge with different "wiring" are required to identify the correct response. For example, the paper [27] describes fourth-grade reading as follows "*reading includes developing a general understanding of written text, thinking about texts, and using texts for different purposes*", while [9] have identified 15 factors (dimensions) related to language ability. In recent years, the study of nervous systems, minds and other complex phenomena through artificial intelligence systems has shown a successful performance and defined state of the art results in many fields [11,24,32]. Furthermore, there is an increasing effort to apply such methods to problems in psychometry and education [18]. However, for the greater part, this attention is focused only on predicting student performance by using classification techniques, such as support vector machines [14,41], or by matrix factorization [36,38], etc. Furthermore, these algorithms often act as a "black box" and do not provide detailed information about the reasons for a given decision.

The field of Explainable AI (XAI) has produced an abundance of explanation techniques, from the simplest, such as intrinsically interpretable models [10], to more sophisticated ones, which assign a local interpretation, according to the specific characteristics of the input. Among them, one of the most prominent is the Layer-wise Relevance Propagation (LRP) [4]. It belongs to the propagation-based class, which assumes that the prediction was produced by a neural network, and takes advantage of this structure to produce an explanation, starting from the last layer and backtracking to the initial [33]. That is, from the weights of the network and its activation functions, the LRP propagates the contribution of each neuron to the input layer. Nevertheless, so far, research on explanation methods has mainly concentrated on supervised learning.

In a technical sense, our methodology can be seen as an extension of 2PL model by considering more interaction layers between latent abilities, modeled by a Deep Boltzmann Machine. Furthermore, it is a new framework that can explain an unsupervised model, such as 2PL, rewritten as neural networks - or "neuralizing it", and then applying a post hoc explanation such as LRP.

3   Proposed Method

Over the past several years, big data methods, including but not limited to the use of Machine Learning (ML), have been very successful in computer science applications, and there is increasing effort to apply such methods to problems in psychometry and education [18]. This includes methods that have a "deep" architecture mimicking the information representation structure in human brains. Recently, Deep Boltzmann Machines (DBMs [31]) have risen to prominence due to their capacity of learning efficient representations of seemingly complex data. DBM is a probabilistic undirected graphical model of interconnected units that learn a joint probability density over these units by adapting connections between them. Formally, a DBM can be used to specify the relation between the level of a latent trait and item responses, as follows

$$P\left(\mathbf{X}, \mathbf{Y}, Z; \mathbf{\Theta}\right) = \frac{e^{-f(\mathbf{x}, \mathbf{y}, z; \mathbf{\Theta})}}{G}, \tag{2}$$

$$G = \int e^{-f(\mathbf{x}, \mathbf{y}, z; \mathbf{\Theta})} \, d\mathbf{x} \, d\mathbf{y} \, dz,$$

where $\mathbf{X}$ are the item responses, $\mathbf{Y}$ is the first hidden layer (m-dimensional) and Z is the second hidden layer (unidimensional). The term in the denominator is called the partition function, which normalizes the probability distribution by integrating over all possible values of $\mathbf{X}$, $\mathbf{Y}$ and Z, and

$$f(\mathbf{x}, \mathbf{y}, z; \mathbf{\Theta}) = \frac{1}{2}(\mathbf{y} - \mathbf{b})^T \mathbf{\Sigma}(\mathbf{y} - \mathbf{b})^T - \mathbf{y}\mathbf{\Sigma}^{1/2}\mathbf{W}\mathbf{x}$$

$$+ \frac{1}{2}\frac{(z - c)^2}{\sigma^2} - \frac{z}{\sigma}\mathbf{V}\mathbf{y} - \mathbf{x}^T\mathbf{a},$$

where $\mathbf{\Theta} = (\mathbf{\Sigma}, \mathbf{W}, \mathbf{V}, \mathbf{a}, \mathbf{b}, c)$ groups the model parameters. $\mathbf{W} \in m \times I$ corresponds to the items discrimination parameters matrix (the slope of the surface). The parameters b and c are the bias into hidden units. $\mathbf{\Sigma}$ is a diagonal matrix of precisions. Figure 2 shows how the variables are interconnected. The standard 2PL model corresponds to having one unique output y_1. The DBM extends this model by allowing more output dimensions (y_2 to y_m) and also by providing a combined view of those dimensions in z.

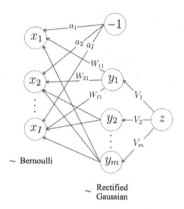

Fig. 2. Relationship between variables in the DBM.

Almost all models for latent traits have unidentified parameters due to scale invariance, this occurs because of overparametrization. To rule out it, we fixed the matrix of precisions to the identity matrix, as in [6]. Many other impressive results fixing those parameters to unity have been published [16,20,30]. The probability of the latent traits is given by:

$$P(Y = y|\mathbf{x}, z; \mathbf{\Theta}) = \mathcal{N}_R\left(\mathbf{y}|\mathbf{W}\mathbf{x} + \frac{\mathbf{V}^T z}{\sigma} + b, \mathbf{I}_m\right) \tag{3}$$

$$P(Z = z|\mathbf{y}; \mathbf{\Theta}) = \mathcal{N}\left(z|\mathbf{V}\mathbf{y} + c, \sigma^2\right) \tag{4}$$

where $\mathcal{N}(\cdot|\mu, \sigma^2)$ denotes the Gaussian probability density function with mean μ and variance σ^2 and $\mathcal{N}_R(\cdot|\mu, \sigma^2)$ denotes the rectified Gaussian distribution. Thus, it is possible to demonstrate that Eq. 2 is structurally similar to the expression for 2PL (Eq. 1). Detailed proofs of these results are given in the Appendix 1. This is a first result presented in this work, that is, that DBMs generalize a 2PL model.

3.1   Explaining DBMs Parameters

In contrast to their very high flexibility, in DBMs it is often difficult to estimate the item parameters in high dimensional spaces and provide a comprehensive explanation along several dimensions of knowledge, limiting their broader applicability. This fact stands a huge drawback in ML models, since their nested non-linear structure makes them highly non-transparent. Not surprisingly, this issue has recently received a lot of attention in the literature [4,5,34]. The goal of the explanation is to assign the importance of each input (item) for the output (latent trait). The multidimensional item discrimination can be interpreted through the $\|W_i\|$, and can be seen as sensitivity analysis. Therefore, the higher the value of $\|W_i\|$, the higher the discrimination power of that item, independently from the assumed underlying latent structure. The multidimensional item difficulty can be interpreted through $d_i = a_i/\|W_i\|$. A limitation of this approach is that it is an explanation of the latent trait variation rather than the value itself. In other words, it answers the question "which items lead to increase/decrease of latent trait when changed?". The aforementioned weakness has led to the development of a more precise technique, that explains "which items contribute much to the latent trait". To achieve this, we use the Layer-wise Relevance Propagation (LRP) [4]. Let j and k be indices for the neurons in layers l and $l + 1$, respectively, and assume that the output of function $f(x)$ has been propagated through from the upper layer to layer $l + 1$. The LRP defines the "message" $R_{j\leftarrow k}$ as the redistribution of relevance R_k to the neurons in the previous layer:

$$R_{j\leftarrow k} = \frac{y_j \cdot w_{jk}}{\sum_j y_j \cdot w_{jk}} R_k.$$

The overall relevance of j is then obtained by calculating $R_j = \sum_k R_{j\leftarrow k}$. Note that, by definition, $\sum_i R_i = \ldots = \sum_j R_j = \sum_k R_k = \ldots = f(x)$, that is, it satisfies the conservation (or completeness) property, since for all $x \in \mathcal{X}$, we have $\mathbf{1}^T \mathcal{E}_f = f(x)$. The process is outlined in Fig. 3. First, the DBMs estimate the latent trait, $f(x)$, based on the input. Then, LRP is applied to explain the prediction, that is, to indicate which items are more or less relevant.

3.2   Recommender System

When a student finishes a test, there are two fundamental and commonly asked questions: "what did I do wrong?" and "how can I improve it?" The former is about identifying specific breakdowns; whereas the latter focuses on creating knowledge. We are asking these two questions to introduce our methodology. The algorithm, at each time-step, reads the input (items responses) and returns an output (student proficiency). Once the current state is outputted, it is aggregated through the recommender system to define the next items to be administered,

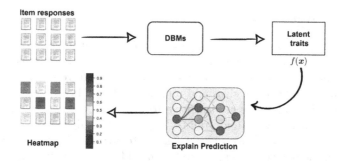

Fig. 3. Explaining predictions of an AI system using LRP.

which are sorted according to items with the same difficulty parameter (vector **d**) and lesser relevance (defined by the LRP) answered by the student, which can be viewed as a content-based recommender system focusing on items where the student needs more attention/learning [3]. Next, the sequence in which the items are presented is considered. The system intentionally incites experiences of like-minded students failure, grading the stimuli to be presented accordingly. In such a setting, the approach comprises collaborative filtering-based methods [35], by performing failure-based learning collaboratively. Various research posits fault-stages as a strategic way to identify possible breakdowns and engender higher-order knowledge [17,21,37]. This, in turn, has pedagogical benefits because it allows students to challenge their existing (incomplete/inaccurate) mental models and activate cognitive mechanisms for deep learning.

Formally, we have a set of latent traits $\mathcal{Z} = \{z_j\}$, $j = 1, \ldots, n$, and a set-based responses $\mathcal{X} = \{\mathbf{x}_i\}$, $i = 1, \ldots, I$. The aim of the system is to create a list of the best l recommendations called stimuli, $\mathcal{S} = \{s_k\}$, $k = 1, \ldots, l$, for an examinee with ability $z_a \in \mathcal{Z}$. The formulation of the algorithm is based on two steps. The first step involves selecting a set of items based on the difficulty parameter (note that the latent trait is in the same unit as the parameter). These items are chosen to be similar to those with lesser relevance (redistributed "message" given z_a) that have been answered by the student, such as:

$$S_a = \{i \ : \ d_i - \delta \leq z_a \leq d_i + \delta\}, \tag{5}$$

where δ is determined by the user and defines the homogeneity of items presented to the student. For the second step, failure is an intentional part. The items are sorted by the average of correct responses based on the k nearest neighbors of z_a. Consequently, we are able to control the examinee's exposure to failure. This strategy allows better opportunities for the student deepen conceptual understanding through items suitable to their proficiency level, and increase the problem-solving capacity through failure-based learning.

4 Simulation Study

To demonstrate our arguments and set up the subsequent discussion, in this section two simulated examples will be presented. The first example will investigate how the true values of latent traits and items are recovered in different situations, involving sample sizes and number of items. In the second example, we show how the test items are sensitive to examinee differences on a number of correlated dimensions (latent traits). The data is designed to mirror a real-life scenario where items have different characteristics and latent traits exhibit multiple dimensions with varying correlations. These initial experiments aim to show the DBM model is indeed a precise estimator of latent characteristics, generalizing the 2PL model to more latent dimensions.

4.1 Example 1

The datasets were generated containing responses of $n = 500, 2000$ and 5000 examinees to $I = 50, 100$ and 250 items. The proficiency parameters are normally distributed, by model assumption. The values of discrimination and difficulty parameters were generated from a uniform distribution such that $W \in (0.4, 1.8)$ and $a \in (-2.4, 2.4)$. As mentioned earlier, the proposed model has a structural similarity with the two-parameter logistic (2PL) Item Response Model [7] if we consider one hidden layer and $m = 1$. Thus, the responses were simulated using the 2PL parameter values. Each implementation involved 10 replications. Using these data, the parameters were estimated based on the DBM model. To carry out comparisons, the average root-mean-squared error and correlation between the true and estimated parameters for the unidimensional models were obtained, and they are summarized in Table 1. It can be seen that the parameter estimates are almost similar and appear to have converged to reasonable values, with high correlation to the ground truth values and low RMSE in all scenarios.

Table 1. Average root-mean-squared error and correlation between the true (2PL's) and estimated parameters (DBM's) based on 10 replications.

Examinees	Items	RMSE			Correlation		
		W	a	y	**W**	a	y
500	50	0.16	0.21	0.34	0.92	0.99	0.94
	100	0.15	0.20	0.25	0.93	0.99	0.96
	250	0.17	0.19	0.19	0.91	0.99	0.98
2000	50	0.10	0.08	0.33	0.97	0.99	0.94
	100	0.12	0.09	0.23	0.96	0.99	0.97
	250	0.13	0.08	0.18	0.96	0.99	0.98
5000	50	0.10	0.07	0.32	0.98	0.99	0.94
	100	0.11	0.05	0.24	0.97	0.99	0.97
	250	0.13	0.06	0.17	0.95	0.99	0.98

4.2   Example 2

For the second example, the response datasets were produced with three levels of correlation between latent trait dimensions ($\rho = 0.2, 0.6$ and 0.8). The data is considered more unidimensional when the value of ρ is closer to 1, since all latent traits will represent the same piece of knowledge. The proficiency parameters were drawn from a multivariate normal distribution. Two datasets with known characteristics are used here. Dataset 1 was designed to represent more realistic conditions where the difficulty parameters were confounded with dimensionality, such that the easier items tend to have high values for W_1 and more difficult items tend to have high values for W_2. Thus, the discrimination parameters, $\mathbf{W}_i = (W_{i1}, W_{i2})$, were ranged from $\mathbf{W}_1 = (1.8, 0.4)$ to $\mathbf{W}_I = (0.4, 1.8)$ and a was ranged from -2.4 (for $W_{i1} = 0.4, W_{i2} = 1.8$) to 2.4 (for $W_{i1} = 1.8, W_{i2} = 0.4$). This pattern of systematic shift in the item's parameters is the same as those also found in [28] and [1]. Dataset 2 was generated according to [2], considering three dimensions in the latent trait to obtain the simulated response. The results are presented in Fig. 4. In contrast to the previous example where the dimension was set at 1, in cases involving multiple dimensions, there are additional discrimination parameters to consider. Consequently, only a correlation metric was employed. Overall, the model recovered the parameters precisely, particularly when the model considered was two-dimensional in dataset 1 and three-dimensional in dataset 2 (matching the number of latent traits used to generate the responses). It is worth noting that when a one-dimensional model was employed with multidimensional data, we observed poor estimates of the discrimination parameters. Furthermore, the results were slightly worse when the correlation between competencies was 0.8, a quite high correlation value.

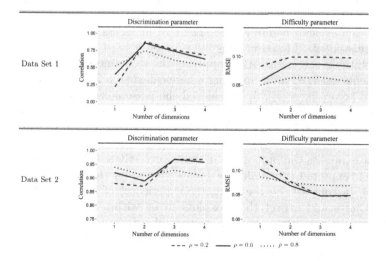

Fig. 4. Results of correlation and RMSE for the Data Sets 1 and 2.

5   Real Data Analysis

In the following, the operation of the recommender system is illustrated by a real example which comprises the Brazilian National High School Examination [13]. This is a non-mandatory, standardized exam, that evaluates high school students in Brazil. Every year, more than 5 million students take the test, which is composed of 180 multiple-choice items, coded as zero (incorrect) or one (correct), on four main content areas referred to as Natural Sciences (NC), Human Sciences (HS), Languages and Codes (LC) and Mathematics (MT). After 2009, its importance has increased since ENEM scores became accepted for admission by more than 23 universities and other institutions as well as for certification for a high school degree. This study selected a sample of 100.000.000 responses of the 2018 ENEM dataset[1] (among those that answered all 180 items of the test). The number of test dimensions (latent traits) was inferred from the presence of a single factor in all items and evidence of distinct clusters of local dependencies formed by other factors, based on the Bayesian Information Criterion. The results indicate that the data set's dimensionality is equal to 4. Thus, we considered a model with four latent traits. Furthermore, the generative p-value for the latent traits was 0.69 and for the overall items it was 0.66. These results suggest that the selected model (in terms of the number of dimensions) is appropriate for the data set. Figure 5a illustrates the distribution of the latent trait according to each of the four dimensions of knowledge. Additionally their correlation is illustrated in Fig. 5b. Note that the distribution of latent traits are complementary (compensatory structure), reinforcing their multidimensional nature. In other words, a high ability on one dimension can compensate for a low ability on another dimension. For instance, having a high level of proficiency in 'reading comprehension' can offset a lower proficiency in 'basic mathematics'. In addition, it is possible to assemble all dimensions into one unidimensional representation as described in Fig. 5c. Thereby, we assess the overall latent trait to compare students. Moreover, this information is used to make recommendations for suitable new items. After applying the recommendation system, many improvements emerged after applying a here called step 2, where items are recommended according to Eq. 5. It achieved the personalized items based on the examinee's neighborhood, dug further into the quality of recommendations to determine the degree of fault-stages, as shown in Fig. 6(left), instead of simply choosing according to $P\left(X_{ij} = 1|\mathbf{y}; \mathbf{\Theta}\right)$, as depicted in Fig. 6(right). On the left, each point expresses the proportion of items correctly answered, chosen according to their expected probability of failure (based on the k nearest neighbors responses). In contrast, on the right, the experiences of like-minded student is ignored. One would expect the box-plots to be within the horizontal dashed lines, that is, if one sets as goal an average of 0.2 (20%) of correct responses, the students responses should be around this value. This clearly happens more effectively when step 2 is used (left of Fig. 6), since the boxplots presents a lower variation and are around the horizontal lines. In practice, this means being able

---

[1] Dataset available at http://portal.inep.gov.br/microdados.

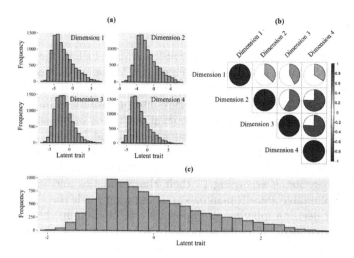

Fig. 5. The distribution of the latent trait according to each dimension (panel a), its correlation (panel b), as well as the assemble unidimensional representation (panel c).

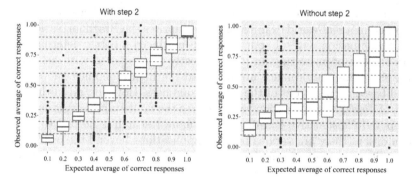

Fig. 6. Results of recommender system step 2. The left panel refers to the distribution of the observed average of correct responses from items chosen according to responses of the nearest k neighbor. The right panel shows the distribution of the observed average of correct responses from items chosen according to $P(X|\mathbf{y};\boldsymbol{\Theta})$.

to recommend items that challenge the student at any level, from items with a low chance of solution (e.g., expected average = 0.2) to high chance of solving (e.g., expected average = 0.8). This information is more refined than simply considering $P(X_{ij} = 1|\mathbf{y};\boldsymbol{\Theta})$, which does not yield good results, especially for items that do not challenge the student (with expected average of correct responses from 0.4 onwards).

6   Conclusion and Discussion

In this paper, we have reformulated the two-parameter logistic model as a neural network, with the Deep Boltzmann Machine (DBM) being defined as its extension. This makes the model explainable based on the underlying interpretation of Item Response Theory parameters. But interpretability is further extended by the use of Layer-wise Relevance Propagation, which propagates the prediction backwards through the neural network to the input features. Based on that model, we have developed an algorithm for individualized education. The approach is continuously adapted based on substantively complex constructs and recommends new items according to their characteristics and students' latent traits. The explainable algorithm allows grasping what made the algorithm arrive at a particular decision, playing an important role of open up the "black box of learning" of both algorithm and student. In the second case, this interpretable feedback is a valuable tool for verifying and improving the items to be recommended, as well as generating higher-order knowledge.

We illustrate our procedure by analyzing two simulated data sets, where the DBM model recovered the IRT parameters precisely in different situations, involving sample sizes, number of items and dimensions. Furthermore, we employ the method in a real-world example which comprises the Brazilian National High School Examination. Overall, our approach produced a better determination of the probability of answering an item correctly as compared with using simply $P(X|\mathbf{y};\boldsymbol{\Theta})$. This performance arose due to the highly detailed subpopulation selected based on responses of the others examinees. However, given the student and set of items, it is not obvious what degree the fault-stages serves as catalysts for deep and constructive learning. Thus, more research is needed to ascertain it. In addition, there are various possible applications beyond the one presented here, particularly in more complex domains, such as massive open online courses and e-commerce.

Acknowledgements. The authors thank the financial support of FAPESP (grant 2021/06870-3) and CNPq.

Appendix 1

6.1   Proof of the Equations

Proof of the Eq. 2:

$$P\left(X_i|\mathbf{y};\boldsymbol{\Theta}\right) = \frac{e^{-f(\mathbf{x},\mathbf{y},z;\boldsymbol{\Theta})}}{\sum\limits_{x} e^{-f(\mathbf{x},\mathbf{y},z;\boldsymbol{\Theta})}}$$

$$= \frac{exp\left\{-\frac{1}{2}(\mathbf{y}-\mathbf{b})^T \boldsymbol{\Sigma}(\mathbf{y}-\mathbf{b})^T + \mathbf{y}\boldsymbol{\Sigma}^{1/2}\mathbf{W}_{.i}x_i - \frac{1}{2}\frac{(z-c)^2}{\sigma^2} + \frac{z}{\sigma}\mathbf{V}\mathbf{y} + x_i a_i\right\}}{\sum\limits_{x} exp\left\{-\frac{1}{2}(\mathbf{y}-\mathbf{b})^T \boldsymbol{\Sigma}(\mathbf{y}-\mathbf{b})^T + \mathbf{y}\boldsymbol{\Sigma}^{1/2}\mathbf{W}_{.i}x_i - \frac{1}{2}\frac{(z-c)^2}{\sigma^2} + \frac{z}{\sigma}\mathbf{V}\mathbf{y} + x_i a_i\right\}}$$

$$= \frac{exp\left\{\mathbf{y}\boldsymbol{\Sigma}^{1/2}\mathbf{W}_{.i}x_i + x_i a_i\right\}}{\sum\limits_{x_i} exp\left\{\mathbf{y}\boldsymbol{\Sigma}^{1/2}\mathbf{W}_{.i}x_i + x_i a_i\right\}} \quad \text{(the only terms that depend on x)}$$

Proof of the Eq. 3:

$$P\left(Y|\mathbf{x},z;\boldsymbol{\Theta}\right) = \frac{e^{-f(\mathbf{x},\mathbf{y},z;\boldsymbol{\Theta})}}{\int_u e^{-f(\mathbf{x},\mathbf{u},z;\boldsymbol{\Theta})}du}$$

$$= \frac{exp\left\{-\frac{1}{2}(\mathbf{y}-\mathbf{b})^T \boldsymbol{\Sigma}(\mathbf{y}-\mathbf{b})^T + \mathbf{y}\boldsymbol{\Sigma}^{1/2}\mathbf{W}\mathbf{x} + \frac{z}{\sigma}\mathbf{V}\mathbf{y}\right\}\mathbb{I}(\mathbf{y}\geq 0)}{\int_\mathbf{u} exp\left\{-\frac{1}{2}(\mathbf{u}-\mathbf{b})^T \boldsymbol{\Sigma}(\mathbf{u}-\mathbf{b})^T + \mathbf{u}\boldsymbol{\Sigma}^{1/2}\mathbf{W}\mathbf{x} + \frac{z}{\sigma}\mathbf{V}\mathbf{u}\right\}\mathbb{I}(\mathbf{u}\geq 0)d\mathbf{u}}$$

$$= \mathcal{N}_R\left(\mathbf{y}|\mathbf{W}\mathbf{x} + \frac{\mathbf{V}^T z}{\sigma} + b, \boldsymbol{\Sigma}\right)$$

Proof of the Eq. 4:

$$P\left(Z|\mathbf{y};\boldsymbol{\Theta}\right) = \frac{e^{-f(\mathbf{x},\mathbf{y},z;\boldsymbol{\Theta})}}{\int_u e^{-f(\mathbf{x},\mathbf{u},z;\boldsymbol{\Theta})}du}$$

$$= \frac{exp\left\{-\frac{1}{2}\frac{(z-c)^2}{\sigma^2} + \frac{z}{\sigma}\mathbf{V}\mathbf{y}\right\}}{\int_\mathbf{u} exp\left\{-\frac{1}{2}\frac{(u-c)^2}{\sigma^2} + \frac{u}{\sigma}\mathbf{V}\mathbf{y}\right\}d\mathbf{u}}$$

$$= \mathcal{N}\left(z|\mathbf{V}\mathbf{y} + c, \sigma^2\right)$$

References

1. Ackerman, T.A.: Unidimensional IRT calibration of compensatory and noncompensatory multidimensional items. Appl. Psychol. Meas. **13**, 113–127 (1989)
2. Ackerman, T.A.: Multidimensional item response theory modeling. In: Contemporary Psychometrics: A Festschrift for Roderick P. McDonald (2013)
3. Ansari, A., Essegaier, S., Kohli, R.: Internet recommendation systems. J. Mark. Res. **37**, 363–375 (2000)
4. Bach, S., Binder, A., Montavon, G., Klauschen, F., Müller, K.R., Samek, W.: On pixel-wise explanations for non-linear classifier decisions by layer-wise relevance propagation. PLoS ONE **10**, e0130140 (2015)
5. Baehrens, D., Schroeter, T., Harmeling, S., Hansen, K., Mueller, K.R.: How to explain individual classification decisions. J. Mach. Learn. Res. **11**, 1803–1831 (2010)
6. Béguin, A.A., Glas, C.A.: MCMC estimation and some model-fit analysis of multidimensional IRT models. Psychometrika **66**, 541–561 (2001). https://doi.org/10.1007/BF02296195
7. Birnbaum, A.: Some latent trait models and their use in inferring an examinee's ability. In: Statistical Theories of Mental Test Scores, pp. 397–479 (1968)
8. Breslow, L., Pritchard, D.E., DeBoer, J., Stump, G.S., Ho, A.D., Seaton, D.T.: Studying learning in the worldwide classroom: research into edX's first MOOC. Res. Pract. Assess. **8**, 13–25 (2013)
9. Carroll, J.B., et al.: Human Cognitive Abilities: A Survey of Factor-Analytic Studies. vol. 1, Cambridge University Press (1993)
10. Craven, M.W., Shavlik, J.W.: Extracting tree-structured representations of trained networks. In: Proceedings of the 8th International Conference on Neural Information Processing Systems, NIPS 1995, pp. 24–30. MIT Press, Cambridge, MA, USA (1995)
11. Dahlkamp, H., Kaehler, A., Stavens, D., Thrun, S., Bradski, G.: Self-supervised monocular road detection in desert terrain, pp. 25–32 (2016)
12. Ehlers, U.D.: Extending the territory: from open educational resources to open educational practices. J. Open Flex. Distance Learn. **15**, 1–8 (2011)
13. ENEM: Exame Nacional do Ensino Médio (2019). https://enem.inep.gov.br/
14. Gray, G., McGuinness, C., Owende, P.: An application of classification models to predict learner progression in tertiary education. In: Souvenir of the 2014 IEEE International Advance Computing Conference, IACC 2014, pp. 549–554 (2014)
15. Hambleton, R.K., Swaminathan, H.: Item Response Theory: Principles and Applications. Springer, Dordrecht (1985). https://doi.org/10.1017/CBO9781107415324.004
16. Hinton, G.E.: A practical guide to training restricted Boltzmann machines. In: Montavon, G., Orr, G.B., Müller, K.-R. (eds.) Neural Networks: Tricks of the Trade. LNCS, vol. 7700, pp. 599–619. Springer, Heidelberg (2012). https://doi.org/10.1007/978-3-642-35289-8_32
17. Kapur, M., Bielaczyc, K.: Designing for productive failure. J. Learn. Sci. **21**, 45–83 (2012)
18. KDD Cup: Educational data minding challenge (2010). http://pslcdatashop.web.cmu.edu/KDDCup/downloads.jsp
19. Kizilcec, R.F., Piech, C., Schneider, E.: Deconstructing disengagement: analyzing learner subpopulations in massive open online courses. In: Proceedings of the Third International Conference on Learning Analytics and Knowledge, LAK 2013, pp. 170–179 (2013)

20. Krizhevsky, A.: Learning multiple layers of features from tiny images. Technical report, Department of Computer Science University of Toronto (2009)
21. Lam, R.: What students do when encountering failure in collaborative tasks. npj Sci. Learn. **4**, 6 (2019). https://doi.org/10.1038/s41539-019-0045-1
22. van der Linden, W.J.: Handbook of Item Response Theory, Three Volume Set. Chapman and Hall/CRC, Boca Raton, FL: CRC Press, 2015 (February 2018). https://doi.org/10.1201/9781315119144. https://www.taylorfrancis.com/books/9781315119144
23. Lord, F.M.: Applications of Item Response Theory to Practical Testing Problems (2014). https://doi.org/10.4324/9780203056615
24. Lu, C., Tang, X.: Surpassing human-level face verification performance on LFW with GaussianFace. In: Proceedings of the 29th AAAI Conference on Artificial Intelligence, pp. 3811–3819 (2015)
25. Pappano, L.: The Year of the MOOC. The New York Times **2**(12) (2012)
26. Pank, C.M.: Online education. AJN Am. J. Nurs. **107**(5), 74–76 (2007)
27. Perie, M., Grigg, W., Donahue, P.: The Nation's Report Card: Reading, 2005. NCES (2006-451). Technical report (2005)
28. Reckase, M.D.: The difficulty of test items that measure more than one ability. Appl. Psychol. Measure. **9**, 401–412 (1985). https://doi.org/10.1177/014662168500900409
29. Romero, C., Ventura, S., Delgado, J.A., De Bra, P.: Personalized links recommendation based on data mining in adaptive educational hypermedia systems. In: Duval, E., Klamma, R., Wolpers, M. (eds.) EC-TEL 2007. LNCS, vol. 4753, pp. 292–306. Springer, Heidelberg (2007). https://doi.org/10.1007/978-3-540-75195-3_21
30. Salakhutdinov, R.: Learning deep generative models. Ph.D. thesis, University of Toronto (2009)
31. Salakhutdinov, R., Hinton, G.: Deep Boltzmann machines. In: van Dyk, D., Welling, M. (eds.) Proceedings of the Twelfth International Conference on Artificial Intelligence and Statistics. Proceedings of Machine Learning Research, Hilton Clearwater Beach Resort, Clearwater Beach, Florida USA, 16–18 April 2009, vol. 5, pp. 448–455. PMLR (2009). https://proceedings.mlr.press/v5/salakhutdinov09a.html
32. Schölkopf, B., Williamson, R., Smola, A., Shawe-Taylor, J., Platt, J.: Support vector method for novelty detection. In: Advances in Neural Information Processing Systems, vol. 12 (2000)
33. Selvaraju, R.R., Cogswell, M., Das, A., Vedantam, R., Parikh, D., Batra, D.: Grad-CAM: visual explanations from deep networks via gradient-based localization. In: 2017 IEEE International Conference on Computer Vision (ICCV), pp. 618–626 (2017). https://doi.org/10.1109/ICCV.2017.74
34. Srinivasan, V., Lapuschkin, S., Hellge, C., Muller, K.R., Samek, W.: Interpretable human action recognition in compressed domain. In: Proceedings of the IEEE International Conference on Acoustics, Speech and Signal Processing (ICASSP), pp. 1692–1696 (2017). https://doi.org/10.1109/ICASSP.2017.7952445
35. Su, X., Khoshgoftaar, T.M.: A survey of collaborative filtering techniques. In: Advances in Artificial Intelligence, pp. 115–129 (2009). https://doi.org/10.1155/2009/421425
36. Sweeney, M., Rangwala, H., Lester, J., Johri, A.: Next-term student performance prediction: a recommender systems approach. J. Educ. Data Min. **8**, 22–51 (2016)

37. Tawfik, A.A., Rong, H., Choi, I.: Failing to learn: towards a unified design approach for failure-based learning. Educ. Technol. Res. Dev. **63**(6), 975–994 (2015). https://doi.org/10.1007/s11423-015-9399-0

38. Thai-Nghe, N., et al.: Multi-relational factorization models for predicting student performance. In: Proceedings of the KDD Workshop on Knowledge Discovery in Educational Data (2011)

39. The White House: Making college affordable (2016). https://obamawhitehouse.archives.gov/issues/education/higher-education/making-college-affordable

40. UNESCO: Global Education First Initiative (2016). http://www.unesco.org/new/en/gefi/about/

41. Yu, H., Lo, H., Hsieh, H.: Feature engineering and classifier ensemble for KDD cup 2010. J. Mach. Learn. Res. **1**, 1–16 (2010)

BLUEX: A Benchmark Based on Brazilian Leading Universities Entrance eXams

Thales Sales Almeida[1,2](✉) , Thiago Laitz[1,3] , Giovana K. Bonás[1] ,
and Rodrigo Nogueira[1,2]

[1] State University of Campinas (UNICAMP), Campinas, Brazil
`g216832@dac.unicamp.br`
[2] Maritaca AI, Campinas, Brazil
`{thales,rodrigo}@maritaca.ai`
[3] NeuralMind AI, Campinas, Brazil
`thiago.laitz@neuralmind.ai`

Abstract. One common trend in recent studies of language models (LMs) is the use of standardized tests for evaluation. However, despite being the fifth most spoken language worldwide, few such evaluations have been conducted in Portuguese. This is mainly due to the lack of high-quality datasets available to the community for carrying out evaluations in Portuguese. To address this gap, we introduce the Brazilian Leading Universities Entrance eXams (BLUEX), a dataset of entrance exams from the two leading universities in Brazil: UNICAMP and USP. The dataset includes annotated metadata for evaluating the performance of NLP models on a variety of subjects. Furthermore, BLUEX includes a collection of recently administered exams that are unlikely to be included in the training data of many popular LMs as of 2023. The dataset is also annotated to indicate the position of images in each question, providing a valuable resource for advancing the state-of-the-art in multimodal language understanding and reasoning. We describe the creation and characteristics of BLUEX and establish a benchmark through experiments with state-of-the-art LMs, demonstrating its potential for advancing the state-of-the-art in natural language understanding and reasoning in Portuguese. The data and relevant code can be found at https://github.com/Portuguese-Benchmark-Datasets/BLUEX.

1 Introduction

Recent advances in Language Models (LMs) have generated significant interest due to their demonstrated capabilities on a wide range of language tasks, including text classification, language translation, and text generation [3,7]. LM performance has been particularly impressive on standardized tests, which present challenging questions requiring high levels of domain-specific knowledge and reasoning. For instance, recent benchmarks on GPT-4 [16] showed that it can achieve human-level performance on a variety of graduate-level benchmarks.

M. C. Naldi and R. A. C. Bianchi (Eds.): BRACIS 2023, LNAI 14195, pp. 337–347, 2023.
https://doi.org/10.1007/978-3-031-45368-7_22

Despite the impressive performance of LMs on standardized tests, few evaluations have been performed in Portuguese [15], partially due to the lack of available datasets in the language. This lack of high-quality, standardized datasets presents a significant challenge for researchers interested in developing and evaluating LMs in Portuguese. To address this gap for Brazilian Portuguese, we introduce BLUEX, a dataset consisting of entrance exams for the two leading universities in Brazil. Our dataset offers a rich source of high-quality high school-level questions annotated with their respective subjects, as well as flags indicating the required capabilities necessary to respond accurately to the questions, such as knowledge of Brazilian culture and the application of mathematical reasoning. These annotations can be used to evaluate the performance of LMs on a variety of subjects and capabilities such as domain-specific knowledge and reasoning. Additionally, BLUEX includes a collection of recently administered entrance exams that are unlikely to be included in the training data of many currently popular LMs.

In anticipation of the emergence of multimodal models that combine text and image understanding, we have annotated BLUEX to indicate the position of images in each question. Additionally, we have included all necessary images with the dataset to facilitate research on multimodal language tasks. We believe that this resource will be essential in evaluating the performance of models that reason with both text and image inputs to solve complex problems.

In this paper, we describe the creation and characteristics of BLUEX and establish a benchmark through experiments with state-of-the-art LMs. Our findings suggest that BLUEX provides a valuable resource for benchmarking and advancing the state-of-the-art in natural language understanding and reasoning in Portuguese. This is particularly relevant since even the current state-of-the-art models, such as GPT-4, still have considerable room for improvement and do not achieve the highest cutoff grades for both universities.

2   Related Work

In the realm of Portuguese Natural Language Processing (NLP) datasets, there appears to be a limited availability.

For question-answering tasks, Faquad [21] is available, which exhibits an extractive style akin to SQuAD [18]. It features questions concerning Brazilian higher education institutions, with documents sourced from a federal university and supplemented by Wikipedia articles. Another option is the Multilingual Knowledge Questions and Answers (MKQA) dataset, which covers 26 languages [12]. This dataset was generated by selecting 10,000 queries from the Natural Questions dataset [10] and acquiring new passage-independent answers for each question. Subsequently, human translators translated the questions and answers into 25 non-English, typologically diverse languages, including Portuguese.

Regarding sentence entailment tasks, ASSIN 1 and 2 [5,19] are available. These datasets encompass Recognizing Textual Entailment (RTE), also referred to as Natural Language Inference (NLI), and Semantic Textual Similarity (STS) tasks. The former involves predicting if a given text (premise) implies another text (hypothesis), while the latter quantifies the semantic equivalence between two sentences.

The Portuguese Language Understanding Evaluation (PLUE) benchmark [6] provides Portuguese translations of the GLUE [26], SNLI [1], and SciTAIL [8] datasets. These translations have been generated using automatic translation tools including Google Translate and OpusMT [24].

The Winograd Schema Challenge (WSC) dataset [9] contains pairs of sentences with minimal differences, featuring an ambiguous pronoun that is resolved divergently between the two sentences. Melo et al. [13] manually translated and adapted this dataset to Portuguese.

For sentiment analysis tasks, the TweetsentBr dataset [2] consists of 15,000 tweets related to the TV show domain, collected between January and July 2017. The tweets were manually annotated by seven annotators into three classes: positive, neutral, and negative.

The Multilingual Amazon Slu resource package (SLURP) for Slot-filling, Intent classification, and Virtual assistant Evaluation (MASSIVE) [4] is a 1M-example dataset containing realistic virtual utterances in 51 languages, including Portuguese. Professional translators translated the dataset from English, and it is annotated for slot (55 classes) and intent (60 classes) prediction tasks.

A dataset more closely related to BLUEX is the ENEM-challenge dataset [22], which includes the editions of the Brazilian national exam, Exame Nacional do Ensino Medio (ENEM), from 2009 to 2017. Additionally, Nunes et al. [15] introduced a dataset containing the ENEM exam of 2022, the same paper evaluated the performance of LMs such as GPT-3.5-Turbo and GPT-4 on both the ENEM-challenge and the ENEM 2022 datasets.

3 The BLUEX Dataset

3.1 Dataset Creation

BLUEX is a dataset comprising more than 1,000 multiple choice questions from the entrance exams of the two leading universities in Brazil, Unicamp and USP, administered between 2018 and 2023. The dataset was created by automatically extracting each question text, alternatives, and related images using scripts, and subsequently each example was manually annotated to correct extraction errors and provide additional metadata such as image positioning.

3.2 Annotated Question Metadata

The annotated metadata is described below.

- **Prior Knowledge (PRK)** - Indicates whether the question requires knowledge from outside of what has been provided in the question, such as familiarity with a particular author's work or a specific mathematical formula.

- **Text Understanding (TU)** - Indicates whether the question requires understanding of a particular text.
- **Image Understanding (IU)** - Indicates whether the question requires understanding of an image. It should be noted that not all questions with images require their understanding to answer the question.
- **Mathematical Reasoning (MR)** - Indicates whether the question requires mathematical reasoning, such as the ability to perform calculations and symbolic manipulations.
- **Multilingual (ML)** - Indicates whether the question requires knowledge of two or more languages, such as questions designed to test English skills of Portuguese speakers.
- **Brazilian Knowledge (BK)** - Indicates whether the question involves knowledge specific to Brazil, such as Brazilian history, literature, geography, or culture.
- **Subjects** - A list of subjects related to the question, such as geography, physics, etc.
- **Related Images** - A list of all the related images for the question.
- **Alternative Type** - Indicates whether the answer choices are presented as text or as images. This is important because some questions may use images as answer choices, which requires different processing techniques than questions with only textual answers.

By providing such annotations along with the questions we aim to facilitate research into language understanding and reasoning in Portuguese for both pure language models and multimodal models. We believe that BLUEX will be a valuable resource for researchers to evaluate and improve the performance of future language models in the context of Portuguese-language standardized tests.

3.3   Image Positioning

Many of the questions in the exams require a contextual or informational understanding of images. Despite active research in the field of multimodal models, models that can adeptly process both text and image data and yield satisfactory results remain scarce in the public domain. We believe that BLUEX can serve as an essential evaluation tool for such models. Anticipating the use of models that will process images and text in an interleaved manner, we also provide precise information regarding the placement of images within the question, as illustrated in Fig. 1.

3.4   Dataset Distribution

The BLUEX dataset covers a wide range of high school subjects, including Mathematics, Physics, Chemistry, Biology, History, Geography, English, Philosophy and Portuguese, as well as multidisciplinary questions that involve two or more subjects. The distribution of questions is shown in Table 1, where we also provide

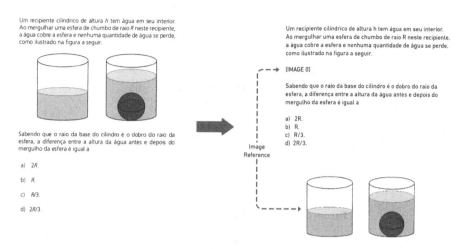

Fig. 1. Example of image annotation in BLUEX.

the distribution for the subset of questions without images, which accounts for approximately 58% of the total dataset.

Furthermore, Table 2 shows the distribution of the dataset across annotated categories, as explained in Sect. 3.2. We observe that the majority of questions require specific knowledge and the ability to comprehend text, two expected capabilities in students taking these exams. Note that any given question can be part of multiple categories.

Table 1. Distribution over subjects.

	biology	chemistry	english	geography	history	mathematics	philosophy	physics	portuguese	multidisciplinary	Total
UNICAMP	60	45	57	51	60	89	1	61	86	46	556
USP	50	63	41	55	63	69	4	63	88	43	539
BLUEX	110	108	98	106	123	158	5	124	174	89	1095
No images											
UNICAMP	35	15	20	22	49	64	1	36	65	31	338
USP	23	15	25	16	52	36	4	26	80	23	300
BLUEX	58	30	45	38	101	100	5	62	145	54	638

4   Results

To enable future comparisons, we evaluated our dataset using several language models, ranging from 6B to 66B parameters, including OpenAI's GPT-4 and GPT-3.5-Turbo models. Our experiments were conducted using large language models with no specific training for this task. Each model was provided with one example in the input and then asked to answer a question from the test set. The example was randomly selected from an exam of the same university

Table 2. Distribution over categories.

	DS	TU	IU	MR	ML	BK
UNICAMP	431	440	160	209	60	69
USP	446	442	203	174	43	63
BLUEX	877	882	363	383	103	132
No Images						
UNICAMP	273	282	0	118	23	46
USP	237	269	0	70	25	43
BLUEX	510	551	0	188	48	89

Table 3. Accuracy in the BLUEX dataset.

Model	BLUEX	UNICAMP	USP	MR	BK
Highest Cutoff Score	0.863	0.855	0.872	-	-
Average Human Score	0.521	0.530	0.511	-	-
Random	0.220	0.250	0.200	0.223	0.228
GPT-4 [16]	**0.748**	**0.749**	**0.747**	**0.447**	**0.854**
Sabiá 65B [17]	0.632	0.615	0.650	0.239	0.775
GPT-3.5-Turbo	0.582	0.580	0.583	0.277	0.764
LLaMA 65B [25]	0.542	0.530	0.557	0.271	0.652
OPT 66B [30]	0.223	0.246	0.197	0.186	0.258
Sabiá 7B [17]	**0.466**	**0.494**	**0.433**	0.25	**0.551**
Alpaca 7B [23]	0.284	0.308	0.257	**0.261**	0.258
BloomZ 7B [14]	0.284	0.275	0.293	0.17	0.326
LLaMA 7B [25]	0.255	0.275	0.233	0.255	0.247
Bertin 6B [20]	0.241	0.293	0.183	**0.261**	0.315
Bloom 7B [29]	0.238	0.302	0.167	0.255	0.281
XGLM 7.5B [11]	0.205	0.219	0.19	0.213	0.202
OPT 6.7B [30]	0.205	0.240	0.167	0.207	0.281
GPT-J 6B [27]	0.197	0.222	0.17	0.186	0.236

as the current question, but from a different year. For example, if the current question is from UNICAMP 2019, the example provided in the prompt would be a question from a UNICAMP exam, but not from 2019. We excluded all questions containing images from our experiments since the language models we used can only process text. This resulted in a total of 638 questions being used, which corresponds to approximately 60% of the dataset.

Table 3 summarizes our experimental findings, including the mean score achieved by exam-taking students, as well as the mean cutoff score of the most competitive major, which is medicine in both universities.[1] The BLUEX column shows the accuracy of the whole subset used in the evaluation, while the UNI-CAMP and USP columns account for only the questions from the respective universities. The MR and BK columns account only for questions that include those categories.

Among the language models tested in the 7B-parameter range, Sabiá [17], a model further pre-trained in Portuguese, consistently outperformed all other models, coming close to matching the average human score. Among the open-source models in the 60B-parameter range, LLaMA 65B [25] significantly outperformed OPT 66B [30] and achieved similar performance to GPT-3.5-Turbo. Sabiá 65B achieved better performance than GPT-3.5-Turbo but still lagged behind GPT-4 by ten points. GPT-4 was by far the best model in our evaluations but did not achieve an average score high enough to pass in medicine, the most competitive major. It is worth noting that the average and cutoff scores provided in Table 3 are computed taking into account the whole exam, including questions with images, while the scores obtained by the language models utilize only the subset of questions with no images.

We also conducted a more detailed analysis of the models' performance by examining their ability to handle specific question types. Table 3 presents the findings for questions that required Mathematical Reasoning (MR) and Brazilian Knowledge (BK). We observe that, with the exception of GPT-4, all models struggled to perform significantly better than random chance in questions that required Mathematical Reasoning. Even GPT-4 only achieved an accuracy of 44% in MR questions. On the other hand, when considering questions that require brazilian knowledge, Sabiá greatly outperformed all the other models in the 7B-parameter range, indicating that the extra pretraining in Portuguese provided the model with additional regional knowledge. In the 60B-parameter range, Sabiá also showed improvement over LLaMA, increasing the accuracy in these questions by 10 points and slightly outperforming GPT-3.5-Turbo. Nevertheless, it could not match the remarkable performance of GPT-4.

Moreover, Fig. 2 displays the performance of the top four models on the exams conducted each year. It can be observed that the models have a small variance between the years, which is expected as the difficulty of each exam and the number of questions in the subset vary across years. A surprising result, however, is the increased performance that all models seem to exhibit in 2023. The average and highest cutoff scores also increased slightly over the years, indicating that the exams became slightly easier in recent years. Since the 2023 exams were very recently administered, it is unlikely that they are part of any of the studied models' training data. Therefore, since the models' performance in the most recent years is comparable to that in older exams, it is reasonable to assume that the models are not merely memorizing the answers for the questions in the dataset.

---

[1] The average and cutoff scores are reported by the entities responsible for administering the exams. The results presented in Table 3 are the average of all the exams contained in the BLUEX dataset.

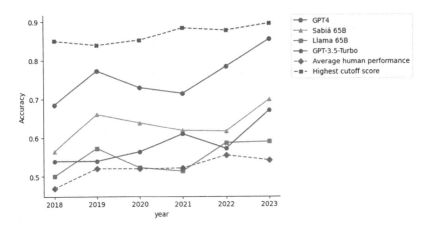

Fig. 2. Accuracy of the best models over the years of the exams.

5 Conclusion

This work introduced BLUEX, a new dataset that consists of 13 college entrance exams applied between 2018 and 2023 from two of the leading Brazilian universities, UNICAMP and USP. Each question of these exams was extensively annotated to help measure different abilities across multiple subjects in Portuguese. Beyond that, by providing images and their corresponding positions within the text, BLUEX is one of the few Portuguese datasets that are ready to evaluate multimodal models. We provide results from multiple LMs as baselines and reference scores based on students performance to facilitate future comparisons. We believe that BLUEX will be a important benchmark in the evaluation of the Portuguese capabilities of future models.

6 Future Work

The models used in this study employed a single in-context example. However, there's room for further investigation, such as determining whether increasing the number of few-shot examples could boost the performance of each model, as well as assessing their zero-shot performance. Furthermore, Nunes et al. [15] showed that GPT-4's performance on ENEM questions was significantly boosted when chain-of-thought prompts [28] were used. Adopting a similar approach here could potentially lead to performance improvement.

Finally, regarding multimodal models, their performance can be assessed utilizing the BLUEX dataset. This provides an opportunity for researchers to investigate the models' capabilities in integrating visual and textual information to address high school level questions.

7   Appendix

7.1   Prompt for Evaluation

The prompt used for all the experiments in this paper is shown in the Fig. 3.

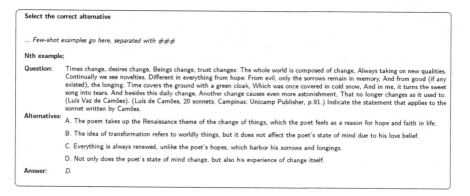

Select the correct alternative

... Few-shot examples go here, separated with ###

Nth example:

Question: Times change, desires change, Beings change, trust changes: The whole world is composed of change, Always taking on new qualities. Continually we see novelties, Different in everything from hope: From evil, only the sorrows remain in memory, And from good (if any existed), the longing. Time covers the ground with a green cloak, Which was once covered in cold snow, And in me, it turns the sweet song into tears. And besides this daily change, Another change causes even more astonishment, That no longer changes as it used to. (Luís Vaz de Camões). (Luís de Camões, 20 sonnets. Campinas: Unicamp Publisher, p.91.) Indicate the statement that applies to the sonnet written by Camões.

Alternatives:
A. The poem takes up the Renaissance theme of the change of things, which the poet feels as a reason for hope and faith in life.

B. The idea of transformation refers to worldly things, but it does not affect the poet's state of mind due to his love belief.

C. Everything is always renewed, unlike the poet's hopes, which harbor his sorrows and longings.

D. Not only does the poet's state of mind change, but also his experience of change itself.

Answer: *D.*

Fig. 3. Example of prompt used in the experiments, the question was translated into English for the convenience of readers. The text in red is the expected output. (Color figure online)

7.2   Benchmark per Subject

Table 4 provides a detailed report of each model achieved accuracy by subject. Questions that were associated with more than one subject contributed to the accuracy of both scores. For example, a question related to mathematics and English will be taken into account when calculating the accuracy of both mathematics and English subjects.

Table 4. Results for each model by subject in BLUEX.

Model	Biology	Chemistry	English	Geography	History	Mathematics	Philosophy	Physics	Portuguese
GPT-4	0.871	0.675	0.918	0.935	0.930	0.389	1.000	0.557	0.805
Sabiá 65B	0.771	0.350	0.837	0.774	0.883	0.278	1.000	0.257	0.755
GPT-3.5-Turbo	0.700	0.350	0.714	0.806	0.805	0.259	0.714	0.329	0.629
LLaMA 65B	0.657	0.350	0.816	0.677	0.719	0.306	0.429	0.286	0.572
OPT 66B	0.229	0.275	0.286	0.161	0.273	0.176	0.286	0.200	0.189
Sabiá 7B	0.514	0.350	0.592	0.565	0.672	0.241	0.571	0.271	0.509
Alpaca 7B	0.286	0.225	0.347	0.306	0.320	0.269	0.143	0.229	0.264
BloomZ 7B	0.243	0.075	0.551	0.371	0.336	0.185	0.143	0.171	0.308
LLaMA 7B	0.229	0.325	0.286	0.210	0.266	0.231	0.000	0.314	0.245
Bertin 6B	0.186	0.225	0.347	0.226	0.234	0.259	0.286	0.243	0.245
Bloom 7B	0.243	0.225	0.327	0.210	0.219	0.259	0.143	0.214	0.239
XGLM 7.5B	0.143	0.300	0.245	0.161	0.164	0.204	0.000	0.171	0.264
OPT 6.7B	0.186	0.250	0.143	0.145	0.234	0.185	0.000	0.257	0.214
GPTJ 6B	0.214	0.200	0.204	0.113	0.227	0.194	0.000	0.200	0.195

References

1. Bowman, S., Angeli, G., Potts, C., Manning, C.D.: A large annotated corpus for learning natural language inference. In: Proceedings of the 2015 Conference on Empirical Methods in Natural Language Processing, pp. 632–642 (2015)
2. Brum, H.B., das Graças Volpe Nunes, M.: Building a sentiment corpus of tweets in Brazilian Portuguese (2017)
3. Chowdhery, A., et al.: Palm: scaling language modeling with pathways (2022)
4. FitzGerald, J., et al.: MASSIVE: a 1 m-example multilingual natural language understanding dataset with 51 typologically-diverse languages (2022)
5. Fonseca, E., Santos, L., Criscuolo, M., Aluisio, S.: ASSIN: Avaliacao de similaridade semantica e inferencia textual. In: 12th International Conference on Computational Processing of the Portuguese Language, Tomar, Portugal, pp. 13–15 (2016)
6. Gomes, J.R.S.: PLUE: Portuguese language understanding evaluation (2020). https://github.com/jubs12/PLUE
7. Hoffmann, J., et al.: Training compute-optimal large language models (2022)
8. Khot, T., Sabharwal, A., Clark, P.: SciTaiL: a textual entailment dataset from science question answering. In: Proceedings of the AAAI Conference on Artificial Intelligence, vol. 32 (2018)
9. Kocijan, V., Lukasiewicz, T., Davis, E., Marcus, G., Morgenstern, L.: A review of Winograd Schema Challenge datasets and approaches. arXiv preprint arXiv:2004.13831 (2020)
10. Kwiatkowski, T., et al.: Natural questions: a benchmark for question answering research. Trans. Assoc. Comput. Linguist. 7, 453–466 (2019)
11. Lin, X.V., et al.: Few-shot learning with multilingual language models (2022)
12. Longpre, S., Lu, Y., Daiber, J.: MKQA: a linguistically diverse benchmark for multilingual open domain question answering. Trans. Assoc. Computat. Linguist. 9, 1389–1406 (2021)
13. de Melo, G., Imaizumi, V., Cozman, F.: Winograd schemas in portuguese. In: Anais do XVI Encontro Nacional de Inteligência Artificial e Computacional, pp. 787–798. SBC (2019)
14. Muennighoff, N., et al.: Crosslingual generalization through multitask finetuning (2022)
15. Nunes, D., Primi, R., Pires, R., Lotufo, R., Nogueira, R.: Evaluating GPT-3.5 and GPT-4 models on Brazilian University admission exams (2023)
16. OpenAI: GPT-4 technical report (2023)
17. Pires, R., Abonizio, H., Almeida, T.S., Nogueira, R.: Sabiá: Portuguese large language models (2023)
18. Rajpurkar, P., Zhang, J., Lopyrev, K., Liang, P.: SQuAD: 100,000+ questions for machine comprehension of text. In: Proceedings of the 2016 Conference on Empirical Methods in Natural Language Processing, pp. 2383–2392 (2016)
19. Real, L., Fonseca, E., Gonçalo Oliveira, H.: The ASSIN 2 shared task: a quick overview. In: Quaresma, P., Vieira, R., Aluísio, S., Moniz, H., Batista, F., Gonçalves, T. (eds.) PROPOR 2020. LNCS (LNAI), vol. 12037, pp. 406–412. Springer, Cham (2020). https://doi.org/10.1007/978-3-030-41505-1_39
20. de la Rosa, J., Ponferrada, E.G., Villegas, P., de Prado Salas, P.G., Romero, M., Grandury, M.: BERTIN: efficient pre-training of a Spanish language model using perplexity sampling (2022)
21. Sayama, H.F., Araujo, A.V., Fernandes, E.R.: FaQuAD: reading comprehension dataset in the domain of Brazilian higher education. In: 2019 8th Brazilian Conference on Intelligent Systems (BRACIS), pp. 443–448. IEEE (2019)

22. Silveira, I.C., Mauá, D.D.: Advances in automatically solving the ENEM. In: 2018 7th Brazilian Conference on Intelligent Systems (BRACIS), pp. 43–48. IEEE (2018)
23. Taori, R., et al.: Stanford Alpaca: an instruction-following LLaMA model (2023). https://github.com/tatsu-lab/stanford_alpaca
24. Tiedemann, J., Thottingal, S.: OPUS-MT - building open translation services for the world. In: Proceedings of the 22nd Annual Conference of the European Association for Machine Translation (EAMT), Lisbon, Portugal (2020)
25. Touvron, H., et al.: LLaMA: open and efficient foundation language models (2023)
26. Wang, A., Singh, A., Michael, J., Hill, F., Levy, O., Bowman, S.R.: GLUE: a multitask benchmark and analysis platform for natural language understanding. In: International Conference on Learning Representations (2019). https://openreview.net/forum?id=rJ4km2R5t7
27. Wang, B., Komatsuzaki, A.: GPT-J-6B: a 6 billion parameter autoregressive language model, May 2021. https://github.com/kingoflolz/mesh-transformer-jax
28. Wei, J., et al.: Chain of thought prompting elicits reasoning in large language models. In: Oh, A.H., Agarwal, A., Belgrave, D., Cho, K. (eds.) Advances in Neural Information Processing Systems (2022). https://openreview.net/forum?id=_VjQlMeSB_J
29. Le Scao, T., et al.: BLOOM: a 176B-parameter open-access multilingual language model (2023)
30. Zhang, S., et al.: OPT: open pre-trained transformer language models (2022)

Agent Systems

Towards Generating P-Contrastive Explanations for Goal Selection in Extended-BDI Agents

Henrique Jasinski[ID], Mariela Morveli-Espinoza$^{(\boxtimes)}$[ID],
and Cesar Augusto Tacla[ID]

Graduate Program in Electrical and Computer Engineering (CPGEI), Federal
University of Technology - Paraná (UTFPR), Curitiba, Brazil
morveli.espinoza@gmail.com, tacla@utfpr.edu.br

Abstract. As agent-based systems have been growing, more and more
the general public has access to them and is influenced by decisions taken
by these systems. This increases the necessity for such systems to be
capable of explaining themselves to a user. The Beliefs-Desires-Intentions
(BDI) is a commonly used agent model that has an two-phase internal
goal selection process (desires and intentions) to decide what goals to
pursue. Belief-based goal processing (BBGP) model is an extended-BDI
model whose selection process consists of four phases. This more-grained
behavior may have relevant consequences for the analysis of what an
intention is and better explain how an intention becomes what it is.
Contrastive explanations are commonly employed by people and can
bring benefits to the explanation exchange process. A Property-contrast
(P-contrast) explanation is a type of contrastive explanation that compares the properties of an explanation object. Thus, we can take a goal
as the object about which we want an explanation and its status in the
selection process (that is, how much it has advanced) as a property. This
work tackles the problem of generating P-contrast explanations and proposes a method to construct them. The method consists of two phases,
which in turn consist of a set of steps. The first step of the second phase
returns a set of beliefs that constitute a future explanation. Thus, this
work focuses on the first phase and the first step of the second phase. We
use a scenario of the cleaner world in order to illustrate the performance
of our proposal.

Keywords: Goal selection · Contrastive explanations · BDI agents ·
BBGP agents

1 Introduction

In intelligent agents, practical reasoning is the process of figuring out what to
do. According to Wooldridge [18], practical reasoning involves two phases: deliberation and means-ends reasoning. Deliberation is decomposed in two parts: (i)

Supported by organization CAPES/Brazil and CNPq Proc. 409523/2021-6.

M. C. Naldi and R. A. C. Bianchi (Eds.): BRACIS 2023, LNAI 14195, pp. 351–366, 2023.
https://doi.org/10.1007/978-3-031-45368-7_23

firstly, the agent generates a set of possible goals and (ii) then, the agent chooses which goals he will be committed to bring about. This process is known as goal selection. In order to make this decision, a goal may have to pass from more than one stage depending on the practical reasoning model. BDI model [1] is a very known and used practical model that has two stages: (i) *desires* represent some end state that the agent deems as desirable and (ii) *intentions* are desires that the agent is committed to, with their respective plans. BBGP model [2] can be seen an extended model of BDI that consider four stages for goal processing: (i) activation, (ii) evaluation, (iii) deliberation, and (iv) checking. Consequently, four different statuses for a goal are defined: (i)active, (ii)pursuable, (iii)chosen, and (iv)executive. Figure 1 shows a parallel between these models. The schema presents the goal reasoning stages and the status of a goal after passing each stage. In [2], the authors claim that chosen goals can be seen as future-directed intentions and executive ones as present-directed intentions. Thus, chosen and executive goals are intentions from a BDI perspective.

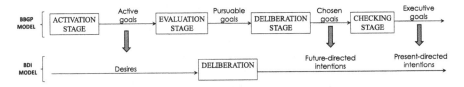

Fig. 1. Schema of the goal reasoning stages in BDI and BBGP models. This figure was extracted from [10].

Contrastive explanations are commonly employed by people and can bring benefits to the exchange process during the explanation. *Contrastive questions* provide an insight into the questioner's mental model, allowing to have a better understanding of what they do not know and *contrastive explanations* usually are more straightforward, more feasible, and less demanding both for the questioner and explainer [8]. In [16], the authors distinguish three types of contrastive questions:

- **(P-contrast)** Why does object o have property p, rather than property p'?
- **(O-contrast)** Why does object o have property p, while object o' has property p'?
- **(T-contrast)** Why does object o have property p at time t, but property p' at time t'?

In this work, we focus on generating explanations for P-contrast questions. We consider goals as the object and the status of goals as properties.

Interest in explainable artificial intelligence (XAI) grounded on social and cognitive sciences studies has grown recently [9] as well as contrastive explanations [15]. Despite such growth, there are relatively few works directed to the end user with such grounding (e.g., [5,7,14]). Since the explanation process is an exchange of information most likely involving a person, it should follow the rules of conversation. Grice [4] presents the *Cooperation Principle* with four categories: quantity, quality, relation, and manner. Each with a set of maxims that two entities in a cooperative dialogue are expected to follow. The "Quantity" category is of particular interest in this work. Its maxims state that one should be as informative as necessary, but not more than that. Thus, we will take into account this category when generating the conditions for future explanations. Regarding explainable agents focused on goal selection, to the best of our knowledge, no contrastive explanation generation approach was found. However, there are some related approaches for generating ontological and causal explanations (e.g., [10,11,17]). Against this background, we can formulate our research questions:

1. *How to generate P-contrastive explanations for goal selection process?*
2. *Considering social and cognitive sciences works, what information is required for that generation?*

In order to address the first question, we propose a method for generating P-contrastive explanations, which are constructed from the agent's execution log. The method does not present a final user-ready explanation. It provides a set of conditions (beliefs) that are required to attribute a causal relation to the P-contrastive question and that are the base for future explanations. Since a goal can advance or go back during goal selection, we generate explanations for each case and also consider the case when a goal changes its status due to a preference order, which happens when there are conflicts between goals and preference is used to determine the status change. Regarding the second question, the method is based on Grice's Cooperation Principles as requirements for the quality of explanations, specifically on the quantity category. In this sense, the resultant set of conditions includes a concise set of beliefs that are directly related to the status change of goals.

The remainder of this paper is structured as follows. In Sect. 2, we present method for generating explanations. In Sect. 3, we focus on the generation of conditions for P-contrastive explanations. Some related work is presented in Sect. 5. Finally, Sect. 6 is devoted to conclusions and future work.

2 Explanation Generation Method

This section presents the agent architecture and the method for generating explanations.

2.1 Building Blocks

We begin by presenting the agent architecture. An agent in the context of this work is an artificial entity that has a set of beliefs, a set of goals, and a set of rules. The temporal aspect of the agent execution is captured by the goals reasoning cycles. A cycle is considered to be a single execution of the selection process.

First of all we present the logical language that will be used. Let $\mathcal{L}$ be a first order logical (FOL) language which will be used to represent the goals, beliefs, and rules of an agent. The following symbols $\wedge, \vee, \rightarrow$, and $\neg$ denote the logical connectives conjunction, disjunction, implication, and negation, and $\vdash$ stands for the inference.

Definition 1 (Agent). *An agent* agt *is a tuple* agt $= \langle \mathcal{KB}, \mathcal{G}, \mathcal{P} \rangle$*, where* $\mathcal{KB} = \mathcal{B} \cup \mathcal{R}$ *is the knowledge base, which is composed of a finite set of beliefs* $\mathcal{B}$ *and a set of finite rules* $\mathcal{R}$*,* $\mathcal{G}$ *is a finite set of goals, and* $\mathcal{P}$ *is a set of plans. A goal* $g \in \mathcal{G}$ *is a tuple* $g = \langle gid, st \rangle$ *where* gid *is a unique atom representing goal* g *and* st *is the current status of goal* g*. A rule* $r \in \mathcal{R}$ *is an expression of the form* $r = \bigwedge b \rightarrow h$*, where* $b \in \mathcal{B}$ *and* $h \in \mathcal{B}$ *or* $h \in \mathcal{G}$*. Let* $p \in \mathcal{P}$ *be a plan such that* $p = \langle g, Gd, Act \rangle$*, where* $g \in \mathcal{G}$*,* Gd *is the set of guard clauses that act as a precondition for the plan execution, and* Act *is an ordered list of actions that allows the agent to achieve the goal* g*.*

$\mathcal{B}$ and $\mathcal{G}$ are subsets of literals[1] from the language $\mathcal{L}$. It also holds that $\mathcal{B}$ and $\mathcal{G}$ are pairwise disjoint. For this method, the nature of the agent's plan library and the actions list are not relevant. In the case of plans, only the goal and the guard clause are required information. A guard clause is a literal in language $\mathcal{L}$.

An explanation can be described as the assignment of causal responsibility, as it presents possible causes for what is being explained [6]. Thus, by using the agent's beliefs and the causal relationship between them and their goals, contrastive explanations can be constructed. Since there is a causal relation between the beliefs and goals, the rules and plans of the agent knowledge base can be used to define it. Thus, a cause is a relation between a set of beliefs and a single event, which is either a belief or a goal.

Definition 2 (Cause). *Let* agt $= \langle \mathcal{KB}, \mathcal{G}, \mathcal{P} \rangle$ *be an agent and* $\mathrm{E} = \mathcal{B} \cup \mathcal{G}$ *be a finite set of events. A cause* c *is a tuple* $c = \langle \mathbf{e}, \mathrm{COND} \rangle$*, where* $\mathbf{e} \in \mathrm{E}$ *is an event and* $\mathrm{COND} \subset \mathcal{B}$ *is a set of beliefs called conditions. If* $\mathcal{KB} \vdash \mathrm{COND}$ *then event* $\mathbf{e}$ *is expected to happen. Let* C *be the set of causes.*

Given a cause $c = \langle \mathbf{e}, \mathrm{COND} \rangle$, we say that it is **activated** if $\forall b \in \mathrm{COND}$ $\mathcal{KB} \vdash b$ otherwise, if $\exists b \in \mathrm{COND}$ such that $\mathcal{KB} \nvdash b$ then the cause c is **deactivated**. When two events share a piece of causal history, they are said to be causally related. Besides, the causal history of an event is called a *causal chain*. Formally, let $c = \langle \mathbf{e}, \mathrm{COND} \rangle$, $c' = \langle \mathbf{e}', \mathrm{COND}' \rangle$, and $c'' = \langle \mathbf{e}'', \mathrm{COND}'' \rangle$ be three causes s.t. $c, c', c'' \in \mathrm{C}$:

– c' and c are in a *causal chain* if $\mathbf{e}' \in \mathrm{COND}$ or if $\mathbf{e}'' \in \mathrm{COND} \wedge \mathbf{e}' \in \mathrm{COND}''$
– c and c' are *causally related* if $\exists c'' \in \mathrm{C}$, such that c'' is in a causal chain with c and c'. These relations are denoted by related$(c, c') \in \{true, false\}$.

---

[1] A literal is either an atomic formula or the negation of an atomic formula. When a literal is an atomic formula, we say that it is a positive literal, and when a literal is the negation of an atomic formula, we say it is a negative literal.

Causes describes a relation of a set of beliefs with another belief or goal. This relation can be extracted from rules and plans. Thus, we convert rules and plans to causes as follows:

- **Plan to cause conversion:** Given a plan $p = \langle g, Gd, Act \rangle$, a cause c from p is $c = \langle g, Gd \rangle$.
- **Rule to cause conversion:.** Given a rule $r = \bigwedge b \rightarrow h$, a cause c from r is $c = \langle h, \bigwedge b \rangle$.

2.2 The Proposed Method

In order to generate a contrastive explanations about a past decision during goal selection, it is necessary to reproduce the agent's knowledge base configuration when the explained events happened. Thus, a memory of the execution needs to be kept.

We propose a method that consists of two phases: Interface and Explanation Generator. Figure 2 shows these phases and the required precedures in each phase. In the Interface, four elements are required: i) the *execution history* is a log-like ordered list of events, ii) the p*reference function* encodes a preference relationship between two events, iii) the *causal function* is responsible for mapping events to the set of conditions that are causally related to it, and iv) the *conflict function* is responsible for identifying if two given goals have any sort of incompatibilities between them. The Explanation Generator has four domain-independent elements: i) the *presumptions functions* that, given a set of beliefs, return the subset containing only the presumptions (that is, beliefs that need no support); ii) the *generate related causes procedure* is responsible for receiving the posed question, retrieving the required information using the functions defined in Interface and the presumption functions, and returns the set of **related conditions**; iii) the *generate possible explanations procedure* takes as input the set of related conditions and outputs sets of explanations, where each possible explication is a subset of the related conditions; and iv) the *select an explanation procedure*, which takes the set of possible explanations and returns one considered the "most adequate" according to the context. These last elements are out of the scope of this work. Thus, we focus on generating a set of related conditions.

2.2.1 Interface

Each entry in the execution history needs to encapsulate, for each reasoning cycle, the set of goals that changed their status and the set of beliefs that were added or removed from the agent's knowledge base. Note that the goals changes happened during the deliberation, but the beliefs changes are updated after the deliberation has finished. Thus, for a cycle i, the reconstructed knowledge base needs to apply the changes only up to cycle $i - 1$.

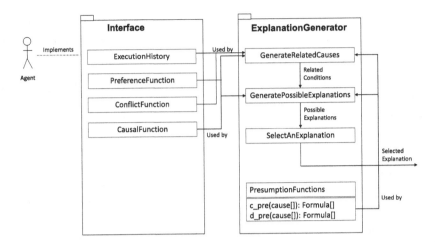

Fig. 2. Schema of the explanation generation method.

Definition 3 (Execution history entry). *Let* $\mathtt{agt} = \langle \mathcal{KB}, \mathcal{G}, \mathcal{P} \rangle$ *be an agent and* STATS *be a list of possible statuses of a goal. An entry for the execution history is a tuple* $hist = \langle i, \mathcal{G}_H, \mathcal{B}_H \rangle$, *where i is the cycle identifier, $\mathcal{G}_H$ is the set of changed goals s.t. a changed goal g_H is a tuple $g_H = \langle g, st \rangle$ where $g \in \mathcal{G}$ and $st \in$ STATS is the new status of goal g, and $\mathcal{B}_H$ is the set of changed beliefs s.t. a changed belief b_H is a tuple $b_H = \langle b, v \rangle$ where $b \in \mathcal{B}$ and $v \in \{\mathtt{ADD}, \mathtt{REM}\}$ represents the beliefs addition or removal, respectively.*

We now define the functions employed in the Interface:

Definition 4 (Causal, Preference, and Conflict Functions). *Let* $\mathtt{agt} = \langle \mathcal{KB}, \mathcal{G}, \mathcal{P} \rangle$ *be an agent,* $\mathbf{E} = \mathcal{B} \cup \mathcal{G}$ *be the set of events, and* $\mathbf{C}$ *the set of causes.*

- **Causal function:** causes : $\mathbf{E} \longrightarrow 2^{\mathbf{C}}$ *maps an event $e \in \mathbf{E}$ to the set of every cause related to it.*
- **Preference function:** preferred : $\mathcal{G}^2 \times \mathbb{N} \longrightarrow \{-1, 0, 1\}$. *This function takes as input two goals and the cycle index where they are conflicting and returns the preference relation between both goals. Thus, -1 represents that the first goal of the pair is less preferred than the second one, 0 represents that both are equally preferred, and 1 denotes that the first goal of the pair is more preferred than the second one.*
- **Conflict function:** conflict : $\mathcal{G}^2 \longrightarrow \{True, False\}$. *This function takes as input two goals and returns if they are conflicting or not. We have that $True$ denotes that there is a conflict between both goals of the pair and $False$ denotes that there is no conflict.*

2.2.2   Explanation Generator

Different patterns can be observed when analyzing how an agent answers questions about its goals. Such patterns represent how the reasoning mechanism deliberates about the selected goal in each step of the goal selection. The three observed patterns are: a) *positive filter beliefs*, which are beliefs that lead to the advance of the targeted goal to its next status; b) *negative filter beliefs*, which are beliefs that impede the advance of a given goal to its next status; and c) *preference filters*, which express the preference relationship between conflicting goals.

The agent's goals always have some associated status. The set of possible statuses is dependent on the agent model. For example, the BDI model defines two statuses: desire and intention; in turn, the BBGP model defines four: active, pursuable, chosen, and executive. The possible statuses of a goal are ordered. Considering that order, a status can be regarded as subsequent to another if it comes afterward in the list. Thus, given a finite ordered list of possible statuses $\texttt{STATS} = \{st_1, ..., st_n\}$ and $st_j, st_k \in \texttt{STATS}$ two possible statuses. We say that:

- st_k is **directly subsequent** to st_j if $k = j + 1$.
- st_k is **subsequent** to st_j if $k > j$.

The expanded knowledge base, denoted by $\mathcal{KB}^+$, is the knowledge base $\mathcal{KB}$ of the agent and everything that entails from it. It contains everything that the agent considers as true.

Definition 5 *(Expanded Knowledge Base)*. *Let $\mathcal{KB}$ be the knowledge base of the agent, the expanded knowledge base $\mathcal{KB}^+ = \{\varphi \mid \mathcal{KB} \vdash \varphi\}$, and $\nexists \varphi \in \mathcal{KB}^+ \mid \mathcal{KB} \vdash \varphi$, where $\varphi \notin \mathcal{KB}^+$.*

The explanations are built using the most fundamental beliefs of the agent, called presumptions, as they need no support to be considered true. Thus, $\forall r \in \mathcal{R}$, if $r : \emptyset \rightarrow h$ then h is a presumption.

There are two types of presumption extraction functions, one when the desired outcome is an activated causal rule and the other when the desired outcome is that a deactivated causal rule. The function **d_pre** maps a set of causes to the corresponding set of presumptions used in them. The following disjunctive form of the function is required for answers that need to deactivate a cause, as it represents the idea that if a single cause were removed, the event would not have happened.

Definition 6 *(Disjunctive Presumptions)*. *Let $\texttt{agt} = \langle \mathcal{KB}, \mathcal{G}, \mathcal{P} \rangle$ be an agent, $\texttt{E}$ be a set of events, $e \in \texttt{E}$ a given event, and $\texttt{C} = \{c_1, ..., c_n\}$ be a set of causes. **d_pre** is defined as follows:*

$$\texttt{d\_pre(C)} = \begin{cases} \bigcup_{i=1}^{n} \texttt{d\_pre}(c_i) & if \mid \texttt{C} \mid > 1 \\ h & if\ \texttt{C} = \{\emptyset \rightarrow h\} \wedge h \in \mathcal{B} \\ \bigcup_{j=1}^{m} \texttt{d\_pre}(\texttt{causes}(b_j)) & if\ \texttt{C} = b_1, ..., b_m \rightarrow e \end{cases} \quad (1)$$

Let us now focus on conjunctive presumptions. Every presumption in a cause of event e forms a single conjunctive formula. There is one formula for each different cause of e. This conjunctive form is used for answers that need to activate a cause, as it represents the idea that if all the beliefs that are missing were to become true, the event would have happened.

Definition 7 (*Conjunctive Presumptions*). *Let* $\mathtt{agt} = \langle \mathcal{KB}, \mathcal{G}, \mathcal{P} \rangle$ *be an agent,* E *be a set of events,* $e \in$ E *a given event, and* C $= \{c_1, ..., c_n\}$ *be a set of causes.* c_pre *is defined as follows:*

$$c\_pre(C) = \begin{cases} \bigcup_{i=1}^{n} c\_pre(c_i) & if \mid C \mid > 1 \\ h & if\ C = \{\emptyset \to h\} \wedge h \in \mathcal{B} \\ \bigwedge_{j=1}^{m} c\_pre(causes(b_j)) & if\ C = b_1, ..., b_m \to e \end{cases} \quad (2)$$

Figure 3 shows an example of applying the presumption functions on the causes for $\neg mop(x, y)$. Rectangles denote the event, ellipses denote causes, dashed lines denote that the causes are directly related to the event, solid lines denote that the causes are in a causal chain. If two solid lines are joined by a line, it denotes that their causes are connected by logic conjunction. In (a) it is possible to see the conjunctive function, where the presumptions for each cause form a single formula. In (b) it is possible to see that the causes are disjoint.

Let us recall that there is a preference relation between goals. This relation is especially important between conflicting goals since it is the reason for selecting one over the other. This preference is denoted by $g > g'$, which means that g is more preferred than g'. Sometimes two sets of goals need to be compared, especially when the goal being explained has a low overall preference and other compatible goals help it to become selected. Note that both sets of goals must be internally compatible. This relation is denoted by $\mathcal{G}' > \mathcal{G}''$, which means that $\mathcal{G}'$ is preferred over $\mathcal{G}''$. Sometimes a single goal has a very high priority, that is, it is preferred over a set of other goals. This is important when the high priority of such goal can explain its selection over any other goals. We denote it by $g \gg \mathcal{G}'$, which means that g is maximally preferred over a set of goals $\mathcal{G}' \subset \mathcal{G}$.

The procedure Generate Related Causes is tackled in the next section because the output depends on the type of contrast explanation.

3   Generating Conditions for P-Contrast Explanations

In this section, we focus on generating a set of conditions for P-contrast explanations.

P-Contrast questions are of the form: "*Why is goal g in status st rather than in status st'?*". The answer is "a set of absent conditions RC that would ensure that g achieves status st'.

Two sub-cases can be defined considering the sequence of the statuses:

1. $st' \succ st$, that is, "Why has not g advanced in the goal reasoning stages?
2. $st \succ st'$, that is, "Why has not g gone back in the goal reasoning stages?

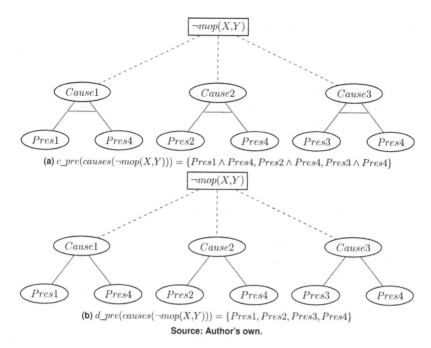

(a) $c\_pre(causes(\neg mop(X,Y))) = \{Pres1 \wedge Pres4, Pres2 \wedge Pres4, Pres3 \wedge Pres4\}$

(b) $d\_pre(causes(\neg mop(X,Y))) = \{Pres1, Pres2, Pres3, Pres4\}$

Source: Author's own.

Fig. 3. Example of presumption functions for goal $\neg mop(x, y)$. (a) shows a conjunction example and (b) shows a disjunction example.

(CASE 1). For $st' \succ st$, each deliberative pattern is defined as follows:

- **Positive:** The conditions for the explanation needs to present the set of presumptions that, if present, would have ensured the goal's progress. This set can be obtained in the following manner: $RC = c\_pre(causes(g)) - \mathcal{KB}$.
- **Negative:** The explanation needs to present a set of presumptions that, if were made absent, would have allowed the goal to progress freely. The set is obtained in the following manner: $RC = d\_pre(causes(\neg g)) \cap \mathcal{KB}$.
- **Preference:** Although no presumption was missing or impeding the goal progression, two cases may occur there was a conflict between the goal that is object of the explanation and another goal and the lack of preference made the goal not to be selected to advance. Thus, the conditions of an explanation are assertions about the preference, identifying why g was not selected: $RC = conflict(g, g') = True, preferred(g', g, i) = True$.

(CASE 2). For $st \succ st'$, each deliberative pattern is defined as follows:

- **Positive:** The conditions for an explanation need to present the set of presumptions that, if present, would have ensured the goal regression. This set can be obtained in the following manner: $RC = d\_pre(causes(\neg g)) \cap \mathcal{KB}$.
- **Negative:** The conditions for an explanation need to present a set of presumptions that, if are in the $\mathcal{KB}$, would have impeded the goal to stay with

status st. The set is obtained in the following manner: RC = c_pre(causes(g)) $- \mathcal{KB}$.

- **Preference:** Although no presumptions forced the goal regression, a conflict of goals was expected, but the preference made the goal to be selected to continue with the status st. Thus, the conditions of an explanation are assertions about the preference, identifying why g was not selected: RC = conflict(g, g') = $True$, preferred(g, g', i) = $True$.

4 Case Study: Cleaner World Scenario

In this section, a case study is presented. We show the mental state of the agent and show how the conditions fora p-contrast explanation are generated.

In the cleaner world scenario, the agent is a robot in charge of cleaning a certain area. In the robot's representation, the area is divided into cells referred to by a coordinate system (x, y). The robot can move along the cells, clean solid and liquid dirt by sweeping and mopping, respectively, recharging itself when necessary, replacing some broken parts, and disposing of the dirt in its internal storage. Lastly, for the robot not to disrupt the passage when there are no jobs at the moment, there is a command for it to rest. There are two special locations for the robot, the workshop and the dumpster. In the workshop, the robot can find replacement parts, recharge, and is where it is assigned to rest. The dumpster is where the dirt is disposed of. To do the mopping, the robot needs a functional mop part, and to sweep, the robot needs a functional vacuum. The mop is a part that the robot can replace by itself, given that there is a replacement available.

On a given day, the following series of events were observed. At time t_0, three cells were dirty, where (2,2) and (3,2) had liquid dirt and (3,4) solid dirt. At t_1, the robot cleaned (2,2). Followed by cell (3,4) at t_2. The robot proceed to go back to the workshop at t_3.

4.1 Mental States

The agent's knowledge base, goals, and plans for the scenario are described next. Let us begin describing the agent's starting **beliefs**, that is, the beliefs it had at t_0 before the first deliberation cycle. Thus, we have $\mathcal{B} = \{workshop(1,1),$ $dumpster(4,1)$, $\neg available(mop)$, $at(1,1)$, $have(battery, 75)$, $have(cargo, 65)$, $solid\_dirt(3,4)$, $liquid\_dirt(3,2)$, $liquid\_dirt(2,2)\}$. The **goals** are: $\mathcal{G} = \{(replace(x), 0)$, $(mop(x,y), 0)$, $(sweep(x,y), 0)$, $(recharge, 0)$, $(rest, 0)$, $(dispose, 0)\}$, where 0 means that the goal do not have a status yet.

Table 1 shows the rules for the scenario. In the BBGP model, there are four types of rules, each type for each stage. Only the activation and evaluation rules are depicted because the rules for deliberation and checking stages are fixed since they are domain-independent[2].

---

[2] For more details about the BBGP model, the reader is referred to [10].

Table 1. Activation and Evaluation Rules.

Activation Rules	Evaluation Rules
$\langle liquid\_dirt(x,y) \rightarrow mop(x,y)\rangle$	$\langle \neg have(battery,40) \wedge liquid\_dirt(x,y) \rightarrow \neg mop(x,y)\rangle$
$\langle solid\_dirt(x,y) \rightarrow sweep(x,y)\rangle$	$\langle have(cargo,80) \wedge liquid\_dirt(x,y) \rightarrow \neg mop(x,y)\rangle$
$\langle \neg have(battery,15) \rightarrow recharge\rangle$	$\langle broken(mop) \wedge liquid\_dirt(x,y) \rightarrow \neg mop(x,y)\rangle$
	$\langle \neg have(battery,30) \wedge solid\_dirt(x,y) \rightarrow \neg sweep(x,y)\rangle$
	$\langle have(cargo,90) \wedge solid\_dirt(x,y) \rightarrow \neg sweep(x,y)\rangle$

The **preference** order of goals is exactly the activation rules order, in decreasing order. That is, $mop(x,y)$ has the highest preference and $recharge$ has the lowest one.

Let us recall that the first phase of the method, that is, Interface, requires four elements to be implemented: the execution history, the causal function, the preference function, and the conflict function. For our scenario, the agent's execution history is depicted in Table 2, where each row is an entry of the log. The log allows for the reconstruction of the agent knowledge base and tracks the goal status changes.

Table 2. Execution History.

Cycle	Changed Goals	Changed Beliefs
0	$\emptyset$	$\emptyset$
1	$\langle mop(2,2), Executive\rangle$ $\langle mop(3,2), Pursuable\rangle$ $\langle sweep(3,4), Pursuable\rangle$	$\langle broken(mop), \text{ADD}\rangle, \langle liquid\_dirst(2,2), \text{REM}\rangle,$ $\langle have(battery,75), \text{REM}\rangle, \langle habe(battery,35), \text{ADD}\rangle,$ $\langle have(cargo,65), \text{REM}\rangle, \langle have(cargo,85), \text{ADD}\rangle,$ $\langle at(2,2), \text{ADD}\rangle, \langle at(1,1), \text{REM}\rangle$
2	$\langle sweep(3,4), Executive\rangle$ $\langle mop(3,2), Active\rangle$	$\langle solid\_dirt(3,4), \text{REM}\rangle,$ $\langle have(battery,35), \text{REM}\rangle, \langle have(battery,5), \text{ADD}\rangle,$ $\langle have(cargo,85), \text{REM}\rangle, \langle have(cargo,95), \text{ADD}\rangle,$ $\langle at(3,4), \text{ADD}\rangle, \langle at(2,2), \text{REM}\rangle,$
3	$\langle recharge, Executive\rangle$	$\langle have(battery,5), \text{REM}\rangle, \langle have(battery,100), \text{ADD}\rangle,$ $\langle at(1,1), \text{ADD}\rangle, \langle at(3,4), \text{REM}\rangle$

The Causal Function: The activation and evaluation rules can be easily converted into causes by attributing the consequent (the goal) as e and the set of beliefs in the antecedent as COND. Table 3 shows the mapping generated. The BBGP model has two more stages: deliberation and checking. The deliberation has no rules to be converted, as it evaluates conflicts and preferences among goals, it will be encoded by the conflict and preference functions.

The Preference Function: In the example, the model defines a preference, which was described with the scenario, as such, using the execution history, the total (decreasing) preference relation for the scenario is as follows: $mop(2,2) > mop(3,2) > sweep(3,4) > recharge$.

The Conflict Function: Given the nature of the scenario and the defined goals, every goal is incompatible with each other. This is the case because the goals are

Table 3. Mapping of the Activation and Evaluation rules to causes.

Input	Causes
$mop(x, y)$	$\langle mop(x, y), \{liquid\_dirt(x, y)\}\rangle$
$\neg mop(x, y)$	$\langle \neg mop(x, y), \{\neg have(battery, 40), liquid\_dirt(x, y)\}\rangle$
	$\langle \neg mop(x, y), \{\neg have(cargo, 80), liquid\_dirt(x, y)\}\rangle$
	$\langle \neg mop(x, y), \{broken(mop), liquid\_dirt(x, y)\}\rangle$
$sweep(x, y)$	$\langle sweep(x, y), \{solid\_dirt(x, y)\}\rangle$
$\neg sweep(x, y)$	$\langle \neg sweep(x, y), \{have(cargo, 90), solid\_dirt(x, y)\}\rangle$
	$\langle \neg sweep(x, y), \{have(battery, 30), solid\_dirt(x, y)\}\rangle$
$recharge$	$\langle recharge, \{\neg have(battery, 15)\}\rangle$
	$\langle recharge, \{\neg have(battery, 80)\}\rangle$

localized in space, and the robot needs to be at the specified place to perform the actions to achieve the goal.

4.2   The Conditions for a P-Contrast Explanation

We can now present how the conditions for generating a P-contrast explanation are constructed. Let us start by posing a question: *Why $mop(3, 2)$ is active, instead of executive?* The rationale of this question is very straightforward: why a goal is not being executed? First, let us identify the information in the question:

- Goal: $mop(3, 2)$
- Status: Active
- Foil status: Pursuable (from Executive)
- Cycle id: 2

Notice that in the posed question, the contrasting statuses are Active and Executive. In this case, Executive is converted to Pursuable, a previous case from Executive and directly subsequent to Active because the goal did not even became Pursuable so we have to know why this happened. Since the question requires a single time frame, which is cycle 2, the knowledge base for that cycle can be reconstituted using the execution history by adding and removing the beliefs for each cycle to the initial beliefs, until the desired cycle. We have that the relevant beliefs are:

$workshop(1, 1)$
$dumpster(4, 1)$
$\neg available(mop)$
$at(2, 2)$
$have(battery, 35)$
$solid\_dirt(3, 4)$
$liquid\_dirt(3, 2)$
$broken(mop)$

Related Causes Procedure: First, it is necessary to evaluate the relation of the statuses. Be $st = active$ the factual status and $st' = pursuable$ the foil status such that $st \prec st'$, as such the relevant patterns are for "not advanced its status".

For an active goal become pursuable, it is assessed in the evaluation stage, that is, the second stage of the BBGP model. In the case of evaluation stage, it uses only negative filters and the way of calculating this filter is: $d\_pre(causes (\neg mop(3,2))) \cap \mathcal{KB}^+$.

For this scenario $\mathcal{KB}^+ = \mathcal{KB}$ because there are no rules with only beliefs that may infer another beliefs. Thus, the calculation is as follows for the negative pattern:

First, the set of causes for $(mop(3,2)$ is calculated. For simplicity, we only write the set of beliefs that are part of COND:

$$causes(mop(3,2)) = \{\{\neg have(battery, 40), liquid\_dirt(3,2)\},$$
$$\{have(cargo, 80), liquid\_dirt(3,2)\}$$
$$\{broken(mop), liquid\_dirt(3,2)\}\}$$

Then, the disjunctive presumptions are calculated:

$$d\_pre(causes(mop(3,2))) = \{\neg have(battery, 40), liquid\_dirt(3,2),$$
$$have(cargo, 80), broken(mop)\}$$

Finally, we obtain the set of conditions:

$$RC = d\_pre(causes(mop(3,2))) \cap \mathcal{KB}^+ = \{\neg have(battery, 40), liquid\_dirt(3,2),$$
$$have(cargo, 80), broken(mop)\}$$

This concludes with the procedure whose output is the set of conditions RC, which constitute future explanations.

5   Related Work

In this section, some articles about contrastive explanations generation and explanations for goal selections are presented. Besides, we compare them with our proposal.

To the best of our knowledge, there are few works where contrastive explanations are used in the context of intelligent agents. In [13], the authors propose an approach for generating contrastive explanations for questions of the for "Why P and not Q?", where P and Q are stand-ins for the current robot plan and the foil one, respectively. An algorithm for generating minimal explanations for users of different levels of expertise is presented. The result is that agents can explain its ongoing or planned behavior in a way that is both tailored to the user's background and is designed to reduce cognitive burden on the user's end. In [3], the authors propose an approach for generating contrastive explanations about agents' preferences in the context of reinforcement learning. The explanations identify the state feature values that certain agents prefers compared to the other agent. In [12], the work focus on generating contrastive explanations in planning, more specifically, for questions such as "Why action A instead of action B?". Thus, the contrastive explanation that compares properties of the original plan containing A against the contrastive plan containing B. We can notice that the contrastive questions tackled in these works have a different nature compared with the P-contrast question of our approach. In our approach, we consider an object of the explanation and a property about it, which gives one more dimension to the generated explanation.

Regarding explanation generation for goal selection. To the best of our knowledge, only [11] and [10] present an explanation method for goal selection. Their approach also considers BBGP-based agents and uses the formal argumentation base for generating explanations. However, there is no a social sciences grounding and it does not account for contrastive explanations. We can see our approach as a complement to these works.

6   Conclusions and Future Work

This article presented a method for generating P-contrastive explanations about the BBGP-based goal selection process. A set of beliefs – called conditions – is returned to the questioner and is the base for future explanations. Although, we present the phases and all the steps for generating a final explanation, the final procedure was not developed; however, all the building blocks for generating an explanations were established. We illustrate how our method works by applying it to the scenario of the cleaner world.

Let us recall that our method used Grice's Cooperation Principles as a guide for how the agent should present its explanations, specifically, we base on the quantity category. This category is concerned with the amount of information provided: "be as informative as required, and no more than that". It is subjective how informative a piece of information is. Not just in the sense that it depends on who receives the information, but it is also not measurable. In a sense, this is also part of the selection problem: how to select the most adequate answer?. We could say that the "most informative" explanations could be a solution to the problem. The method addresses this requirement by avoiding repetitive information, that is, if a condition c_1 can answer all the required events, there is no need to include a second condition.

As mentioned above, an important steps of the method were not tackled in this work. Thus, the immediate future work is to generate the set of possible explanations. Besides, we plan to study the selection problem, which focuses on deciding which single explanation will be returned as final answer.

As presented in Introduction, there are other types of contrastive question and explanations. We aim to generate the set of conditions and explanations for these types of explanations. Finally, another possible improvement is to include special questions that can better explain internal behaviors of the agent that are beyond the standard BBGP model. A good example would be a special question type to explain trust deliberation in an agent.

References

1. Bratman, M.: Intention, Plans, and Practical Reason (1987)
2. Castelfranchi, C., Paglieri, F.: The role of beliefs in goal dynamics: Prolegomena to a constructive theory of intentions. Synthese **155**, 237–263 (2007)
3. Gajcin, J., Nair, R., Pedapati, T., Marinescu, R., Daly, E., Dusparic, I.: Contrastive explanations for comparing preferences of reinforcement learning. In: AAAI Conference on Artificial Intelligence (2022)
4. Grice, H.P.: Logic and conversation. In: [Cole, P., Morgan, J.L. (eds.) Syntax and Semantics, vol. 3, Speech Acts (1975)]
5. Harbers, M., van den Bosch, K., Meyer, J.J.: Design and evaluation of explainable BDI agents. In: 2010 IEEE/WIC/ACM International Conference on Web Intelligence and Intelligent Agent Technology, vol. 2, pp. 125–132. IEEE (2010)
6. Josephson, J.R., Josephson, S.G.: Abductive Inference: Computation, Philosophy, Technology. Cambridge University Press (1996)
7. Kaptein, F., Broekens, J., Hindriks, K., Neerincx, M.: The role of emotion in self-explanations by cognitive agents. In: 2017 Seventh International Conference on Affective Computing and Intelligent Interaction Workshops and Demos (ACIIW), pp. 88–93. IEEE (2017)
8. Miller, T.: Contrastive explanation: a structural-model approach. Knowl. Eng. Rev. **36**, e14 (2021)
9. Miller, T., Howe, P., Sonenberg, L.: Explainable AI: beware of inmates running the asylum or: how i learnt to stop worrying and love the social and behavioural sciences. arXiv preprint arXiv:1712.00547 (2017)
10. Morveli-Espinoza, M., Nieves, J.C., Tacla, C.A., Jasinski, H.M.: An argumentation-based approach for goal reasoning and explanations generation. J. Logic Comput. **105**, 1–26 (2022)
11. Morveli-Espinoza, M., Possebom, A.T., Tacla, C.A.: Argumentation-based agents that explain their decisions. In: 2019 8th Brazilian Conference on Intelligent Systems (BRACIS), pp. 467–472. IEEE (2019)
12. Sarwar, M., Ray, R., Banerjee, A.: A contrastive plan explanation framework for hybrid system models. ACM Trans. Embed. Comput. Syst. **22**(2), 1–51 (2023)
13. Sreedharan, S., Srivastava, S., Kambhampati, S.: Using state abstractions to compute personalized contrastive explanations for AI agent behavior. Artif. Intell. **301**, 103570 (2021)
14. Stange, S., Kopp, S.: Effects of a social robot's self-explanations on how humans understand and evaluate its behavior. In: Proceedings of the 2020 ACM/IEEE International Conference on Human-Robot Interaction, pp. 619–627 (2020)
15. Stepin, I., Alonso, J.M., Catala, A., Pereira-Fariña, M.: A survey of contrastive and counterfactual explanation generation methods for explainable artificial intelligence. IEEE Access **9**, 11974–12001 (2021)

H. Jasinski et al.

16. Van Bouwel, J., Weber, E.: Remote causes, bad explanations? J. Theor. Soc. Behav. **32**(4), 437–449 (2002)
17. Winikoff, M.: Debugging agent programs with "why?" questions (2017)
18. Wooldridge, M.: Reasoning About Rational Agents. MIT Press (2003)

Applying Theory of Mind to Multi-agent Systems: A Systematic Review

Michele Rocha[1], Heitor Henrique da Silva[1], Analúcia Schiaffino Morales[1],
Stefan Sarkadi[2], and Alison R. Panisson[1(✉)]

[1] Department of Computing, UFSC, Florianópolis, Brazil
{analucia.morales,alison.panisson}@ufsc.br
[2] Department of Informatics, KCL, London, UK
stefan.sarkadi@kcl.ac.uk

Abstract. Life in society requires constant communication and coordination. These abilities are efficiently achieved through sophisticated cognitive processes in which individuals are able to reason about the mental attitudes and actions of others. This ability is known as Theory of Mind. Inspired by human intelligence, the field of Artificial Intelligence aims to reproduce these sophisticated cognitive processes in intelligent software agents. In the field of multi-agent systems, intelligent agents are defined not only to execute reasoning cycles inspired by human reasoning but also to work similarly to human society, including aspects of communication, coordination, and organisation. Consequently, it is essential to explore the use of these sophisticated cognitive processes, such as Theory of Mind, in intelligent agents and multi-agent systems. In this paper, we conducted a literature review on how Theory of Mind has been applied to multi-agent systems, and summarise the contributions in this field.

Keywords: Artificial Intelligence · Multi-Agent Systems · Theory of Mind · Systematic Review

1 Introduction

Theory of Mind (ToM) refers to the ability of individuals to attribute mental states to themselves and others, including beliefs, desires, intentions, and emotions. It is an important cognitive skill that enables us to understand and predict the behaviour of others. Additionally, it plays a fundamental role in communication and social interaction [24].

A Multi-Agent System (MAS) is a system composed of multiple autonomous agents that interact with each other to achieve a common goal. These agents can be humans or software, and the system's goals are accomplished through collaboration, negotiation, and coordination among them [68]. The conceptualisation of MAS takes inspiration from by human society, and considers approaches of modelling not only autonomous intelligent agents as individuals, but the aspects concerning the societies made up of these agents. Furthermore, the field of MAS also explores hybrid systems, where humans are considered as part of the MAS, which aligns MAS with emerging concepts such as hybrid intelligence [1].

M. C. Naldi and R. A. C. Bianchi (Eds.): BRACIS 2023, LNAI 14195, pp. 367–381, 2023.
https://doi.org/10.1007/978-3-031-45368-7_24

Although there have been fascinating achievements in the field of MAS, including models for describing and implementing sophisticated reasoning mechanisms and communication methods, there has always been a continuous exploration of incorporating human cognitive capabilities, such as ToM, within the field of MAS. By integrating ToM into autonomous intelligent agents and, consequently into MAS composed of these agents, there is a promise of achieving sophisticated capabilities that were previously seen only in humans, human interactions, and human societies. This integration also opens up new possibilities for systems where software agents and humans can work together synergistically, leading to the development of systems that align with the concept of hybrid intelligence [1].

In this paper, we present a systematic review which aims to give a brief, but comprehensive, account of the state-of-the-art in applying ToM in the area of MAS. Our goal is to identify what is lacking to achieve the ambitious vision of an AI-human society. It is crucial to evaluate current approaches and techniques in artificial and cognitive intelligence, as well as to address the limitations and challenges involved in reaching these goals.

There are four sections in the paper. Section 2 presents an overview of the systematic review methodology, including the identification of research gaps, the search strategies used, and the selection and exclusion criteria used for articles. In Sect. 3, the findings are discussed in relation to the specific points of interest outlined earlier. Finally, the paper concludes with a few final remarks and references.

2   Methodology

In order to conduct the systematic review, Preferred Reporting Items for Systematic Reviews and Meta-Analyses (PRISMA) guidelines [41] were adhered to. The objective of the study was to fill a few research gaps regarding ToM and MAS employing the SPIDER as a strategy defining the following aspects [15]: **Sample (S)**, **Phenomenon of Interest (P)**, **Design (D)**, **Evaluation (E)**, and **Research type (R)**. This strategy is useful for systematic reviews involving studies of different designs and types of interventions. Below, each aspect for this strategy is detailed:

- **Sample**: Studies focused on the ToM and MAS context.
- **Phenomenon of Interest**: Theoretical and empirical research on applying ToM to MAS.
- **Interest/Information Source**s: Peer-reviewed articles, conference proceedings, and book chapters published in English, available online.
- **Design**: Qualitative and quantitative studies, theories, and case studies.
- **Evaluation**: Studies that examine the relationship between ToM and MAS, or applied ToM to MAS.
- **Research type**: Primary research studies, including experimental and non-experimental designs. And the applications areas (health, education, games, personal assistant).

The following research questions were used to identify such gaps:

(RQ1) What theory underlies the use of the theory of mind? What is the reference work? (here, we may have multiple works grounded in the same theory, which could be from psychology, philosophy, etc.).

(RQ2) Which mental attitudes are modelled by the work? (for example beliefs, memories, intentions, emotions, etc.).

(RQ3) How does the system model knowledge related to the theory of mind? What is the form of representation for information about the theory of mind?

(RQ4) Does the work model Theory of Mind of human users or only computational entities (software agents)?

(RQ5) What is the application area of the work? (for example, health, games, personal assistants, theoretical works, etc.)

(RQ6) What technologies were used to develop the work? (for example, modelling and simulation platforms, algorithms, AI techniques, etc.)

(RQ7) What are the main results achieved?

(RQ8) What are the main challenges encountered?

(RQ9) What are the limitations pointed out by the work?

2.1 Databases

Four databases were utilised for this study, each of which was carefully selected based on their relevance to our research topic. Specifically, we focused on databases that were most closely aligned with the scope of our study, and that had a proven track record of returning a significant number of relevant papers. Our aim was to ensure that we had access to the most comprehensive and up-to-date literature on the subject, in order to provide a rigorous and thorough analysis. In this regard, the databases we selected were instrumental in providing us with the necessary information to carry out our review. We examined PubMed, Web of Science, Scopus, and IEEE Xplore databases. Grey literature is not considered in this study.

Table 1. Inclusion and Exclusion Criteria.

Stage	Inclusion Criteria	Exclusion Criteria
Selection	Search terms at title	Short papers (shorter than 4 pages)
	Search terms at abstract	Workshop papers
	Search terms at keywords	Out of scope (outside of computer science)
Reading	Papers about ToM and MAS	Out of scope (outside of computer science)
	Paper that answered (partially) the RQs	Papers that did not applied/modelled ToM in software agents

2.2   Adopted Criteria and Selection Procedures

This study employed an accurate search approach, utilising four electronic databases (see Sect. 2.1) to identify relevant scientific articles on ToM and MAS. After the search process, it was applied a two-stage selection process. In the first stage, the researchers screened the titles, abstracts, and keywords of each article to identify potentially relevant studies. Subsequently, studies that did not meet the inclusion criteria were excluded from the analysis. In cases where there was uncertainty during the initial evaluation, a second evaluation of the full text was conducted, in which we were able to evaluate if those studies were part of the scope of this literature review. The inclusion and exclusion criteria for all studies are outlined in Table 1.

Across all databases, a string with the following keywords was investigated.

```
( multiagent systems OR multi-agent systems OR autonomous agents
  OR personal assistants OR chatbots OR chatterbots
  OR distributed artificial intelligence )
AND ( theory of mind ).
```

Note that this search string focus on searching articles that use synonyms for 'agent' in the context of software agents, but also including the concept of 'Theory of Mind' specifically, which is the interest of this study.

2.3   Selection Process

The selection process began with 698 articles, including 144 from PUBMED, 184 from WoS, 311 from Scopus database, and 58 from IEEE Xplore. Out of the 698 articles, 126 were removed due to duplication. From the remaining 572 articles, 456 were excluded in the initial stage as they did not meet the selection criteria. Five reviewers participated in the first stage, with each assigned to read the title and abstract of all 572 articles that remained after removing the duplicates.

The review was conducted in a blinded manner using the Rayyan tool[1] to minimise the influence of individual researchers' decisions. Subsequently, the blinding review was removed, allowing for a comparison of the researchers' decisions. In this stage, a set of rules was established to facilitate the selection process. Articles receiving three or more votes for inclusion were selected for full reading, while those receiving four or more votes for exclusion were directly excluded. In cases where conflicting votes occurred, the reviewers convened to make a final decision. Applying these criteria, a total of 116 articles were selected for full reading and subsequently categorized into a separate section.

With the final set of 116 articles for full reading, each researcher was tasked with answering the questions created before the reading. In this part of the research, three researchers participated. After reading the articles, 22 were excluded because they were out of scope (mostly related to the study of ToM in humans), 16 were excluded because they were only published in workshops,

---

[1] https://www.rayyan.ai/.

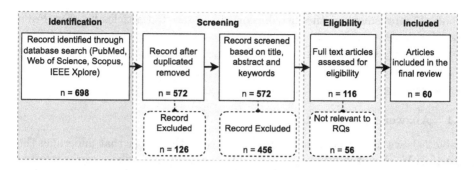

Fig. 1. PRISMA Diagram.

11 were excluded because they were short papers (extended abstracts), 4 were excluded because they were books, 2 articles were excluded because they could not be found online for free, and 1 article was excluded because it was a survey.

Figure 1 presents the PRISMA diagram of the systematic review, which illustrates the eligible investigation process.

2.4 Selected Articles

In the end, 60 articles remained and were included in this study. The selected studies were published between 2002 and 2022. We observe an increase in the popularity of the topic from 2019 to 2022. However, there is a slight decrease in publications from 2020 to 2022, which could be attributed to external factors, such as the COVID-19 pandemic. The pandemic could have limited the participation of researchers in events and access to university and research labs as a

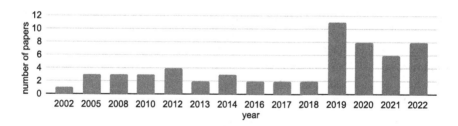

Fig. 2. Papers by year of publication.

Table 2. Main references used by the papers.

Main Reference	References
Premack and Woodruff (1978) [46]	[4, 5, 8, 11, 16–19, 30, 32, 39, 45, 62]
Baron-Cohen and Leslie (1985) [6, 7]	[3, 27, 28, 31, 56, 57, 69]
Whiten and Byrne (1991) [67]	[22, 26, 38, 48–51, 59]
Goldman (2012) [24]	[42, 54, 63]

whole. Figure 2 displays the distribution of the selected articles by year. From the 60 selected articles, 26 of them are from academic journals and 34 from conference proceedings.

3   Results

3.1   Answer to RQ1 (Main References)

Table 2 shows those works that were referenced as the theory that underlines the use of ToM, or as the primary reference to the work. Only those works referred to by multiple sources are included. Furthermore, in 25% of the articles included, the main reference could not be identified since the authors did not provide this information explicitly. Of the remaining articles (approximately 19%), while they did provide this information, none of those main references were referenced by multiple articles.

3.2   Answer to RQ2 (Modelled Mental Attitudes)

Table 3 summarises the mental attitudes (or states) modelled by the selected articles. It is notable that most of the research focuses on mental attitudes commonly present in agent architectures, including the well-known BDI (Beliefs-Desires-Intentions) architecture [10]. Additionally, some works focus on modelling memories (which are slightly stronger than beliefs), emotions, and goals.

Table 3. Mental attitudes explicitly modelled.

Mental Attitude	References
Beliefs	$[3–5, 8, 9, 11–14, 16–22, 25–28, 30, 31, 33, 35, 36, 38, 40, 42, 44, 45, 47–49, 51, 52, 54, 55, 57–66, 69]$
Desires	$[9, 12, 13, 16, 26–28, 31, 40, 42, 44, 51–55, 57, 61, 63, 64, 69]$
Intentions	$[2, 5, 12, 13, 16, 20, 22, 26–29, 32, 34, 39, 40, 42, 44, 45, 51, 52, 54, 55, 58, 60–66]$
Memories	$[20, 50, 56, 61, 65]$
Emotions	$[22, 23, 52, 65]$
Goals/Motivations	$[9, 14, 38, 57]$

Table 4. Knowledge representation.

Knowledge Representation	References
Formal (abstract) Representation	$[2, 4, 5, 8, 9, 12, 13, 16–21, 29, 31–36, 40, 42, 44, 45, 47, 50, 52–55, 57–59, 61–65, 69]$
Predicates (FOL, LISP, PDDL)	$[3, 11, 13, 14, 19, 25–28, 33, 40, 42, 51, 54, 55, 65]$
Propositional Logic and Text	$[22, 23, 30, 64, 66]$
Ontologies	$[56]$
Vector of characteristics	$[38, 39, 48, 49, 60]$

3.3 Answer to RQ3 (Knowledge Representation)

Table 4 summarises the distribution of the selected articles according to how they represent the knowledge related to the ToM. A formal representation is presented in approximately 65% of the works, including epistemic logic, Bayesian modeling, state transition systems, and MDPs. In addition, many of the works utilise symbolic representations, which are similar to agent-oriented programming languages. Furthermore, some works use a representation based on vectors of characteristics, which is typically used in machine learning techniques. Finally, there are works that use natural language representation, and one work differs from others by using semantic databases (ontologies) to represent ToM.

3.4 Answer to RQ4 (Modelled Entities)

Table 5 summarises the distributions of the selected articles based on whether they modelled ToM for other software agents, for humans (e.g., human users), or for both. Most of the research focuses on modelling ToM for other software agents and studying the diverse phenomena that can emerge from agents with this capability. Some works focus on modelling ToM for both software agents

Table 5. Entities modelled.

Entity	References
Humans	[4, 13, 19, 23, 27, 28, 32, 39, 56, 60, 62]
Agents	[2, 3, 5, 9, 12, 14, 16, 18, 20, 21, 25, 30, 31, 33–36, 38, 40, 44, 45, 47–53, 58, 59, 61, 63–66, 69]
Both	[8, 11, 17, 22, 26, 29, 42, 54, 55, 57]

Table 6. Application areas.

Application Area	References
General	[2, 12, 25, 31, 35, 36, 40, 42, 44, 52, 54, 58, 61, 64]
Dishonest Attitudes	[55]
Personal Assistants	[13, 26, 29, 49]
Games	[4, 5, 11, 14, 17, 21, 28, 33, 34, 45, 62, 63, 66]
Negotiation	[16–18, 50, 60]
Robotics	[19, 20, 56]
False Beliefs	[8, 69]
Cooperative Agents	[65]
Education	[3]
Human-Agent Interaction	[23, 30, 32, 39, 53, 56, 57, 59]
Agent-Based Simulation	[9, 22, 27, 38, 47–49, 51]

and humans, often introducing abstract approaches and logic that can incorporate the modelling and reasoning about ToM. Finally, there are some works that focus on modelling ToM for humans, particularly related to personal assistants. In these works, an software agent aims to improve its decision-making and interaction by using ToM.

3.5   Answer to RQ5 (Application Areas)

Table 6 summarises the distributions of the selected articles according to their application domains. It can be noted that most of the research focus on providing approaches without any particular application domain, providing theoretical proofs and definitions over the proposed logic or framework.

Regarding the works that apply the proposed approach and explore applications, there is a notable interest in the areas of games, human-agent interaction and agent-based simulations. In the area of games, most works explore how ToM can provide an advantage to agents with this capability, studying whether they outperform agents without this capability. Additionally, some works focus on providing a more realistic experience for players and/or improving the naturalness of Non-Playable Characters (NPCs), with some also contrasting human-agent interaction in games. In the area of human-agent interaction, most works explore how ToM can improve the software agent communication by anticipating humans thoughts, intentions, etc. and remembering important information such as their beliefs, goals, and preferences. In the area of agent-based simulation, most works focus on studying social phenomena in which individuals model ToM. Those studies include emergent behaviours, agents profiles, and other related factors.

Furthermore, there are works focusing on how ToM can be applied to: (i) identify false beliefs, (ii) deceive others (and similar dishonest behvaiours), (iii) train and/or implement better negotiators, (iv) achieve cooperation, (v) achieve better results in educational platforms, (vi) infer mental attitudes applied to robotics, and (vii) create better personal assistants.

Table 7. Main technologies.

Technology	References
AOP Languages and Platforms	[12,13,55] (JaCaMo), [26] (2APL), [26] (JADE)
Machine Learning (NN, Deep Learning, etc.)	[5,18,23,33,34,40,45,50,58,61]
Probabilistic Approaches (Bayesian, (PO)MDPs, etc.)	[2,8,16,17,19,21,23,23,31,32,39,59,63]
2D/3D Simulation Platforms, or Tools with Interfaces	[3,4,11,19,20,22,23,25,28,30,38,39,47–49,51,60,62,64,65,69]
Planning	[44,66]
Robotics and Software Integration (ROS, etc.)	[56,57]
Formal	[9,27,35,36,42,54]

3.6 Answer to RQ6 (Technologies)

Table 7 presents the main technologies employed in the selected articles for this literature review. It is evident that the majority of the works concentrate on studies that necessitate a human-computer interface, enabling interaction between humans and agents, or showcasing emergent behaviours through agent-based simulations. Additionally, a significant number of articles employ probabilistic approaches and machine learning as their primary technologies. A few studies employ robotics, planning, and agent-oriented programming languages as their main technologies. Lastly, some works remain focused on theoretical aspects without utilising any specific technology.

3.7 Answer to RQ7 (Main Results)

Most of the selected works focus on providing formal properties, semantics, and definitions regarding their approaches for modelling and reasoning with ToM [11, 13,25,35,36,42,51,54,55,69] or aim to provide proof of concepts [3,9,14,16–19, 22,28–30,39,44,45,47–49,62–64]. For example, some studies demonstrate that ToM improves the performance of agents, especially when they have access to more information. They also show that agents with ToM outperform those who are not able to model and reason about ToM [16,63].

Furthermore, many works aim to provide a computational model for ToM, which can address various challenges, such as (i) dealing with uncertainty [21, 32,54]; (ii) handling preferences [13]; (iii) based on machine learning models [34, 40,50,65]; (iv) using planning [58,66]; (v) incorporating emotions [23,52,53,59]; (vi) using ontologies [56]; among others [2,8,12,20,26,27,31,38,57,61].

Finally, one work provides a datasets as one of the main results [4]. Some works provide empirical results, such as comparing the ability of humans to model ToM with those modeled by agents [5], comparing the performance of agents with and without ToM [33,60], among others [50].

3.8 Answer to RQ8 (Main Challenges)

More than half of the selected papers do not point out any challenges in developing their approaches. Most of the works that did mention challenges highlight the difficulty of dealing with the dynamism of real-world environments, which also creates a highly dynamic mental state of agents. For instance, in [4], the authors describe how changes in the physical world, which is very dynamic, affect agents' beliefs, consequently the ToM they have about each other, making it a challenging issue. In [18], the authors also describe the unpredictability of the environment as one of the main challenges. In [20], the authors explain the challenge of robots representing knowledge about their interactions with the environment and other agents. In [23], the authors discuss about the difficulty of capturing real-time gestures from human users and creating a more general model applicable to other domains. In [31], the authors emphasise the complexity of applying achieved results in real-world scenarios. In [12], the authors describe challenges associated with selecting the correct plan agents should use to achieve user-delegated goals, considering the dynamics of the environment and their individual state. Also, in [39], the authors state that scalability is one of the main challenges, making it hard to consider more factors to make inferences related to ToM. Further, in [47], the authors describe the great challenge of including humans in the loop when considering ToM.

Furthermore, in [2], the authors describe how challenging it can be for agents to decide on initial interactions when their behaviors are unknown. In [25], the authors point out the difficulty of developing cognitive models capable of making inferences about "common sense". In [36], the authors describe the challenge of finding a balance between representing the explicit and implicit beliefs of others. In [22], the authors find it challenging to collect and use knowledge from crowdsourcing and to evaluate different narratives based on the diversity of phrases, actions, and sentiments. In [57], the authors describe the challenge of implementing complex social capabilities in humanoid robots. In [64], the authors highlight the difficulty of extending the proposed approach to real and more complex case studies, such as legal cases. In [52], the authors describe the challenge of integrating different emotional and cognitive theories into a single agent architecture. In [21], the authors describe the challenge of learning parameters from collected data and exploring different learning techniques. In [60], the authors describe the difficulty of training and measuring skills related to modeling and reasoning about ToM.

Finally, in [11], the authors point out the inadequacy of mathematical frameworks from computer science to model ToM. In [63], the authors describe that higher-order ToM only gains an advantage over lower-order ToM or other simple strategies when enough information is available, which is often hard to achieve in many scenarios. In [33], the authors describe the difficulty of dealing with wrong and uncertain ToM. Lastly, in [66], the authors describe the challenges related to debugging the system, particularly in their case, when debugging narrative planning problems.

3.9   Answer to RQ9 (Main Limitations)

The works that applied subjective or quantitative evaluations supported by human users, such as [13,28], indicated that the number of human users in the experiments was limited. Also, some works have described the need to extend their approach with other mental attitudes and knowledge representations. For instance, in [19], the authors describe that they intends to extend their work to model desires and intentions in addition to beliefs, while in [66], the authors explain that their approach does not allow the representation of uncertainty, which may be necessary for certain applications.

Furthermore, in [58], the authors note that their approach only explored a limited number of agents. In [23,45], the authors describe the specificity of their approaches as limitations. In [23], the authors highlight the difficulty of learning the user's model, while in [36], the authors discuss inconsistencies that can arise from using propositional logic in their approach. Additionally, in [31], the authors acknowledge that their simplified model for agents and environment could not capture all complexities of social interaction as they occur. Moreover, in [39], the authors state that their approach is limited to inferring the intents of only one user and considers a limited number of intents. In [22,30], the authors describe limitations regarding the technologies used, such as robotics and crowdsourcing. In [21], the authors describe limitations regarding the data used and consequently the generated model. In [63], the authors mention that they only explore two levels of ToM, which could be a limitation of their work. In [33,44], the authors assume perfect knowledge about other agents' policies, which could also be a limitation of their work. In [25], the authors describe that they intend to explore other formal properties based on possible worlds.

Many articles also point out common limitations related to validation and evaluation methods, such as [12,17,28,48,52,53,60]. They often highlight limitations resulting from (i) using controlled environments for experiments that may not reflect the real world, and (ii) using a limited number of parameters and/or fixed instances of the problem, among others. Finally, note that the majority of the articles do not mention limitations (their limitations, if any, could not be accessed for this literature review).

4   Conclusion

In this paper we have presented a short and systematic literature review of the approaches in MAS used to apply ToM, that is the concept from Cognitive Science understood as the ability of humans to model the minds of other agents. As future work, we plan to situate these approaches w.r.t. the 'flavours' of ToM in Cognitive Science [24].

There seems to be a representational equivalence between the flavours of ToM from Cognitive Science and types of AI modelling approaches. For instance *Theory-Theory* of Mind (TT) would be the equivalent of symbolic AI, *Simulation Theory* of Mind (ST) of subsymbolic AI, and the *Hybrid Theory* of Mind would be something similar to Neurosymbolic AI. While TT represents the type of ToM where agents already have an understanding to some extent of the other agents' minds, e.g., a set or a knowledge base that consists of beliefs of others' mental attitudes, ST represents a process for using the already known set of others' mental attitudes to simulate their minds in order to predict their behaviour. HT is a 'high-level' ToM that combines TT and ST in a practical reasoning process in the way its target (the other agent who's mind is being mentalised) would.

Perhaps the most obvious application of HT in AI is to enable language, communication, and explanation [37], a component that is foundational for MAS where agents must communicate in order to interact successfully with other artificial agents, or humans. This could enable future research directions in MAS, in which simulation of intelligent agents are showed to humans, and humans are asked to estimate their mental attitudes, that means, humans building a ToM about the mental attitudes of intelligent agents [43].

However, as we have seen from answering the literature review questions in this paper, there are quite a few limitations regarding the application of ToM in

MAS, and quite a few challenges still remain to be addressed in order for ToM in MAS to showcase the promises regarding realistic human-like mentalisation in agent-agent interactions. Furthermore, there might be even more limitations which more than half of the papers that presented applications of ToM to MAS did not even consider to report.

References

1. Akata, Z., et al.: A research agenda for hybrid intelligence: augmenting human intellect with collaborative, adaptive, responsible, and explainable artificial intelligence. Computer **53**(08), 18–28 (2020)
2. Albrecht, S.V., Crandall, J.W., Ramamoorthy, S.: Belief and truth in hypothesised behaviours. Artif. Intell. **235**, 63–94 (2016)
3. Aylett, R., et al.: Werewolves, cheats, and cultural sensitivity. In: International Conference on Autonomous Agents and Multi-agent Systems (2014)
4. Bara, C.P., CH-Wang, S., Chai, J.: MindCraft: theory of mind modeling for situated dialogue in collaborative tasks. arXiv preprint arXiv:2109.06275 (2021)
5. Bard, N., et al.: The Hanabi challenge: a new frontier for AI research. Artif. Intell. **280**, 103216 (2020)
6. Baron-Cohen, S.: Mindblindness: An Essay on Autism and Theory of Mind. MIT Press (1997)
7. Baron-Cohen, S., Leslie, A.M., Frith, U.: Does the autistic child have a "theory of mind?". Cognition **21**(1), 37–46 (1985)
8. Berthiaume, V.G., Shultz, T.R., Onishi, K.H.: A constructivist connectionist model of transitions on false-belief tasks. Cognition **126**(3), 441–458 (2013)
9. Boella, G., Van der Torre, L., et al.: From the theory of mind to the construction of social reality. In: Proceedings of the Annual Conference on the Cognitive Science Society, vol. 5, pp. 298–303 (2005)
10. Bratman, M.: Intention, Plans, and Practical Reason. Harvard University Press, Cambridge (1987)
11. Bringsjord, S., et al.: Toward logic-based cognitively robust synthetic characters in digital environments. Front. Artif. Intell. Appl. **171**, 87 (2008)
12. Cantucci, F., Falcone, R.: Towards trustworthiness and transparency in social human-robot interaction. In: 2020 IEEE International Conference on Human-Machine Systems (ICHMS), pp. 1–6. IEEE (2020)
13. Cantucci, F., Falcone, R.: Collaborative autonomy: human-robot interaction to the test of intelligent help. Electronics **11**(19), 3065 (2022)
14. Chang, H.M., Soo, V.W.: Simulation-based story generation with a theory of mind. In: Proceedings of the AAAI Conference on Artificial Intelligence and Interactive Digital Entertainment, vol. 4(1), pp. 16–21 (2008)
15. Cooke, A., Smith, D., Booth, A.: Beyond PICO: the SPIDER tool for qualitative evidence synthesis. Qual. Health Res. **22**(10), 1435–1443 (2012)
16. de Weerd, H., Verbrugge, R., Verheij, B.: Higher-order theory of mind in negotiations under incomplete information. In: Boella, G., Elkind, E., Savarimuthu, B.T.R., Dignum, F., Purvis, M.K. (eds.) PRIMA 2013. LNCS (LNAI), vol. 8291, pp. 101–116. Springer, Heidelberg (2013). https://doi.org/10.1007/978-3-642-44927-7_8
17. De Weerd, H., Verbrugge, R., Verheij, B.: Negotiating with other minds: the role of recursive theory of mind in negotiation with incomplete information. Auton. Agent. Multi-Agent Syst. **31**, 250–287 (2017)

18. De Weerd, H., Verbrugge, R., Verheij, B.: Higher-order theory of mind is especially useful in unpredictable negotiations. Auton. Agent. Multi-Agent Syst. **36**(2), 30 (2022)

19. Dissing, L., Bolander, T.: Implementing theory of mind on a robot using dynamic epistemic logic. In: IJCAI, pp. 1615–1621 (2020)

20. Djerroud, H., Chérif, A.A.: VICA: a vicarious cognitive architecture environment model for navigation among movable obstacles. In: ICAART, vol. 2, pp. 298–305 (2021)

21. Doshi, P., Qu, X., Goodie, A.S., Young, D.L.: Modeling human recursive reasoning using empirically informed interactive partially observable Markov decision processes. IEEE Trans. Syst. Man Cybern. Part A Syst. Hum. **42**(6), 1529–1542 (2012)

22. Feng, D., Carstensdottir, E., El-Nasr, M.S., Marsella, S.: Exploring improvisational approaches to social knowledge acquisition. In: 18th International Conference on Autonomous Agents and MultiAgent Systems, AAMAS 2019 (2019)

23. Gebhard, P., Schneeberger, T., Baur, T., André, E.: MARSSI: model of appraisal, regulation, and social signal interpretation. In: International Conference on Autonomous Agents and Multi-agent Systems (2018)

24. Goldman, A.I.: Theory of Mind. Oxford University Press, United Kingdom (2012)

25. Gouidis, F., Vassiliades, A., Basina, N., Patkos, T.: Towards a formal framework for social robots with theory of mind. In: ICAART, vol. 3, pp. 689–696 (2022)

26. Harbers, M., van den Bosch, K., Meyer, J.J.C.: Agents with a theory of mind in virtual training. In: Multi-agent Systems for Education and Interactive Entertainment: Design, Use and Experience, pp. 172–187. IGI Global (2011)

27. Hoogendoorn, M., Merk, R.-J.: Action selection using theory of mind: a case study in the domain of fighter pilot training. In: Jiang, H., Ding, W., Ali, M., Wu, X. (eds.) IEA/AIE 2012. LNCS (LNAI), vol. 7345, pp. 521–533. Springer, Heidelberg (2012). https://doi.org/10.1007/978-3-642-31087-4_54

28. Hoogendoorn, M., Soumokil, J.: Evaluation of virtual agents utilizing theory of mind in a real time action game. In: Proceedings of the 9th International Conference on Autonomous Agents and Multiagent Systems, vol. 1, pp. 59–66 (2010)

29. Hu, Q., Lu, Y., Pan, Z., Gong, Y., Yang, Z.: Can AI artifacts influence human cognition? The effects of artificial autonomy in intelligent personal assistants. Int. J. Inf. Manage. **56**, 102250 (2021)

30. Husemann, S., Pöppel, J., Kopp, S.: Differences and biases in mentalizing about humans and robots. In: 2022 31st IEEE International Conference on Robot and Human Interactive Communication (RO-MAN), pp. 490–497. IEEE (2022)

31. Kaufmann, R., Gupta, P., Taylor, J.: An active inference model of collective intelligence. Entropy **23**(7), 830 (2021)

32. Kelley, R., Tavakkoli, A., King, C., Nicolescu, M., Nicolescu, M., Bebis, G.: Understanding human intentions via hidden Markov models in autonomous mobile robots. In: Proceedings of the 3rd ACM/IEEE International Conference on Human Robot Interaction, pp. 367–374 (2008)

33. Lerer, A., Hu, H., Foerster, J., Brown, N.: Improving policies via search in cooperative partially observable games. In: Proceedings of the AAAI Conference on Artificial Intelligence, vol. 34, no. 05, pp. 7187–7194 (2020)

34. Lin, B., Bouneffouf, D., Cecchi, G.: Predicting human decision making with LSTM. In: 2022 International Joint Conference on Neural Networks (IJCNN), pp. 1–8. IEEE (2022)

35. Lorini, E.: Rethinking epistemic logic with belief bases. Artif. Intell. **282**, 103233 (2020)

36. Lorini, E., Jimenez, B.F.R.: Decision procedures for epistemic logic exploiting belief bases. In: 18th International Conference on Autonomous Agents and MultiAgent Systems, AAMAS 2019, pp. 944–952 (2019)

37. Malle, B.F.: The relation between language and theory of mind in development and evolution. In: The Evolution of Language Out of Pre-language, vol. 18, pp. 265–284 (2002)

38. Marsella, S.C., Pynadath, D.V.: Modeling influence and theory of mind. In: Virtual Social Agents, p. 199 (2005)

39. Narang, S., Best, A., Manocha, D.: Inferring user intent using Bayesian theory of mind in shared avatar-agent virtual environments. IEEE Trans. Vis. Comput. Graph. **25**(5), 2113–2122 (2019)

40. Nguyen, D., Nguyen, P., Le, H., Do, K., Venkatesh, S., Tran, T.: Learning theory of mind via dynamic traits attribution. arXiv preprint arXiv:2204.09047 (2022)

41. Page, M.J., et al.: The PRISMA 2020 statement: an updated guideline for reporting systematic reviews. BMJ **372**, n71 (2021)

42. Panisson, A.R., et al.: On the formal semantics of theory of mind in agent communication. In: Lujak, M. (ed.) AT 2018. LNCS (LNAI), vol. 11327, pp. 18–32. Springer, Cham (2019). https://doi.org/10.1007/978-3-030-17294-7_2

43. Pantelis, P.C., et al.: Agency and rationality: adopting the intentional stance toward evolved virtual agents. Decision **3**(1), 40 (2016)

44. Persiani, M., Hellström, T.: Inference of the intentions of unknown agents in a theory of mind setting. In: Dignum, F., Corchado, J.M., De La Prieta, F. (eds.) PAAMS 2021. LNCS (LNAI), vol. 12946, pp. 188–200. Springer, Cham (2021). https://doi.org/10.1007/978-3-030-85739-4_16

45. Pöppel, J., Kahl, S., Kopp, S.: Resonating minds-emergent collaboration through hierarchical active inference. Cogn. Comput. **14**(2), 581–601 (2022)

46. Premack, D., Woodruff, G.: Does the Chimpanzee have a theory of mind? Behav. Brain Sci. **1**(4), 515–526 (1978)

47. Pynadath, D.V., et al.: Disaster world: decision-theoretic agents for simulating population responses to hurricanes. Comput. Math. Organ. Theor. **29**(1), 84–117 (2022)

48. Pynadath, D.V., Marsella, S.C.: PsychSim: modeling theory of mind with decision-theoretic agents. In: IJCAI, vol. 5, pp. 1181–1186 (2005)

49. Pynadath, D.V., Marsella, S.C.: Socio-cultural modeling through decision-theoretic agents with theory of mind. In: Advances in Design for Cross-Cultural Activities, pp. 417–426 (2012)

50. Pynadath, D.V., Rosenbloom, P.S., Marsella, S.C.: Reinforcement learning for adaptive theory of mind in the sigma cognitive architecture. In: Goertzel, B., Orseau, L., Snaider, J. (eds.) AGI 2014. LNCS (LNAI), vol. 8598, pp. 143–154. Springer, Cham (2014). https://doi.org/10.1007/978-3-319-09274-4_14

51. Pynadath, D.V., Si, M., Marsella, S.C.: Modeling theory of mind and cognitive appraisal with decision-theoretic agents. In: Social Emotions in Nature and Artifact: Emotions in Human and Human-Computer Interaction, pp. 70–87 (2011)

52. Reisenzein, R., et al.: Computational modeling of emotion: toward improving the inter-and intradisciplinary exchange. IEEE Trans. Affect. Comput. **4**(3), 246–266 (2013)

53. Rumbell, T., Barnden, J., Denham, S., Wennekers, T.: Emotions in autonomous agents: comparative analysis of mechanisms and functions. Auton. Agent. Multi-Agent Syst. **25**, 1–45 (2012)

54. Sarkadi, Ş, Panisson, A.R., Bordini, R.H., McBurney, P., Parsons, S.: Towards an approach for modelling uncertain theory of mind in multi-agent systems. In: Lujak, M. (ed.) AT 2018. LNCS (LNAI), vol. 11327, pp. 3–17. Springer, Cham (2019). https://doi.org/10.1007/978-3-030-17294-7_1

55. Sarkadi, S., Panisson, A.R., Bordini, R.H., McBurney, P., Parsons, S., Chapman, M.: Modelling deception using theory of mind in multi-agent systems. AI Commun. **32**(4), 287–302 (2019)

56. Sarthou, G., Clodic, A., Alami, R.: Ontologenius: a long-term semantic memory for robotic agents. In: 2019 28th IEEE International Conference on Robot and Human Interactive Communication (RO-MAN), pp. 1–8. IEEE (2019)

57. Scassellati, B.: Theory of mind for a humanoid robot. Auton. Robot. **12**, 13–24 (2002)

58. Shum, M., Kleiman-Weiner, M., Littman, M.L., Tenenbaum, J.B.: Theory of minds: understanding behavior in groups through inverse planning. In: Proceedings of the AAAI Conference on Artificial Intelligence, vol. 33(1), pp. 6163–6170 (2019)

59. Si, M., Marsella, S.C., Pynadath, D.V.: Modeling appraisal in theory of mind reasoning. Auton. Agent. Multi-Agent Syst. **20**, 14–31 (2010)

60. Stevens, C., de Weerd, H., Cnossen, F., Taatgen, N.: A metacognitive agent for training negotiation skills. In: Proceedings of the 14th International Conference on Cognitive Modeling, ICCM 2016 (2016)

61. Tekülve, J., Schöner, G.: Neural dynamic concepts for intentional systems. In: CogSci, pp. 1090–1096 (2019)

62. Veltman, K., de Weerd, H., Verbrugge, R.: Training the use of theory of mind using artificial agents. J. Multimodal User Interfaces **13**, 3–18 (2019)

63. Von Der Osten, F.B., Kirley, M., Miller, T.: The minds of many: Opponent modeling in a stochastic game. In: IJCAI, pp. 3845–3851 (2017)

64. Walton, D.: Using argumentation schemes to find motives and intentions of a rational agent. Argument Computat. **10**(3), 233–275 (2019)

65. Wang, R.E., Wu, S.A., Evans, J.A., Tenenbaum, J.B., Parkes, D.C., Kleiman-Weiner, M.: Too many cooks: coordinating multi-agent collaboration through inverse planning. In: Proceedings of the 19th International Conference on Autonomous Agents and MultiAgent Systems, pp. 2032–2034 (2020)

66. Ware, S.G., Siler, C.: Sabre: a narrative planner supporting intention and deep theory of mind. In: Proceedings of the AAAI Conference on Artificial Intelligence and Interactive Digital Entertainment, vol. 17, no. 1, pp. 99–106 (2021)

67. Whiten, A., Byrne, R.: Natural Theories of Mind: Evolution, Development and Simulation of Everyday Mindreading. B. Blackwell Oxford, UK (1991)

68. Wooldridge, M.: An Introduction to Multiagent Systems. Wiley, Hoboken (2009)

69. Yousefi, Z., Heinke, D., Apperly, I., Siebers, P.-O.: An agent-based model for false belief tasks: belief representation systematic approach (BRSA). In: Thomson, R., Dancy, C., Hyder, A., Bisgin, H. (eds.) SBP-BRiMS 2018. LNCS, vol. 10899, pp. 111–126. Springer, Cham (2018). https://doi.org/10.1007/978-3-319-93372-6_14

A Spin-off Version of Jason for IoT and Embedded Multi-Agent Systems

Carlos Eduardo Pantoja[1,2(✉)] , Vinicius Souza de Jesus[1] ,
Nilson Mori Lazarin[1,2] , and José Viterbo[1]

[1] Institute of Computing, Fluminense Federal University (UFF), Niterói, RJ, Brazil
{vsjesus,nlazarin}@id.uff.br, viterbo@ic.uff.br
[2] Federal Center for Technological Education Celso Suckow da Fonseca (Cefet/RJ),
Rio de Janeiro, RJ, Brazil
pantoja@cefet-rj.br

Abstract. Embedded artificial intelligence in IoT devices is presented as
an option to reduce connectivity dependence, allowing decision-making
directly at the edge computing layer. The Multi-agent Systems (MAS)
embedded into IoT devices enables, in addition to the ability to perceive and act in the environment, new characteristics like pro-activity,
deliberation, and collaboration capabilities to these devices. A few new
frameworks and extensions enable the construction of agent-based IoT
devices. However, no framework allows constructing them with hardware
control, adaptability, and fault tolerance, besides agents' communicability and mobility. This work presents an extension of the Jason framework
for developing Embedded MAS with BDI agents capable of controlling
hardware, communicating, and moving between IoT devices capable of
dealing with fault tolerance. A case study of an IoT solution with a smart
home, a monitoring center, and an autonomous vehicle is presented to
demonstrate the framework's applicability.

Keywords: Internet of Things · Multi-agent Systems · Edge
Computing

1 Introduction

The Internet of Things (IoT) promotes the use of devices that sense physical environments in various application domains. However, they generate a considerable
amount of raw data stream that needs to be transmitted to be processed [2].
Pervasive Computing allows the development of distributed cognitive devices
capable of extracting relevant information and making decisions directly at the
edge computing layer, reducing the dependence on connectivity [10].

One of the fields of Distributed Artificial Intelligence (DAI) is Multi-agent
Systems (MAS), which are composed of multiple autonomous and proactive entities with decision-making capacity and social abilities that can collaboratively
interact to achieve a common goal for the system [28]. In this research area, an

© The Author(s), under exclusive license to Springer Nature Switzerland AG 2023
M. C. Naldi and R. A. C. Bianchi (Eds.): BRACIS 2023, LNAI 14195, pp. 382–396, 2023.
https://doi.org/10.1007/978-3-031-45368-7_25

integration of hardware and software that allow agents to sense and act in a real-world environment using sensors and actuators is named Embedded MAS [3]. These systems can also contribute to reduce dependency on connectivity since a cognitive agent embedded in an IoT device can process the raw data received from sensors and act immediately, thus accelerating decision-making [9].

One of the most well-known cognitive agent architecture, the Belief-Desire-Intention (BDI) model [7], is based on the knowledge that an agent can have from the environment (perceptions), other agents, or itself [21]. When applied embedded in devices, it provides decision-making at edge level by using perceptions and beliefs captured from the real world and other devices [1,5,9]. We performed a mapping review, where we found some works that present a framework [8,12,15,25] or provide some features to construct IoT agent-based devices using BDI frameworks, such as: extension to provide interoperable between cyber-physical and IoT systems using fuzzy logic [16]; approaches to reconfiguring agent's goals on the fly [11] and developing MAS with IoT objects [6], or architecture to Ambient Intelligence with IoT [23]. However, none of these solutions simultaneously meets all the needs of an agent-based IoT device, such as the hardware control, the communicability with other MAS or devices, the mobility of the agents between different MAS or devices, fault tolerance, and adaptability.

This paper presents the Jason Embedded – a spin-off extended version of the Jason framework [4] – that provides autonomy, pro-activity, social ability, adaptability, and fault-tolerance to IoT devices. It allows programming agents dedicated to exchanging KQML messages [14] between different devices and agents dedicated to controlling sensors and actuators and capable of deciding when to perceive the environment or yet can define a strategy to gather perceptions from sensors. Besides, allow the programming of Open MAS [26], where agents can move from one system to another using an IoT network.

For this, the reasoning cycle of a standard Jason agent was modified to create two novel specific types of BDI agents to program Embedded MAS. In the first, the agent can perceive or act in the environment and monitor the connection status with wired microcontrollers, allowing fault tolerance and adaptability. In the second, agents can exchange messages with other MAS and handle agents' arrival and departure into its MAS. Then, we integrated and adapted the ad-hoc solutions [17,19,24,27] in a single distribution. Despite the modifications, all agents maintain the original features of Jason's agents. Jason Embedded allows the designer to abstract some aspects of hardware interfacing and communication, focusing only on programming the MAS. The contribution of this work is an extended framework to allow hardware controlling, communicability, and agent mobility to be integrated into a single framework. To demonstrate this, we build and present a case study that implemented an IoT solution for home monitoring integrated with a central control and an unmanned autonomous vehicle.

This paper is structured as follows: an analysis of related works is presented in Sect. 2; in Sect. 3 we present the Jason Embedded and the newly available behavior of agents; we demonstrate an Embedded MAS in a case study in Sect. 4; finally, a discussion and future work are presented in Sect. 5.

2   Related Works

A mapping review was conducted to find works that use the BDI framework for the development of IoT systems based on cognitive agents. We followed these steps: first was to define the search string[1] for Google Scholar which returned 246 results; next, the results were filtered, ignoring duplicity and results that were not a thesis or a paper, remaining 215; the third step considered only works published after 2017, remaining 171 works; the fourth step ignored surveys or reviews, remaining 153 works. After this, 56 works with the words *framework, tool, architecture, library, hardware, IoT, cyber-physical, things, or ubiquitous* in the title were considered for the next step.

Finally, it was rejected the works that were not directly related to IoT and BDI (i.e., simulation, data mining, and others were discarded); finally, 11 works remained. Three works [1,5,9] only present an IoT implementation based on cognitive agents. Four works [6,11,16,23] feature an extension to a BDI framework, however, they do not meet all the needs of an agent-based IoT device, such as the hardware control, the communicability with other MAS or devices, the mobility of the agents between different MAS or devices, fault tolerance, and adaptability. Thus, only the below four works [8,12,15,25] present a framework for programming IoT agents.

The first [8] try to provide connectivity to the Web of Things using an agent-oriented visual programming IDE using a framework endowed with a REST endpoint. The IDE is a web-based solution for reducing that expertise in adopting the BDI. The solution allows agents to discover and connect to Things using internal structures, but it does not offer embedded solutions to direct control hardware, being dependent on the IoT infrastructure. Our solution employs specialized agents to control hardware and access IoT gateways in an extended version of Jason. The hardware control is independent of the IoT infrastructure, which is used for the communicability and mobility of agents.

Similarly, the second [12] provides a declarative language in a BDI-like style for programming MAS for the IoT using microcontrollers in Python. However, microcontrollers are limited in processing capacity and memory and need other components to be connected to any IoT Gateway. Withal allows high-level agents to connect and control these microcontrollers if only they are available and connected to the IoT. The PHIDIAS is an improved version of PROFETA, and it presents ways of developing intelligence in devices, including IoT, but does not allow agent mobility between different MAS or Environments. PHIDIA allows communication using HTTP. The communication between microcontrollers and agents is performed using wireless devices. The microcontrollers, sensors, and actuators are not part of the MAS. Besides, one agent is responsible for each edge device. The third [25] try reduce the gap between theory and practical applications in cybersecurity using BDI agents and MAS. In their framework, agents can control IoT devices (remotely or hosted in the device), utilizing API to

---

[1] *("embedded" OR "embodied") AND ("multiagent system" OR "multi-agent system") AND "belief-desire-intention" AND "framework" AND "internet of things".*

connect to IoT gateways or simulated software. In both, the dependence on IoT gateways avoid agents from properly working even if embedded. The proposed in this paper allows agents to directly control hardware using serial communication (without network connection) and access the IoT network to exchange information with other MAS.

The fourth [15] try to bring together academic development and industry by offering ways of developing MAS using recent technologies (e.g., Node.JS). Similarly, our approach aims to provide and facilitate ways of developing embedded and IoT systems to be adopted in domains such as education, academia, and industry. In addition to the mapping review, the SPADE 3 [22], in its most recent version, uses XMPP and has an extension for BDI agents. However, it does not allow agent mobility between different MAS. The authors even argue that mobility between different physical or logical nodes is an interesting and desired functionality, however not contemplated by SPADE.

As previously presented, none of the works embed a MAS in IoT devices with BDI agents capable of communicating and moving between different Embedded MAS and also adapting to some faults that may arise in hardware malfunctioning, for example. Therefore, our spin-off version of the Jason framework presented in this work has customized architectures of BDI agents able to control hardware, communicate, and move between MAS embedded in devices using an IOT network. Besides, the agents that interface hardware can be fault tolerant if the hardware becomes unavailable and adapt themselves to find another way to comply with their goals.

3   Jason Embedded

Jason Embedded[2] is an extended version of Jason [4] that supports the development of Embedded MAS to provide autonomy and communicability to IoT devices and mobility and adaptability to agents. In this paper, we define autonomy as a characteristic an Embedded MAS provides to a device that does not need external architectures to control and manage it. Communicability is the agent's ability to communicate with other agents in its MAS or another MAS. We define mobility as the agents' ability to move from one MAS to another. Finally, Fault Tolerance and Adaptability are defined as the ability of an agent to identify if a physical resource is absent or not answering commands and to overcome this situation by modifying its goals.

Jason Embedded also provides new types of agents to control hardware devices and to communicate with agents hosted in other MAS. Besides, our extended version allows agent mobility between Embedded MAS, where one agent or a group of agents can move from one system to another. Then, it specializes the MAS by creating agents dedicated to certain functionalities without modifying the core of Jason, maintaining all of its original functionalities. Jason Embedded is an initiative employed by our research group in embedded solutions

---

[2] https://jasonembedded.chon.group/.

to facilitate the development and teaching of practical MAS. The overall picture of our framework is depicted in Fig. 1.

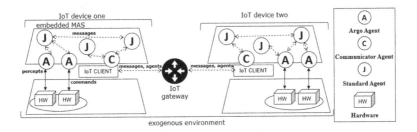

Fig. 1. The overall picture of the framework architecture.

In this approach, the developer can employ Standard, Argo, and Communicator agents to develop the Embedded MAS. Each type of agent has specific abilities to deal with hardware interfacing, communicability, and mobility using an IoT network. Argo agents use hardware interfacing to capture perceptions from the exogenous environment and send commands to activate actuators. The Communicator agent has an IoT client middleware for connecting to an IoT gateway to send messages or agents from one IoT device to another. Standard agents can communicate and exchange information only with agents from their system.

For example, in a scenario concerning three different IoT devices hosting Embedded MAS, one is responsible for managing resources in a house, one is responsible for a monitoring center, and the last is an emergency vehicle. An Argo agent monitors the house's sensors, and if something is detected, it informs the house's Communicator agent to send an alert to the monitoring center. Consequently, the Communicator agent from the monitoring center searches for an available vehicle and sends a Standard agent to help the Argo agent to drive the vehicle to attend the house's emergency. The three devices are distributed, autonomous, and controlled by an Embedded MAS. They can communicate by sending messages using an IoT gateway available, and Argo agents perform all sensing in each device. Finally, agents can move from one system to another.

The new types of agents are based on the standard Jason's agent reasoning cycle. Their reasoning cycle is modified at specific points to create new behaviors without modifying any existing function. In this way, every extended agent is still a Jason agent. The agent's reasoning cycle comprises ten steps that complete a BDI decision loop [4]. In the first step, the agent senses the simulated environment using the Perceive method to gather all available perceptions. At this point, the agent can only perceive a simulated environment if it exists, and if it needs to interface with real sensors, this environment must be properly coded. The last step executes an intention by performing actions from a selected plan. These two steps represent how the agent can perceive and modify the environment, and agents could use them to interface sensors and actuators.

Then, the first step could allow agents to gather perceptions directly from sensors and process them as beliefs. Since this mechanism works as passive perception, the agent is aware of all available information, even if it does not need all of them to accomplish a goal. To overcome this issue, the agent could also decide whether or not to gather available perceptions and filter undesired perceptions at runtime. It allows the agent to indirectly perform active perception since it can decide when and what to perceive, but it always does it in the first step of its reasoning cycle. Similarly, the last step could allow agents to act upon actuators by adopting a protocol that runs over the serial port by sending activation commands [24]. The agent does not need to use Jason's simulated environment—the endogenous representation of the system's environment—since it can directly activate or deactivate actuators in the real world—the exogenous environment. This approach reduces the use of abstractions and layers to reflect actions in the device's hardware where the Embedded MAS is hosted.

In the third step of the reasoning cycle, an agent checks its mail, looking for messages that other agents from its MAS could have sent to it. Since Jason Embedded aims to provide a framework for programming autonomous and embedded solutions using agents, it is worthwhile to think of prototypes communicating with other prototypes. Considering that each prototype has an Embedded MAS, some agents could work as a communicator, receiving and sending messages to other communicator agents. Otherwise, the Embedded MAS will be autonomous and proactive but without social ability. Then, these agents could check an alternative mailbox with messages from other Embedded MAS. In the last step, agents could send messages to agents within and out of its system by addressing other communicators using an IoT middleware, for instance.

As sending a message occurs in the last step, it could also allow specific agents to move agents from one Embedded MAS to another. In Embedded MAS, agent mobility is based on bio-inspired protocols [27]. The protocols simulate natural behaviors that could be explored in collaborative tasks using IoT devices. One or more agents or even the whole MAS could move from one device to another to take control of the destination device (predation) or to use it as a temporary non-hazardous relationship (mutualism and inquilinism) if both parties accept the protocol. For communicability and mobility, agents must connect to an IoT middleware to redirect the messages and agents to an agent of the target Embedded MAS. The description of all new internal actions, their behaviors, and requested parameters can be seen in Table 1.

Below are described the new characteristics allowed by the framework presented in this work. Jason Embedded offers two types of extended agents: one for interfacing hardware named Argo and another for communication named Communicator, responsible for all communications and mobility from outside its MAS. Besides, Jason Embedded maintains the standard agents present in the Jason. It is important to remark that the option for specializing agents leads to well-defined responsibilities, allowing agents to focus on their purpose to minimize some drawbacks. For example, when interfacing hardware, it is interesting that the agent can deliberate, considering the perceptions available at that

Table 1. The Jason Embedded's internal actions

Agent	New Action	Description
Argo	.port(S);	Defines a serial communication port with an IoT device
	.limit(N);	Defines an interval for the cycle of environmental perception
	.percepts(open|lose);	Listens or not the environmental perceptions
	.filter(add | remove, c, P, C);	Defines or not an environmental perception filter
	.act(O);	Sends an order to the microcontroller to execute
Communicator	.connectCN(T,G,E,U,K);	Joins an IoT network
	.sendOut(D,f,M);	Dispatches a message to another MAS
	.moveOut(D,b,A);	Carries over the agents to other MAS
	.disconnectCN;	Leaves the IoT network

Where:

A is one, all, or a set of agents (i.e., all, agent or $[agent_1, agent_2, agent_n]$).
C is an optional field representing the necessary context for applying the filter.
D It is a literal that represents the identification of the recipient MAS.
E is a number that represents the network port of an IoT gateway.
G is a literal that represents the FQDN or the network address of an IoT gateway.
K is an optional field representing the device's credentials in communication technology.
M is a literal that represents the message.
N is a positive number ($N > 0$) that represents an interval in milliseconds.
O is a literal that represents an order for the microcontroller to execute.
P is one or a set of environmental perceptions (i.e., perception or $[perception_1, perception_2, perception_n]$).
S is a literal that represents a serial port (i.e., ttyACM0).
T is a literal that represents the communication technology.
U is a literal that represents the identification of the device in the IoT network.
V is an optional field that represents a context to apply the filter.
b is a bio-inspired protocol (inquilinism \| mutualism \| predation).
c is a filter criterion (all \| comply \| except \| only).
f is a illocutionary force (tell \| untell \| achieve \| unachieve).

moment. Suppose the agent is compromised to some other task as communication or trying to achieve another intention. So, it could affect the agent's reaction in some applications and domains where the response time is essential—i.e., object deviation.

3.1   Controlling Hardware

Argo is a customized agent architecture built on the Jason framework that extends Jason's standard agents by adding the ability to control microcontrollers [24]. Argo agents interact with the physical environment, capturing information from the environment using sensors and acting by sending commands to the microcontroller to activate the actuators. Sensors' information is automatically processed as perceptions in the agent's belief base. Argo agents have four internal actions to control microcontrollers: port, limit, percepts, act, and filter. These actions define which port the agent is accessing, a time limit for accessing them, if they open or close the flow of perceptions, activate actuators, or filter the incoming percepts, respectively. Argo architecture was modified to work properly along with all existing architectures since it was initially designed to be an ad-hoc solution. Furthermore, the behavior of swapping resources [17] at runtime was added to avoid stopping the Embedded MAS during its execution.

3.2   Communicability

The Communicator agent is another customized agent architecture built on Jason's Standard agent by adding the ability to communicate with agents from other MAS, which also have Communicator agents [23]. It allow agents from different MAS to interact, exchange knowledge and even collaborate between them. The Communicator agent can interact by exchanging knowledge and even collaborating between them using KQML performatives [14]. To use these new capabilities, the communicator agent has specific internal actions: connect, disconnect, and sendOut. These actions connects and disconnects from the IoT Gateway, and send a message to another Embedded MAS. The Communicator agent is a client instance of an IoT middleware [13] that needs an IoT Gateway to communicate. The agent can send a message to another MAS using this IoT middleware even if the target is disconnected. Once it connects, the message is redirected.

In this case, the original communicator architecture was totally modified to allow the connection in the IoT Gateway generically and work with Argo agents in the same Embedded MAS. The connect and disconnect are new functionalities once the agent sometimes can deliberate whether to be offline.

3.3   Mobility

Subsequently, the Communicator architecture was extended with bioinspired protocols [27] following ecological relations concepts that allow agents to be transferred between different MAS. They can be used to preserve the MAS knowledge if the physical device is damaged since it is subject to the unpredictability of the real world. The bioinspired protocols currently have three ecological relations implemented: Predation, Inquilinism, and Mutualism: in Predation, all agents are transferred to the target MAS to prey and dominate. It eliminates all agents of the target and the origin MAS (after moving them) to prevent unwanted access to any residue of its knowledge; in the Inquilinism, it sends all its agents to the target MAS to preserve its knowledge, but they do not interfere in its activities and existence. However, similar to the predation protocol, the origin MAS is eliminated to prevent unwanted access to any residue of knowledge; the Mutualism sends an agent, a group of agents, or the entire MAS to interact, learn, and transmit knowledge to a target MAS. Agents using this protocol can return to the origin MAS or move to any other if allowed.

The Communicator has the moveOut internal action for the mobility behavior. It uses the IoT middleware instance to move agents from one system to another. In this version, Mutualism was modified to send all agents except the Communicator since the agents need a way to return to the origin MAS. In the Inquilinism, we modified the protocol to drop all desires in the target MAS of the moving agents since they cannot interfere with the MAS functioning. We modified the protocols to guarantee the departure and arrival of agents to avoid communication problems and that agents do not get lost during the agents' transference.

4   Case Study

In the case study, we consider a scenario, shown in Fig. 2, that integrates a smart home, a central control, and an unmanned vehicle, all controlled by Multi-agent Systems. The MAS running in the central control has a Communicator to forward service calls and two Jason agents for decision-making in case of calls. The vehicle has an Argo agent for driving and a Communicator agent for receiving agents to use the vehicle. The MAS running in the house has a Communicator agent responsible for interacting with the central and an Argo agent controlling the sensors and actuators.

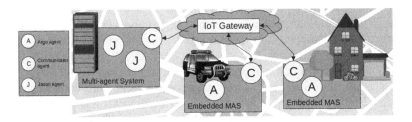

Fig. 2. The scenario of the case study proposed.

In this scenario, when the Argo agent in the House's Embedded MAS perceives a change in a specific sensor, it will inform the Communicator agent to send a message to central control. In the central control, the message is forwarded to a coordinator agent, who will send an agent to the house to analyze it. After completion, the sent agent will send a message to the coordinator requesting a patrol. The coordinator agent will migrate to an unmanned vehicle to execute the mission. When in the vehicle, the coordinator will inform the driving agent about the path.

To fulfill the proposed scenario, we built three Embedded MAS: the first – Batcave project – represents the house, hosted in a Raspberry Pi with an Arduino Board and has an LDR (light dependent resistor) sensor and a LED (light-emitting diode) as the actuator; the second – WayneMansion project – represents the control center, hosted in a Raspberry Pi; the third – BatMobile project – represents the unmanned vehicle, hosted in a ChonBot 2WD [18] prototype.

When the Argo agent in the house MAS (batCave.mas2j) notices a change in the environment (batSignal(true)) it activates an actuator (led(red)) and requests the Communicator agent to forward an alert to the center control MAS (wayneMansion.mas2j). The Fig. 3 shown the code of the BatCave project.

When this alert arrives at the destination, it is forwarded to the Jason agent responsible for handling occurrences (bruce.asl); when analyzing the alert, the agent Bruce decides to designate other Jason agent (alfred.asl) to verify the situation (at MAS BatCave). The agent Alfred requests the Communicator agent of WayneMansion MAS to transfer itself to BatCave MAS; after this, the Communicator of BatCave MAS receives the agent Alfred in the system. Alfred requests

```
≡ argoAgent.asl ×     ≡ comm.asl
batCave > multi-agentProject > agt >  ≡ argoAgent.asl
   1    /* Agent ARGO in batCave.mas2j - Initial beliefs */
   2    mySerialPort(ttyACM0).
   3
   4    /* Plans */
   5    +batSignal(S): S≡true <-     !led(red); .broadcast(tell,batSignal(S));
   6         .send(comm,achieve,sendExternalMessageForAllKnown(tell,batSignal(S))).
   7
   8    +!led(Act)[source(A)] : light(S) & S\==Act <-    .act(Act).
   9
  10    { include("../../../_commonAgents/argo.asl") }
```
```
≡ argoAgent.asl      ≡ comm.asl ×
batCave > multi-agentProject > agt >  ≡ comm.asl
   1    // Agent comm in project batCave.mas2j
   2    where(batCave).
   3
   4    { include("../../../_commonAgents/communicator.asl") }
```

Fig. 3. The implementation of the BatCave project. (Color figure online)

the Argo agent for a new check of sensors and to change the alert state to yellow (led(yellow)); finally, Alfred requests to Communicator of BatCave MAS to send a message to agent Bruce. The Fig. 4 shown the code of the WayneMansion project.

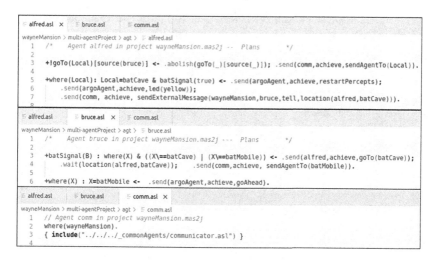

```
≡ alfred.asl ×    ≡ bruce.asl      ≡ comm.asl
wayneMansion > multi-agentProject > agt >  ≡ alfred.asl
   1    /*     Agent alfred in project wayneMansion.mas2j -- Plans     */
   2
   3    +!goTo(Local)[source(bruce)] <- .abolish(goTo(_)[source(_)]); .send(comm,achieve,sendAgentTo(Local)).
   4
   5    +where(Local): Local=batCave & batSignal(true) <- .send(argoAgent,achieve,restartPercepts);
   6         .send(argoAgent,achieve,led(yellow));
   7         .send(comm, achieve, sendExternalMessage(wayneMansion,bruce,tell,location(alfred,batCave))).
   8
```
```
≡ alfred.asl      ≡ bruce.asl ×     ≡ comm.asl
wayneMansion > multi-agentProject > agt >  ≡ bruce.asl
   1    /*    Agent bruce in project wayneMansion.mas2j --- Plans     */
   2
   3    +batSignal(B) : where(X) & ((X\==batCave) | (X==batMobile)) <- .send(alfred,achieve,goTo(batCave));
   4         .wait(location(alfred,batCave));    .send(comm,achieve, sendAgentTo(batMobile)).
   5
   6    +where(X) : X=batMobile <-   .send(argoAgent,achieve,goAhead).
```
```
≡ alfred.asl      ≡ bruce.asl      ≡ comm.asl ×
wayneMansion > multi-agentProject > agt >  ≡ comm.asl
   1    // Agent comm in project wayneMansion.mas2j
   2    where(wayneMansion).
   3    { include("../../../_commonAgents/communicator.asl") }
   4
```

Fig. 4. The implementation of the WayneMansion project. (Color figure online)

The Communicator of WayneMansion MAS forwards the message to Bruce, who decides to patrol near the house. In this way, it requests the Communicator to transport itself to the autonomous unmanned vehicle (batMobile.mas2j); once it arrives in the BatMobile MAS, the agent Bruce requests to Argo agent to pilot the vehicle (goAhead). Figure 5 shown the implementation of the BatMobile project. Finally, all projects' Argo and Communicator agents have standard plans, beliefs, and intentions shown in Fig. 6.

argoAgent.asl × | comm.asl

batMobile > multi-agentProject > agt > argoAgent.asl

```
/* Agent ARGO in batMobile.mas2j -- Initial beliefs */
mySerialPort(ttyACM0).

/* Plans */
+!goAhead <-    .act(ledOn);    .act(goAhead).

{ include("../../../_commonAgents/argo.asl") }
```

argoAgent.asl | comm.asl ×

batMobile > multi-agentProject > agt > comm.asl

```
// Agent comm in project batMobile.mas2j
where(batMobile).

{ include("../../../_commonAgents/communicator.asl") }
```

Fig. 5. The implementation of the BatMobile project.

communicator.asl × | argo.asl

_commonAgents > communicator.asl

```
/* Generic Communicator Agent -- Initial beliefs */
ioTGateway("skynet.chon.group",5500).
masDirectory(batCave,       "99194818-4993-4388-4567-816391351812").
masDirectory(batMobile,     "996b6f17-c7e7-4400-982b-d5f4c91fbee6").
masDirectory(wayneMansion, "9916c2df-51be-42ae-8a39-1cb39b8d7727").

/* Generic Communicator Agent -- Initial goals */
!connect.

/* Generic Communicator Agent --  Plans */
+!connect: where(OurMAS) & masDirectory(MAS,UUID) & MAS=OurMAS  & ioTGateway(Server,Port) <-
    .connectCN(Server,Port,UUID);  .broadcast(tell,where(OurMAS)).

+!sendAgentTo(Destination)[source(Agent)]: where(OurMAS) & masDirectory(MAS,UUID) & Destination=MAS <-
    .send(Agent,untell,where(OurMAS));  .moveOut(UUID,mutualism,Agent); .sendOut(UUID, achieve, welcome(Agent)).

+!welcome(Agent)[source(X)]: where(OurMAS) <-   .send(Agent,tell,where(OurMAS)).

+!sendExternalMessage(MAS,Receiver,Force,Message)[source(Source)]: masDirectory(A,UUID) & MAS=A  <-
    .sendOut(UUID, achieve, internalUnicastMessage(Receiver,Force,Message[source(Source)])).

+!internalUnicastMessage(Destination,Force,Message)[source(X)]<-    .send(Destination,Force,Message[source(X)]).

+!sendExternalMessageForAllKnown(Force,Message)[source(Source)]: where(OurMAS) <-
    for((masDirectory(OtherMAS,UUID)) & (OtherMAS\==OurMAS)){
        .sendOut(UUID, achieve, internalBroadcastMessage(Force,Message[source(Source)]));
    }.

+!internalBroadcastMessage(Force,Message)[source(X)] <- .broadcast(Force,Message[source(X)]).
```

communicator.asl | argo.asl ×

_commonAgents > argo.asl

```
/*          Generic ARGO Agent Initial goals          */
!startPercept.

/* Plans */
+!startPercept: mySerialPort(SP) <- .port(SP); .percepts(open); .limit(1000).

+!stopPercept <-    .percepts(close);   .abolish(_[source(percept)]).

+!restartPercepts[source(A)] <- !stopPercept; !startPercept.
```

Fig. 6. The common source of the Argo and Communicator agents.

Aiming to enable the reproducibility of what is presented in this work, the framework and its source code, the implementation of the case study, and a demonstration video using prototypes representing the house, the control center, and the unmanned vehicle of the proposed scenario are available[3].

[3] http://bracis2023.chon.group/.

5   Discussion

This paper presented an extended version of the Jason framework named Jason Embedded to program autonomous devices using BDI agents. It provides two new types of agents for allowing hardware interfacing, IoT-based communication, and mobility between different Embedded MAS. Besides, Jason's agents can deal with the MAS's internal issues where it lives. We also presented a study case to cover the new functionalities provided by the extended framework.

Using an Embedded MAS could bring advantages compared to adopting just one agent to interface all prototype's functionalities. One agent could be overloaded depending on how much information it gathers or which goals it is pursuing. The decision to use a framework that allows specialized agents to deal with certain functionality leads to an internal distributed solution to optimize the response to stimuli. If one agent has to deal with both perception gathering and external mail checking, it can eventually get overloaded. A prototype embedded with a MAS is autonomous and proactive by its capabilities of perceiving and acting upon the real world. However, when dealing with prototypes, it is important that these autonomous devices could collaborate somehow. Jason Embedded allows these prototypes to exchange messages using the Communicator agents. Then, devices with different Embedded MAS can negotiate, coordinate and exchange information without being part of the same physical architecture or network. They only need the Internet to connect to the IoT server.

Every Embedded MAS is an Open MAS if it adopts at least one Communicator agent. An Open MAS allows agents to enter and leave the system whenever necessary. So, in our extended version, the Communicator is responsible for moving agents (and itself) to other systems by invoking bio-inspired protocols. Depending on the domain, agents can move and dominate the destiny system or co-exist within the new one. In some cases, the agents can come and go from a system (mutualism). These two contributions now allow communicability and mobility in BDI agents using Jason by using some new internal actions since all background technologies are abstracted from the designer.

Jason Embedded could run on any platform or IDE that the designer chooses. But to exploit the most of it, it can work along the ChonOS and ChonIDE, which are an OS distribution and an IDE, to support the embedding process and develop a MAS using Jason Embedded. The OS comprises all the necessary technological dependencies in a single distribution, including the IDE to program the firmware and the MAS, start and stop the Embedded MAS, verify the prototype's outputs (logs), and inspect the mind of any agent. We remark that it has been an initiative of our research group to develop embedded solutions during the past years. In this paper, we consolidate several solutions that used to work alone into a single framework to facilitate their use and adoption.

In future works, we will provide an alternative to program artifacts that interfaces hardware using a serial interface and CArtAgO [20]. Then, the designer can choose between Argo agents and artifacts. The former is one agent dedicated to interface sensors and actuators, while the latter could be available to all agents existing in the system. Furthermore, the Communicator agent needs a different format for identifying itself in the system since it uses a non-intuitive hexadecimal number. One possibility would be to use an email address to identify the Communicator agent. Besides, two new protocols could help in the development process of Embedded MAS: cloning could be used to send a copy of one or more agents to another system without killing these agents in origin; the cryogenics could be used to dump all agents into files and then restart them in the future. From the technology point of view, we intend to match the Jason Embedded version with the most recent Jason framework. This version uses version 1.4.1 since some single-board computers only run Java 8, and the recent Jason framework uses Java 17. Another point is to exploit alternative communication infrastructures such as SMTP/POP3 or XMPP to offer more reliability and privacy since ContextNet uses UDP, and the message is transmitted without cryptography.

References

1. Akhtar, S.M., Nazir, M., Saleem, K., Mahfooz, H., Hussain, I.: An ontology-driven IoT based healthcare formalism. Int. J. Adv. Comput. Sci. Appl. **11**(2) (2020). https://doi.org/10.14569/IJACSA.2020.0110261
2. Baccour, E., et al.: Pervasive AI for IoT applications: a survey on resource-efficient distributed artificial intelligence. IEEE Commun. Surv. Tut. **24**(4), 2366–2418 (2022). https://doi.org/10.1109/COMST.2022.3200740
3. Barnier, C., Aktouf, O.E.K., Mercier, A., Jamont, J.P.: Toward an embedded multiagent system methodology and positioning on testing. In: 2017 IEEE International Symposium on Software Reliability Engineering Workshops (ISSREW), pp. 239–244 (2017). https://doi.org/10.1109/ISSREW.2017.57
4. Bordini, R.H., Hübner, J.F., Wooldridge, M.: Programming Multi-Agent Systems in AgentSpeak Using Jason. Wiley Series in Agent Technology. Wiley, Hoboken (2007). https://dl.acm.org/doi/10.5555/1197104
5. Brandao, F., Nunes, P., de Jesus, V.S., Pantoja, C.E., Viterbo, J.: Managing natural resources in a smart bathroom using a ubiquitous multi-agent system. In: Proceedings of the 11th Workshop-School on Agents, Environments, and Applications, WESAAC 2017, pp. 101–112. FURG, São Paulo (2017)
6. Brandão, F.C., Lima, M.A.T., Pantoja, C.E., Zahn, J., Viterbo, J.: Engineering approaches for programming agent-based iot objects using the resource management architecture. Sensors **21**(23), 8110 (2021). https://doi.org/10.3390/s21238110
7. Bratman, M.: Intention, Plans, and Practical Reason. Harvard University Press, Cambridge, MA (1987)
8. Burattini, S., et al.: Agent-oriented visual programming for the web of things (2022). https://www.alexandria.unisg.ch/handle/20.500.14171/109205
9. Souza de Castro, L.F., et al.: Integrating embedded multiagent systems with urban simulation tools and IoT applications. RITA **29**(1), 81–90 (2022). https://doi.org/10.22456/2175-2745.110837
10. Chander, B., Pal, S., De, D., Buyya, R.: Artificial Intelligence-based Internet of Things for Industry 5.0. In: Pal, S., De, D., Buyya, R. (eds.) Artificial Intelligence-Based Internet of Things Systems. Internet of Things. Springer, Cham (2022). https://doi.org/10.1007/978-3-030-87059-1_1

11. Ciortea, A., Mayer, S., Michahelles, F.: Repurposing manufacturing lines on the fly with multi-agent systems for the web of things. In: Proceedings of the 17th International Conference on Autonomous Agents and MultiAgent Systems, AAMAS 2018, pp. 813–822. International Foundation for Autonomous Agents and Multiagent Systems, Richland, SC (2018). https://dl.acm.org/doi/10.5555/3237383.3237504

12. D'Urso, F., Longo, C.F., Santoro, C.: Programming intelligent IoT systems with a Python-based declarative tool. In: Proceedings of the 1st Workshop on Artificial Intelligence and Internet of Things co-located with the 18th International Conference of the Italian Association for Artificial Intelligence, AI*IA 2019. CEUR Workshop Proceedings, Rende (CS), Italy, vol. 2502, pp. 68–81. CEUR-WS.org (2019)

13. Endler, M., et al.: ContextNet: context reasoning and sharing middleware for large-scale pervasive collaboration and social networking. In: Proceedings of the Workshop on Posters and Demos Track. PDT 2011, Association for Computing Machinery, New York, NY, USA (2011). https://doi.org/10.1145/2088960.2088962

14. Finin, T., Fritzson, R., McKay, D., McEntire, R.: KQML as an agent communication language. In: Proceedings of the Third International Conference on Information and Knowledge Management, CIKM 1994, pp. 456–463. Association for Computing Machinery, New York, NY, USA (1994). https://doi.org/10.1145/191246.191322

15. Kampik, T., Nieves, J.C.: JS-son - a lean, extensible JavaScript agent programming library. In: Dennis, L.A., Bordini, R.H., Lespérance, Y. (eds.) Engineering Multi-Agent Systems, vol. 12058, pp. 215–234. Springer, Cham (2020). https://doi.org/10.1007/978-3-030-51417-4_11

16. Karaduman, B., Tezel, B.T., Challenger, M.: Enhancing BDI agents using fuzzy logic for CPS and IoT interoperability using the JaCa platform. Symmetry **14**(7) (2022). https://doi.org/10.3390/sym14071447

17. Lazarin., N., Pantoja., C., Viterbo., J.: Swapping physical resources at runtime in embedded multiagent systems. In: Proceedings of the 15th International Conference on Agents and Artificial Intelligence - Volume 1: ICAART, pp. 93–104. INSTICC, SciTePress (2023). https://doi.org/10.5220/0011750700003393

18. Lazarin, N., Pantoja, C., Viterbo, J.: Towards a toolkit for teaching AI supported by robotic-agents: proposal and first impressions. In: Anais do XXXI Workshop sobre Educação em Computação, pp. 20–29. SBC, Porto Alegre, RS, Brasil (2023). https://doi.org/10.5753/wei.2023.229753

19. Lazarin, N.M., Pantoja, C.E.: A robotic-agent platform for embedding software agents using Raspberry PI and Arduino boards. In: Proceedings of the 11th Workshop-School on Agents, Environments, and Applications, WESAAC 2015, pp. 13–20. Niteroi (2015)

20. Manoel, F., Pantoja, C.E., Samyn, L., de Jesus, V.S.: Physical artifacts for agents in a cyber-physical system: a case study in oil & gas scenario (EEAS). In: The 32nd International Conference on Software Engineering and Knowledge Engineering, SEKE 2020, pp. 55–60. KSI Research Inc. (2020)

21. Michel, F., Ferber, J., Drogoul, A.: Multi-agent systems and simulation: a survey from the agents community's perspective. In: Danny Weyns, A.U. (ed.) Multi-Agent Systems: Simulation and Applications, p. 47. Computational Analysis, Synthesis, and Design of Dynamic Systems, CRC Press - Taylor & Francis, May 2009

22. Palanca, J., Terrasa, A., Julian, V., Carrascosa, C.: SPADE 3: supporting the new generation of multi-agent systems. IEEE Access **8**, 182537–182549 (2020). https://doi.org/10.1109/ACCESS.2020.3027357

23. Pantoja, C., Soares, H.D., Viterbo, J., Seghrouchni, A.E.F.: An architecture for the development of ambient intelligence systems managed by embedded agents. In: The 30th International Conference on Software Engineering & Knowledge Engineering, pp. 215–249. KSI Research Inc., San Francisco Bay, July 2018. https://doi.org/10.18293/SEKE2018-110

24. Pantoja, C.E., Stabile, M.F., Lazarin, N.M., Sichman, J.S.: ARGO: an extended Jason architecture that facilitates embedded robotic agents programming. In: Baldoni, M., Müller, J.P., Nunes, I., Zalila-Wenkstern, R. (eds.) EMAS 2016. LNCS (LNAI), vol. 10093, pp. 136–155. Springer, Cham (2016). https://doi.org/10.1007/978-3-319-50983-9_8

25. Rafferty, L.: Agent-based modeling framework for adaptive cyber defence of the Internet of Things. Ph.D. Thesis, Faculty of Business and IT, University of Ontario Institute of Technology, Oshawa, Ontario, Canada (2022)

26. da Rocha Costa, A.C., Hübner, J.F., Bordini, R.H.: On entering an open society. In: XI Brazilian Symposium on Artificial Intelligence, vol. 535, p. 546 (1994)

27. Souza de Jesus., V., Pantoja., C.E., Manoel., F., Alves., G.V., Viterbo., J., Bezerra., E.: Bio-inspired protocols for embodied multi-agent systems. In: Proceedings of the 13th International Conference on Agents and Artificial Intelligence - Volume 1: ICAART, pp. 312–320. INSTICC, SciTePress (2021). https://doi.org/10.5220/0010257803120320

28. Wooldridge, M.: An Introduction to MultiAgent Systems. Wiley, Hoboken (2009)

Explainability

Hybrid Multilevel Explanation: A New Approach for Explaining Regression Models

Renato Miranda Filho[1,2(✉)] and Gisele L. Pappa[1]

[1] Universidade Federal de Minas Gerais, Belo Horizonte, Minas Gerais, Brazil
renato.miranda@dcc.ufmg.br
[2] Instituto Federal de Minas Gerais, Sabará, Minas Gerais, Brazil

Abstract. Regression models are commonly used to model the associations between a set of features and an observed outcome, for purposes such as prediction, finding associations, and determining causal relationships. However, interpreting the outputs of these models can be challenging, especially in complex models with many features and nonlinear interactions. Current methods for explaining regression models include simplification, visual, counterfactual, example-based, and attribute-based approaches. Furthermore, these methods often provide only a global or local explanation. In this paper, we propose a hybrid multilevel explanation (Hybrid Multilevel Explanation - HuMiE) method that enhances example-based explanations for regression models. In addition to a set of instances capable of representing the learned model, the HuMiE method provides a complete understanding of why an output is obtained by explaining the reasons in terms of attribute importance and expected values in similar instances. This approach also provides intermediate explanations between global and local explanations by grouping semantically similar instances during the explanation process. The proposed method offers a new possibility of understanding complex models and proved to be able to find examples statistically equal to or better than the main competing methods and to provide a coherent explanation with the context of the explained model.

Keywords: Regression · Explanation · Multilevel hybrid explanation

1 Introduction

Regression models are widely used in various fields, including medicine, engineering, finance, marketing, among others. They are used to generate associations between a set of features and an observed outcome, and can be used for: *(i)*

This work was supported by FAPEMIG (through the grant no. CEX-PPM-00098-17), MPMG (through the project Analytical Capabilities), CNPq (through the grant no. 310833/2019-1), CAPES, MCTIC/RNP (through the grant no. 51119) and IFMG - Campus Sabará.

M. C. Naldi and R. A. C. Bianchi (Eds.): BRACIS 2023, LNAI 14195, pp. 399–414, 2023.
https://doi.org/10.1007/978-3-031-45368-7_26

prediction—when, given a prediction value, we want to interpolate or extrapolate from observations to estimate the outcome, or *(ii)* to find associations between the independent features and the outcome and how strong these associations are.

When the regression models can account for non-linear data relationships, the produced models cannot be directly interpreted by analyzing their coefficients, as it is usual in linear regression. As these models can be complex, have many features and nonlinear interactions, having methods able to understand the reasons that lead to a regression model's output is important, especially in critical areas that can affect human life.

Current methods for explaining regression models provide different approaches, including simplification, visual, counterfactual, example-based, and feature-based explanations [3, 9, 16]. Each of these approaches provides complementary explanations for the model's output, but often they do not provide a complete understanding of why an output occurs. Additionally, these approaches usually provide a global explanation of the overall model behavior or a local explanation about the output of a specific instance.

In this paper we propose Hybrid Multilevel Explanation (HuMiE), a method for explaining regression models that build up on example-based explanations. Example-based explanations have the advantage of being similar to the human way of explaining things by similarity to past events, but they often do not provide a complete understanding of why an output occurs. Our proposed method, on the other hand, provides a complete understanding of the reasons why an output is obtained, in terms of feature importance and expected values in similar instances.

For each selected example, HuMiE explain the reasons why the output is obtained in terms of feature importance and expected values in similar instances. Furthermore, the method provides intermediate explanations between the possibilities of global and local explanations, grouping semantically similar instances during the explanation process and bringing new possibilities of output understanding to the user.

The remainder of this paper is organized as follows. In Sect. 2, we review the existing literature on explanation methods for regression models. In Sect. 3, we introduce HuMiE. In Sect. 4, we report the experimental results. Finally, in Sect. 5, we report our conclusions and possibilities of future work.

2   Related Work

As the use of models created by machine learning algorithm is increasingly used to make critical decisions in our daily lives, so does the concern to understand the internal mechanisms that lead such models to obtain their outputs.

Machine learning models can be divided into two categories: those that create intrinsically interpretable models and those that are considered black-boxes. Intrinsically interpretable models are easily understandable by humans, and include linear models, and decision trees. In black box models, in turn, a human cannot easily understand the internal mechanisms that led to the output [13]. It is fair to assume that currently the most accurate models are precisely of this type [9]. Examples include Support Vector Regressors (SVR) and Neural Network Regression (NNR).

The eXplainable Artificial Intelligence (XAI) literature encompasses methods generated for explaining black-box models, with an emphasis on classification tasks [12]. More recently we have seen various initiatives aimed to explain regression models, where the output is represented by a continuous value instead of a categorical one.

Among the main explanation approaches described in the literature, we have: i) simplification, where mathematical substitutions are made, generating more trivial algebraic expressions; ii) visual, where the model is explained through graphical elements [1, 23]; iii) counterfactual, methods demonstrate that minimal modifications made to the input data can significantly modify the output [8, 17]; iv) feature-based, analyze the characteristics of the data-including, for example, statistical summaries and feature importance- that cause a given black-box model to perform a prediction [20, 22]; and v) example-based, where representative instances of the model are created or selected from the training set (and are also called prototypes) to generate a global explanation [2, 10, 11, 15], and local explanations are based on the distance of these prototypes to the instance to be explained.

In this context, we observe that example-based explanation approaches, although working in a similar to humans in the sense of justifying events based on analogies by similarity, end up not fully explaining why a similar instance resulted in a specific output. Therefore, we propose a hybrid multilevel approach that will fill this gap, providing feature-based explanations that will demonstrate how their output was obtained for each selected example.

One of the few works that has already developed hybrid approaches in the literature was MAME (Model Agnostic Multilevel Explanations) [19]. As per MAME, little attention has been given to obtaining insights at an intermediate level of model explanations. MAME is a meta-method that builds a multilevel tree from a given method of local explanation. The tree is built by automatically grouping local explanations, without any hybrid character combining different types of explanation. The intermediate levels are based on this automatic grouping of the local explanations. Furthermore, the local explanation methods evaluated in MAME were not specifically built to work with regression problems, which can cause some damage when we start to have a continuous range of possible values for each instance output. Finally, MAME did not make the source code available and neither did they make available the tree generated for the only regression dataset used in their work (Auto-MPG), which makes any kind of comparison between the approaches difficult.

3   Hybrid Multilevel Explanations

The section introduces Hybrid Multilevel Explanation (HuMiE), an explanation method that combines both example-based and feature-base explanations.

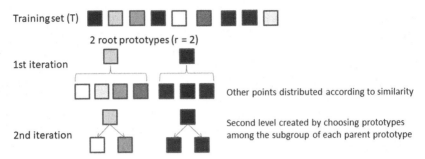

Fig. 1. Example of the multilevel tree built by HuMiE. In this figure, the squares represent instances, and the colors of the instances the feature values used to calculate similarities.

HuMiE produces a tree where each node represents a prototype selected from the training set. After prototypes are selected, a feature-based approach is used to provide further details on the models decision.

The dynamics of creating this explanation tree is exemplified in Fig. 1. In the figure, each square represents an instance from the training set (T), and the colors of the squares reflect the values of the features of the instances. We calculate the similarity between instances using these features. In this example, HuMiE first selects two prototypes as the root of the tree (r). After that, the distance from the remaining instances to each prototype is calculated, and each instance associated with its most similar prototype, generating 2 subgroups. For each of these subgroups, again two children prototypes are chosen. This process is repeated until it reaches the height (h) of the tree defined by the user.

Having a tree of prototypes, for each node of the tree, HuMiE shows the most relevant features considering the prototype associated with that node, together with the range of values assumed by the instances of the subgroup represented by that prototype. Finally, HuMiE allows for a local explanation of a test instance by selecting the leaf-level prototype most similar to the instance to be explained.

Different algorithms can be used to select both the prototypes and to determine features importance. In this paper, we use M-PEER (Multiobjective Prototype-basEd Explanation for Regression) [7] to select prototypes and DELA (Dynamic Explanation by Local Approximation) [5] to determine feature importance, although they could be replaced by other algorithms.

3.1   Prototype Selection

The prototype selection is done by M-PEER. M-PEER is based on the SPEA-2 evolutionary algorithm and simultaneously optimizes three metrics: fidelity of global prototypes, stability of global prototypes, and Root Mean Squared Error (RMSE).

The fidelity and stability measures were originally proposed in the context of classification [14,18] and adapted in [7] for regression. The fidelity metric assesses how close the output of the chosen prototypes is from the output of the instances to be explained, and is defined in Eq. 1.

$$\overline{GF}(f, PR_t, t) = median_{\forall t \in T}(LF(f(t), PR(t))) \tag{1}$$

$$LF(f(t'), PR(t')) = dist(y_i, y_j) = (y_i - y_j)^2 \tag{2}$$

The second metric, global stability, measures whether the characteristics of the prototypes are similar to the characteristics of the instances to be explained, as defined in Eq. 3.

$$\overline{GS}(t, PR_t) = median_{\forall t \in T}(LS(t, PR_t)) \tag{3}$$

$$LS(x_i, x_j) = dist_{Mink}(x_i, x_j) = (\sum_{a=1}^{d} |x_{ia} - x_{ja}|^{0.3})^{1/0.3} \tag{4}$$

where x_i and x_j are the features of instances to be compared, d is the number of dimensions, and $dist_{Mink}$ is the Minkowski distance. In addition, $\overline{GS}$ and $\overline{GF}$ are normalized using min-max.

In M-PEER, the authors observed that the obtained results were better when the $\overline{GS}$ and $\overline{GF}$ metrics are used together through the GFS combination, shown in Eq. 5. By using GFS we make the similarity take into account both the features of the instance and the output predicted by the model.

$$GFS(f, PR_t', t') = \sqrt{\left[\frac{1}{\log(\overline{GF}(f, PR_t', t'))}\right]^2 + \left[\frac{1}{\log(\overline{GS}(t', PR_t'))}\right]^2} \tag{5}$$

Finally, the error is used to ensure that the selected set of prototypes is as close as possible to the error of the black-box model.

3.2 Feature Importance

After selecting the prototypes, we want to obtain a feature-based explanation showing what motivated the output obtained for the specific instance. For this, we use the DELA explanation method.

DELA is an improvement on the work presented by ELA (Explanation by Local Approximation) [6]. By analyzing the coefficients found in local linear regressions, this method provides a global explanation of the model, using a stacked area chart where the user can check the feature importance for different output values, and also provides local explanations for specific instances through the local importance of the features, expected value ranges, and the coefficients of the local explanation. Additionally, DELA considers the characteristics of the data sets used in the model training stage, chooses the most appropriate distance measure to be used, dynamically defines the number of neighbors that will be used for local explanation according to the local data density, and calculates feature importance based on the location of the test instance.

DELA calculates the most relevant features of a prototype using Eq. 6:

$$Importance_{x_i} = \frac{|LevelImportance_i| \times 100}{\sum_{a=1}^{d} |LevelImportance_a|} \tag{6}$$

$LevelImportance_i$ is defined as:

$$LevelImportance_i = b_i \times mean(V_i) \tag{7}$$

where b_i is the coefficient returned for feature i and V_i the vector of values found in the neighboring examples (set of closest instances) for feature i - and the minimum and maximum values of each feature, extracted considering the prototype and the instances in its subgroup.

3.3   Algorithm

The method is illustrated in Algorithm 1. It receives as input the training (T) and test (T') sets, together with the height h of the explanation tree to be built and the number r of root nodes present in the tree, where the root nodes represent the global explanations of the model.

In line 1, we create the regression model to be explained. Note that HuMiE is model agnostic, meaning that it has the ability to explain any regression model. To ensure semantically valid results, HuMiE works with continuous features only. Any categorical feature should be pre-processed. Next, we initialize some important variables for building the multilevel tree. In line 2, the tree is initialized as an array. This array will be filled with tuples containing three pieces of information: the selected prototype, the prototype's depth level in the tree, and the prototype at the previous level (parent) of the selected prototype. In line 3, a dictionary is initialized to store, for each selected prototype, the minimum and maximum values of each feature found in the subset of instances from where the prototype was chosen. This information will serve in the future to show the minimum and maximum values of each feature within the subset that the prototype will be representing. In line 4, an array is initialized to store the prototypes already chosen during the tree construction process. Finally, on line 5, we initialize an array to store the set of prototypes selected at the top level. As the explainer is built on a top-down fashion (i.e., starting from the root level), it has no higher-level prototype, and this variable is initialized with -1. In line 6 it is defined that the number of prototypes of the root level is the value of r received as a parameter in the algorithm.

After that, we carry out the selection of the prototypes that will belong to the tree in the loop between lines 7 and 29. Note that the construction of this tree will start from the root, which will represent the global explanation of the model. After that, for each root level prototype selected, we expand it up to the level at limit h, defined by the user. In line 10 it is defined that for intermediate levels of

Algorithm 1. Hybrid Multilevel Explanation (HuMiE)

Require: T (train), T' (test), h (height of explanation tree), r (number of tree root prototypes)

1: f = Regression model(T)
2: $tree = [\,]$
3: $rangeAtt = \{\}$
4: $selectedPR = [\,]$
5: $prevLevelPR = [-1]$
6: $numPR = r$
7: **for** each $depth \in$ range(h) **do**
8:    $nearestPR = [\,]$
9:    **if** $depth > 0$ **then**
10:      $numPR = 2$
11:      **for** $t \in T$ **do**
12:         $nearestPR$.append(selectPR(t, $prevLevelPR$, f))
13:      **end for**
14:    **end if**
15:    **for** $previousPR \in prevLevelPR$ **do**
16:      $validIndexes = (T \notin selectedPR) \wedge (T \in nearestPR(previousPR))$
17:      **if** $depth == 0$ **then**
18:         $validIndexes = T$
19:      **end if**
20:      $prototypes =$ MPEER($validIndexes$, $numPR$, f)
21:      $tree$.append($prototypes$, $depth$, $previousPR$)
22:      $rangeAtt[previousPR] =$ findMinMaxFeatures($validIndexes$)
23:      $prevLevelPR = prototypes$
24:      $selectedPR$.append($prototypes$)
25:      **if** $depth = h$ **then**
26:         $leaves = prototypes$
27:      **end if**
28:    **end for**
29: **end for**
30: $topFeatures = \{\}$
31: **for** $PR \in selectedPR$ **do**
32:    $topFeatures[PR] =$ DELA(PR)
33: **end for**
34: $localExplanation = \{\}$
35: **for** $t \in T'$ **do**
36:    $localExplanation[t] =$ selectPR(t, $leaves$, f)
37:    $topFeatures[t] =$ DELA(t)
38: **end for**
39: drawHybridTree($tree$, $topFeatures$, $rangeAtt$, $LocalExplanation$)

---

the explanation tree we will have a binary division of the prototypes. In the loop of lines 11 to 13, we check, for all instances of the training set, which prototype of the previous level is the most similar to represent it. We are redistributing the training points to identify the most suitable prototypes in a subset. Note that if we are at the first level this array will be an empty set (line 8). To measure the similarity between the instances we used the L2 of the normalized values of LF and LS measure (Eq. 8).

$$L2FS(f, PR_{t'}, t') = \sqrt{\left[\frac{1}{\log(\overline{LF}(f(PR_{t'}), f(t')))}\right]^2 + \left[\frac{1}{\log(\overline{LS}(t', PR_{t'}))}\right]^2}$$

(8)

$\overline{LF}$ and $\overline{LS}$ are normalized values by the min-max of the original metrics.

In the loop between lines 15 and 28 we find the prototypes that represent each of the subsets divided by the prototypes in the previous iteration. To accomplish this, in line 20, we check the indexes of valid training points, i.e., all training points explained by the same prototype in the previous tree level, excluding instances already selected as prototypes. The conditional performed on line 17 verifies whether the chosen prototypes are from the root level, if so, all instances of the training set make up the list of valid indexes. In line 20, the M-PEER method is called for this subset of instances and the chosen prototypes are added to the tree in line 21. In line 22, the minimum and maximum values of all instances used to select each of the prototypes are checked, and in line 24 the set of selected prototypes is added to the array of previously selected prototypes. Finally, the condition in line 25 tests whether we have reached the last level h defined by the user, which corresponds to the leaf nodes of the tree that will be later associated with test instances.

The selection of the most relevant attributes and the range of expected values for each attribute of similar instances is performed in the loop between lines 31 and 33 for selected prototype points at all levels of the tree. The same procedure of feature analysis is performed in the loop between lines 35 and 38 for the set of test instances. This is done by DELA. In line 36, we check which leaf node is closest to the test point to associate the two in the final view of the tree.

Finally, on line 39, a function is called to draw the tree, showing the most relevant features and the range of expected features for the selected prototype instance and for the test instances.

4   Experiments

We perform both a quantitative and a qualitative evaluation of the prototypes chosen by HuMiE and their characteristics.

4.1   Experimental Setup

In the quantitative evaluation, we compare the prototypes chosen by HuMiE for local explanation with those chosen by other methods of explanation based on examples: ProtoDash [10] and the standalone version of M-PEER. ProtoDash is an example-based explanation method that selects prototypes by comparing the

Table 1. Dataset used in the experiments of the HuMiE explanation method.

Id	Dataset	# Attributes	# Train	# Test
1	cholesterol	14	209	88
2	auto-mpg	8	279	119
3	sensory	12	403	173
4	strike	7	437	188
5	day_filter	12	510	221
6	qsar_fish	7	636	272
7	concrete	9	721	309
8	music	70	741	318
9	house	80	1022	438
10	wineRed	12	1279	320
11	communities	103	1395	598
12	crimes	103	1550	664
13	abalone	8	2924	1253
14	wineWhite	12	3918	980
15	cpu_act	22	5734	2458
16	bank32nh	33	5734	2458
17	puma32H	33	5734	2458
18	compactiv	21	5734	2458
19	tic	85	6876	2947
20	ailerons	41	9625	4125
21	elevators	19	11619	4980
22	california	8	14448	6192
23	house_16H	17	15949	6835
24	fried	11	28538	12230
25	mv	10	28538	12230

distribution of training data and the distribution of candidate prototypes, i.e., it does not take into account the model output.

For that, we use the 25 datasets presented in Table 1. To measure the quality of the prototypes, we use the classic root mean squared error (RMSE) and also the metrics GF, GS and GFS presented in Sect. 3.1. We also performed a qualitative analysis on the classic Auto-MPG dataset, which portrays the fuel consumption of automobiles (miles per gallon - MPG) according to their construction characteristics.

4.2 Results

This section first presents a quantitative experimental analysis of the explainability method proposed. The models to be explained were built using a Random Forest Regression algorithm, implemented with sklearn[1]. All parameters were used with their default values.

Taking as input the model built using each of the datasets shown in Table 1, we run the explanation methods ProtoDash, M-PEER and HuMiE to compare

---

[1] https://scikit-learn.org/stable/modules/generated/sklearn.ensemble.
RandomForestRegressor.html.

Table 2. Result of the quality of the prototypes chosen by different methods in the local explanation of the test instances.

Id	ProtoDash				MPEER				HuMiE			
	GS	GF	GFS	RMSE	GS	GF	GFS	RMSE	GS	GF	GFS	RMSE
1	0.06	433.10	0.78	0.98	0.11	463.83	0.81	0.69	0.08	**354.21**	0.75	**0.61**
2	0.02	117.60	0.60	**0.56**	0.07	96.17	0.63	0.81	0.05	**73.77**	0.60	0.68
3	0.14	478.57	0.85	1.26	0.19	465.81	0.86	**0.71**	0.10	**337.01**	0.77	0.72
4	0.02	153.61	0.59	0.61	0.03	171.31	0.61	**0.53**	0.02	**118.43**	0.57	0.59
5	0.11	409.93	0.82	0.85	0.09	608.41	0.88	0.73	0.03	**396.44**	0.77	0.68
6	0.04	68.37	**0.56**	0.75	0.10	69.00	0.60	0.71	0.05	**59.99**	0.56	**0.60**
7	0.10	**282.32**	0.74	**0.77**	0.11	307.73	0.76	1.06	0.06	283.02	0.72	0.79
8	0.11	**1.86e+5**	1.25	0.82	0.13	1.97e+5	1.29	0.88	0.09	1.95e+5	1.27	0.99
9	0.05	9.58e+4	0.74	0.78	0.07	9.64e+4	0.75	0.84	0.05	**8.45e+4**	0.72	0.68
10	0.02	1656.46	**0.43**	0.73	0.06	1391.73	0.48	0.68	0.06	**1230.08**	0.48	**0.60**
11	0.10	2.08e+6	1.43	0.94	0.13	1.91e+6	1.39	0.66	0.10	**1.70e+6**	1.30	**0.62**
12	0.00	1.56e+6	1.16	**0.35**	0.00	1.66e+6	1.20	0.35	0.00	**1.54e+6**	1.16	0.36
13	0.05	525.57	0.42	0.93	0.06	386.61	0.43	0.70	0.01	**338.52**	0.35	**0.60**
14	0.01	1831.79	**0.40**	0.68	0.05	1544.18	0.45	**0.67**	0.05	**1415.63**	0.45	0.66
15	0.00	4972.93	0.40	**0.94**	0.01	**2978.10**	0.40	0.98	0.00	3076.94	0.38	1.25
16	0.00	7.05e+4	**0.53**	**0.81**	0.02	7.01e+4	0.55	0.93	0.05	**6.76e+4**	0.58	0.96
17	0.02	**8.21e+4**	**0.57**	1.01	0.21	8.37e+4	0.66	**1.00**	0.05	8.51e+4	0.59	1.01
18	0.01	4035.01	0.41	**0.91**	0.01	3106.38	0.40	0.97	0.00	**2945.77**	0.38	1.04
19	0.00	1.13e+5	0.50	**0.72**	0.00	1.01e+5	0.49	**0.72**	0.00	**8.37e+4**	0.47	0.72
20	0.02	2.40e+4	0.47	**0.75**	0.05	2.26e+4	0.49	0.95	0.01	**2.16e+4**	0.46	0.86
21	0.01	4277.07	0.38	**0.84**	0.02	**3706.47**	0.40	0.89	0.01	3763.58	0.38	0.88
22	0.02	670.03	0.42	0.91	0.07	478.54	0.48	1.08	0.02	**470.18**	0.41	**0.83**
23	0.01	3627.19	0.37	**0.75**	0.00	2745.08	0.36	0.98	0.00	**2646.99**	0.36	0.81
24	**0.28**	1253.83	**1.56**	**0.85**	0.39	1357.11	1.65	0.94	0.29	**1248.61**	1.56	0.97
25	0.04	525.34	1.00	**0.81**	0.06	590.88	1.06	1.00	0.02	**504.68**	0.98	0.87

the quality of the selected prototypes. For ProtoDash, we selected 6 global explanation prototypes and assigned each of the test instances to the most similar. For the M-PEER method we selected 3 global prototypes that are used for local explanation according to the proximity of the test instances. Finally, in HuMiE we select 3 global prototypes (the same as in the M-PEER method) to generate a sublevel of 6 prototypes that were then used to explain the test instances.

The results obtained for each dataset are shown in Table 2, where the best ones are highlighted in bold. In addition, we show the critical diagrams of each of the metrics in the graphs in Fig. 2. These diagrams were generated after carrying out an adapted Friedman test followed by a Nemenyi post-hoc with a significance level of 0.05. As we are comparing multiple approaches, Bonferroni's correction was applied to all tests [4]. In these diagrams, the main line shows the average ranking of the methods, i.e., how well one method did when compared to the others. This ranking takes into account the absolute value obtained by each method according to the evaluated metric. The best methods are shown on the left (lowest rankings).

As we can see, the creation of a prototype explanation intermediate level managed to keep the same RMSE as its competitors, as not method was statistically superior than the others. Furthermore, HuMiE was statistically better than the explanation with only one M-PEER prototype level in relation to Global Stability and GFS. HuMiE was also statistically better than all competitors in terms of the Global Fidelity of the chosen prototypes.

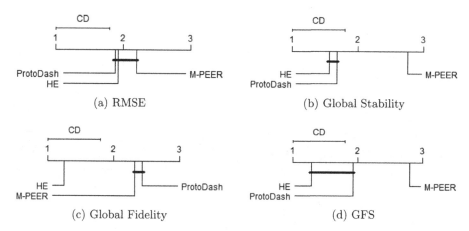

Fig. 2. Critical diagrams of the choice of prototypes for local explanation of test instances according to RMSE metrics (a), Global Stability (b), Global Fidelity (c) and GFS (d). From left to right, methods are ranked according to their performance, and those connected by a bold line present no statistical difference in their results.

Next, we show a qualitative analysis for Auto-MPG, which has 398 instances described by 8 features. To evaluate the model error, we divided the data into two subsets, the first being the training set with 70% of the instances (279) and the test set with the remaining 30% (119). Note that, in this case, the model error is used only to identify how good the explanation model is related to the original regression model to be explained. As our focus is the explanation provided and not minimizing the error, we do not perform any further evaluation, for example, through cross-validation.

The Auto-MPG dataset portrays a classic regression problem whose objective is to find the consumption of a car having as features constructive characteristics of the vehicle, they are: cylinders, displacement, horsepower, weight, acceleration, model year and origin. To evaluate the explanation provided by our HuMiE approach we build a model using all available features in a Random Forest algorithm. The RMSE error found on the test set was 0.36.

Figure 3 shows the explanation provided for the model. In the explanations provided in this work we defined the height h of the tree as 2 and the total number of prototypes in the global explanation of the root as 3. Two test instances, one with a low output value and another with a high output value, were previously selected for local explanation.

In the explanation, each node highlighted in gray represents a prototype selected from the training set and the nodes, highlighted in red represent two test instances selected for local explanation. In addition, within each of the nodes, the actual output value of the instance (y) and the respective value predicted by the regressor model (y') are presented in the first line. The other lines of the tree node represent the 5 main features identified for that instance with their

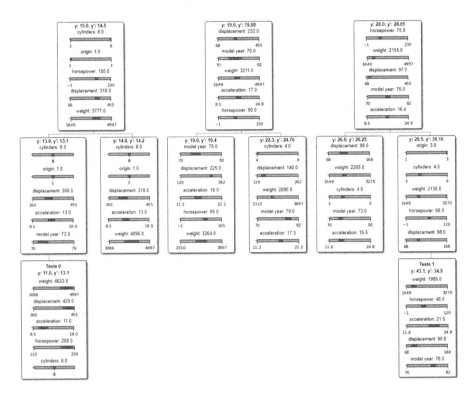

Fig. 3. Hybrid tree explanation for the Auto-MPG dataset. (Color figure online)

respective value (the most relevant number of features to be shown can easily be changed if the user wants to).

The bars represent the observed threshold values between all valid closest points when finding the prototype. In the case of the first upper level (root of the tree) the limit is imposed by the values of all instances of the training set. In the case of the test node the limit is imposed by all the closest features that originated the prototype local explanation leaf (leaf node of the hybrid explanation tree). The red dot represents the feature-specific value for the instance. The green range represents the allowable variation for creating a similar instance, determined by the DELA method. Specifically, it represents the range of feature values found among instances that deviate no more than 10% from the output predicted by the model. When hovering over one of the features, the green range is shown numerically with a highlight balloon for the user.

Note that the global explanation prototypes managed to cover the solution space well, with the first prototype being a high fuel consumption car (14.5 MPG), the second an intermediate consumption car (19.08 MPG) and the last one a low fuel consumption car of fuel (28.05 MPG). In addition, among the most relevant features at this level, we find those that domain experts

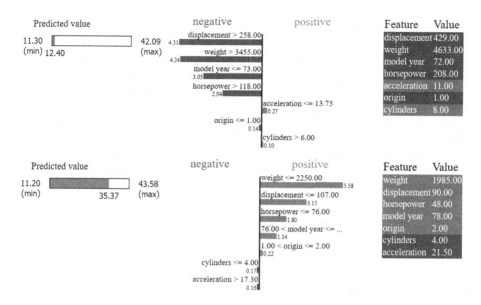

Fig. 4. LIME Explanation for the Auto-MPG Dataset

traditionally portray as determining the consumption of a car, we can highlight: weight, displacement, horsepower and acceleration.

In the second level (intermediate explanation) we observe that the prototypes really represent subgroups of the prototypes of the global explanation. This observation can be made both in terms of the output returned by the model and the range for each of the features. In the output we can observe that the values obtained do not extrapolate to values of the side prototypes of the previous level. In the range of features, we observed, for example, that in the first prototype of the second level the displacement feature varied between 260 and 455 while in its parent prototype these values varied between 68 and 455, portraying a specification of the group.

Finally, we can also observe that the L2FS similarity metric was adequate when associating the leaf prototypes of the explanation tree with the test instances, since both the values returned by the regressor model and the feature intervals represent the best possible association in each specific scenario.

Note that an explanation provided by the M-PEER for this dataset would only consist of the 3 instances represented at the root level of the explanation tree. In this context, we would only have information about the values for each of the features for these instances.

We can also compare our explanation with what we could obtain from a popular feature-based explanation method called LIME (Local Interpretable Model-Agnostic Explanations) [21]. LIME explains an observation locally by generating a linear model (considered interpretable by humans) using randomly generated instances in the neighborhood of the instance of interest. The linear model is then used to portray the feature's contribution to the output of the specific instance based on the model's coefficients.

Using the same two test instances shown in Fig. 3, we have the explanations provided by LIME in Fig. 4. As we can observe, LIME would end up showing semantically invalid interpretations to the user, such as portraying a positive contribution of the "weight" feature for cars weighing below 2250.0, without a lower limit. Thus, one could interpret, for example, the possibility of a negative weight for the car. Furthermore, with our hybrid multilevel approach, we have access to both similar instances in the training set and the characteristics that make these instances relevant, providing an explanation that brings new information about the model.

5   Conclusions and Future Work

HuMiE is a method capable of creating multilevel hybrid explanations for regression problems. By hybrid we mean that the provided explanation encompasses as many elements as explanations based on examples, through the selection of prototypes from the training set, as well as the presentation of explanations based on features such as, for example, the most relevant features and the expected values of features for the region of a given instance. By multilevel, we refer to the fact that HuMiE presents the model's global explanations in a tree format, with local explanations for specific test instances being at the leaves of this tree, global explanations at the root but also intermediate explanations, portraying semantically similar subgroups. HuMiE uses the M-PEER approach as a auxiliar method for choosing prototypes and the DELA approach for features analysis.

Experiments with real-world datasets quantitatively showed that the prototypes chosen by HuMiE were better than all competitors (M-PEER and Proto-Dash) in relation to the fidelity metric and better than M-PEER with a single level in relation to the stability metrics and GFS. Qualitatively, we showed that HuMiE is able to diversify the choice of prototypes according to the characteristics of the presented dataset, both in terms of model output and features. In addition, HuMiE was able to find subgroups of instances that are similar, providing an intermediate interpretation between the local and global explanations which might be useful for the user to understand the moded's rationale. In this context, HuMiE was able to identify which features are most important according to the subgroup the instance belongs to.

As future work, we propose to explore additional metrics to evaluate HuMiE's performance and the robustness of explanations, in order to gain a more comprehensive view of the method's capabilities. Furthermore, combining other explanation approaches, such as counterfactuals, during the construction of the hybrid explanation could contribute to an even more thorough understanding of the model.

References

1. Adadi, A., Berrada, M.: Peeking inside the black-box: a survey on explainable artificial intelligence (XAI). IEEE Access **6**, 52138–52160 (2018)
2. Bien, J., Tibshirani, R.: Prototype selection for interpretable classification. Ann. Appl. Stat. **5**(4), 2403–2424 (2011)
3. Carvalho, D.V., Pereira, E.M., Cardoso, J.S.: Machine learning interpretability: A survey on methods and metrics. Electronics **8**(8), 832 (2019)
4. Demšar, J.: Statistical comparisons of classifiers over multiple data sets. J. Mach. Learn. Res. **7**, 1–30 (2006)
5. Filho, R.M.: Explaining Regression Models Predictions. Ph.D. thesis, Universidade Federal de Minas Gerais (2023)
6. Filho, R.M., Lacerda, A., Pappa, G.L.: Explaining symbolic regression predictions. In: 2020 IEEE Congress on Evolutionary Computation (CEC), pp. 1–8 (2020)
7. Filho, R.M., Lacerda, A.M., Pappa, G.L.: Explainable regression via prototypes. ACM Trans. Evol. Learn. Optim. **2**(4) (2023)
8. Grari, V., Lamprier, S., Detyniecki, M.: Adversarial learning for counterfactual fairness. Mach. Learn. **11**, 1–23 (2022)
9. Guidotti, R., Monreale, A., Ruggieri, S., Turini, F., Giannotti, F., Pedreschi, D.: A survey of methods for explaining black box models. ACM Comput. Surv. **51**(5), 9:31-93:42 (2018)
10. Gurumoorthy, K.S., Dhurandhar, A., Cecchi, G.A., Aggarwal, C.C.: Efficient data representation by selecting prototypes with importance weights. In: ICDM 2019 (2019)
11. Kim, B., Khanna, R., Koyejo, O.O.: Examples are not enough, learn to criticize! criticism for interpretability. In: Lee, D., Sugiyama, M., Luxburg, U., Guyon, I., Garnett, R. (eds.) Advances in Neural Information Processing Systems, vol. 29. Curran Associates, Inc. (2016)
12. Letzgus, S., Wagner, P., Lederer, J., Samek, W., Muller, K.R., Montavon, G.: Toward explainable artificial intelligence for regression models: a methodological perspective. IEEE Sig. Process. Mag. **39**(4), 40–58 (2022)
13. Loyola-González, O.: Black-box vs. white-box: understanding their advantages and weaknesses from a practical point of view. IEEE Access **7**, 154096–154113 (2019)
14. Melis, D.A., Jaakkola, T.: Towards robust interpretability with self-explaining neural networks. In: NeurIPS 2018, pp. 7775–7784 (2018)
15. Ming, Y., Xu, P., Qu, H., Ren, L.: Interpretable and steerable sequence learning via prototypes. In: KDD 2019, pp. 903–913 (2019)
16. Molnar, C.: Interpretable Machine Learning, 2nd edn. Lulu.com (2022). https://christophm.github.io/interpretable-ml-book
17. Mothilal, R.K., Sharma, A., Tan, C.: Explaining machine learning classifiers through diverse counterfactual explanations. In: Proceedings of the 2020 Conference on Fairness, Accountability, and Transparency, FAT* 2020, pp. 607–617. Association for Computing Machinery, New York, NY, USA (2020)
18. Plumb, G., Al-Shedivat, M., Xing, E.P., Talwalkar, A.: Regularizing black-box models for improved interpretability. CoRR (2019)
19. Ramamurthy, K.N., Vinzamuri, B., Zhang, Y., Dhurandhar, A.: Model agnostic multilevel explanations. In: Larochelle, H., Ranzato, M., Hadsell, R., Balcan, M., Lin, H. (eds.) Annual Conference on Neural Information Processing Systems 2020, NeurIPS 2020. Advances in Neural Information Processing Systems 33, 6–12 December 2020, Virtual (2020)

20. Ribeiro, M.T., Singh, S., Guestrin, C.: "why should i trust you?": explaining the predictions of any classifier. In: Proceedings of the 22Nd ACM SIGKDD International Conference on Knowledge Discovery and Data Mining, KDD 2016, pp. 1135–1144. ACM, New York, NY, USA (2016)
21. Ribeiro, M.T., Singh, S., Guestrin, C.: "Why should i trust you?" Explaining the predictions of any classifier. In: KDD 2016, pp. 1135–1144 (2016)
22. Schwab, P., Karlen, W.: CXPlain: causal explanations for model interpretation under uncertainty. In: NeurIPS 2019, pp. 10220–10230 (2019)
23. Zhao, Q., Hastie, T.: Causal interpretations of black-box models. J. Bus. Econ. Stat. **39**(1), 272–281 (2021)

Explainability of COVID-19 Classification Models Using Dimensionality Reduction of SHAP Values

Daniel Matheus Kuhn[1]([⊠]), Melina Silva de Loreto[2],
Mariana Recamonde-Mendoza[1], João Luiz Dihl Comba[1],
and Viviane Pereira Moreira[1]

[1] Institute of Informatics, UFRGS, Porto Alegre, Brazil
{daniel.kuhn,mrmendoza,comba,viviane}@inf.ufrgs.br
[2] Hospital de Clínicas de Porto Alegre, Porto Alegre, Brazil

Abstract. The critical scenario in public health triggered by COVID-19 intensified the demand for predictive models to assist in the diagnosis and prognosis of patients affected by this disease. This work evaluates several machine learning classifiers to predict the risk of COVID-19 mortality based on information available at the time of admission. We also apply a visualization technique based on a state-of-the-art explainability approach which, combined with a dimensionality reduction technique, allows drawing insights into the relationship between the features taken into account by the classifiers in their predictions. Our experiments on two real datasets showed promising results, reaching a sensitivity of up to 84% and an AUROC of 92% (95% CI, [0.89–0.95]).

Keywords: Mortality prediction · Classification Models · Explainable AI · Dimensionality Reduction · SHAP values

1 Introduction

The development of predictive models for the diagnosis and prognosis of patients has been recently investigated by numerous works [17]. The increasing interest in clinical predictive models may be explained by factors such as the growing capacity of data collection, the positive impact on decision-making and clinical care, and the great advances in machine learning (ML). These models aim to assist in the various challenges faced by health professionals while treating patients, optimizing clinical management processes, and allocating hospital resources, ultimately promoting better patient care.

The COVID-19 pandemic was a critical public health emergency worldwide, given the sudden need for many hospital beds and increased demand for medical equipment and health professionals. This scenario intensified the demand for resources and, consequently, the interest and need in using predictive models to support decision-making for diagnosis and prognosis in the context of COVID-19.

© The Author(s), under exclusive license to Springer Nature Switzerland AG 2023
M. C. Naldi and R. A. C. Bianchi (Eds.): BRACIS 2023, LNAI 14195, pp. 415–430, 2023.
https://doi.org/10.1007/978-3-031-45368-7_27

As discussed by previous works [2], a considerable number of papers related to COVID-19 were published in a short period, including papers proposing ML applications for COVID-19 prognosis prediction.

Among the models for prognosis prediction, we can cite predictive models that aim to anticipate the patient's outcome (*e.g.*, hospital discharge or death), the severity of the disease, the hospitalization length, the evolution of the patient's clinical condition during hospitalization, the need for intensive care unit (ICU) admission or interventions such as intubation, and the development of complications (*e.g.*, cardiac problems, thrombosis, acute respiratory syndrome) [16,17]. In this work, we are particularly interested in mortality prediction models.

Mortality prediction models help recognize patients with higher risks of poor outcomes at the time of admission or during hospitalization, providing support to more effective triage and treatment of ill patients and reducing death rates. Although previous studies have shown that ML algorithms achieved promising results in the mortality prediction task [2,9,18], the lack of explainability of the features taken into account by the predictive models has been pointed out as a limiting factor for their application in the clinical routine [9].

This work aims to evaluate several ML algorithms for predicting COVID-19 mortality and explore state-of-the-art explainability approaches to understand the factors considered during prediction. We developed a visualization technique based on SHAP values [11] combined with a dimensionality reduction method that draws insights into the relationship between the features taken into account by the predictive models and their correct and incorrect predictions. To assess our approach, we conducted a retrospective study of patients diagnosed with COVID-19, analyzing data collected during hospitalization in two hospitals. We also analyze the ability of classifiers to identify patients with a high probability of dying at different time intervals. Our main contributions can be summarized as follows: (*i*) evaluation of six ML algorithms for the mortality prediction task; (*ii*) proposal of a visualization technique based on SHAP values to assist in explainability; and (*iii*) experiments with real data from two hospitals.

Our findings showed that, in general, deceased patients were older. In addition, age, oxygen saturation, heart disease, and hemogram information also tended to have high contributions in the predictions. Logistic Regression, AdaBoost, and XGBoost models recurrently achieved the highest sensitivity scores among the evaluated classifiers.

2   Related Work

This section discusses existing works that addressed mortality prediction for COVID-19 patients based on the information available at hospital admission. We considered works evaluating classical risk scores (including those designed especially for COVID-19 patients and those that precede the pandemic) and ML classifiers. Several techniques have been employed to address mortality prediction, such as using risk scores based on classical statistical approaches and popular classification algorithms [5,9,20].

Covino *et al.* [5] evaluated six physiological scoring systems recurrently used as early warning scores for predicting the risk of hospitalization in an ICU and the risk of mortality—both tasks considering up to seven days. The Rapid Emergency Medicine Score obtained the best results with an AUROC of 0.823 (95% CI, [0.778–0.863]). Zhao *et al.* [20] developed specific risk scores for predicting mortality and ICU admission. The scores were developed based on logistic regression and reached an AUROC of 0.83 (95% CI, [0.73–0.92]). Another score specifically designed to predict mortality risk was the 4C Mortality Score [9]. It uses generalized additive models to categorize continuous feature values and pre-select predictors, followed by training a logistic regression model. The 4C score yielded an AUROC of 0.767 (95% CI, [0.760–0.773]), which was outperformed by the XGBoost algorithm by a small margin. Still, the authors claim that their score is preferable due to the difficulty in interpreting the predictions made by XGBoost. Paiva *et al.* [13] also evaluated classification algorithms for predicting mortality, including a transformer (FNet), convolutional neural networks (CNN), boosting algorithms (LightGBM), Random-Forest (RF), support vector machines (SVM), and K-Nearest-Neighbors (KNN). In addition, an ensemble meta-model was trained to combine all the aforementioned classifiers. The ensemble model achieved the best results, with an AUROC of 0.871 (95% CI [0.864–0.878]). To explain the predictive models, the study contemplated the application of SHapley Additive exPlanations (SHAP) *et al.* [11]. Araújo *et al.* [3] developed risk models to predict the probability of COVID-19 severity. To perform the study, the authors collected blood biomarkers data from COVID-19 patients. The models were developed with a LightGBM-based implementation, which achieved an average AUROC of 0.91 ± 0.01 for mortality prediction. The authors of this work apply a visualization technique similar to the one proposed in our work. The main difference is that they use it only to assess the model's ability to separate the targeted classes. Conversely, in our work, we use visualization to analyze the relationship between the features taken into account by the classifiers and their correct and incorrect predictions.

The datasets used in the studies typically range from a single hospital unit with a few hundred patients to datasets comprising thousands of hospitalization records from hundreds of hospital units in several countries. Covino *et al.* [5] and Zhao *et al.* [20] carried out their studies based on datasets containing 334 and 641 patients, respectively, from a single hospital unit. Paiva *et al.* [13] used a dataset with 5,032 patients from 36 hospitals in 17 cities in 5 Brazilian states, and Araújo *et al.* [3] used a dataset with 6,979 patients from two Brazilian hospitals. To develop the 4C Mortality Score, Knight *et al.* [9] used a large dataset with 57,824 patients collected from 260 hospitals in England, Scotland, and Wales. Although the datasets used in the different studies do not contain the same features, there is a recurrence among the features considered most relevant by the predictive models. Age is the most frequently cited variable. Other features include oxygen saturation, respiratory rate, comorbidities, heart failure, lactate dehydrogenase (LDH), C-reactive protein (CRP), and consciousness level metrics (*e.g.*, Glasgow Coma Scale).

Fig. 1. Pipeline employed in this work.

Our study complements the aforementioned works by evaluating six algorithms in new datasets from two hospitals. In this work, in addition to applying the explainability approach, we developed a visualization technique that allows drawing insights into the relationship between the features taken into account by the classifiers and their correct and incorrect predictions.

3   Materials and Methods

This work aims to predict the risk of COVID-19 mortality during patient hospitalization based on information available at the time of admission. We approach this task through the use of ML classification algorithms.

Our method comprises five steps, depicted in Fig. 1. Our input dataset goes through preprocessing and cleaning, feature selection, then classification. The results of the classifiers are evaluated and submitted to the explainability approach. The next sections describe our data and each step in greater detail.

3.1   Patient Records and Preprocessing

The study involved patients hospitalized in two institutions: HMV and HCPA. HMV is a private hospital. HCPA is a public tertiary care teaching hospital academically linked to a University. The project was approved by the institutional Research Ethics Committee of both institutions[1]. Informed consent was waived due to the retrospective nature of the study. The collected data did not include any identifiable information to ensure patient privacy.

Patients included in the study are adults (age ≥ 18) diagnosed with COVID-19 using the reverse transcriptase polymerase chain reaction (RT-PCR) test. The features include clinical and demographic data: age, sex, ethnicity, comorbidities (hypertension, obesity, diabetes, *etc.*), lab tests (arterial blood gas, lactate, blood count, platelets, C-reactive protein, d-dimer, creatine kinase (CK), creatinine, urea, sodium, potassium, bilirubin), need for ICU admission, type of ventilatory support during hospitalization and ICU period (O2 catheter, non-invasive ventilation, high-flow nasal cannula, invasive mechanical ventilation), start and end date of hospitalization, as well as hospitalization outcome. We performed the Shapiro-Wilk test on continuous features. The results showed that our features are not normally distributed ($\alpha = 0.01$).

---

[1] Ethics committee approval: HCPA 32314720.8.0000.5327, HMV 32314720.8.3001.5330.

Patients without information on the outcome of hospitalization (dependent variable) were not considered in this study. At the HCPA hospital, 318 patients had no record of outcome due to transfer to other hospitals and 6 due to withdrawal or evasion of treatment. For the HMV hospital, patients without outcome information were disregarded during data collection. Variables with missing values greater than 40% were also not considered. In the HMV dataset the variable brain natriuretic peptide was discarded, as well as the variables weight, height and blood type for HCPA. The variables troponin from HMV and fibrinogen from HCPA were the only exceptions to this rule, because although they had a missing value rate of 44.67% and 41.78%, respectively, there is previous evidence regarding their importance as indicators of severity for COVID-19. In HMV, 23 patients (5 deaths and 18 discharges) were disregarded because they had a missing value index greater than 30%, while in HCPA, 258 (93 deaths and 165 discharges) were not considered.

In HMV and HCPA, 7 and 19 patients, respectively, had two hospital admissions (readmissions). Among these, 7 patients from HMV and 13 patients from HCPA were readmitted within 30 days of the discharge from the index hospitalization. For these patients, we used the admission records collected at the index hospitalization with the readmission outcome as the label (patients who were discharged, returned within 30 days, and died, were considered deaths). For the other 6 patients from HCPA who were readmitted after 30 days, we used only the record of the index hospitalization.

HMV Datasets. contain the anonymized records of patients confirmed for COVID-19 admitted at HMV between March 2020 and June 2021. There are 1,526 records, among which 58.78% refer to male patients. The dataset covers both ward patients and patients admitted to the ICU. The data collection process was conducted by health professionals, who manually transcribed the information contained in the patients' medical records into a form containing the features previously established for the study (clinical information available at hospital admission). Patients without outcome information were not considered during the data transcription process. Thus, our study did not include hospitalization records in which patients were transferred to other hospitals and/or dropout cases. Automatic verification processes were applied to identify inconsistencies. We developed classifiers considering two scenarios. In the first scenario, data of all patients are considered, including ward and ICU patients (HMV_{ALL}). In the second scenario, only patients admitted to the ICU are considered (HMV_{CTI}). In both cases, we focus on information available during admission.

$HCPA_{CTI}$ Dataset contains anonymized records of 2,269 patients with confirmed COVID-19 admitted to the ICU of HCPA between March 2020 and August 2021. Male patients represent 54.69% of the dataset. The data was collected directly from the hospital database.

The outcome of interest was mortality during hospitalization, thus, records were categorized as *alive* or *deceased* according to data provided by the hospitals. Details of the datasets after the preprocessing step are shown in Table 1.

Table 1. Statistics of the Datasets

	HMV$_{ALL}$	HMV$_{CTI}$	HCPA$_{CTI}$
Records	1,526	458	2,269
Deceased	182	143	849
Alive	1,344	315	1,420

3.2 Feature Selection

Each raw instance was represented by roughly 60 and 70 features for HMV and HCPA datasets, respectively. Features with more than 40% of missing values were discarded. To pre-select features with good predictive power and avoid multicollinearity among predictive features, we applied the Correlation-based Feature Selection approach (CFS). This approach evaluates subsets of features prioritizing features highly correlated with the classes and uncorrelated with each other [7]. Feature selection was performed only over the training instances to avoid data leakage.

3.3 Classification

We applied six classification algorithms: (*i*) Logistic Regression (LR), three tree-based algorithms, namely (*ii*) J48, (*iii*) RandomForest (RF), and (*iv*) Gradient Boosting Tree - XGBoost (XGB), (*v*) a boosting-based ensemble (AdaBoostM1), and (*vi*) Multi Layer Perceptron (MLP).

In supervised classification approaches, class imbalance refers to the inequality of the proportions of instances for each class. Class imbalance is frequently observed in healthcare data, where the class of interest is often the minority (*e.g.*, disease occurrence or poor diagnosis). When training classification algorithms on imbalanced data without the necessary care, these algorithms tend to present biased results, prioritizing the performance of the majority class. In our datasets, the outcome variable presents, to some degree, an imbalance between the classes (*alive* and *deceased*). The greatest imbalance is observed in the HMV$_{ALL}$ dataset, with approximately 7.38 patients with an *alive* outcome for each patient with a *deceased* outcome. Considering only patients admitted to the ICU, the imbalance is 2.2 and 1.67 alive for each deceased, for the HMV$_{CTI}$ and HCPA$_{CTI}$ datasets, respectively.

To deal with class imbalance, we applied the Cost-sensitive matrix approach, in which the classification errors (false-negatives (FN) or false-positives (FP)) are weighted depending on the task to be addressed [14]. Since the *outcome = deceased* is our positive class, we applied a cost matrix that increased the penalty for FN. The cost matrices applied considered the imbalance between classes so that the penalty for false negatives was defined as $pen = \frac{nc\_maj}{nc\_min}$ where nc_maj is the number of instances of the majority class and nc_min is the number of instances of the minority class.

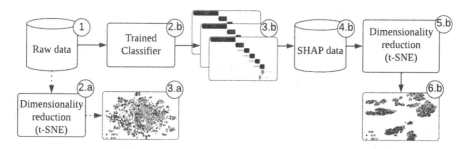

Fig. 2. Steps applied in explaining and interpreting the results returned by the predictive models.

3.4   Evaluation

Metrics. With the aim of evaluating the classification models, we calculated several metrics, namely: sensitivity (also known as recall or true positive rate), specificity (or true negative rate), positive predictive value (PPV or precision), negative predictive value (NPV), F1 (both for the positive class ($F1_+$) and macro-averaged (ma-F1) for both classes), Cohen's kappa, Area Under Receiver Operating Characteristic Curve (AUROC), and Area Under Precision-Recall Curve (AUPRC). Using different metrics is important to interpret the results from different perspectives.

Resampling. We used the resampling technique *Bootstrap* to evaluate the predictive models. For each dataset, we performed 100 iterations of resampling with replacement. For each iteration, several instances—equal to the number of instances in the dataset—were randomly selected to compose the training set, and the remaining instances to constitute the test set. Thus, for each of the 100 iterations, the classifiers were trained on the training set and evaluated on the test set. At the end of the process, we take the means of the evaluation metrics.

3.5   Explainability of Classification Results

This section describes the proposed approach to help explain the features taken into account by the underlying model during inference. The goal is to extract insights into the relationship between the features utilized by the predictive models and their correct and incorrect predictions. The proposed approach corresponds to step four, shown in Fig. 1 and its components are shown in Fig. 2. The idea is similar to supervised clustering [10], in which SHAP values (not raw feature values) are used to cluster instances. However, instead of clustering, we use them to generate visual representations to inspect predictions.

First, starting from a generic dataset D, which in this case, could be any of our datasets, we use dimensionality reduction to understand aspects of linear separability of the data. It is expected that in the raw data, there are no clear separability patterns. Rectangles *2.a* and *3.a* represent the visualization of the raw data using a t-SNE projection [12]. We can see that the arrangement of instances does not constitute clusters, suggesting little linearity in the data.

Starting from the same dataset D, the proposed approach consists of the following steps: (*2.b*) training a classification algorithm and (*3.b*) applying the SHAP explanation over predictions. The training process in *2.b* follows standard practices, such as handling missing values, applying techniques to deal with class imbalance, model evaluation, and calibration. In *3.b*, we select instances from the test set and submit them to the SHAP approach. We obtain, for each feature value v of the instance i, their contribution values (SHAP values). In our experiments, we used KernelSHAP to estimate SHAP values, but other methods could also have been used (*i.e.*, TreeSHAP, Linear SHAP [11], Kernel SHAP method to handle dependent features [1], *etc.*). With the SHAP values, we create an instance i', which consists of the calculated SHAP values instead of the raw feature values. In the next step (*4.b*), we create a dataset C constituted by the instances i'.

Finally, dimensionality reduction is applied to the SHAP-transformed instances in C to generate a 2D plot that gives insights into the separability of instances. This procedure is represented by step *5.b*. We applied the t-SNE dimensionality reduction (DR) technique in our tests, but other DR techniques could be used. The representation in *6.b* is the result of the projection into 2D generated by t-SNE. We observe that the instances may form well-defined clusters in the projection. For instance, certain regions may have a prevalence of a specific class, while others may show higher uncertainty. Moreover, the projected points can be colored based on the feature values, which makes it possible to obtain important insights, such as the relationship between the features taken into account by the predictive models and their correct and incorrect predictions.

4 Experimental Evaluation

In this section, we describe the experiments performed to answer the following research questions:

RQ1 Is it possible to identify, among patients with COVID-19, who is more likely to die from the disease?

RQ2 Which features are taken into account by the classifiers for making predictions and how do they relate to correct/incorrect classifications?

RQ3 Is the performance obtained with multivariable classifiers superior to the performance obtained with single-variable classifiers using only age as independent variable?

4.1 Experimental Setup and Executions

Data preprocessing was performed using *Python v3.5* and the libraries *Pandas* and *Numpy*. The classifiers were implemented using the Weka v3.9.6 API through the *wrapper* python-weka-wrapper3 v0.2.9[2]—with the exception of the *Gradient*

---

[2] https://fracpete.github.io/python-weka-wrapper3.

Boosting Tree, for which used the XGBoost implementation[3]. XGBoost can handle missing values, so no data imputation was needed. For the other algorithms, the feature values were imputed with the median value from the training set. Categorical features were imputed with mode values. The weights of the cost matrix were calculated as shown in Sect. 3.3, considering the imbalance rate of the training set of each bootstrap iteration. To interpret the generated models, we use the SHAP library[4]. SHAP relies on *Shapley Values* to explain the contribution that each variable makes to the output predicted by the model. Specifically, we used the KernelSHAP algorithm, as it is model-agnostic. The perplexity parameter for the t-SNE projection considered values between 10 and 40. Empirical tests were performed in multiple executions to set the appropriate perplexity value.

4.2   Results for Mortality Prediction (*RQ1*)

Table 2 displays the scores achieved by each classifier. Algorithms have better/worse performances across all metrics, with no clear winner. In our analysis, we give emphasis to sensitivity results, as the consequences of not identifying a patient with a high probability of dying are much more severe than wrongly predicting a patient into the *alive* class when a patient has a high probability of dying. Considering only sensitivity, Logistic Regression is the winner, leading all scenarios. Only in HCPA$_{CTI}$ XGBoost reaches the same top score as LR. AdaBoostM1 has the second highest sensitivity values for HMV$_{ALL}$ and HMV$_{CTI}$, and the third highest sensitivity value for HCPA$_{CTI}$. In all cases, J48 has the lowest sensitivity values.

Overall, the sensitivity values we obtain demonstrate the classifier's promising ability to predict mortality based on the information available at the time of admission (answer to *RQ1*). The average sensitivity for the classification algorithms is higher for the HMV$_{CTI}$ dataset (0.74) when compared to the HCPA$_{CTI}$ dataset (0.65). This difference may be explained by the individual characteristics of patients admitted to each hospital and the features available in the datasets for each hospital. A similar discussion is mentioned in Futoma *et al.* [6], which warns about the risks of excessively seeking the development of predictive models that generalize to various hospitals with different clinical protocols and units.

Our best AUROC is higher than the ones reported in related work [5,9,13,20] (see Table 2). Yet a direct comparison among approaches is not possible as they were applied to different datasets.

---

[3] https://github.com/dmlc/xgboost.
[4] https://github.com/slundberg/shap.

Table 2. Results for mortality prediction

Dataset	Algorithm	Sen	Spe	PPV	NPV	F1$_+$	ma-F1	Kappa	AUROC	AUPRC
HMV$_{ALL}$	LR	**.84**	.87	.46	**.98**	.59	.76	.52	.92 [.89–.95]	.60 [.48–.69]
	J48	.54	.91	.46	.94	.50	.71	.42	.73 [.67–.79]	.44 [.31–.56]
	RF	.57	**.95**	**.60**	.94	.58	.76	.53	**.93 [.90–.95]**	**.61 [.50–.72]**
	AB	.78	.85	.41	.97	.54	.72	.45	.89 [.86–.92]	.52 [.40–.64]
	MLP	.62	.92	.51	.95	.56	.74	.49	.87 [.80–.92]	.57 [.48–.67]
	XGB	.72	.91	.52	.96	**.60**	**.80**	**.54**	.92 [.90–.94]	.60 [.50–.69]
HCPA$_{CTI}$	LR	**.68**	.73	.60	**.79**	**.64**	**.70**	.39	.77 [.74–.79]	.65 [.60–.68]
	J48	.56	.70	.53	.73	.54	.63	.25	.62 [.58–.67]	.53 [.48–.59]
	RF	.62	.77	**.62**	.77	.62	.70	**.40**	**.78 [.75–.81]**	.66 [.61–.71]
	AB	.67	.71	.58	.78	.62	.69	.37	.76 [.74–.79]	.64 [.59–.69]
	MLP	.66	.69	.57	.78	.61	.67	.34	.75 [.72–.78]	.61 [.55–.66]
	XGB	**.68**	**.74**	.61	**.79**	**.64**	**.70**	**.40**	**.78 [.75–.81]**	**.67 [.61–.71]**
HMV$_{CTI}$	LR	**.80**	.82	.66	**.90**	**.72**	**.79**	**.58**	**.87 [.83–.92]**	**.73 [.63–.84]**
	J48	.68	.80	.60	.84	.63	.73	.46	.73 [.62–.83]	.58 [.42–.72]
	RF	.75	**.83**	**.67**	.88	.70	.78	.56	**.87 [.81–.91]**	.69 [.54–.81]
	AB	.77	.80	.64	.89	.70	.77	.54	.86 [.80–.91]	.69 [.53–.82]
	MLP	.70	.82	.64	.86	.66	.75	.50	.83 [.76–.89]	.69 [.57–.81]
	XGB	.73	**.83**	.66	.87	.69	.76	.54	**.87 [.81–.92]**	.71 [.57–.83]

4.3 Analysis of Feature Importance and Explainability ($RQ2$)

We use the results of applying dimensionality reduction over SHAP values to help understand the factors considered by classifiers during predictions and answer $RQ2$. Figure 3 presents the global view of how features contribute to mortality prediction using the LR classifier on HMV$_{ALL}$. Each line refers to the contribution of a feature. Each dot is an instance. Low values for the feature are in blue and high values are in red. The contribution of the feature value is given by the horizontal displacement. The further the instance is positioned to the right, the greater the contribution to the *deceased* class. The further to the left, the greater the contribution of the feature value to the prediction of the *alive* class. For example, high values for age are associated with increased probability for the *deceased* class. The ten features with the highest contribution to the predictions were age, erythrocytes, bilirubin, oxygen saturation, neutrophils, troponin, heart disease, hemoglobin, creatinine, and central nervous system disease. Using the LR classifier on HCPA$_{CTI}$, the ten features with the highest contribution were age, urea, red cell distribution width, D-dimer, mean corpuscular volume (MCV), lymphocytes, platelets, creatinine, erythrocytes, and troponin.

Figure 4 shows the t-SNE plots generated using SHAP values. Each point corresponds to a patient. The points are colored based on the values of different features to aid in interpreting the generated projections. Each row is composed of three plots. The first plot displays the correct and incorrect predictions (true positive (TP), true negative (TN), false positive (FP) and false negative(FN)). The second and third plots show points colored by different features.

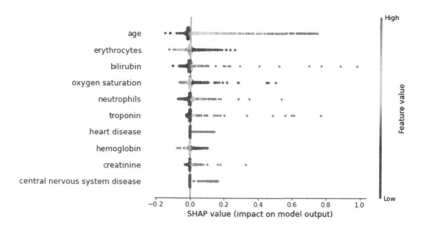

Fig. 3. Global explanations based on LR classifier using SHAP for HMV_{ALL}

The t-SNE plots reinforce the high importance attributed by the predictive models to the age feature. When analyzing the plots with the classifications and the colored plots based on the age of the patients (Figs. 4b, 4e 4h, and 4k), it is possible to observe that recurrent FP errors (purple dots) and TP occur in the regions with a high concentration of older patients. On the other hand, FN errors (red dots) and TN are in regions with younger patients.

Figure 4c also shows a trend of moderate and low erythrocyte values for TP and FP patients. These findings are supported by previous studies, which showed that anemia is associated with poor prognosis in critically ill patients.

Figures 4d, 4e, and 4f refer to LR on the HMV_{CTI} dataset. In Fig. 4f, we can observe that patients with heart disease tend to be older. Both features predispose those patients to be classified in the *deceased* class. This finding is in agreement with the related literature – studies have shown that heart disease is a risk factor for patient mortality [4].

Comparing the predictions for HCPA_{CTI} using XGB (Figs. 4g, 4h, and 4i) and LR (Figs. 4j, 4k, and 4l) we notice that XGB tends to present a greater overlap between FN and FP. This may be due to the influence of the t-SNE visualization method, which may favor the visualization for the LR algorithm.

Figures 4i and 4l, generated based on the SHAP values provided by the XGB and LR classifiers, show regions of high concentrations of patients with high values of D-dimers. Both younger and older patients are found in those regions. The pattern is more evident in the visualization obtained with LR, in which TP and FP classifications prevail. High D-dimer values are associated with poor prognosis in patients with COVID-19 infection [15], and in both cases (XGB and LR), models tend to assign these patients to the *deceased* class.

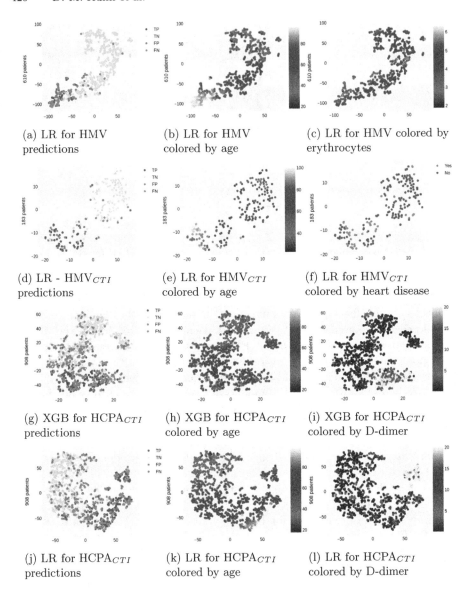

(a) LR for HMV predictions

(b) LR for HMV colored by age

(c) LR for HMV colored by erythrocytes

(d) LR - HMV$_{CTI}$ predictions

(e) LR for HMV$_{CTI}$ colored by age

(f) LR for HMV$_{CTI}$ colored by heart disease

(g) XGB for HCPA$_{CTI}$ predictions

(h) XGB for HCPA$_{CTI}$ colored by age

(i) XGB for HCPA$_{CTI}$ colored by D-dimer

(j) LR for HCPA$_{CTI}$ predictions

(k) LR for HCPA$_{CTI}$ colored by age

(l) LR for HCPA$_{CTI}$ colored by D-dimer

Fig. 4. Visualizations of the model's predictions based on SHAP values – the x-axis corresponds to t-SNE dimension 1 and the y-axis to t-SNE dimension 2.

4.4 Single vs. Multiple Variable Classifiers ($RQ3$)

As seen in the analysis of Sect. 4.3, as well as recurrently pointed out by works in the related literature, age was frequently flagged as the attribute with the greatest contribution during mortality prediction tasks. This research question aims to compare the performance between classifiers with multiple independent variables

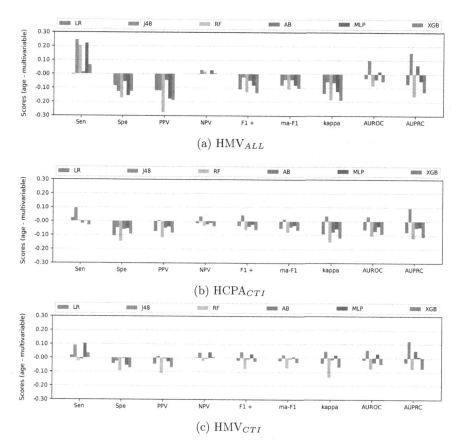

(a) HMV$_{ALL}$

(b) HCPA$_{CTI}$

(c) HMV$_{CTI}$

Fig. 5. Multivariable and single-variable classifiers (age) for mortality prediction

and their respective versions with only one independent variable. For this, we compare the performance of the classifiers presented in Sect. 4.2. Figure 5 shows the differences between the performances obtained by the classifiers trained using only age as independent variable.

For most algorithms, classification models that used only age as independent variable did not show degradation in sensitivity, and in some cases, higher levels of sensitivity were found. Although this positive effect on sensitivity levels was observed, the reduction in PPV and specificity levels was also recurrent. As a consequence, in most cases, no gains were observed for the metrics F1+, ma-F1, Kappa, AUROC and AUPRC. The LR algorithm, which showed the best results in the experiments in Sect. 4.2, tended not to show substantial decrease in sensitivity levels. For the HMV$_{ALL}$ dataset, considerable increases in sensitivity levels occur for some algorithms. In contrast, there were considerable reductions in the levels of PPV and specificity, mainly in the HMV$_{ALL}$ dataset and in the HCPA$_{CTI}$ dataset.

Algorithm J48, which presented the lowest sensitivity scores in the experiment in Sect. 4.2, was recurrently the one that obtained the highest sensitivity gains among the classifiers that used only age as an independent variable. When considering the HMV$_{CTI}$ dataset, in addition to the J48 algorithm, the RF and MLP algorithms stood out for their increased sensitivity. However, despite the increase in sensitivity, due to the losses in the levels of PPV and specificity, no gains were observed for the metrics F1+, ma-F1 and Kappa. Finally, J48 and MLP were the only ones that obtained AUROC gains. In HCPA$_{CTI}$ (Fig. 5b), only the J48 algorithm obtained an increase in the other metrics, in addition to sensitivity. In HMV$_{CTI}$ (Fig. 5c), J48 and MLP achieved the highest sensitivity gains. In this scenario, both algorithms showed a smaller reduction in terms of PPV and specificity, resulting in an increase in the levels of other metrics.

Essentially, the classifiers that use only age as independent variable reach similar or higher sensitivity scores compared to their multivariable versions. However, these scores were achieved at the cost of lower Spe and PPV levels, showing that other variables are useful in improving these metrics. The full version of Table 2 with CI (RQ1) and significance tests performed on RQ3 can be accessed at https://github.com/danimtk/covid-19-paper.

5   Conclusion

This work evaluated several ML algorithms for predicting COVID-19 mortality. The classifiers showed a promising ability to predict mortality based on the information available at the time of admission, reaching a sensitivity of up to 84% and an AUROC of 92% (95% CI, [0.89–0.95]). Overall, besides age, oxygen saturation, heart disease, hemogram information also tended to have high contributions to the predictions. We applied a visualization technique based on SHAP values, which helps to corroborate the high importance attributed to the age feature by the classifiers in their predictions. The visualizations also showed that some groups of patients with low erythrocytes and high D-dimer (risk factors for mortality) tend to be recognized by the classifiers with a higher risk of dying. We believe that the visualization technique employed here can also be used in other prediction tasks.

One limitation of our work is that SHAP is relatively time-consuming when applied to datasets with a large number of features [19]. This limitation can be mitigated using optimized SHAP algorithms (like TreeSHAP or [19]). Although the visualizations obtained are intuitive and seem to represent already known relationships (such as the high risk for older patients, as well as those with comorbidities), it should be noted that approaches that use Shapley values, as well as approximations of these values (such as SHAP), may generate misconceptions regarding the actual contribution of features [8]. Thus, we emphasize the importance for specialists to use other evaluation techniques concomitantly to assist in the interpretation of predictive models, avoiding misinterpretations.

In future work, we plan to experiment with other dimensionality reduction methods and classifiers. We also plan to investigate whether there is a set of attributes (and how many) would be needed to maintain sensitivity levels without considerable PPV losses.

Acknowledgment. This work has been financed in part by CAPES Finance Code 001 and CNPq/Brazil.

References

1. Aas, K., Jullum, M., Løland, A.: Explaining individual predictions when features are dependent: accurate approximations to Shapley values. Artif. Intell. **298**, 103502 (2021)
2. Alballa, N., Al-Turaiki, I.: Machine learning approaches in Covid-19 diagnosis, mortality, and severity risk prediction: a review. Inf. Med. Unlocked **24**, 100564 (2021)
3. Araújo, D.C., Veloso, A.A., Borges, K.B.G., das Graças Carvalho, M.: Prognosing the risk of Covid-19 death through a machine learning-based routine blood panel: a retrospective study in brazil. IJMEDI **165**, 104835 (2022)
4. Broberg, C.S., Kovacs, A.H., Sadeghi, S., et al.: Covid-19 in adults with congenital heart disease. JACC **77**(13), 1644–1655 (2021)
5. Covino, M., Sandroni, C., Santoro, M., et al.: Predicting intensive care unit admission and death for Covid-19 patients in the emergency department using early warning scores. Resuscitation **156**, 84–91 (2020)
6. Futoma, J., Simons, M., Panch, T., et al.: The myth of generalisability in clinical research and machine learning in health care. Lancet Digit. Health **2**(9), 489–492 (2020)
7. Hall, M.A.: Correlation-based feature selection for machine learning. Ph.D. thesis, The University of Waikato (1999)
8. Huang, X., Marques-Silva, J.: The inadequacy of Shapley values for explainability. arXiv preprint arXiv:2302.08160 (2023)
9. Knight, S.R., Ho, A., Pius, R., et al.: Risk stratification of patients admitted to hospital with Covid-19 using the ISARIC WHO clinical characterisation protocol: development and validation of the 4C mortality score. BMJ **370**, m3339 (2020)
10. Lundberg, S.M., Erion, G.G., Lee, S.I.: Consistent individualized feature attribution for tree ensembles. arXiv preprint arXiv:1802.03888 (2018)
11. Lundberg, S.M., Lee, S.I.: A unified approach to interpreting model predictions. In: Advances in Neural Information Processing Systems, vol. 30 (2017)
12. Van der Maaten, L., Hinton, G.: Visualizing data using t-SNE. JMLR **9**(11), 2579–2605 (2008)
13. Miranda de Paiva, B.B., Delfino-Pereira, P., de Andrade, C.M.V., et al.: Effectiveness, explainability and reliability of machine meta-learning methods for predicting mortality in patients with COVID-19: results of the Brazilian COVID-19 registry. medRxiv (2021)
14. Qin, Z., Zhang, C., Wang, T., Zhang, S.: Cost sensitive classification in data mining. In: Advanced Data Mining and Applications, pp. 1–11 (2010)
15. Rostami, M., Mansouritorghabeh, H.: D-dimer level in COVID-19 infection: a systematic review. Exp. Rev. Hematol. **13**(11), 1265–1275 (2020)
16. Subudhi, S., Verma, A., Patel, A.B.: Prognostic machine learning models for Covid-19 to facilitate decision making. IJCP **74**(12), e13685 (2020)
17. Wynants, L., Van Calster, B., Collins, G.S., et al.: Prediction models for diagnosis and prognosis of COVID-19: systematic review and critical appraisal. bmj **369**, m1328 (2020)

18. Yadaw, A.S., Li, Y., Bose, S., et al.: Clinical features of COVID-19 mortality: development and validation of a clinical prediction model. Lancet Digt. Health **2**(10), E516–E525 (2020)
19. Yang, J.: Fast TreeSHAP: accelerating SHAP value computation for trees. arXiv preprint arXiv:2109.09847 (2021)
20. Zhao, Z., Chen, A., Hou, W., et al.: Prediction model and risk scores of ICU admission and mortality in COVID-19. PLoS ONE **15**(7), e0236618 (2020)

An Explainable Model to Support the Decision About the Therapy Protocol for AML

Jade M. Almeida[1], Giovanna A. Castro[1], João A. Machado-Neto[2], and Tiago A. Almeida[1(✉)]

[1] Department of Computer Science (DComp-So), Federal University of São Carlos (UFSCar), Sorocaba, São Paulo 18052-780, Brazil
jade.almeida@dcomp.sor.ufscar.br, giovannacastro@estudante.ufscar.br, talmeida@ufscar.br
[2] Institute of Biomedical Sciences, The University of São Paulo (USP), São Paulo 05508-000, Brazil
jamachadoneto@usp.br

Abstract. Acute Myeloid Leukemia (AML) is one of the most aggressive types of hematological neoplasm. To support the specialists' decision about the appropriate therapy, patients with AML receive a prognostic of outcomes according to their cytogenetic and molecular characteristics, often divided into three risk categories: favorable, intermediate, and adverse. However, the current risk classification has known problems, such as the heterogeneity between patients of the same risk group and no clear definition of the intermediate risk category. Moreover, as most patients with AML receive an intermediate-risk classification, specialists often demand other tests and analyses, leading to delayed treatment and worsening of the patient's clinical condition. This paper presents the data analysis and an explainable machine-learning model to support the decision about the most appropriate therapy protocol according to the patient's survival prediction. In addition to the prediction model being explainable, the results obtained are promising and indicate that it is possible to use it to support the specialists' decisions safely. Most importantly, the findings offered in this study have the potential to open new avenues of research toward better treatments and prognostic markers.

Keywords: Decision support system · Explainable artificial intelligence · Acute Myeloid Leukemia · Knowledge discovery and pattern recognition

1 Introduction

Acute Myeloid Leukemia (AML) is one of the most aggressive types of hematological neoplasm, characterized by the infiltration of cancer cells into the bone marrow. AML has decreasing remission rates regarding the patient's age, and its average overall survival rate is just 12 to 18 months [21].

Supported by CAPES, CNPq, and FAPESP grant #2021/13325-1.

In 2010, the European LeukemiaNet (ELN) published recommendations for diagnosing and treating AML [5], which became a field reference. A significant update to these recommendations was published in 2017 [6] and 2022 [7], incorporating new findings concerning biomarkers and subtypes of the disease combined with a better understanding of the disease behavior.

For a diagnosis of AML, at least 10% or 20% of myeloblasts must be present in the bone marrow or peripheral blood, depending on the molecular subtype of the disease [1]. This analysis is performed according to the Classification of Hematopoietic and Lymphoid Tissue Tumors, published and updated by the World Health Organization.

In addition to the diagnosis, the patient with AML receives a prognostic of outcomes, often divided into three risk categories: favorable, intermediate, and adverse. Cytogenetic and molecular characteristics define such stratification [25]. The cytogenetic characteristics come from certain chromosomal alterations. In turn, the molecular ones are determined according to mutations in the *NPM1*, *RUNX1*, *ASXL1*, *TP53*, *BCOR*, *EZH2*, *SF3B1*, *SRSF2*, *STAG2*, and *ZRSR2* genes. Specialists commonly use the ELN risk classification to support critical decisions about the course of each treatment, which can directly impact patients' quality of life and life expectancy.

Patients with a favorable risk prognosis generally have a good response to chemotherapy. On the other hand, those with adverse risk tend not to respond well to this therapy, needing to resort to other treatments, such as hematopoietic stem cell transplantation [25]. The problem with the current risk prognosis is the high rate of heterogeneity between patients of the same risk group. In addition, there is no clear definition regarding the intermediate risk since these patients do not show a response pattern to treatments.

Most patients with AML receive an intermediate-risk classification [5]. Unfortunately, this makes specialists demand more information, such as the results of other tests and analyses, to support their decisions regarding the most appropriate treatment, even with little or no evidence of efficacy. This process can result in delayed initiation of treatment and consequent worsening of the patient's clinical condition.

To overcome this problem, this study presents the result of a careful analysis of real data composed of clinical and genetic attributes used to train an explainable machine-learning model to support the decision about the most appropriate therapy protocol for AML patients. The model is trained to identify the treatment guide that maximizes the patient's survival, leading to better outcomes and quality of life.

2   Related Work

The decision on therapy for patients with AML is strongly based on the prediction of response to treatment and clinical outcome, often defined by cytogenetic factors [9]. However, the current risk classification can be quite different among patients within the same risk groups, in which the result can range from decease within a few days to an unexpected cure [5].

Since the mid-1970s, the standard therapy for patients with AML has been chemotherapy, with a low survival rate. However, with advances, various data on mutations and gene expressions began to be collected, analyzed, and made available, accelerating the development of therapeutic practices.

In 2010, the European LeukemiaNet (ELN) proposed a risk categorization based on cytogenetic and molecular information, considering the severity of the disease [5]. This classification comprises four categories: favorable, intermediate I, intermediate II, and adverse.

In 2017, a significant update to the ELN's risk classification was published [6]. The updated risk classification grouped patients into three categories (favorable, intermediate, and adverse) and refined the prognostic value of specific genetic mutations. Since then, specialists have commonly used this stratification to support important decisions about the course of each treatment, which can directly impact the patient's quality of life and life expectancy.

In 2022, the ELN's risk classification was updated again. The main change provided is related to the expression of the *FLT3-ITD* gene. All patients with high expression but without any other characteristics of the adverse group are classified as intermediate risk. Another significant change is that mutations in *BCOR*, *EZH2*, *SF3B1*, *SRSF2*, *STAG2*, and *ZRSR2* genes are related to the adverse risk classification [7].

Specialists often rely on the ELN risk classification to define the treatment guidelines given to the patient shortly after diagnosis. Patients with a favorable risk generally present a positive response to chemotherapy. In contrast, patients with an adverse risk tend not to respond well to this therapy, requiring other treatments, such as hematopoietic stem cell transplantation [25]. However, there is no clear definition regarding the therapeutic response of AML patients with intermediate risk.

The problem with using the current risk classifications as a guide for deciding the most appropriate treatment is that there can be significant variability of patients in the same risk group, with different characteristics such as age and gender. For example, patients under 60 tend to respond better to high-dose chemotherapy. On the other hand, patients over 60 years old tend to have a low tolerance to intense chemotherapy and may need more palliative therapies [14]. Several studies suggest that age is a relevant factor when deciding the treatment for a patient, a fact that is not considered by the current risk classification. However, as most patients with AML receive the intermediate risk, specialists often require additional information, such as the results of other tests and analyses, to decide the most appropriate treatment, even with little or no evidence of efficacy [5]. This process can lead to a delay at the start of treatment and worsen the patient's clinical condition.

Studies have emphasized the significance of analyzing mutations and gene expression patterns in families of genes to determine the therapeutic course in AML. Over 200 genetic mutations have been identified as recurrent in AML patients through genomic research [25]. With genetic sequencing, the patient profile for AML has transitioned from cytogenetic to molecular [15]. However, due to the heterogeneity of the disease, it is difficult to manually analyze the various genetic alterations that may impact the course of the disease. To overcome these challenges, recent studies have sought to apply machine learning (ML) techniques to automatically predict the outcome after exposure to specific treatments and complete remission of the disease.

For example, [10] trained supervised ML models with data extracted from RNA sequencing and clinical information to predict complete remission in pediatric patients with AML. The k-NN technique obtained the best performance, with an area under the ROC curve equals to 0.81. The authors also observed significant differences in the gene expressions of the patients concerning the pre-and post-treatment periods.

Later, [19] used clinical and genetic data to train a random forest classifier capable of automatically predicting the survival probability. According to the authors, the three most important variables for the model were patient age and gene expression of the *KDM5B* and *LAPTM4B* genes, respectively. The authors concluded that applying ML techniques with clinical and molecular data has great predictive potential, both for diagnosis and to support therapeutic decisions.

In the study of [17], a statistical decision support model was built for predicting personalized treatment outcomes for AML patients using prognostic data available in a knowledge bank. The authors have found that clinical and demographic data, such as age and blood cell count, are highly influential for early death rates, including death in remission, which is mainly caused to treatment-related mortality. Using the knowledge bank-based model, the authors concluded that roughly one-third of the patients analyzed would have their treatment protocol changed when comparing the model's results with the ELN treatment recommendations.

The success reported in these recent studies is an excellent indicator that recent ML techniques have the potential to automatically discover patterns in vast amounts of data that specialists can further use to support the personalization and recommendation of therapy protocols. However, one of the main concerns when applying machine learning in medicine is that the model can be explainable, and experts can clearly understand how the prediction is generated [4].

In this context, this study presents the result of a careful analysis of real data composed of clinical and genetic attributes used to train an explainable machine-learning model to support the decision about the most appropriate therapy protocol for AML patients. Our main objective is to significantly reduce the subjectivity involved in the decisions specialists must make and the time in the treatment decision processes. This can lead to robust recommendations with fewer adverse effects, increasing survival time and quality of life.

3   Materials and Methods

This section details how the data were obtained, processed, analyzed, and selected. In addition, we also describe how the predictive models were trained.

3.1   Datasets

The data used to train the prediction models come from studies by *The Cancer Genome Atlas Program* (TCGA) and *Oregon Health and Science University* (OHSU). These datasets are known as *Acute Myeloid Leukemia* [25,27] and comprise clinical and genetic data of AML patients. Both are real and available in the public domain at https://www.cbioportal.org/. We used three sets with data collected from the same patients: one with clinical information (CLIN), another with gene mutation data (MUT), and another with gene expression data (EXP). Table 1 summarizes these original data.

Table 1. Amount of original data in each database. Each database is composed of three sets of features: clinical information (CLIN), gene mutation data (MUT), and gene expression data (EXP)

	Samples	Patients	Attributes		
			CLIN	MUT	EXP
TCGA	200	200	31	25,000	25,000
OHSU	672	562	97	606	22,825

3.2   Data Cleaning and Preprocessing

Since the data comes from two sources, we have processed them to ensure consistency and integrity. With the support of specialists in the application domain, we removed the following spurious data:

1. Samples not considered AML in adults observed by (*i*) the age of the patient, which must not be less than 18 years, and (*ii*) the percentage of blasts in the bone marrow, which should be greater or equal to 20%;
2. Samples without information on survival elapsed time after starting treatment (*Overall Status Survival*);
3. Duplicate samples; and
4. Features of patients in only one of the two databases.

We used the 3-NN method to automatically fill empty values in clinical data features (CLIN). We used the features with empty values as the target attributes and filled them using the value predicted from the model trained with other attributes. Nevertheless, we removed the features of 37 genes with no mutations.

Subsequently, we kept only the samples in which all the variables are compatible, observing data related to the exams and treatment received by the patients, as these affect the nature of the clinical, mutation, and gene expression data. Of the 872 initial samples in the two databases, 272 were kept at the end of the preprocessing and data-cleaning processes. Of these, there are 100 samples from patients who remained alive after treatment and 172 who died before, during, or after treatment. Cytogenetic information was normalized and grouped by AML specialists. Moreover, the same specialists analyzed and grouped the treatments in the clinical data into four categories according to the intensity of each therapy:

1. *Target therapy* – therapy that uses a therapeutic target to inhibit some mutation/AML-related gene or protein;
2. *Regular therapy* – therapy with any classical chemotherapy;
3. *Low-Intensity therapy* – non-targeted palliative therapy, generally recommended for elderly patients; and
4. *High-Intensity therapy* – chemotherapy followed by autologous or allogenic hematopoietic stem cell transplantation.

Finally, the specialists checked and validated all the data.

3.3 Feature Selection

This section describes how we have analyzed and selected the features used to represent clinical, gene mutation, and gene expression data.

Clinical Data. Among the clinical attributes common in the two databases, specialists in the data domain selected the following 11 according to their relevance for predicting clinical outcomes. In Table 2, we briefly describe all selected clinical features, and Table 3 summarizes the main statistics of those with a continuous nature. Figures 1 and 2 summarize their main statistics.

Table 2. Clinical features description

Feature	Description
Diagnosis age	Patient age when diagnosed with AML
Bone marrow blast %	Percentage of blasts in the bone marrow
Mutation count	Number of genetic mutations observed
PB blast %	Percentage of blasts in peripheral blood
WBC	White blood cell count
Gender	Patient gender
Race	Whether the patient is white or not
Cytogenetic info	Cytogenetic information the specialist used in diagnosing the patient
ELN risk classification	ELN risk groups (favorable, intermediate, and adverse)
Treatment intensity classification	The intensity of treatment received by the patient (target, regular, low-intensity, or high-intensity therapy)
Overall survival status	Patient survival status (living or deceased)

Table 3. Main statistics of clinical features with a continuous nature

Feature	Minimum	Maximum	Median	Mean
Diagnosis age	18	88	58	55.11
Bone marrow blast %	20	100	72	68.13
Mutation count	1	34	9	9.54
PB blast %	0	99.20	39.10	40.99
WBC	0.4	483	39.40	19.85

Among the clinical attributes, in line with several other studies, the only noticeable highlight is that the patient's age seems to be a good predictor of the outcome. The older the patient, the lower the chances of survival. All other attributes showed similar behavior for both classes, with subtle differences.

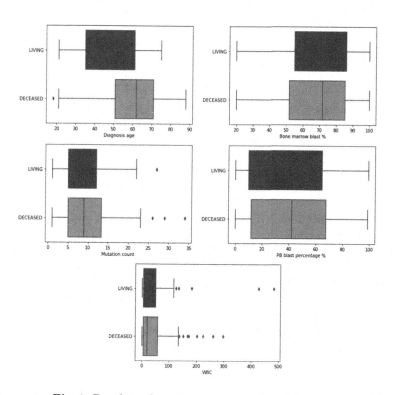

Fig. 1. Boxplots of continuous nature clinical features

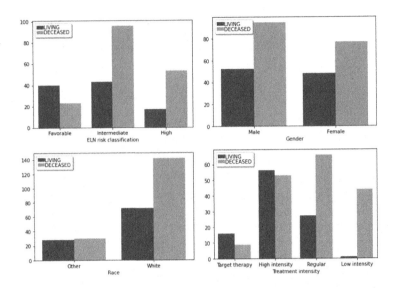

Fig. 2. Bar plots of categorical nature clinical features

Gene Mutation Data. After cleaning and preprocessing the data, 281 gene mutation features remained. Then, we employed the χ^2 statistical method to select a subset of these features. We chose to use the χ^2 test because it has been widely used in previous studies to analyze the correlation between genetic mutations and certain types of cancer [23]. We defined the following hypotheses: H0 – patient survival is independent of gene mutation; and H1 – both groups are dependent. Using $p < 0.05$, only two features were selected: *PHF6* and *TP53* gene mutations.

The *TP53* mutation is the best known among the two gene mutations selected. Several studies show the relationship between *TP53* mutation with therapeutic response and prognosis. The *TP53* gene is considered the guardian of genomic stability, as it controls cell cycle progression and apoptosis in situations of stress or DNA damage, and mutations in this gene are found in approximately half of the cancer patients [12]. Although mutations in *TP53* are less common in AML patients (about 10%), they predict a poor prognosis [11].

The mutation in the *PHF6* gene has been identified as a genetic alteration associated with hematologic malignancies [13]. *PHF6* is a tumor suppressor gene, and several studies have shown a high mutation frequency in the adverse risk group of AML [8]. These observations suggest that *PHF6* mutations may have a significant role in the development and progression of AML and may serve as a potential prognostic marker for the disease [28].

To further investigate the potential of gene mutation data on outcome prediction, we have enriched the set of gene mutation features with well-known genes already highlighted in studies in the literature [3,14,22] and used by the ELN [7].

The literature features used were: *FLT3, NPM1, DNMT3A, IDH1, IDH2, TET2, ASXL1, RUNX1, CEBPA, NRAS, KRAS, SF3B1, U2AF1, SRSF2.*

Gene Expression Data. After data cleaning and preprocessing, 14,712 gene expression features remained. To select the most relevant features for outcome prediction, we have employed a method similar to Lasso Regression [26]: we have trained an SVM model with L1 regularization. This method estimates the relevance of the features by assigning a weight coefficient to each of them. When a feature receives a zero coefficient, it is irrelevant enough for the problem the model was trained for. As a consequence, these features are not selected.

The method was trained with all 14,712 gene expression features, from which 22 were selected.

The final datasets we have used to train and evaluate the outcome prediction models are publicly available at https://github.com/jdmanzur/ml4aml_databases. It is composed of 272 samples (patient data) consisting of 11 clinical features (CLIN), 22 gene expression features (EXP), and 16 gene mutation features (MUT). Table 4 summarizes each of these datasets.

Table 4. Final datasets used to train and evaluate the outcome prediction models

Dataset	#Features	Features
Clinical (CLIN)	11	Diagnosis age, Bone marrow blast (%), Mutation count, PB blast (%), WBC, Gender, Race, Cytogenetic info, ELN risk classification, Treatment intensity classification, Overall survival status (class)
Gene expression (EXP)	24	*CCDC144A, CPNE8, CYP2E1, CYTL1, HAS1, KIAA0141, KIAA1549, LAMA2, LTK, MICALL2, MX1, PPM1H, PTH2R, PTP4A3, RAD21, RGS9BP, SLC29A2, TMED4, TNFSF11, TNK1, TSKS, XIST* Treatment intensity classification, Overall survival status (class)
Gene mutation (MUT)	18	*FLT3, NPM1, DNMT3A, IDH1, IDH2, TET2, ASXL1, RUNX1, CEBPA, NRAS, KRAS SF3B1, U2AF1, SRSF2, PHF6, TP53,* Treatment intensity classification, Overall survival status (class)

3.4 Training the Outcome Prediction Models

Since interpretability is a crucial pre-requisite for machine-learning models in medicine [4], we have employed the well-known Explainable Boosting Machine (EBM) technique [2].

EBM is a machine learning approach that combines the strengths of boosting techniques with the goal of interpretability. It is designed to create accurate and easily understandable models, making it particularly useful in domains where interpretability and transparency are important.

EBM extends the concept of boosting by incorporating a set of interpretable rules. Instead of using complex models like neural networks as weak learners, EBM employs a set of rules defined by individual input features. These rules are easily understandable and can be represented as "if-then" statements.

During training, EBM automatically learns the optimal rules and their associated weights to create an ensemble of rule-based models. The weights reflect the importance of each rule in the overall prediction, and the ensemble model combines their predictions to make a final prediction.

The interpretability of EBM comes from its ability to provide easily understandable explanations for its predictions. Using rule-based models, EBM can explicitly show which features and rules influenced the outcome, allowing AML specialists to understand the underlying decision-making process.

EBM has been applied successfully in various domains, such as predicting medical conditions, credit risk assessment, fraud detection, and predictive maintenance, where interpretability and transparency are paramount [20].

We have used the EBM classification method from the InterpretML library[1] to train seven outcome prediction models: one per dataset (CLIN, MUT, EXP) and four using all possible combinations (CLIN+MUT, CLIN+EXP, MUT+EXP, CLIN+MUT+EXP).

3.5   Performance Evaluation

We evaluated the performance of the prediction models using holdout [18]. For this, we have divided the data into three parts 80% was randomly separated for training the models, 10% of the remaining data was randomly selected for model and feature selection, and the remaining 10% was used to test. The data separation was stratified; therefore, each partition preserves the class balance of the original datasets. We must highlight we performed the feature selection processes using only training and validation partitions.

We calculated the following well-known measures to assess and compare the performance obtained by the prediction models: accuracy, recall (or sensitivity), precision, F1-Score, and the Area Under the ROC Curve (AUC).

4   Results and Discussion

First, we trained the outcome prediction models using only the best-known genes consolidated by studies in the literature, both for the expression and mutation contexts. These genes are *FLT3, NPM1, DNMT3A, IDH1, IDH2, TET2, ASXL1,*

---

[1] InterpretML is a Python library that provides a set of tools and algorithms for interpreting and explaining machine learning models. The documentation is available at https://interpret.ml/docs.

RUNX1, CEBPA, NRAS, KRAS, SF3B1, U2AF1, and *SRSF2.* Table 5 presents the prediction performance obtained.

Table 5. Results achieved by the outcome prediction models. The genes from MUT and EXP were selected according to consolidated studies in the literature

Model	F1-Score	AUC	Accuracy	Precision	Recall
CLIN	0.57	0.53	0.57	0.57	0.57
MUT	0.64	**0.70**	0.64	**0.76**	0.64
EXP	0.66	0.62	**0.68**	0.66	**0.68**
CLIN+MUT	**0.67**	0.64	**0.68**	0.67	**0.68**
CLIN+EXP	0.57	0.53	0.57	0.57	0.57
MUT+EXP	0.63	0.59	0.64	0.63	0.64
CLIN+MUT+EXP	0.54	0.51	0.54	0.55	0.54

The model that achieved the best result was the one that combined clinical and genetic mutation data. When analyzing the models trained with individual datasets, the ones based on gene mutation and expression showed the best performances. However, the overall results obtained are low and unsatisfactory for predicting the outcomes of AML patients. Surprisingly, the genes most known in the literature seem not strongly associated with outcomes prediction.

We then trained the outcome prediction models using the data resulting from the pre-processing, data analysis, and feature selection process described in Sect. 3 (Table 4). Table 6 shows the results obtained.

Table 6. Results achieved by the outcome prediction models. The genes from MUT were selected using χ^2-test + the genes selected according to the literature. The genes from EXP were selected using LASSO

Model	F1-Score	AUC	Accuracy	Precision	Recall
CLIN	0.57	0.53	0.57	0.57	0.57
MUT	0.65	0.63	0.64	0.66	0.64
EXP	**0.86**	**0.84**	**0.86**	**0.86**	**0.86**
CLIN+MUT	0.67	0.64	0.68	0.67	0.68
CLIN+EXP	0.78	0.74	0.79	0.78	0.79
MUT+EXP	0.82	0.79	0.82	0.82	0.82
CLIN+MUT+EXP	0.78	0.74	0.79	0.78	0.79

The performance of the model trained only with the mutation data deteriorated slightly compared to the one obtained only with the genes highlighted in the literature. However, the performance of the model trained only with the expression data showed a remarkable improvement since all performance measures were up about 30%, and figuring as the best model we achieved. This strong increase in model performance is probably due to the careful KDD (Knowledge Discovery in Databases) process performed on the data and the new genes discovered to be good predictors.

Since gene expression data are expensive to obtain, they are usually absent on the first visit with specialists [6]. In this case, the outcome prediction model trained with clinical data and genetic mutations can be used as an initial guide to support the first therapeutic decisions.

The main advantage of using EBMs is that they are highly intelligible because the contribution of each feature to an output prediction can be easily visualized and understood. Since EBM is an additive model, each feature contributes to predictions in a modular way that makes it easy to reason about the contribution of each feature to the prediction. Figure 3 shows the local explanation for two test samples correctly classified as positive and negative using the classification model trained with the EXP feature set.

Figure 4 presents the top-15 attributes according to their importance in generating the prediction of outcome using gene mutation (Fig. 4a), gene expression (Fig. 4b), and clinical data (Fig. 4c), respectively. The attribute importance scores represent the average absolute contribution of each feature or interaction to the predictions, considering the entire training set. These contributions are weighted based on the number of samples within each group.

The four most influential clinical features are (*i*) when low-intensity treatment is chosen by the specialist; (*ii*) the patient's age; (*iii*) when high-intensity treatment is chosen; and (*iv*) the ELN risk classification. It is well-known that the age at diagnosis and the ELN risk classification can potentially impact the patient's outcome [1,7]. Considering that specialists often do not have access to the most suitable treatment intensity during model prediction, the predictions are automatically generated for the four categorized treatment types (Sect. 3.1), and the one that best optimizes the patient's survival time is selected as the recommended therapy.

Regarding genetic mutation data, the mutations in the *TP53* and *PHF6* genes are ranked as the most influential, followed by the gene mutations already well-known in the literature. If, on the one hand, the mutation in the *TP53* gene was already expected, to the best of our knowledge, there are no studies in the literature associating the *PHF6* gene with predicting outcomes in the context of AML. Therefore, laboratory tests should be performed to confirm whether this gene may serve as a potential prognostic marker.

Among the most influential genetic expression features for model prediction, the following stand out *KIAA0141*, *MICALL2*, and *SLC9A2*. Unlike the other genes, such as *PPM1* and *LTK*, which are already related in several AML studies, as far as we know, there is no study in the literature relating any of the three genes mentioned in the context of AML. In particular, the gene *KIAA0141*, also known as *DELE1*, has been recently identified as a key player [24]. In a

Local Explanation (Actual Class: 1 | Predicted Class: 1
Pr(y = 1): 0.981)

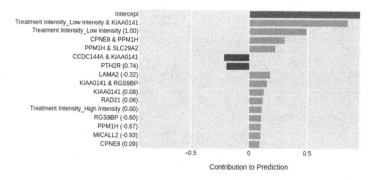

Contribution to Prediction

Local Explanation (Actual Class: 0 | Predicted Class: 0
Pr(y = 0): 0.840)

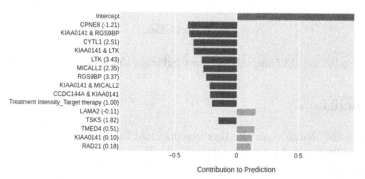

Contribution to Prediction

Fig. 3. Local explanation showing how much each feature contributed to the prediction for a single sample using the classification model trained with the EXP feature set. The intercept reflects the average case presented as a *log* of the base rate (e.g., -2.3 if the base rate is 10%). The 15 most important terms are shown.

pan-cancer analysis, *MICALL2* was highly expressed in 16 out of 33 cancers compared to normal tissues [16]. The role of *SLC9A2* in cancer is still an area of active research, and the exact relationship between *SLC9A2* and cancer development or progression is not fully understood. However, some studies have suggested potential associations between *SLC9A2* and certain types of cancer, such as colorectal, breast, and gastric cancer.

The findings presented in this paper suggest that the biological role of these genes in the pathogenesis and progression of AML deserves future functional studies in experimental models and may provide insights into the prognosis and the development of new treatments for the disease.

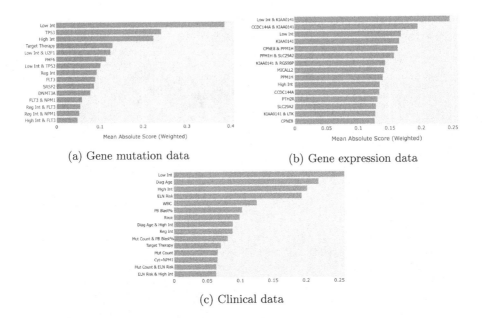

(a) Gene mutation data

(b) Gene expression data

(c) Clinical data

Fig. 4. Top-15 features that most influence the models' prediction

5 Conclusion

To support the decision on the therapy protocol for a given AML patient, specialists usually resort to a prognostic of outcomes according to the prediction of response to treatment and clinical outcome. The current ELN risk stratification is divided into favorable, intermediate, and adverse. Despite being widely used, it is very conservative since most patients receive an intermediate risk classification. Consequently, specialists must require new exams, delaying treatment and possibly worsening the patient's clinical condition.

This study presented a careful data analysis and explainable machine-learning models trained using the well-known Explainable Boosting Machine technique. According to the patient's outcome prediction, these models can support the decision about the most appropriate therapy protocol. In addition to the prediction models being explainable, the results obtained are promising and indicate that it is possible to use them to support the specialists' decisions safely.

We showed that the prediction model trained with gene expression data performed best. In addition, the results indicated that using a set of genetic features hitherto unknown in the AML literature significantly increased the prediction model's performance. The finding of these genes has the potential to open new avenues of research toward better treatments and prognostic markers for AML.

For future work, we suggest collecting more data to keep the models updated regarding the disease variations over time. Furthermore, the biological role of the genes *KIAA0141*, *MICALL2*, *PHF6*, and *SLC92A* in the pathogenesis and progression of AML deserves functional studies in experimental models.

References

1. Arber, D.A., et al.: International consensus classification of myeloid neoplasms and acute leukemias: integrating morphologic, clinical, and genomic data. Blood **140**(11), 1200–1228 (2022)
2. Caruana, R., et al.: Intelligible models for healthcare: predicting pneumonia risk and hospital 30-day readmission. In: Proceedings of the 21th ACM SIGKDD International Conference on Knowledge Discovery and Data Mining, pp. 1721–1730. ACM, Sydney NSW Australia (2015)
3. Charrot, S., et al.: AML through the prism of molecular genetics. Br. J. Haematol. **188**(1), 49–62 (2020)
4. Combi, C., et al.: A manifesto on explainability for artificial intelligence in medicine. Artif. Intell. Med. **133**, 102423 (2022)
5. Döhner, H., et al.: Diagnosis and management of acute myeloid leukemia in adults: recommendations from an international expert panel, on behalf of the European LeukemiaNet. Blood **115**(3), 453–474 (2010)
6. Döhner, H., et al.: Diagnosis and management of AML in adults: 2017 ELN recommendations from an international expert panel. Blood **129**(4), 424–447 (2017)
7. Döhner, H., et al.: Diagnosis and management of AML in adults: 2022 recommendations from an international expert panel on behalf of the ELN. Blood **140**(12), 1345–1377 (2022)
8. Eisa, Y.A., et al.: The role of PHF6 in hematopoiesis and hematologic malignancies. Stem Cell Rev. Rep. **19**(1), 67–75 (2023)
9. Estey, E.A.: Acute myeloid leukemia: 2019 update on risk-stratification and management. Am. J. Hematol. **93**(10), 1267–1291 (2019)
10. Gal, O., et al.: Predicting complete remission of acute myeloid leukemia: machine learning applied to gene expression. Cancer Inf. **18**, 1–5 (2019)
11. Grob, T., et al.: Molecular characterization of mutant TP53 acute myeloid leukemia and high-risk myelodysplastic syndrome. Blood **139**(15), 2347–2354 (2022)
12. Kastenhuber, E.R., Lowe, S.W.: Putting p53 in context. Cell **170**(6), 1062–1078 (2017)
13. Kurzer, J.H., Weinberg, O.K.: PHF6 mutations in hematologic malignancies. Front. Oncol. **11**, 704471 (2021)
14. Lagunas-Rangel, F.A., et al.: Acute myeloid leukemia-genetic alterations and their clinical prognosis. Int. J. Hematol. Oncol. Stem Cell Res. **11**, 328–339 (2017)
15. Ley, T.J., et al.: DNA sequencing of a cytogenetically normal acute myeloid leukaemia genome. Nature **456**(7218), 66–72 (2008)
16. Lin, W., et al.: Identification of MICALL2 as a novel prognostic biomarker correlating with inflammation and T cell exhaustion of kidney renal clear cell carcinoma. J. Cancer **13**(4), 1214–1228 (2022)
17. Gerstung, M., et al.: Precision oncology for acute myeloid leukemia using a knowledge bank approach. Nat. Genet. **49**, 332–340 (2017)
18. Mitchell, T.M.: Machine Learning. McGraw-Hill, New York (1997)
19. Mosquera Orgueira, A., et al.: Personalized survival prediction of patients with acute myeloblastic leukemia using gene expression profiling. Front. Oncol. **11**, 657191 (2021)
20. Nori, H., Caruana, R., Bu, Z., Shen, J.H., Kulkarni, J.: Accuracy, interpretability, and differential privacy via explainable boosting. In: Meila, M., Zhang, T. (eds.) Proceedings of the 38th International Conference on Machine Learning. Proceedings of Machine Learning Research, vol. 139, pp. 8227–8237. PMLR (2021)

21. Pelcovits, A., Niroula, R.: Acute myeloid leukemia: a review. R I Med. J. **103**(3), 38–40 (2020)
22. Pimenta, R.J.G., et al.: Genome-wide approaches for the identification of markers and genes associated with sugarcane yellow leaf virus resistance. Sci. Rep. **11**(1), 15730 (2021)
23. Rahman, M.M., et al.: Association of p53 gene mutation with helicobacter pylori infection in gastric cancer patients and its correlation with clinicopathological and environmental factors. World J. Oncol. **10**(1), 46–54 (2019)
24. Sharon, D., et al.: DELE1 loss and dysfunctional integrated stress signaling in TP53 mutated AML is a novel pathway for venetoclax resistance [abstract]. Can. Res. **83**, 2530 (2023)
25. The Cancer Genome Atlas Research Network: Genomic and epigenomic landscapes of adult de novo acute myeloid leukemia. N. Engl. J. Med. **368**(22), 2059–2074 (2013)
26. Tibshirani, R.: Regression shrinkage and selection via the lasso. J. Royal Stat. Soc. Ser. B (Methodological) **58**(1), 267–288 (1996)
27. Tyner, J.W., et al.: Functional genomic landscape of acute myeloid leukaemia. Nature **562**(7728), 526–531 (2018)
28. Van Vlierberghe, P., et al.: PHF6 mutations in adult acute myeloid leukemia. Leukemia **25**(1), 130–134 (2011)

IA Models

Bayes and Laplace Versus the World: A New Label Attack Approach in Federated Environments Based on Bayesian Neural Networks

Pedro H. Barros[1]($\boxtimes$), Fabricio Murai[2], and Heitor S. Ramos[1]

[1] Department of Computer Science, Federal University of Minas Gerais, Belo Horizonte, Brazil
`{pedro.barros,ramosh}@dcc.ufmg.br`
[2] Department of Computer Science, Worcester Polytechnic Institute, Worcester, USA
`fmurai@wpi.edu`

Abstract. Federated Learning (FL) is a decentralized machine learning approach developed to ensure that training data remains on personal devices, preserving data privacy. However, the distributed nature of FL environments makes defense against malicious attacks a challenging task. This work proposes a new attack approach to poisoning labels using Bayesian neural networks in federated environments. The hypothesis is that a label poisoning attack model trained with the marginal likelihood loss can generate a less complex poisoned model, making it difficult to detect attacks. We present experimental results demonstrating the proposed approach's effectiveness in generating poisoned models in federated environments. Additionally, we analyze the performance of various defense mechanisms against different attack proposals, evaluating accuracy, precision, recall, and F1-score. The results show that our proposed attack mechanism is harder to defend when we adopt existing defense mechanisms against label poisoning attacks in FL, showing a difference of 18.48% for accuracy compared to the approach without malicious clients.

Keywords: Federated Learning · Model Attack · Bayesian Neural Network

1 Introduction

The Internet of Things (IoT) has become increasingly impactful, empowering diverse applications [14]. Approximately 5.8 billion IoT devices are estimated to be in use this year [6]. Moreover, privacy issues are becoming increasingly relevant for distributed applications, as seen recently in General Data Protection Regulation (GDPR). A decentralized machine learning approach called Federated Learning (FL) was proposed to guarantee that training data remains on personal devices and facilitates complex models of collaborative machine learning on distributed devices.

M. C. Naldi and R. A. C. Bianchi (Eds.): BRACIS 2023, LNAI 14195, pp. 449–463, 2023.
https://doi.org/10.1007/978-3-031-45368-7_29

The training of a federated model is performed in a distributed fashion where data remains on users' local devices while the model is updated globally [15]. During training, the global model is sent to users' devices, which evaluate the model with local estimates. Furthermore, local updates are sent back to the server, aggregating them into a central model. Data privacy tends to be preserved as no actual data is shared, only model updates.

With the increasing use of machine learning algorithms in federated environments, it is necessary to ensure user data privacy and security and the constructed models' reliability. However, the distributed nature of these environments and the heterogeneity of user data make this process challenging. Furthermore, these techniques still present vulnerabilities, as the model is trained based on data from multiple users, including potential attackers.

Developing defense strategies against malicious attacks in federated environments becomes crucial to ensure the security and privacy of users. A promising approach to this issue is the construction of malicious models that can be used to attack other models and, therefore, test the robustness of the federation. Furthermore, these malicious models are built to exploit the vulnerabilities of models in federated environments, making them helpful in evaluating the effectiveness of defense techniques.

In federated learning, neural networks are commonly used as a machine learning model due to their ability to learn and generalize complex patterns in large datasets. However, despite their promising results, neural networks have limitations that can restrict their applications, such as difficulty in model calibration and overconfidence in predictions, especially when the data distribution changes between training and testing [10].

This overconfidence problem is where neural networks exhibit excessively high confidence levels in their predictions, even when these predictions are incorrect. This happens because neural networks are trained to maximize the accuracy of their predictions without considering the uncertainty associated with the input data. As a result, neural networks may exhibit overconfidence in their predictions, even when the input data is ambiguous or noisy.

Bayesian neural networks (BNNs) are models capable of quantifying uncertainty and using it to develop more accurate learning algorithms [9]. In addition, BNNs tend to mitigate the problem of neural networks' overconfidence [8]. This approach allows neural networks to produce a probability distribution for the output rather than just a single output, considering the uncertainty associated with the input data. This allows the network to assess its confidence in the prediction and make more informed decisions, improving its generalization and making it more robust and reliable. Although implementing Bayesian neural networks may be more complex, the advantages in terms of accuracy and confidence in their predictions are significant [11]. For example, we can see an illustration in Fig. 1, an example of the decision region of neural networks (Bayesian). In this example, we observe that BNNs present smoother decision regions, which implies more realistic confidence about predictions. Furthermore, neural networks are typically confident even in regions where the uncertainty is high, such as on the borders of regions.

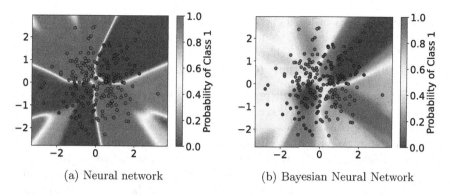

(a) Neural network   (b) Bayesian Neural Network

Fig. 1. Illustration of decision region of (bayesian) neural network. The dataset used in this experiment is synthetic and consists of 200 samples. Each sample has two features (2D dimension) randomly generated from a normal distribution. The output variable is determined by applying the XOR logical operation to the two input features. The goal is to train a (Bayesian) neural network model to make accurate predictions.

This work proposes a **novel attack approach** to poisoning labels using Bayesian neural networks in federated environments. Label poisoning attacks change training data labels, deviating the model from its original goal. As BNNs incorporate the principle of parsimony (Occam's razor), we hypothesize that a label poisoning attack model trained using the uncertainty quantification provided by BNNs can maximize the adherence of malicious data to the model while also estimating a poisoned model with lower complexity, making it difficult to detect attacks in federated environments. This work aims to propose and evaluate this approach, presenting experimental results that demonstrate its effectiveness in generating poisoned models in federated environments.

We organized this paper as follows: Sect. 2 presents the related works to security in FL environments; Sect. 3 describes our proposal and some notations review for a good understanding of our proposal; Sect. 4 describes the experimental setup used to analyze the data; Sect. 5 presents the main results and discussions; and Sect. 6 concludes this work.

2   Related Work

The security issues in machine learning systems, and consequently federated learning, have been extensively studied [17,21,27]. In traditional machine learning, the learning phase is typically protected and centralized in a unique system [13]. Specifically, the approaches in literature usually consider that malicious clients act during inference, i.e., the attacked model is already in production [12]. In FL, malicious clients usually exploit the vulnerability of the models during the learning phase [3]. In general, malicious clients attacking an FL model have one of two adversarial goals:

Case I. Reconstruct or learn client/model information based on data transmitted in the federated training process, and

Case II. Force the model to behave differently than intended, invalidate or train it for a specific purpose (e.g., poisoning attack).

In this proposal, we focus on the problems related to attacks that aim to degrade the performance of the aggregated model (type II attacks), especially in the untargeted attack. Among these types of attacks, data poisoning (i.e., poisoning in the training dataset) is one of the most common forms of poisoning attack.

Several works in the literature propose a poisoning attack method to FL models to degrade the training process. Zhang et al. [25] propose using generative adversarial networks (GANs) to generate examples without any assumption on accessing the participants' training data. Similarly, Zhang et al. [26] then inserts adversarial poison samples assigned with the wrong label to the local training dataset to degrade the aggregate model. Sun et al. [19] studies the vulnerability of federated learning models in IoT systems. Thus, the authors use a bilevel optimization consideration, which injects poisoned data samples to maximize the deterioration of the aggregate model. In addition, Defense mechanisms for distributed poisoning attacks typically draw ideas from robust estimation and anomaly detection [1,18]. Some works are based on aggregation functions robust to outliers, such as median [22], mean with exclusion [23], geometric mean [16], and clustering of nearby gradients [4].

Additionally, assuming scenarios, where all clients train the network model but use their data is paramount. More often than not, we observe that data models differ across clients. Therefore, the resulting models will be abstractions of different real-world conditions. Ultimately, we need to propose solutions that handle such situations bearing in mind the challenges of accounting for such differences without considering them malicious, especially when only a small subset of clients present data models apart from the majority.

3   Our Proposal

3.1   Bayesian Neural Network

A BNN is a neural network $f(\mathbf{x}, \mathbf{w})$ that maps inputs $\mathbf{x}$ to outputs y, where $\mathbf{w} \in \mathbb{R}^M$ is a vector of random variables representing the weights and biases of the network. The BNN assumes a prior distribution over $\mathbf{w}$, denoted $p(\mathbf{w})$, and learns a posterior distribution $p(\mathbf{w} \mid \mathcal{D})$ over the weights and biases given the training data $\mathcal{D} = \{(\mathbf{x}_i, y_i)\}_{i=1}^N$. Thus, for a dataset $\mathcal{D}$, we have $p(\mathcal{D} \mid \mathbf{w}) = \Pi_{i=1}^N p(y_i \mid f(\mathbf{x}_i, \mathbf{w}))$. Bayesian inference techniques such as Markov Chain Monte Carlo (MCMC), variational inference, and Laplace approximation can approximate the posterior distribution.

Given the posterior distribution, the network can make predictions by computing the predictive distribution $p(y \mid \mathbf{x}, \mathcal{D})$ over the output given the input and the training data. The BNN approach provides a probabilistic framework for modeling uncertainty in the network's predictions and can be particularly useful in applications where knowing the level of uncertainty is essential. In addition, the regularization effect of the prior distribution can also prevent overfitting and improve the generalization performance of the network.

3.2 Laplace Approximation

In Bayesian inference, we are interested in computing the posterior distribution of the model parameters $\mathbf{w}$ given the observed data $\mathcal{D}$, which is given by

$$p(\mathbf{w} \mid \mathcal{D}) = \frac{p(\mathcal{D} \mid \mathbf{w})p(\mathbf{w})}{\int p(\mathcal{D} \mid \mathbf{w})p(\mathbf{w})d\mathbf{w}},$$

where $p(\mathcal{D} \mid \mathbf{w})$ is the data likelihood, $p(\mathbf{w})$ is the prior distribution of the parameters, and the integral in the denominator is the normalization constant.

Unfortunately, $\int p(\mathcal{D} \mid \mathbf{w})p(\mathbf{w})d\mathbf{w}$ is often intractable for complex models (as neural networks), and we need to resort to approximate inference methods. One such method is the Laplace approximation, which approximates the posterior distribution with a Gaussian distribution centered at the mode of the posterior, $\mathbf{w}_{MAP}$ (maximum a posterior), and with a covariance matrix given by the inverse Hessian matrix evaluated at $\mathbf{w}_{MAP}$.

To derive the Laplace approximation, we start by manipulating the integral in the denominator $\int p(\mathcal{D} \mid \mathbf{w})p(\mathbf{w})d\mathbf{w} = \int \exp(\log p(\mathcal{D}, \mathbf{w}))d\mathbf{w}$. Next, we use the second-order Taylor expansion of $\log p(\mathcal{D}, \mathbf{w})$ around $\mathbf{w}_{MAP}$:

$$\log p(\mathcal{D}, \mathbf{w}) \approx \log p(\mathcal{D}, \mathbf{w}_{MAP}) - g_{MAP}^\top(\mathbf{w} - \mathbf{w}_{MAP})$$
$$- \frac{1}{2}(\mathbf{w} - \mathbf{w}_{MAP})^\top \mathbf{H}_{MAP}(\mathbf{w} - \mathbf{w}_{MAP}) \qquad (1)$$
$$= \log p(\mathcal{D}, \mathbf{w}_{MAP}) - \frac{1}{2}(\mathbf{w} - \mathbf{w}_{MAP})^\top \mathbf{H}_{MAP}(\mathbf{w} - \mathbf{w}_{MAP}),$$

where $g_{MAP} = -\nabla \log p(\mathbf{w} \mid \mathcal{D}) \mid_{\mathbf{w}=\mathbf{w}_{MAP}}$ is the gradient of $\log p(\mathcal{D}, \mathbf{w})$ evaluated at $\mathbf{w}_{MAP}$, and $\mathbf{H}_{MAP} = -\nabla\nabla \log p(\mathbf{w} \mid \mathcal{D}) \mid_{\mathbf{w}=\mathbf{w}_{MAP}}$ is the Hessian matrix evaluated at $\mathbf{w}_{MAP}$. Using the Laplace approximation, we approximate the marginal likelihood $p(\mathcal{D})$ as

$$p(\mathcal{D}) = \int \exp(\log p(\mathcal{D}, \mathbf{w}))d\mathbf{w}$$
$$\approx \int \exp(\log p(\mathcal{D}, \mathbf{w}_{MAP}) - \frac{1}{2}(\mathbf{w} - \mathbf{w}_{MAP})^\top \mathbf{H}_{MAP}(\mathbf{w} - \mathbf{w}_{MAP}))d\mathbf{w}.$$

Since the integral is Gaussian, we can solve it analytically and obtain

$$p(\mathcal{D}) \approx p(\mathcal{D}, \mathbf{w}_{MAP})(2\pi)^{\frac{M}{2}}|\mathbf{H}_{MAP}|^{-\frac{1}{2}} \qquad (2)$$

where M is the dimensionality of $\mathbf{w}$. Using this approximation, we can obtain the posterior distribution $p(\mathbf{w} \mid \mathcal{D}) \approx \mathcal{N}(\mathbf{w} \mid \mathbf{w}_{MAP}, \mathbf{H}_{MAP}^{-1})$, where $\mathcal{N}(\mathbf{w} \mid$

$\mathbf{w}_{MAP}, \mathbf{H}_{MAP}^{-1})$ denotes a multivariate Gaussian distribution with mean $\mathbf{w}_{MAP}$ and covariance matrix $\mathbf{H}_{MAP}^{-1}$ [11]. Therefore, the Laplace approximation allows us to approximate the posterior distribution of the weights as a Gaussian distribution, which can be more computationally efficient than the full posterior distribution.

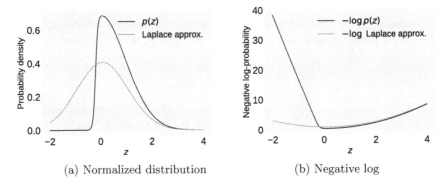

(a) Normalized distribution   (b) Negative log

Fig. 2. Illustration of the Laplace approximation applied to the distribution $p(z) \propto \exp\left(-z^2/2\right)\sigma(20z+4)$ where $\sigma(z)$ is the logistic sigmoid function defined by $\sigma(z) = (1+e^{-z})^{-1}$. Figure (a) shows the normalized distribution $p(z)$, together with the Laplace approximation centered on the mode z_0 of $p(z)$. Figure (b) shows the negative logarithms of the corresponding curves. (Color figure online)

3.3   Occam Razor and Marginal Likelihood

The marginal likelihood automatically encapsulates a notion of Occam's Razor. To illustrate this, we estimate Laplace approximation to find a Gaussian approximation to a probability density defined over a set of continuous variables. We can see a toy example for Laplace approximation in Fig. 2.

Figure 2 (a), the normalized distribution $p(z)$ is shown alongside the Laplace approximation centered on the mode z_0 of $p(z)$. The Laplace approximation, depicted by the orange curve, is a Gaussian distribution that closely matches the original distribution near the mode z_0. This approximation is commonly employed to simplify calculations and provide a more manageable representation of complex problems.

Figure 2 (b) presents the negative logarithms of the corresponding curves. The logarithmic scale enhances subtle differences between the curves. Notably, the Laplace approximation (orange line) exhibits a similar fit to the original curve (blue line) near the mode z_0. This Figure effectively showcases the application of Laplace approximation for estimating a Gaussian distribution that effectively approximates a complex probability distribution.

We can consider the log of the Laplace Marginal Likelihood (LML) in Eq. 2 as

$$\log p(\mathcal{D}) \propto \log p(\mathcal{D}, \mathbf{w}_{\mathrm{MAP}}) + \underbrace{\frac{M}{2} \log(2\pi) - \frac{1}{2} \log |\mathbf{H}_{MAP}|}_{\text{Occam factor}}. \tag{3}$$

The relationship between Occam's Razor and Laplace Approximation is deeply rooted in the theory of Bayesian inference. Laplace Approximation is a technique that allows us to approximate a probability distribution with a Gaussian distribution around the maximum a posteriori point. The maximum a posteriori point is often interpreted as the most likely solution to a given modeling problem.

Occam's Razor, on the other hand, is a philosophical principle that states that if there are several possible explanations for a given set of observations, the simplest explanation is the most likely. In Bayesian inference theory, this translates into the fact that when making inferences about a model, we should prefer simpler and less complex models unless there is clear evidence on the contrary.

When we consider Eq. 3 as our loss function for training the neural network, we realize that maximizing the fit of the model's marginal likelihood corresponds to increasing the value $\log p(\mathcal{D}, \mathbf{w}_{\mathrm{MAP}})$ and minimizing the complexity term $\log |\mathbf{H}_{MAP}|$. The complexity term depends on the log determinant of the Laplace posterior covariance. Therefore, if $\log |\mathbf{H}_{MAP}|$ is large, the model strongly correlates with the training data [11]. So, maximizing the Laplace $\log p(\mathcal{D})$ requires maximizing the data fit while minimizing the training sample correlation.

The Laplace Approximation implements Occam's Razor as we make the simplest possible assumption about the posterior distribution. Furthermore, the Laplace Approximation has been used in many applications, including training machine learning models in federated environments where data privacy is critical. In these scenarios, it is essential to have a model that can generalize well from a small dataset. The Laplace Approximation can help achieve this goal by providing a way to regularize the model and avoid overfitting.

3.4 Federated Learning

In federated learning, users collaboratively train a machine learning model without sharing their raw data. The process consists of three main steps, as can see in Fig. 3:

- (Step 1) **Task initialization:** The system initializes the local models and necessary hyperparameters for the learning task. Each user prepares their local model using their respective dataset.
- (Step 2) **Local model training and update**: Each user independently trains their local model using their local data. The goal is to find the optimal parameters that minimize the loss function specific to their dataset. This step ensures that each user's model is tailored to their data and captures their local patterns.

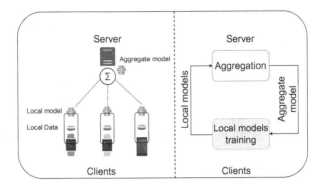

Fig. 3. Left: Conventional (client-server) federated learning scheme. Right: Training of a simple data flow model in federated learning.

– (Step 3) **Model aggregation and update**: The server aggregates the local models from selected participants and generates an updated global model, often referred to as the federated model. The aggregation process typically involves combining the model parameters of the local models . The global model is then sent back to the users for further iterations.

Steps 2 and 3 are repeated iteratively until the aggregate loss function converges or reaches a desired training metric. The iterative nature of the process allows the global model to improve over time by leveraging the collective knowledge from all participating users while maintaining data privacy.

3.5 Model Attack

Federated learning, a distributed learning paradigm, allows multiple clients to collaboratively train a global model without sharing their raw data. However, this collaborative nature makes federated learning vulnerable to malicious attacks.

Before launching our attack on federated learning, an attacker must go through several steps: (i) Firstly, in the attack preparation phase, the attacker injects poisoned labels into the corrupted local data, introducing inaccurate label assignments to the training examples; (ii) Secondly, each local malicious client trains a local malicious model using the corrupted dataset. The malicious model is trained by minimizing a loss function that incorporates the maximization of marginal likelihood using the Laplace approximation. This approach takes advantage of the over-parameterized nature of modern classifiers, enabling them to fit the corrupted data easily [2]; and (iii) Finally, during the global model update process, the local models are aggregated to form the updated global model. In the aggregation step, honest and malicious updates construct the aggregated model. By understanding these steps, we can learn how malicious models exploit the federated learning framework to undermine its integrity and compromise the global model's performance and reliability. We can describe our proposal as follows:

Attack Preparation. The attacker injects poisoned labels into the corrupted local data, introducing inaccurate label assignments to the training examples:

Step (a) Let $\mathcal{D}_i = \{(\mathbf{x}_1, y_1), (\mathbf{x}_2, y_2), \cdots, (\mathbf{x}_n, y_n)\}$ be the local dataset of client i, where x_j is an input and y_j is the true label.

Step (b) Let $\alpha \in [0, 1]$ represent the fraction of random poisoned samples in local malicious data $\mathcal{D}_i'$.

Step (c) The attacker replaces a fraction α of the true labels y_j with random poisoned labels y_j'.

Malicious Model. Each local malicious client trains a local malicious model using the corrupted dataset $\mathcal{D}_i'$:

Step (a) Let $f(\mathbf{x}, \mathbf{w}_i)$ denote the local model of client i, parameterized by $\mathbf{w}_i$. The malicious model is trained by minimizing a loss function that incorporates the maximization of marginal likelihood using the Laplace approximation. Also, note that honest clients are trained using the usual cross-entropy loss.

Step (b) The malicious loss function used is the Laplace Marginal Likelihood loss (Eq. 3), defined as

$$\mathcal{L}_{\mathrm{LML}}(\mathbf{w}_i) = \log p(\mathcal{D}_i, \mathbf{w}_{\mathcal{D}_i}^*) + \frac{M_i}{2} \log(2\pi) - \frac{1}{2} \log |\mathbf{H}_{\mathcal{D}_i}^*|,$$

where $\mathcal{D}_i$ is the corrupted local dataset, $\log p(\mathcal{D}_i, \mathbf{w}_i)$ is the Laplace approximation (Eq. 1), $\mathbf{w}_{\mathcal{D}_i}^*$ refers to the maximum a posterior (MAP), M_i is the dimensionality of $\mathbf{w}_i$, and $\log |\mathbf{H}_{\mathcal{D}_i}^*|$ is the logarithm of the determinant of the Hessian matrix.

Step (c) Maximizing the marginal likelihood promotes the fitting of the corrupted data, while the regularization term controlled by $\log |\mathbf{H}_{\mathcal{D}_i}^*|$ limits the model's complexity. Modern classifiers are heavily over-parameterized [2]. Thus, these models easily fit a random poisoning labeling of the training data and, consequently, present a high Memorization gap [24].

Aggregation. During the global model update process in federated learning, the local models are aggregated to form the updated global model:

step (a) The local models are combined using an aggregation technique; for example, the weighted average of the local model parameters as

$$\mathbf{w}_{Fed} \leftarrow \mathbf{w}_{Fed} + \eta \underbrace{\sum_{u_i \in S} \left[p_i(\mathbf{w}_{\mathcal{D}_i}^* - \mathbf{w}_{Fed}) \right]}_{\text{Honest updates}} + \eta \underbrace{\sum_{m_i \in S'} \left[p_i(\mathbf{w}_{\mathcal{D}_i'}^* - \mathbf{w}_{Fed}) \right]}_{\text{Malicious updates}},$$

where S' and S denote the set of selected malicious and benign clients for training, respectively; $\mathbf{w}_{\mathcal{D}_i'}^*$ and $\mathbf{w}_{\mathcal{D}_i}^*$ refers to the maximum a posterior (MAP) estimate of the malicious and honest model parameters, respectively; and $\mathbf{w}_{Fed}$ is the aggregated or global model that is constructed by combining the local model updates from multiple clients.

4   Methodology

4.1   Dataset

To analyze the federated performance, we used the EMNIST dataset, a collection of handwritten characters with over 800,000 images. This unique dataset combines two popular datasets, the MNIST dataset and the NIST Special Database 19, to create a more extensive and diverse collection of handwritten characters. The dataset contains handwritten characters from 62 classes, including uppercase and lowercase letters, digits, and symbols. It is a valuable resource for researchers and developers interested in optical character recognition (OCR) and related applications.

The images in the EMNIST dataset are grayscale and have a 28 by 28 pixels resolution. To prepare the data for training, we performed a min-max normalization on all input features to scale the values between 0 and 1, ensuring that no individual feature would dominate the learning process.

4.2   Model Evaluation

We use the framework Flower to develop solutions and applications in federated learning. We perform a non-iid data distribution among the users in this experiment. We randomly distribute the data in a non-uniform way (quantity-based label imbalance) [15].

We employ a server and 100 clients to evaluate our model, and we trained our method with an NVIDIA Quadro RTX 6000 GPU (24 GB) for a total of 100 epochs (server). For each training round, the server selects five clients to train the local model, i.e., each model is trained using only the local data. Finally, the models are aggregated on the central server, which forwards the aggregated model to the clients.

For simplicity, we consider uncorrelated noise for the poisoning label attack (all labels are equally likely to be corrupted). We use 10% as the system noise level (ratio of noisy clients). Each malicious client generates $\alpha = 50\%$ of corrupt label data, i.e., even a malicious client produces 50% of correct labels.

4.3   Network Architecture

The neural network architecture consists of two main parts: a convolutional layer and a fully connected layer. The convolutional layer is responsible for extracting features from the input data. It consists of four convolutional layers with ReLU activation functions and max pooling layers, which reduces the spatial dimensions of the output. The first convolutional layer has 32 filters, followed by a layer with 64 filters. The subsequent two layers have 128 filters each. The fully connected layer takes the output of the convolutional layer and maps it to the output classes. It consists of two linear layers, with 1024 and 26 neurons, respectively. The model uses Stochastic Gradient Descent (SGD) with momentum optimization with a learning rate of 0.01 and a momentum of 0.9. This architecture was used to classify characters in the EMNIST dataset.

5 Results

In this work, in order to perform a quantitative comparison, we compared our approach in different evaluation scenarios. Therefore, we considered six experiments with five techniques for mitigating attacks in federated environments and FedAvg aggregation. More specifically, we have the following:

- FedAvg [15] is a method that uses the weighted average of local models to obtain a more accurate global model. The FedAvg aggregation is a widespread technique in FL due to its effectiveness and simplicity of implementation.
- FedEqual [5] equalizes the weights of local model updates in FL, allowing most benign models to counterbalance malicious attackers' power and avoid excluding local models.
- Geometric [16] uses the geometric mean of the local models to obtain a global model that is more robust to outliers. Geometric aggregation is an alternative technique to FedAvg aggregation, which can be more effective in specific applications.
- Krum [4] is a robust aggregation rule that uses a Euclidean distance approach to select similar model updates. This method calculates the sum of squared distances between local model updates and then selects the model update with the lowest sum to update the parameters of the global model.
- Norm [20] consists of calculating a normalization constant for model updates by clients to normalize the contribution of any individual participant. So, all model updates are averaged to update the joint global model.
- Trimmed [23] can be defined as the average of the local models but removes a proportion of the local models, including outliers, that may negatively affect the aggregation. Trimmed aggregation is a technique that can improve the effectiveness of model aggregation in FL systems, especially when there is a significant variation in the local data.

On the other hand, besides our proposed attack, we evaluated our federation using four attack proposals in federated environments in addition to the standard proposal without malicious clients for each of the six experiments:

- None: Approach without any malicious client.
- FedAttack [22]: Uses a contrastive approach to group corrupted samples into similar regions. The authors propose to use globally hardest sampling as a poisoning technique. However, the technique involves malicious clients using their local user embedding to retrieve globally hardest negative samples and pseudo "hardest positive samples" to increase the difficulty of model training.
- Label [22]: Label flipping attack approach that randomly flips the labels of the malicious client's dataset and trains on the contaminated dataset using cross-entropy loss.
- Stat-opt [7]: Attack that consists of adding constant noise with opposite directions to the average of benign gradients.

Note que the malicious clients in FedAttack, Label, and Our proposal only modify the input samples and their labels. At the same time, the model gradients are directly computed on these samples without further manipulation. Finally, Table 1 displays the performance of the described defense mechanisms against various attacks in federated learning. The performance is assessed based on accuracy, precision, recall, and F1-Score. The best performance is highlighted in bold in each category. It is worth noting that the proposed attack model aims to maximize its success rate. Thus a lower performance is better in comparison to other defense proposals.

Table 1. The table shows the performance of various defense mechanisms against different attack proposals in a federated learning setting. The evaluation metrics include accuracy, precision, recall, and F1-score. The best values were shown in **bold** (lower is better).

Def proposal	Atk proposal	Accuracy	Precision	Recall	F1-Score
FedAvg	None	0.863	0.861	0.857	0.859
	FedAttack	0.648	0.653	0.649	0.655
	Label	0.783	0.775	0.769	0.772
	Stat-opt	**0.584**	**0.593**	**0.581**	**0.587**
	Our proposal	0.603	0.632	0.607	0.595
FedEqual	None	0.853	0.859	0.857	0.859
	FedAttack	0.783	0.786	0.789	0.782
	Label	0.801	0.812	0.807	0.809
	Stat-opt	**0.767**	**0.759**	0.762	**0.761**
	Our proposal	0.771	0.764	**0.760**	0.762
Geometric	None	0.860	0.859	0.857	0.857
	FedAttack	0.803	0.799	0.795	0.797
	Label	0.807	0.806	0.805	0.806
	Stat-opt	0.791	0.796	0.791	0.793
	Our proposal	**0.701**	**0.707**	**0.703**	**0.705**
Krum	None	0.838	0.841	0.837	0.839
	FedAttack	0.751	0.759	0.756	0.757
	Label	0.793	0.799	0.795	0.798
	Stat-opt	0.784	0.789	0.781	0.786
	Our proposal	**0.745**	**0.762**	**0.752**	**0.754**
Norm	None	0.822	0.829	0.818	0.824
	FedAttack	0.817	0.812	0.808	0.809
	Label	0.786	0.781	0.776	0.780
	Stat-opt	0.709	0.712	0.701	0.709
	Our proposal	**0.689**	**0.681**	**0.704**	**0.698**
Trimmed	None	0.849	0.853	0.849	0.852
	FedAttack	0.812	0.816	0.809	0.814
	Label	0.813	0.819	0.813	0.816
	Stat-opt	0.782	0.787	0.786	0.786
	Our proposal	**0.725**	**0.729**	**0.728**	**0.729**

Initially, we verified that the "Label" approach is a simplified version of our proposed attack, using the usual cross-entropy loss in a label attack. In contrast, our proposal uses a Bayesian neural network trained with marginal likelihood. The results show that our attack proposal is more effective than the Label approach. In all defense scenarios, our proposal had a lower performance, i.e., our proposal was able to reduce accuracy, precision, recall, and F1-score significantly. The results also show that the performance of defense mechanisms varies according to the attack proposal and the type of defense used. For example, the "FedAvg" defense was less effective against label attacks than other defense proposals. In general, our attack proposal proved more effective in all scenarios, regardless of the type of defense used compared to the Label attack approach.

For the other approaches, compared to the FedAvg defense proposal, the results show that the Stat-opt proposal had the best attack performance, with an accuracy of 0.584, precision of 0.593, recall of 0.581, and F1-Score of 0.587. In the second place, we observed that Our proposal had slightly better performance than the "FedAttack" model, with an accuracy of 0.603, precision of 0.632, recall of 0.607, and F1-Score of 0.595. A similar conclusion was observed for the FedEqual defense strategy. Finally, however, we noticed that for the Recall metric, our proposal had better performance (being more effective in degrading this metric in the federation result).

For the Geometric defense strategy, we observed that our approach achieved the highest degradation of the model. For the accuracy metric, our model obtained 0.701, showing a difference of 18.48% when compared to the approach without malicious clients. The second-best attack proposal achieved a degradation of 8.02% compared to the model without any attack (None). We observed that this behavior is consistent for the other metrics, where our approach achieved a precision of 0.707, recall of 0.703, and F1-Score of 0.705. We also observed similar behavior for the Krum, Norm, and Trimmed experiments, indicating that the attack strategy significantly impacted the defense mechanisms' performance.

Based on the results, our proposed approach was superior in four scenarios, while the stat-opt proposal was better in only two. This conclusion happens because the Stat-opt proposal manipulates the gradient to attack the model, while our approach only performs a label change attack. Manipulating the gradient allows the stat-opt approach to have finer control over changes made to model weights, making the attack more effective. Furthermore, the stat-opt approach can exploit the model's internal structure, which can lead to more sophisticated and challenging to detect attacks, unlike our approach, which only performs a label change attack. However, our proposed approach outperformed the other defense mechanisms in the remaining scenarios. This indicates that our approach is more effective for attacking federated learning models than the label attack in the stat-opt proposal. Overall, our approach is a powerful tool for evaluating the robustness of federated learning models and improving their defenses against label change attacks.

6 Conclusion

This paper proposes a new labeling attack model for federated environments using Bayesian neural networks. The approach is based on a model trained with the marginal likelihood loss function that maximizes the adherence of malicious data to the model while also estimating a poisoned model with lower complexity, making it difficult to detect attacks in federated environments. In federated environments, it is essential to ensure the security and privacy of user data and the reliability of built models. However, machine learning techniques in federated environments still present vulnerabilities since the model is trained based on data from multiple users, including possible attackers.

The evaluation of different defense mechanisms against various attacks in federated learning showed that the performance of defense mechanisms varies according to the attack proposal and the type of defense used. However, in general, our proposed attack was more effective in all scenarios. The results also showed that the proposed attack method was highly influential in degrading the model's performance. Therefore, it is crucial to consider the proposed attack model when designing defense mechanisms for federated learning systems. Furthermore, the findings suggest that future research should focus on developing more effective defense mechanisms to mitigate the risks associated with the proposed attack.

References

1. Alistarh, D., Allen-Zhu, Z., Li, J.: Byzantine stochastic gradient descent. In: Advances in Neural Information Processing Systems, vol. 31. Curran Associates, Inc. (2018)
2. Bansal, Y., et al.: For self-supervised learning, rationality implies generalization, provably. In: International Conference on Learning Representations (ICLR) (2020)
3. Bhagoji, A.N., Chakraborty, S., Mittal, P., Calo, S.: Analyzing federated learning through an adversarial lens. In: International Conference on Machine Learning (ICML), vol. 97 (2019)
4. Blanchard, P., El Mhamdi, E.M., Guerraoui, R., Stainer, J.: Machine learning with adversaries: Byzantine tolerant gradient descent. In: Advances in Neural Information Processing Systems (NeurIPS), vol. 30 (2017)
5. Chen, L.Y., Chiu, T.C., Pang, A.C., Cheng, L.C.: Fedequal: defending model poisoning attacks in heterogeneous federated learning. In: 2021 IEEE Global Communications Conference (GLOBECOM), pp. 1–6 (2021)
6. Dao, N.N., et al.: Securing heterogeneous IoT with intelligent DDoS attack behavior learning. IEEE Syst. J. **16**(2), 1974–1983 (2022)
7. Fang, M., Cao, X., Jia, J., Gong, N.Z.: Local model poisoning attacks to byzantine-robust federated learning. In: Proceedings of the 29th USENIX Conference on Security Symposium (SEC) (2020)
8. Gal, Y., Ghahramani, Z.: Dropout as a Bayesian approximation: representing model uncertainty in deep learning. In: Proceedings of the 33rd International Conference on Machine Learning (ICML) (2016)
9. Ghahramani, Z.: Probabilistic machine learning and artificial intelligence. Nature **521**(7553), 452–459 (2015)

10. Guo, C., Pleiss, G., Sun, Y., Weinberger, K.Q.: On calibration of modern neural networks. In: Proceedings of the 34th International Conference on Machine Learning, ICML 2017 (2017)
11. Immer, A., et al.: Scalable marginal likelihood estimation for model selection in deep learning. In: International Conference on Machine Learning (ICML), vol. 139, pp. 4563–4573 (2021)
12. Jagielski, M., Oprea, A., Biggio, B., Liu, C., Nita-Rotaru, C., Li, B.: Manipulating machine learning: poisoning attacks and countermeasures for regression learning. In: 2018 IEEE Symposium on Security and Privacy (SP), pp. 19–35 (2018)
13. Lamport, L., Shostak, R., Pease, M.: The byzantine generals problem. ACM Trans. Programm. Lang. Syst. **4**(3), 382–401 (1982)
14. Li, T., Sahu, A.K., Talwalkar, A., Smith, V.: Federated learning: challenges, methods, and future directions. IEEE Signal Process. Mag. **37**, 50–60 (2020)
15. McMahan, B., Moore, E., Ramage, D., Hampson, S., Arcas, B.A.: Communication-efficient learning of deep networks from decentralized data. In: Proceedings of the International Conference on Artificial Intelligence and Statistics (AISTATS), pp. 1273–1282 (2017)
16. Pillutla, K., Kakade, S.M., Harchaoui, Z.: Robust aggregation for federated learning. IEEE Trans. Signal Process. **70**, 1142–1154 (2022)
17. Rodríguez-Barroso, N., Martínez-Cámara, E., Luzón, M.V., Herrera, F.: Dynamic defense against byzantine poisoning attacks in federated learning. Future Gener. Comput. Syst. **133**, 1–9 (2022)
18. Shejwalkar, V., Houmansadr, A., Kairouz, P., Ramage, D.: Back to the drawing board: a critical evaluation of poisoning attacks on production federated learning. In: 2022 IEEE Symposium on Security and Privacy (SP), pp. 1354–1371 (2022)
19. Sun, G., Cong, Y., Dong, J., Wang, Q., Lyu, L., Liu, J.: Data poisoning attacks on federated machine learning. IEEE Internet Things J. **9**(13), 11365–11375 (2022)
20. Sun, Z., Kairouz, P., Suresh, A.T., McMahan, H.B.: Can you really backdoor federated learning? (2019)
21. Wang, H., et al.: Attack of the tails: yes, you really can backdoor federated learning. In: Proceedings of the 34th International Conference on Neural Information Processing Systems (NeurIPS). Red Hook, NY, USA (2020)
22. Wu, C., Wu, F., Qi, T., Huang, Y., Xie, X.: Fedattack: effective and covert poisoning attack on federated recommendation via hard sampling. In: Proceedings of the 28th ACM SIGKDD Conference on Knowledge Discovery and Data Mining (KDD), pp. 4164–4172, New York, NY, USA (2022)
23. Yin, D., Chen, Y., Kannan, R., Bartlett, P.: Byzantine-robust distributed learning: towards optimal statistical rates. In: Proceedings of the International Conference on Machine Learning, vol. 80, pp. 5650–5659 (2018)
24. Zhang, C., et al.: Understanding deep learning (still) requires rethinking generalization. Commun. ACM **64**(3), 107–115 (2021)
25. Zhang, J., Chen, B., Cheng, X., Binh, H.T.T., Yu, S.: PoisonGAN: generative poisoning attacks against federated learning in edge computing systems. IEEE Internet Things J. **8**(5), 3310–3322 (2021)
26. Zhang, J., Chen, J., Wu, D., Chen, B., Yu, S.: Poisoning attack in federated learning using generative adversarial nets. In: 2019 18th IEEE International Conference on Trust, Security and Privacy in Computing and Communications (TrustCom/BigDataSE), pp. 374–380 (2019)
27. Zhao, M., An, B., Gao, W., Zhang, T.: Efficient label contamination attacks against black-box learning models. In: Proceedings of the Twenty-Sixth International Joint Conference on Artificial Intelligence (IJCAI), pp. 3945–3951 (2017)

MAT-Tree: A Tree-Based Method for Multiple Aspect Trajectory Clustering

Yuri Santos[1,2]([⊠]) [iD], Ricardo Giuliani[1], Vania Bogorny[2] [iD],
Mateus Grellert[1,2] [iD], and Jônata Tyska Carvalho[1,2] [iD]

[1] Universidade Federal de Santa Catarina, Florianópolis, Brazil
[2] INE, Programa de Pós-Graduação em Ciência da Computação, Florianópolis, Brazil
`yuri.nassar@posgrad.ufsc.br`

Abstract. Multiple aspect trajectory is a relevant concept that enables
mining interesting patterns and behaviors of moving objects for different
applications. This new way of looking at trajectories includes a semantic
dimension, which presents the notion of aspects that are relevant facts
of the real world that add more meaning to spatio-temporal data. Given
the inherent complexity of this new type of data, the development of
new data mining methods is needed. Despite some works have already
focused on multiple aspect trajectory classification, few have focused on
clustering. Although the literature presents several raw trajectory clustering algorithms, they do not deal with the heterogeneity of the semantic
dimension. In this paper, we propose a novel hierarchical clustering algorithm for multiple aspect trajectories using a decision tree structure that
chooses the best aspect to branch and group the most similar trajectories
according to different criteria. We ran experiments using a well-known
benchmark dataset extracted from a location-based social network and
compared our clustering results with a state-of-the-art clustering approach over different internal and external validation metrics. As a result,
we show that the proposed method outperformed the baseline, where it
revealed a formation of more cohesive and homogeneous clusters in 88%
of the clusters, being five times more precise according to the external
metrics.

Keywords: Multiple Aspect Trajectory · Trajectory clustering · Tree
approach

1 Introduction

Mining hidden and useful patterns from data is an activity that has been strongly
performed and researched for at least three decades [3]. The development of
techniques for analyzing trajectory data is growing rapidly [8,24,27], leveraged
by the popularization of GPS-equipped devices which is allowing the collection
of spatial data from moving objects in unprecedented volumes. Moreover, the
advancement of the Internet of Things has allowed the extraction ofnumerous

ⓒ The Author(s), under exclusive license to Springer Nature Switzerland AG 2023
M. C. Naldi and R. A. C. Bianchi (Eds.): BRACIS 2023, LNAI 14195, pp. 464–478, 2023.
https://doi.org/10.1007/978-3-031-45368-7_30

data other than spatio-temporal called aspects. The aspects contribute to enriching the semantic dimensions of mobility data. Furthermore, the enriched dimension can be associated with a moving object, the entire trajectory, or a single trajectory point. Besides, this enriched dimension can contain any data format, from simple labels to sophisticated objects. For example, a certain point of a trajectory that is found in a restaurant may contain, in addition to the aspect that defines the type of place visited, aspects such as price range, user evaluation, and opening hours, among others. This complex type of trajectory is called Multiple Aspect Trajectory (MAT) [13].

Regarding the mobility data domain, several data mining techniques have been proposed for trajectory clustering [2,7,16,20,30,32]. Such technique allows us to extract patterns [15], and detect common and outlying moving objects behaviors [14]. Furthermore, clustering can be used to find how moving objects are similar to each other with respect to their spatio-temporal trajectories, or trajectories of the same user are similar to each other [9]. Some clustering applications can include carpooling services based on common trajectories (e.g., Uber), profiling of users for transportation and route planning, and collaborative filtering and other association rules techniques to find which locations are often associated by their trajectories [18]. Trajectory clustering methods for space [7], spatio-temporal [31], or semantic dimensions [10] have achieved very solid results so far. Meanwhile, few works have focusing on clustering MATs [25], which is a challenging task since it requires the capability of dealing with heterogeneous data and an even bigger data volume. This means that there are opportunities for the development of new techniques capable of grouping and describing multiple semantic aspects together. Thus, it is important that clustering methods for MATs adopt multidimensional similarity metrics or other strategies that allow capturing such data heterogeneity [29].

We propose a divisive hierarchical clustering algorithm that considers the three dimensions of trajectories (space, time, and semantics) by using a decision tree-based approach. Regarding the frequency at which each of the aspects appears in the trajectories of a given dataset, the proposal seeks to iteratively select the ideal aspect to divide the set of trajectories in two by using a threshold (mean or median of aspect frequency). It leads to the formation of hierarchical trajectory clusters that are naturally more similar to each other, i.e., clusters that present a higher frequency for certain aspects and a lower frequency for others. It is noteworthy that since the proposed method uses a tree approach, there is a hierarchy between the formed clusters. Therefore, in the bottom of the tree, at the leaf nodes, we have more specific and detailed clusters, while closer to the top of the tree, clusters have more general information, being the root node the original dataset. Thus, we seek to provide a new method that supports answering questions related to trajectory similarities regarding different aspects and at different levels of abstraction.

For validation purposes, the proposed method is tested on the Foursquare dataset [29], the same data used by state-of-the-art for MAT clustering [25]. This dataset contains user trajectories (spatio-temporal) enriched with other characteristics such as: the type of place visited, rating, price tier, and weather condition. We use internal and external validation metrics to evaluate clustering results. The experimental result shows that our proposal is quantitatively and qualitatively better than the previous state-of-the-art method [25]. The clusters generated by our method were 88% more cohesive and more separable than the baseline method [25]. Regarding the external validation metrics, the proposed method was five times more precise than the baseline. From the qualitative point of view, our proposal provides different options for clustering and visualizations, being a valuable tool for data analysts performing exploratory analysis on multiple aspect trajectory data.

The remainder of this paper is organized as follows. In Sect. 2, we present the basic concepts of multiple aspect trajectories and related works. In Sect. 3 we describe the proposed approach to cluster MAT by decision trees. In Sect. 4 we discuss the experimental results and evaluation of the proposed method. Finally, the conclusion and further research directions of our work are presented in Sect. 5.

2   Basic Concepts and Related Works

In this section we present the basic concepts to guide the reader throughout this paper and a brief review of MAT clustering approaches.

2.1   Multiple Aspect Trajectory

Multiple aspect trajectories are defined by their three-dimensional nature, i.e., the sequences of points composed by space and time, in addition to the semantic dimension [13]. The concept of semantic dimension is the representation of any context information or relevant meanings that are of fundamental importance for understanding the data obtained in a trajectory [21]. The first approach that brought semantic data enrichment to trajectory data was stops and moves [1]. Moves are made of sample locations between stops which could also be at the beginning or end of a trajectory. Stops are groups of sample points close in space and time that reflect interesting spatial places known as Point of Interest (POI). Besides, every stop has a beginning and ending time, a spatial position and a minimum duration.

More recently, the semantic dimension started to represent the vast set of characteristics that each point of a trajectory can present, which is called aspect, thus bringing the idea of multiple aspect trajectories [13]. An aspect can be described as a real-world fact relevant to the analysis of moving object data. Figure 1 illustrates several points on a trajectory that can contain many aspects. It shows the trajectory of a given person where the POIs are represented in circles while different data are collect, such as: weather information, heart rate, emotional status while working at the office, the ticket price, the genre of the film and its rating in a cinema session, and a restaurant with aspects representing its reviews, opening hours, price range and the restaurant type. Thus, the multiple aspect trajectories can present numerous aspects that enrich the semantic dimension.

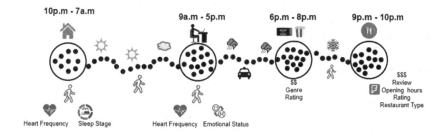

Fig. 1. Example of a Multiple Aspect Trajectory.

2.2   Related Works

Many works focus on raw trajectory analysis, and clustering due to the high availability of GPS tracking devices, enabling the tracking of moving objects such as vehicles, planes, and humans. Besides, when using clustering for MAT, it may extract useful patterns or detect interesting outliers. Nevertheless, the majority of trajectory clustering works in the literature [20,25,26,28,30,32] take into account the space or spatio-temporal dimensions, which employ classical clustering approaches (e.g., k-means [2] and DBSCAN [22]) with adapted measures from these dimensions.

In the work of Hung et al. [7], the group proposed a framework to explore the spatio dimension by clustering and aggregating clues of trajectories to find routes. Nanni and Pedreschi [16] proposed a modified density-based clustering algorithm to explore the temporal dimension to improve the quality of trajectory clustering. Wang et al. [26] proposed a trajectory clustering method based on HDBSCAN, which adaptively clusters trajectories with their shape characteristics using the Hausdorff distance on the spatio dimension to compute similarities. Chen et al. [2] used the DBSCAN method for clustering trajectories. It first divide the trajectories into a set of subtrajectories considering the spatial dimension, then it computes the similarity between trajectories using the Hausdorff distance, and finally the DBSCAN clustering algorithm is applied for clustering. Sun and Wang [22] used the DBSCAN method combined with Minimum Bounding Rectangle and buffer similarity over the ship trajectories using the spatio dimension to improve trajectories clustering and reduce computation time.

Yuan et al. [31] proposed a density-based clustering algorithm that employs an index tree for spatio-temporal trajectories. It works by partitioning the trajectories into trajectory segments, storing them in the index tree and then the segments are clustered based on the density strategy. Yao et al. [30] proposed an RNN-based auto-encoder model to encode the spatio-temporal movement pattern of trajectory to improve similarity computation by learning the low-dimensional representation and then applied the classic k-means algorithm. Liu et al. [11] proposed a new clustering method by extending the k-means algorithm with a time layer to take into account both spatio and temporal proprieties for clustering flight trajectories.

Liu and Guo [10] proposed a semantic trajectory clustering method to capture global relationships among trajectories based on community detection from the perspective of the network. Xuhao et al. [28] proposed a semantic-based trajectory clustering method for arrival aircraft based on k-means and DBSCAN. In the work of Santos et al. [20], the researches proposed a co-clustering method for mining semantic trajectories by clustering trajectories without to test all their attributes by focusing on high frequent ones.

Different of the presented works, Varlamis et al. [25] proposed a MAT similarity measure with hierarchical clustering by including a multi-vector representation of MATs that enables performing cluster analysis on them. The vector representation embeds all trajectory dimensions (space, time, and semantic) into a low-dimension vector which allows clustering multiple aspect trajectories. The authors compared their method with other MAT similarity measures using traditional clustering methods and the results outperformed the baseline methods. Note that such comparison is performed given the gaps in this recent multiple aspect trajectory clustering task. Thus, there is a lack of new methods enabling trajectory cluster analysis using semantic dimension with its multiple aspects and the spatio-temporal dimension simultaneously. We aim to contribute to the reduction of this gap in the literature by proposing a novel MAT clustering method.

3   The Multiple Aspect Trajectory Tree Approach

In this section we present a new method named *MAT-Tree* (Multiple Aspect Trajectory Tree) for finding clusters in a multiple aspect trajectory dataset by using a hierarchical strategy. The main idea is that the *MAT-Tree* groups similar trajectories in the same cluster based on aspects that occurs frequently. *MAT-Tree* aims to identify the most relevant aspects while clustering the trajectories.

Figure 2 illustrates an example of a tree generated by the clustering algorithm using Sankey diagram representation. The vertical bars indicate the clusters generated by the division through the chosen aspect, noting that the leftmost bar represents the complete dataset with all trajectories and the rightmost bars represent the leaf nodes. Figure 2 illustrates tree levels with aspect *RAIN* as the most relevant to start the splitting where other six aspects contributed to identify four clusters (leaf nodes). This result can support analysts to identify moving patterns, for instance, during raining days there are more people that go to indoor events (e.g. movie theaters and comedies) using public transportation than night parties with stops at restaurants.

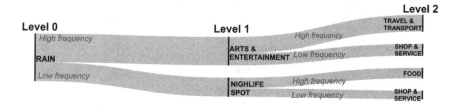

Fig. 2. Example of multiple aspect trajectory clustering tree.

The detailed description of *MAT-Tree* can be seen on Algorithm 1. The main parts and characteristics of the algorithm are: (i) the construction of the frequency matrix; (ii) the criterion chosen for splitting the dataset at each node iteration; (iii) the criterion chosen for evaluating the quality of the split tested; (iv) the stop criterion. *MAT-Tree* receives as input the multiple aspect trajectories *MAT* of dataset *D*, the statistical metric that will be used for dividing the trajectories *stat_met*, the criterion for choosing the aspect *asp_criterion*, the maximum height desired for the tree *max_tl* and the minimum number of trajectories for a leaf node *min_t*.

The first node of the tree is the root and then *MAT-Tree* executes all the process recursively. *MAT-Tree* starts by creating the tree's node structure (line 2) that will store the trajectories received as a parameter of the function (line 3). Next, both child nodes are initialized empty, called left_node (line 4) and right_node (line 5), which may contain the trajectories resulting from the division process. *MAT-Tree* creates the clustering tree by exploring frequency matrices regarding the occurrences of the multiple aspects of each trajectory. It means that, for each new node generated in the tree, the occurrence of each aspect in the trajectories is used to build a frequency matrix for this new node (line 6). Table 1 depicts an example of a frequency matrix showing the number of occurrences of four aspects from three MATs.

Table 1. Frequency matrix of three trajectories with four aspects.

Traj. ID	Nightlife Spot	Travel & Transport	Rain	Arts & Entertainment
tid_1	3	0	12	10
tid_2	14	12	14	11
tid_3	12	3	17	12

We start by selecting a split criterion that is used to identify which aspect is better for partitioning the set of trajectories and how such partitioning will occur. Regarding this partitioning task, *MAT-Tree* uses a statistical metric *stat_met*, e.g. average, to identify the split point. In this scenario, *MAT-Tree* computes the average occurrence of each aspect in the set of trajectories (lines 8–11). After an aspect has been chosen for the division (line 24), the trajectories that present

Algorithm 1. MAT-Tree

Input:	MAT: multiple aspect trajectories of dataset D
	$stat\_met$: statistical metric
	asp_crit: aspect selection criterion
	max_tl: maximum tree level
	min_t: minimum number of trajectories per node
Output:	MAT clustering tree

1: MAT-Tree($MAT, stat\_met, asp\_crit, max\_tl, min\_t$)
2: $node \leftarrow NewTreeNode()$
3: $node.data \leftarrow MAT$
4: $node.left\_node \leftarrow NULL$
5: $node.right\_node \leftarrow NULL$
6: $freq\_matrix \leftarrow CALC\_ASPECT\_FREQUENCY(node.data)$
7: $aspect\_threshold \leftarrow emptyList()$
8: **for all** aspect $a_i \in freq\_matrix$ **do**
9:     $asp \leftarrow CALC\_ASPECT\_STATISTICS(a_i, stat\_met)$
10:     $aspect\_threshold.add(asp)$
11: **end for**
12: $left\_group \leftarrow emptyMatrix()$
13: $right\_group \leftarrow emptyMatrix()$
14: $node.aspect\_chosen \leftarrow \emptyset$
15: **for all** aspect $a_i \in freq\_matrix$ **do**
16:     **for all** trajectory $t_i \in trajectories$ **do**
17:         **if** $freq\_matrix[t_i] < aspect\_threshold[a_i]$ **then**
18:             $left\_group[a_i][t_i]$
19:         **else**
20:             $right\_group[a_i][t_i]$
21:         **end if**
22:     **end for**
23: **end for**
24: $node.aspect\_chosen \leftarrow ASP\_SELECTION(asp\_crit, MAT.aspects, left\_group, right\_group)$
25: $tree\_level \leftarrow current\_level + 1$
26: **if** $size\_of(left\_group) < min\_t \vee tree\_level > max\_tl$ **then**
27:     $node.left\_node \leftarrow NULL$
28: **else**
29:     $node.left\_node \leftarrow MAT\_TREE(left\_group, stat\_met, asp\_crit, max\_tl, min\_t)$
30: **end if**
31: **if** $size\_of(right\_group) < min\_t \vee tree\_level > max\_tl$ **then**
32:     $node.right\_node \leftarrow NULL$
33: **else**
34:     $node.right\_node \leftarrow MAT\_TREE(right\_group, stat\_met, asp\_crit, max\_tl, min\_t)$
35: **end if**

a frequency higher than the *stat_met* in the aspect are grouped together, i.e. *node.right_node* (lines 31–35), while the trajectories that present a frequency lower than the *stat_met* are grouped in another cluster, *node.left_node* (lines 26–30). It is noteworthy that other statistical metrics, such as the median instead of the average, can be used in this step.

We propose four evaluation criteria (*asp_crit*) to select the best aspect, namely: (i) binary division (BD), (ii) minimum variance (MV), (iii) maximum variance reduction (MVR) and (iv) maximum variance reduction considering the largest reduction average among all the aspects set (LRA). The idea is to provide different approaches that better fit the data distribution and the application at hand. Therefore, *MAT-Tree* tests the trajectory partition for all aspects according to the chosen evaluation criterion to identify which aspect is the best choice (line 24). We provide details about each aspect evaluation criterion in the following.

BD will analyze the absolute difference in the number of trajectories between the subsets formed after partitioning. The aspect that generates the smallest difference value is selected as the best. **MV** evaluates the mean variance of an aspect in both trajectory partitions, selecting the one with the minimum mean variance. In addition, **MVR** also considers the mean variance of the aspect in both trajectory partitions. However, it evaluates the variance reduction considering the previous aspect value, that is, the variance of the parent node. The selected aspect using MVR is the one that generates the maximum reduction value compared to the previous node. Last but not least, **LRA** is similar to MVR, but instead of considering the max reduction of the aspects individually, LRA considers all aspects to compute the average reduction. Thus, LRA selects the aspect that generates the largest reduction average of variance regarding the whole aspects set.

We highlight that the variance represents the dispersion of the aspects trajectory distribution, that is, how far the trajectories are from the mean. Considering a low variance, we are interested in more homogeneous trajectory partitions regarding the frequency of the aspects. The proposed hierarchical tree algorithm identifies more detailed clusters regarding the tree depth, i.e., the last levels of the tree characterize clusters with more aspects on its path. It should be noted that, by design, neighboring nodes will have more similar trajectories, while distant nodes will have less similar trajectories.

In conclusion, *MAT-Tree* stops and returns the generated tree (line 36) when the conditions of the minimum number of trajectories or maximum tree level (or both) are satisfied. Thus, these hyperparameters may vary, and they depend on the application domain and the data distribution at hand.

4   Experimental Evaluation

We carried out the experiments[1] using the Foursquare NY dataset [17] to evaluate the performance of different multiple aspect trajectory clustering of several users. Regarding the main hyperparameters, we use the average occurrence of the aspects frequency as the split-point strategy, while the stop criterion with the minimum number of trajectories was set to 25, where it is application-dependent. The minimum number of trajectories needs to be defined empirically for an eventual optimal value. For a fair comparison with the baseline method, we employed the default hyperparameters and set the number of clusters to be equal to the number of different users (i.e. 193 users). Moreover, we run every clustering algorithm 10 times and reports the average of internal and external evaluation scores. We performed the experiments in a machine with a processor Intel i7-7700 3.6 GHz, 16 GB of memory, and OS Windows 10 64 bits.

Petry et al. [17] enriched the original Foursquare datase [29] with semantic information to evaluate multiple aspect trajectory similarity measures. The dataset contains the trajectories of 193 different users, who accounted for a total

---

[1] https://github.com/bigdata-ufsc/mat_tree.

of 66962 points in 3079 trajectories with an average length of 22 points (check-ins) per trajectory and an average of 16 trajectories per user. The aspects contained in the dataset after the enrichment are: i) the geographic coordinates (latitude and longitude, being a numerical attribute); ii) the time (numeric) at which the user checked in; iii) the day of the week (nominal); iv) the point of interest (POI), which can be understood as the name of the establishment such as Starbucks or McDonald's (nominal); v) the type of POI (nominal), such as coffee house or fast food; vi) check-in category (nominal), called the root type (for example, Food, Outdoors & Recreation); vii) the rating assigned by the user in the application (ordinal) and viii) the weather condition (nominal) at the time of check-in.

Clustering evaluation is well-known in the literature due to the fact that clustering is an unsupervised method and we do not usually have a ground truth to compare with. However, in the trajectory clustering application, we can assume that trajectories of the same user are likely to belong to the same cluster, as indicated by Gonzalez et al. [5] and already used in the state-of-the-art [12]. Therefore, the external evaluation of the clustering method is based on this ground truth. For the internal clustering validity metrics, we assume that the best clusters are those that are well separated and compact, as described by Rendón et al. [19]. The external (supervised) and internal (unsupervised) clustering validity metrics [6] comprise: i) homogeneity score (external); ii) completeness score (external); iii) v-measure score (external); iv) adjusted mutual info (Mut. Info) score (external); v) adjusted rand (Adj. Rand) score (external); vi) Fowlkes Mallows (FM) score (external); vii) silhouette (S) score (internal); viii) Calinski Harabaz (CH) score (internal); ix) Davies-Bouldin (DB) Index (separation) (internal).

We performed the experiments using the most strongly correlated aspects as similar as the experiments conducted by Valarmis et al. [25] for comparison purpose. According to this, the aspects used are the weather, day of the week, and POI category. However, *MAT-Tree* is very flexible and it allows any other combination of aspects. In addition, the aspect combination also depends on the application and what kind of patterns the analyst want to mine.

Considering the aspect evaluation criterion, MAT-Tree builds four trees, one tree for each criteria. The result shows numerical similarities in terms of number of nodes, leaf nodes and height. Regarding the number of nodes, on average (rounded-down), the trees generated 388 nodes, while in terms of leaf nodes the average was 194. In addition, the heights of the trees vary between 8 and 14, as shown in Table 2. Besides, we noted that the trees structure are different among them, it indicates that the method can identify different trajectory behavior depending on the chosen criteria.

Figure 3 shows the clustering results from two different aspect selection criterion using the dendogram to visualize them. Figure 3(a) shows the tree generated by the MV aspect choice criterion, while Fig. 3(b) shows the result using the MVR criterion. Comparing both dendograms, it is noted that the clustering on Fig. 3(a) captured an atypical behavior of the trajectories as can be seen in the

Table 2. The AVG number of generated groups (leaf-nodes) and the height of the modeled tree for each aspect selection criterion.

Selection criteria	Total nodes	Leaf-nodes	Tree Height
BD	385	193	8
MV	387	194	14
LRA	389	195	10
MVR	393	195	10

extreme right branch. In this experiment, MAT-Tree selected the *Event* aspect of the root_type attribute as the root node, and grouped all the trajectories of the users who checked-in for this type of aspect. This group (C1) contains 20 users out of 193 with a total of 25 trajectories. The *Event* aspect occurs 26 times in the universe of 66962 check-ins, which can reveal a trajectory behavior practiced only in this group.

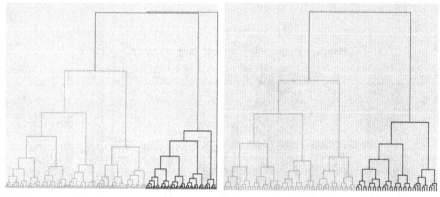

(a) Aspect selection criteria using the minimum variance (MV).

(b) Aspect selection criteria using the maximum variance reduction (MVR).

Fig. 3. Clustering results generated from different aspect selection criterion.

Figure 4 illustrates the exploratory frequency analysis from the aforementioned group C1. The bar graphs show the relative frequency of aspects for *root_type* and *day* category, respectively. It means the number of occurrences of aspects in this specific cluster considering the whole set of trajectories. Furthermore, it can be seen in Fig. 4(a) that the relative frequency for *Event* was equivalent to its absolute frequency. That is, all trajectories that checked in the *Event* aspect belong to the same cluster. Regarding the days of the week, Fig. 4(b) shows that Saturday was the most frequent day in cluster C1. This behavior is expected because the events usually take place on weekends. Regarding the other aspects in Fig. 4, we noted that there are few points referring to *College & University*, which may indicate that the majority of users are not part of the university community. On the other hand, the users in this cluster are more related to activities involving *Art & Entertainment* and *Nightlife*, with *Friday* and *Saturday* more frequent than the other days.

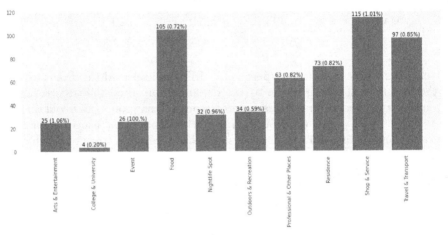

(a) Relative frequencies for *root type* category aspects.

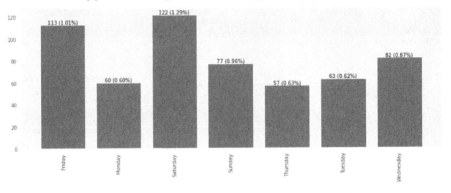

(b) Relative frequencies for *day* category aspects.

Fig. 4. Exploratory frequency analysis of cluster C1.

We quantitatively evaluate the clustering result by employing internal and external validation metrics to verify the goodness of a clustering structure. Furthermore, we used the MSM [4] and MUITAS [17] in *MAT-Tree* as similarity metrics to build the similarity matrix between trajectories. We compare TraFoS [25] results with *MAT-Tree* because it is the state-of-the-art for MAT clustering, and TraFos outperformed the baselines on its experiments. Table 3 shows the clustering evaluation results where *MAT-Tree* obtained promising results. We highlight the best result in bold, while the second one is underscored for each evaluation metric. It can be seen in Table 3 that, in general, the internal and external validation scores for *MAT-Tree* are better than TraFoS, indicating that *MAT-Tree* identified better cluster structures. Additionally, the best approach for selecting an aspect to partition the set of trajectories was the maximum reduction of variance (MVR) using MUITAS.

Table 3. Internal and External Clustering Validation results.

Method	S	CH	DB	Homogeneity	Completeness	V-measure	Mut. Info	Adj. Rand	FM
MAT-Tree$_{BD\_MUITAS}$	−0.305	14.411	<u>0.369</u>	0.557	0.556	0.557	0.168	0.065	0.070
MAT-Tree$_{MV\_MUITAS}$	**−0.284**	5.199	0.449	**0.630**	0.568	**0.597**	0.122	0.057	0.063
MAT-Tree$_{LRA\_MUITAS}$	−0.336	4.408	0.464	<u>0.590</u>	0.561	0.575	0.144	<u>0.066</u>	0.071
MAT-Tree$_{MVR\_MUITAS}$	−0.360	**23.656**	**0.292**	0.587	<u>0.581</u>	<u>0.584</u>	<u>0.210</u>	**0.085**	**0.090**
MAT-Tree$_{BD\_MSM}$	<u>−0.289</u>	12.785	0.476	0.557	0.556	0.557	0.168	0.065	0.070
MAT-Tree$_{MV\_MSM}$	−0.309	1.640	0.509	**0.630**	0.568	**0.597**	0.122	0.057	0.063
MAT-Tree$_{LRA\_MSM}$	−0.305	3.321	0.566	<u>0.590</u>	0.561	0.575	0.144	<u>0.066</u>	0.071
MAT-Tree$_{MVR\_MSM}$	−0.317	<u>18.440</u>	0.413	0.587	<u>0.581</u>	<u>0.584</u>	<u>0.210</u>	**0.085**	**0.090**
TraFoS$_{mean\_SL}$	−0.94	0.99	1.01	0.08	**0.72**	0.14	0.14	0.00	<u>0.08</u>
TraFoS$_{max\_SL}$	−0.94	0.99	1.01	0.08	**0.72**	0.14	0.14	0.00	<u>0.08</u>
TraFoS$_{thr\_SL}$	−0.94	0.99	1.01	0.08	**0.72**	0.14	0.14	0.00	<u>0.08</u>
TraFoS$_{mean\_AL}$	−0.95	0.96	3.66	0.24	0.32	0.27	**0.27**	0.00	0.01
TraFoS$_{max\_AL}$	−0.95	0.96	3.67	0.24	0.32	0.27	**0.27**	0.00	0.01
TraFoS$_{thr\_AL}$	−0.95	0.96	3.67	0.24	0.32	0.27	**0.27**	0.00	0.01

TraFoS tested the hierarchical agglomerative clusteting with single (SL) and average (AL) linkage using the binary partition. In addition, TraFos evaluated three different variations for each clustering strategy such as: i) the first considers the average similarity (TraFoS$_{mean}$), ii) the second takes the maximum similarity (TraFoS$_{max}$) and iii) the third sets a threshold on the average similarity (TraFoS$_{thr}$). Comparing the silhouette scores, TraFoS obtained −0.94 and −0.95 respectively, while MAT-Tree obtained −0.284 and −0.289 respectively. It is important to note that the negative value in all cases is due to the use of similarity matrix instead of distance [25]. This means that the trajectories are better clustered together, i.e., clusters are more cohesive than those found in the previous method. Regarding external validation, in general, our results are better than the baseline, denoting that the resulting clusters are more homogeneous and complete, as shown in Table 2 and Table 3. Thus, *MAT-Tree* obtained a result of 88% better than TraFoS looking at Silhouette and MUITAS, while for V-measure the MAT-Tree was five times more precise.

5   Conclusion and Future Works

The volume and the variety of the big data era we live in require a high computational power which evidences the necessity of new approaches for analyzing complex data. Multiple aspect trajectories bring a lot of opportunities in the data mining domain, where the nature of a sequence, the high dimensionality, heterogeneity, and data volume pose new challenges. Even though a number of methods have been proposed for MAT classification only a few works focus on MAT clustering. Regarding this, we proposed *MAT-Tree*, a novel method for multiple aspect trajectory clustering based on the frequency of occurrence of the aspects of the trajectories, and using a decision tree-based approach. The proposed method presented a result of 88% better than the baseline considering internal clustering evaluation metrics and five times more precise.

The main contribution of *MAT-Tree* is a new approach that allows clustering trajectories considering all their dimensions and semantic aspects. It is noteworthy that studies on the semantic dimension of trajectories are recent and tailored. *MAT-Tree* results outperformed the state-of-the-art for multiple aspect trajectory clustering, it indicates that *MAT-Tree* can identify more cohesive, compact, and connected clusters. Furthermore, *MAT-Tree* offers different options for clustering and visualizations, providing a flexible tool for exploratory data analysis and applications that can adapt to the task at hand. Thus, once clustering is a data mining task that is inherently highly application-dependent and exploratory [23, chapter 1], the flexibility of *MAT-Tree* is an important characteristic.

As future works, it would be interesting to examine other aspect selection strategies that allow *MAT-Tree* to adapt automatically to different applications. The investigation of different split aspect strategies and evaluation criteria is a promising direction because each criterion shows to be suitable for different applications and analyses. We noted that the BD evaluation criterion is well-suited for generating clusters almost of the same size. Nonetheless, the MV criterion seems to easily find outliers trajectories (i.e., trajectories presenting aspects that rarely appear on other trajectories). Therefore, experiments designed specifically for validating these characteristics and uses of these criteria are desirable.

Acknowledgements. This work has been partially supported by CAPES (Finance code 001), CNPQ, FAPESC (Project Match - co-financing of H2020 Projects - Grant 2018TR 1266), and the European Union's Horizon 2020 research and innovation programme under GA N. 777695 (MASTER). We also thank the reviewers who contributed to the improvement of this work. The views and opinions expressed in this paper are the sole responsibility of the author and do not necessarily reflect the views of the European Commission.

References

1. Alvares, L.O., Bogorny, V., Kuijpers, B., de Macedo, J.A.F., Moelans, B., Vaisman, A.: A model for enriching trajectories with semantic geographical information. In: Proceedings of the 15th Annual ACM International Symposium on Advances in Geographic Information Systems, pp. 1–8 (2007)
2. Chen, J., Wang, R., Liu, L., Song, J.: Clustering of trajectories based on Hausdorff distance. In: 2011 International Conference on Electronics, Communications and Control (ICECC), pp. 1940–1944 (2011)

3. Fayyad, U.M., Piatetsky-Shapiro, G., Smyth, P., et al.: Knowledge discovery and data mining: towards a unifying framework. In: KDD, vol. 96, pp. 82–88 (1996)
4. Furtado, A.S., Kopanaki, D., Alvares, L.O., Bogorny, V.: Multidimensional similarity measuring for semantic trajectories. Trans. GIS **20**(2), 280–298 (2016)
5. Gonzalez, M.C., Hidalgo, C.A., Barabasi, A.-L.: Understanding individual human mobility patterns. Nature **453**(7196), 779–782 (2008)
6. Halkidi, M., Batistakis, Y., Vazirgiannis, M.: On clustering validation techniques. J. Intell. Inf. Syst. **17**, 107–145 (2001)
7. Hung, C.-C., Peng, W.-C., Lee, W.-C.: Clustering and aggregating clues of trajectories for mining trajectory patterns and routes. VLDB J. **24**, 169–192 (2015)
8. Zheng, K., Zheng, Y., Yuan, N.J., Shang, S.: On discovery of gathering patterns from trajectories. In: Proceedings of the IEEE International Conference on Data Engineering, Washington, DC. IEEE (2013)
9. Khoroshevsky, F., Lerner, B.: Human mobility-pattern discovery and next-place prediction from GPS data. In: Schwenker, F., Scherer, S. (eds.) MPRSS 2016. LNCS (LNAI), vol. 10183, pp. 24–35. Springer, Cham (2017). https://doi.org/10.1007/978-3-319-59259-6_3
10. Liu, C., Guo, C.: STCCD: semantic trajectory clustering based on community detection in networks. Expert Syst. Appl. **162**, 113689 (2020)
11. Liu, G., Fan, Y., Zhang, J., Wen, P., Lyu, Z., Yuan, X.: Deep flight track clustering based on spatial-temporal distance and denoising auto-encoding. Expert Syst. Appl. **198**, 116733 (2022)
12. May Petry, L., Leite Da Silva, C., Esuli, A., Renso, C., Bogorny, V.: MARC: a robust method for multiple-aspect trajectory classification via space, time, and semantic embeddings. Int. J. Geogr. Inf. Sci. **34**(7), 1428–1450 (2020)
13. Mello, R.D.S., et al.: MASTER: a multiple aspect view on trajectories. Trans. GIS **23**(4), 805–822 (2019)
14. Meng, F., Yuan, G., Lv, S., Wang, Z., Xia, S.: An overview on trajectory outlier detection. Artif. Intell. Rev. **52**, 2437–2456 (2019)
15. Morris, B., Trivedi, M.: Learning trajectory patterns by clustering: experimental studies and comparative evaluation. In: 2009 IEEE Conference on Computer Vision and Pattern Recognition, pp. 312–319 (2009)
16. Nanni, M., Pedreschi, D.: Time-focused clustering of trajectories of moving objects. J. Intell. Inf. Syst. **27**(3), 267–290 (2006)
17. Petry, L.M., Ferrero, C.A., Alvares, L.O., Renso, C., Bogorny, V.: Towards semantic-aware multiple-aspect trajectory similarity measuring. Trans. GIS **23**(5), 960–975 (2019)
18. Poushter, J., et al.: Smartphone ownership and internet usage continues to climb in emerging economies. Pew Res. Center **22**(1), 1–44 (2016)
19. Rendón, E., Abundez, I., Arizmendi, A., Quiroz, E.M.: Internal versus external cluster validation indexes. Int. J. Comput. Commun. **5**(1), 27–34 (2011)
20. Santos, Y., Carvalho, J.T., Bogorny, V.: SS-OCoClus: a contiguous order-aware method for semantic trajectory co-clustering. In: 2022 23rd IEEE International Conference on Mobile Data Management (MDM), pp. 198–207 (2022)
21. Spaccapietra, S., Parent, C., Damiani, M.L., de Macedo, J.A., Porto, F., Vangenot, C.: A conceptual view on trajectories. Data Knowl. Eng. **65**(1), 126–146 (2008)
22. Sun, M., Wang, J.: An approach of ship trajectory clustering based on minimum bounding rectangle and buffer similarity. In: IOP Conference Series: Earth and Environmental Science, vol. 769 (2021)
23. Tan, P.N., Steinbach, M., Kumar, V.: Introduction to Data Mining. Pearson, London (2018)

24. Tortelli Portela, T., Tyska Carvalho, J., Bogorny, V.: HiPerMovelets: high-performance movelet extraction for trajectory classification. Int. J. Geograph. Inf. Sci. **36**(5), 1012–1036 (2022)
25. Varlamis, I., et al.: A novel similarity measure for multiple aspect trajectory clustering. In: Proceedings of the 36th Annual ACM Symposium on Applied Computing, pp. 551–558 (2021)
26. Wang, L., Chen, P., Chen, L., Mou, J.: Ship AIS trajectory clustering: an HDBSCAN-based approach. J. Marine Sci. Eng. **9**(6), 566 (2021)
27. Wu, S.X., Wu, Z., Zhu, W., Yang, X., Li, Y.: Mining trajectory patterns with point-of-interest and behavior-of-interest. In: 2021 IEEE International Conference on Communications Workshops (ICC Workshops), pp. 1–6. IEEE (2021)
28. Xuhao, G., Junfeng, Z., Zihan, P.: Trajectory clustering for arrival aircraft via new trajectory representation. J. Syst. Eng. Electron. **32**(2), 473–486 (2021)
29. Yang, D., Zhang, D., Zheng, V.W., Yu, Z.: Modeling user activity preference by leveraging user spatial temporal characteristics in LBSNs. IEEE Trans. Syst. Man Cybern.: Syst. **45**(1), 129–142 (2014)
30. Yao, D., Zhang, C., Zhu, Z., Huang, J., Bi, J.: Trajectory clustering via deep representation learning. In: 2017 International Joint Conference on Neural Networks (IJCNN), pp. 3880–3887. IEEE (2017)
31. Yuan, G., Xia, S., Zhang, L., Zhou, Y., Ji, C.: An efficient trajectory-clustering algorithm based on an index tree. Trans. Inst. Meas. Control. **34**(7), 850–861 (2012)
32. Yuan, G., Sun, P., Zhao, J., Li, D., Wang, C.: A review of moving object trajectory clustering algorithms. Artif. Intell. Rev. **47**, 123–144 (2017)

Author Index

M. C. Naldi and R. A. C. Bianchi (Eds.): BRACIS 2023, LNAI 14195, pp. 479–483, 2023.
https://doi.org/10.1007/978-3-031-45368-7

Printed in the United States
by Baker & Taylor Publisher Services